Concept Check exercises focus on mathematical thinking and conceptual understanding. Mathematics involves more than notation, calculations, and formulas; pausing to think about the underlying concepts will help you remember and apply them correctly when the time comes.

Concept Check   *Refer to the summary box on solving a trigonometric equation. Decide on the appropriate technique to begin the solution of each equation. Do not solve the equation.*

1. $2 \cot x + 1 = -1$ **2.** $\sin x + 2 = 3$

3. $5 \sec^2 x = 6 \sec x$ **4.** $2 \cos^2 x - \cos x = 1$

5. $9 \sin^2 x - 5 \sin x = 1$ **6.** $\tan^2 x - 4 \tan x + 2 = 0$

7. $\tan x - \cot x = 0$ **8.** $\cos^2 x = \sin^2 x + 1$

Additional complete, step-by-step solutions are now provided at the back of the text for selected exercises. These are solutions to exercises that extend the skills and concepts presented in the examples—actually providing you with a pool of examples to different and/or more challenging problems.

31. Solve $2 \sin \theta = 2 \cos 2\theta$.

31.

$$2 \sin \theta = 2 \cos 2\theta$$
$$\sin \theta = \cos 2\theta \qquad \text{Divide by 2.}$$
$$\sin \theta = 1 - 2 \sin^2 \theta$$
$$\qquad\qquad \text{Cosine double-angle identity}$$
$$2 \sin^2 \theta + \sin \theta - 1 = 0$$
$$(2 \sin \theta - 1)(\sin \theta + 1) = 0$$
$$2 \sin \theta - 1 = 0 \quad \text{or} \quad \sin \theta + 1 = 0$$
$$\sin \theta = \frac{1}{2} \quad \text{or} \quad \sin \theta = -1$$

Over the interval $[0°, 360°)$, the equation $\sin \theta = \frac{1}{2}$ has two solutions, 30° and 150°. Over the same interval, the equation $\sin \theta = -1$ has one solution, 270°.

Solution set: $\{30°, 150°, 270°\}$

Chapter 1 Summary

KEY TERMS

1.1 line line segment (or segment) ray angle initial side terminal side vertex of an angle positive angle	negative angle degree acute angle right angle obtuse angle straight angle complementary angles supplementary angles minute	second angle in standard position quadrantal angle coterminal angles 1.2 vertical angles parallel lines transversal similar triangles	congruent triangles 1.3 sine cosine tangent cotangent secant cosecant

NEW SYMBOLS

θ Greek letter theta	′ minute
° degree	″ second

QUICK REVIEW

CONCEPTS

1.3 Trigonometric Functions

Definitions of the Trigonometric Functions
Let (x, y) be a point other than the origin on the terminal side of an angle θ in standard position. Let $r = \sqrt{x^2 + y^2}$, the distance from the origin to (x, y). Then

$$\sin \theta = \frac{y}{r} \qquad \cos \theta = \frac{x}{r} \qquad \tan \theta = \frac{y}{x} \ (x \neq 0)$$

$$\csc \theta = \frac{r}{y} \ (y \neq 0) \ \sec \theta = \frac{r}{x} \ (x \neq 0) \ \cot \theta = \frac{x}{y} \ (y \neq 0).$$

See the summary table of trigonometric function values for quadrantal angles on page 24.

1.4 Using the Definitions of the Trigonometric Functions

Reciprocal Identities

$$\sin \theta = \frac{1}{\csc \theta} \qquad \cos \theta = \frac{1}{\sec \theta} \qquad \tan \theta = \frac{1}{\cot \theta}$$

$$\csc \theta = \frac{1}{\sin \theta} \qquad \sec \theta = \frac{1}{\cos \theta} \qquad \cot \theta = \frac{1}{\tan \theta}$$

EXAMPLES

If the point $(-2, 3)$ is on the terminal side of angle θ in standard position, then $x = -2$, $y = 3$, $r = \sqrt{4 + 9} = \sqrt{13}$, and

$$\sin \theta = \frac{3\sqrt{13}}{13}, \qquad \cos \theta = -\frac{2\sqrt{13}}{13}, \qquad \tan \theta = -\frac{3}{2},$$

$$\csc \theta = \frac{\sqrt{13}}{3}, \qquad \sec \theta = -\frac{\sqrt{13}}{2}, \qquad \cot \theta = -\frac{2}{3}.$$

The preceding examples given for Section 1.3 satisfy these identities. For example,

$$\tan \theta = \frac{1}{\cot \theta} = \frac{1}{-\frac{2}{3}} = -\frac{3}{2}.$$

Extensive end-of-chapter summaries help you prepare for quizzes and tests. Each end-of-chapter Summary includes a section-by-section list of Key Terms and/or New Symbols, and a Quick Review of Concepts illustrated by a multitude of Examples. This comprehensive review is a hallmark feature of the Lial/Hornsby/Schneider systematic approach to student success.

Trigonometry

Eighth Edition

Margaret L. Lial
American River College

John Hornsby
University of New Orleans

David I. Schneider
University of Maryland

PEARSON
Addison
Wesley

Boston San Francisco New York
London Toronto Sydney Tokyo Singapore Madrid
Mexico City Munich Paris Cape Town Hong Kong Montreal

Publisher: Greg Tobin

Senior Acquisitions Editor: Anne Kelly

Editorial Assistant: Cecilia Fleming

Associate Project Editor: Joanne Ha

Senior Production Supervisor: Karen Wernholm

Production Coordination and Text Design: Elm Street Publishing Services, Inc.

Senior Marketing Manager: Becky Anderson

Marketing Coordinator: Carolyn Buddeke

Senior Author Support/Technology Specialist: Joe Vetere

Associate Media Producer: Sara Anderson

Software Development: Kathleen Bowler and Malcolm Litowitz

Senior Manufacturing Buyer: Evelyn Beaton

Rights and Permissions Advisor: Dana Weightman

Cover Design: Barbara T. Atkinson

Composition: Beacon Publishing Services

Illustrations: Techsetters, Inc.

Cover Photo: © Getty/National Geographics

Photo Credits: p. 1, 26, 92, 126, 132, 211, 386 Corbis RF; **p. 36, 45, 75, 85, 117, 129, 181, 216, 284, 319 (T) (B), 432** PhotoDisc; **p. 44** Bob Kramer Studio, Courtesy Museum of Science, Boston; **p. 65** PictureQuest/Brand X Pictures (RF); **p. 72** SAU; **p. 93, 121, 234** NASA; **p. 131, 151** A.J. Sisco/Corbis; **p. 144** Courtesy Joseph A. Dellinger; **p. 179, 235, 253** Thinkstock (RF); **p. 193** © Jutta Klee/Corbis; **p. 245, 259, 274, 387, 398** PhotoDisc Blue; **p. 275, 281** Corbis TL; **p. 330** Artville (RF); **p. 331, 358** University of Waterloo; **p. 385** Susan Ragan/ Reuters NewMedia Inc./Corbis; **p. 414, 450** Beth Anderson; **p. 430** © AFP/Corbis

Library of Congress Cataloging-in-Publication Data

Lial, Margaret L.

   Trigonometry.—8th ed./Margaret L. Lial, John Hornsby, David I. Schneider.

      p. cm.

    Includes indexes.

    ISBN 0-321-22736-0

    1. Trigonometry.   I. Hornsby, E. John.   II. Schneider, David I.   III. Title.

   QA531.L5   2005

   516.24—dc22

                           2004044494

1 2 3 4 5 6 7 8 9 10—QWT—07060504

Contents

6 Inverse Circular Functions and Trigonometric Equations   *235*

7 Applications of Trigonometry and Vectors   *275*

8   Complex Numbers, Polar Equations, and Parametric Equations   *331*

9   Exponential and Logarithmic Functions   *387*

Preface

In the eighth edition of *Trigonometry,* we continue our ongoing commitment to providing the best possible text to help instructors teach and students succeed. To help students develop both the conceptual understanding and the analytical skills necessary to experience success in mathematics, we present each mathematical topic in the text using a systematic approach designed to actively engage students in the learning process. We have tried to address the diverse needs of today's students through a more open design, updated figures and graphs, helpful features, careful explanations of topics, and a comprehensive package of supplements and study aids.

Although we assume that students have had at least one course in algebra, we include four new appendices that review algebraic prerequisites and we reference these appendices throughout the text. Additional algebraic and geometric concepts are reviewed as needed. Students planning to continue their study of mathematics in calculus, statistics, or other disciplines, as well as those taking this as their final mathematics course, will benefit from the text's student-oriented approach. We believe instructors will particularly welcome the new *Annotated Instructor's Edition,* which provides answers in the margins to almost all exercises, plus helpful Teaching Tips.

New or Enhanced Features

We are pleased to offer the following new or enhanced features.

Chapter Openers   These have been updated and streamlined and provide a quick preview of the chapter topics, plus a reference to a motivating application within the chapter.

Examples   We have added even more examples in this edition. Step-by-step solutions to examples are now easily identified using a **Solution** head. We have carefully polished all solutions and incorporated more side comments and explanations, including helpful section and appendix references to previously covered and prerequisite material. Selected examples continue to provide graphing calculator solutions alongside traditional algebraic solutions. ***The graphing calculator solutions can be easily omitted if desired.***

Now try Exercises   To actively engage students in the learning process, each example now concludes with a reference to one or more parallel odd-numbered exercises from the corresponding exercise set. In this way, students are able to immediately apply and reinforce the concepts and skills presented in the examples.

Exercise Sets   We have taken special care to respond to the suggestions of users and reviewers and have added new exercises to this edition based on their feedback. As a result, the text includes more problems than ever to provide students with ample opportunities to practice, apply, connect, and extend concepts and skills. We have included writing exercises 📝 and optional graphing calculator problems ⊞, as well as multiple-choice, matching, true/false, and completion problems. *Concept Check* problems, which focus on mathematical thinking and conceptual understanding, were well received in the previous edition and have been expanded in this edition.

Solutions to Selected Exercises   Exercise numbers enclosed in a blue circle, such as ⑪., indicate that a complete solution for the problem is included at the back of the text. These new solutions are given for selected exercises that extend the skills and concepts presented in the section examples—actually providing students with a pool of examples to different and/or more challenging problems.

Summary Exercises   These new sets of exercises in Chapters 4 and 5 give students the all-important *mixed* review problems they need to synthesize concepts and select appropriate solution methods.

Function Boxes   Special function boxes offer a comprehensive, visual introduction to each circular and inverse circular function and also serve as an excellent resource for student reference and review throughout the course. Each function box includes a table of values alongside traditional and calculator graphs, as well as the domain, range, and other specific information about the function.

Figures and Photos   Today's students are more visually oriented than ever. As a result, we have made a concerted effort to include mathematical figures, diagrams, tables, and graphs whenever possible. Photos accompany selected applications in examples and exercises.

Chapter Reviews   Each chapter ends with an expanded Summary, featuring a section-by-section list of Key Terms, New Symbols, and a Quick Review of important Concepts, presented alongside corresponding all-new Examples. A comprehensive set of Review Exercises and a Chapter Test are also provided.

Quantitative Reasoning   Now appearing at the end of each chapter, these problems enable students to apply trigonometric concepts to life situations, such as the relationship between utility costs and average annual temperatures or financial planning for retirement. A photo highlights each problem.

Glossary   As an additional student study aid, a comprehensive glossary of key terms from throughout the text is provided at the back of the book.

Continuing Features

We have retained the popular features of previous editions of the text, some of which follow.

Real-Life Applications   We are always on the lookout for interesting data to use in real-life applications. As a result, we have incorporated new or updated applied examples and exercises that show the relevance of trigonometry to daily life. Applications that feature mathematical modeling are labeled with a *Modeling* head. All applications are titled, and a comprehensive Index of Applications is included at the back of the text.

Use of Technology   As in the previous edition, we have integrated the use of graphing calculators where appropriate, although graphing technology is not a central feature of this text. We continue to stress that graphing calculators are an aid to understanding and that students must master the underlying mathematical concepts. We have included graphing calculator solutions for selected examples and continue to mark all graphing calculator notes and exercises that use graphing calculators with an icon ⊟ for easy identification and added flexibility. ***This graphing calculator material is optional and can be omitted without loss of continuity.***

Cautions and Notes   We often give students warnings of common errors and emphasize important ideas in C A U T I O N and N O T E comments that appear throughout the exposition.

Looking Ahead to Calculus   These margin notes offer glimpses of how the trigonometric topics currently being studied are used in calculus.

Connections   Connections boxes continue to provide connections to the real world or to other mathematical concepts, historical background, and thought-provoking questions for writing, class discussion, or group work.

Relating Concepts Exercises   Appearing in selected exercise sets, these sets of problems help students tie together topics and develop problem-solving skills as they compare and contrast ideas, identify and describe patterns, and extend concepts to new situations. These exercises make great collaborative activities for pairs or small groups of students.

Content Changes

A primary focus of this revision of the text was to polish and enhance individual presentations of topics, and we have worked hard to do this throughout the book. Additional content changes you may notice include the following:

- In addition to reviewing algebraic and geometric concepts as needed throughout the text, we have included a thorough review of algebraic prerequisites in new Appendices A–D. As a result, Chapter 1 begins immediately with former Section 1.2 on angles.

- Sections 4.1, 4.3, and 6.1 include new function boxes to highlight important information about the circular and inverse circular functions.

- Chapter 4 concludes with a new section on harmonic motion.

- Section 5.5 now includes product-to-sum and sum-to-product identities.

- Section 8.5 features increased coverage of polar forms of lines and circles.

Supplements

For a comprehensive list of the supplements and study aids that accompany *Trigonometry,* Eighth Edition, see pages xii and xiii.

Acknowledgments

Previous editions of this text were published after thousands of hours of work, not only by the authors, but also by reviewers, instructors, students, answer checkers, and editors. To these individuals and all those who have worked in some way on this text over the years, we are most grateful for your contributions. We could not have done it without you. We especially wish to thank the following individuals who provided valuable input into this edition of the text.

Richard Andrews, *Florida A&M University*
Sandra Arman, *Motlow State Community College*
Melissa Berta, *Saddleback College*
Steven Bogart, *Shoreline Community College*
Larry Bouldin, *Roane State Community College*
Brian Carter, *San Diego City College*
Susan Duke, *Meridian Community College*
Diane Ellis, *Mississippi State University*
Sam Evers, *University of Alabama Tuscaloosa*
Jon Freedman, *Skyline College*
Jeffrey Hughes, *Hinds Community College*
Lynne Kendall, *Metropolitan State College of Denver*
Maria Maspons, *Miami-Dade Community College*
Amy McLanahan, *Foothill College—Los Altos Hills*
Charles Odion, *Houston Community College*
Jane Roads, *Moberly Area Community College*
Alex Rolon, *Northampton Community College*
Mahmoud Shagroni, *Houston Community College*
Virginia Starkenburg, *San Diego City College*
Lewis Walston, *Methodist College*

Over the years, we have come to rely on an extensive team of experienced professionals. Our sincere thanks go to these dedicated individuals at Addison-Wesley, who worked long and hard to make this revision a success: Greg Tobin, Anne Kelly, Becky Anderson, Karen Guardino, Karen Wernholm, Barbara Atkinson, Joanne Ha, Cecilia Fleming, and Joe Vetere.

Terry McGinnis continues to provide invaluable behind-the-scenes guidance—we have come to rely on her expertise during all phases of the revision process. Thanks are due Gina Linko and Elm Street Publishing Services for their excellent production work. Perian Herring, Kitty Pellissier, and Abby Tanenbaum did an outstanding job checking the answers to all exercises, and Abby Tanenbaum also wrote the new solutions to selected exercises that appear at the back of the book. We especially thank Paul Van Erden for the many indexes he has so diligently prepared for us over the years. Joanne Still prepared the index for this text, and Becky Troutman compiled the comprehensive Index of Applications. Special thanks to Patricia Nelson, Richard Watkins, and Margaret Westmoreland for accuracy-checking page proofs.

As an author team, we are committed to the goal stated earlier in the Preface—to provide the best possible text to help instructors teach and students succeed. As we continue to work toward it, we would welcome any comments or suggestions you might have via e-mail to *math@aw.com*.

Margaret L. Lial—John Hornsby—David I. Schneider

Student Supplements	Instructor Supplements

Student's Solutions Manual

- By Heidi Howard, *Florida Community College at Jacksonville*
- Provides detailed solutions to all odd-numbered text exercises
  ISBN: 0-321-22743-3

Graphing Calculator Manual

- By Darryl Nester, *Bluffton College*
- Provides instructions and keystroke operations for the TI-83/83 Plus, TI-84 Plus, TI-85, TI-86, and TI-89
  ISBN: 0-321-22738-7

Videotape Series

- Features an engaging team of lecturers
- Provides comprehensive coverage of each section and topic in the text
  ISBN: 0-321-22741-7

Digital Video Tutor

- Complete set of digitized videos for student use at home or on campus
- Ideal for distance learning or supplemental instruction
  ISBN: 0-321-23734-X

New! Additional Skill & Drill Manual

- By Cathy Ferrer, *Valencia Community College*
- Provides additional practice and test preparation for students
  ISBN: 0-321-23831-1

A Review of Algebra

- By Heidi Howard, *Florida Community College at Jacksonville*
- Provides additional support for students needing further algebra review
  ISBN: 0-201-77347-3

Addison-Wesley Math Tutor Center

- Provides tutoring through a registration number packaged with a new textbook or purchased separately
- Staffed by college mathematics instructors
- Accessible via toll-free telephone, toll-free fax, e-mail, and the Internet
  www.aw-bc.com/tutorcenter

New! Annotated Instructor's Edition

- Special edition of the text
- Provides answers in the margins to almost all text exercises, plus helpful Teaching Tips
  ISBN: 0-321-22740-9

Instructor's Solutions Manual

- By Heidi Howard, *Florida Community College at Jacksonville*
- Provides complete solutions to all text exercises
  ISBN: 0-321-22737-9

Instructor's Testing Manual

- By Christopher Mason, *Community College of Vermont*
- Includes diagnostic pretests, chapter tests, and additional test items, grouped by section, with answers provided
  ISBN: 0-321-22739-5

TestGen with Quizmaster

- Enables instructors to build, edit, print, and administer tests
- Features a computerized bank of questions developed to cover all text topics
- Available on a dual-platform Windows/Macintosh CD-ROM
  ISBN: 0-321-22742-5

New! Adjunct Support Manual

- Includes resources to help new and adjunct faculty with course preparation and classroom management
- Provides helpful teaching tips correlated to the sections of the text
  ISBN: 0-321-23827-3

New! PowerPoint Lecture Presentation

- Classroom presentation software geared specifically to this textbook sequence
- Available within MyMathLab or on the Supplements Central Web site at http://suppscentral.aw.com

New! Adjunct Support Center

- Offers consultation on suggested syllabi, helpful tips on using the textbook support package, assistance with content, and advice on classroom strategies
- Available Sunday–Thursday evenings from 5 P.M. to midnight EST; telephone: 1-800-435-4084; e-mail: AdjunctSupport@aw.com; fax: 1-877-262-9774

MyMathLab®

MyMathLab is a series of text-specific, easily customizable online courses for Addison-Wesley textbooks in mathematics and statistics. MyMathLab is powered by CourseCompass™—Pearson Education's online teaching and learning environment—and by MathXL®—our online homework, tutorial, and assessment system. MyMathLab gives you the tools you need to deliver your course online, whether your students are in a lab setting or working from home. MyMathLab provides a rich and flexible set of course materials, featuring free-response exercises that are algorithmically generated for unlimited practice and mastery. Students can also use online tools, such as video lectures, animations, and a multimedia textbook, to independently improve their understanding and performance. Instructors can use MyMathLab's homework and test managers to select and assign online exercises correlated directly to the textbook, and they can import TestGen tests into MyMathLab for added flexibility. MyMathLab's online gradebook—designed specifically for mathematics and statistics—automatically tracks students' homework and test results and gives the instructor control over how to calculate final grades. MyMathLab is available to qualified adopters. For more information, visit our Web site at www.mymathlab.com.

MathXL®

MathXL is a powerful online homework, tutorial, and assessment system that accompanies your Addison-Wesley textbook in mathematics or statistics. With MathXL, instructors can create, edit, and assign online homework and tests using algorithmically generated exercises correlated to your textbook. All student work is tracked in MathXL's online gradebook. Students can take chapter tests in MathXL and receive personalized study plans based on their test results. The study plan diagnoses weaknesses and links students directly to tutorial exercises for the topics they need to study and retest. Students can also access supplemental animations and video clips directly from selected exercises. MathXL is available to qualified adopters. For more information, visit our Web site at www.mathxl.com.

MathXL Tutorials on CD® (ISBN 0-321-26801-6)

This interactive tutorial CD-ROM provides algorithmically generated practice exercises that are correlated to the exercises in the textbook. Every practice exercise is accompanied by an example and a guided solution designed to involve students in the solution process. Selected exercises may also include a video clip to help students visualize concepts. The software tracks student activity and scores and can generate printed summaries of students' progress.

InterAct Math® Tutorial Web Site   www.interactmath.com

Get practice and tutorial help online! This interactive tutorial Web site provides algorithmically generated practice exercises that correlate directly to the exercises in the text. A detailed worked-out example and guided solution accompany each practice exercise. The Web site recognizes student errors and provides feedback.

1

Trigonometric Functions

There has been an increasing interest in recent years in finding and utilizing sources of renewable energy. Some of these sources include wind, water, and sun. Photovoltaic solar cells convert sunlight directly into electricity. The simplest cells power watches, calculators, and such, while more complex systems light houses and provide power to the energy grid. (*Source:* U.S. Department of Energy.)

The amount of energy a solar cell produces depends on its area. You are asked to solve a problem like this in Exercise 61, Section 1.3.

1.1 | Angles

Basic Terminology ▪ Degree Measure ▪ Standard Position ▪ Coterminal Angles

Line *AB*

Segment *AB*

Ray *AB*

Figure 1

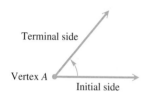

Terminal side

Vertex *A*

Initial side

Figure 2

Basic Terminology Two distinct points *A* and *B* determine a line called **line *AB*.** The portion of the line between *A* and *B*, including points *A* and *B* themselves, is **line segment *AB*,** or simply **segment *AB*.** The portion of line *AB* that starts at *A* and continues through *B*, and on past *B*, is called **ray *AB*.** Point *A* is the *endpoint of the ray.* (See Figure 1.)

An **angle** is formed by rotating a ray around its endpoint. The ray in its initial position is called the **initial side** of the angle, while the ray in its location after the rotation is the **terminal side** of the angle. The endpoint of the ray is the **vertex** of the angle. Figure 2 shows the initial and terminal sides of an angle with vertex *A*. If the rotation of the terminal side is counterclockwise, the angle is **positive.** If the rotation is clockwise, the angle is **negative.** Figure 3 shows two angles, one positive and one negative.

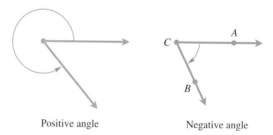

Positive angle

Negative angle

Figure 3

An angle can be named by using the name of its vertex. For example, the angle on the right in Figure 3 can be called angle *C*. Alternatively, an angle can be named using three letters, with the vertex letter in the middle. Thus, the angle on the right also could be named angle *ACB* or angle *BCA*.

Degree Measure The most common unit for measuring angles is the **degree.** (The other common unit of measure, called the *radian*, is discussed in Section 3.1.) Degree measure was developed by the Babylonians, 4000 yr ago. To use degree measure, we assign 360 degrees to a complete rotation of a ray.* In Figure 4, notice that the terminal side of the angle corresponds to its initial side when it makes a complete rotation. One degree, written 1°, represents $\frac{1}{360}$ of a rotation. Therefore, 90° represents $\frac{90}{360} = \frac{1}{4}$ of a complete rotation, and 180° represents $\frac{180}{360} = \frac{1}{2}$ of a complete rotation. An angle measuring between 0° and 90° is called an **acute angle.** An angle measuring exactly 90° is a **right angle.** An angle measuring more than 90° but less than 180° is an **obtuse angle,** and an angle of exactly 180° is a **straight angle.** See Figure 5, where we use the Greek letter θ (theta)** to name each angle.

A complete rotation of a ray gives an angle whose measure is 360°.

$\frac{1}{360}$ of a complete rotation gives an angle whose measure is 1°.

Figure 4

---

*The Babylonians were the first to subdivide the circumference of a circle into 360 parts. There are various theories as to why the number 360 was chosen. One is that it is approximately the number of days in a year, and it has many divisors, which makes it convenient to work with. Another involves a roundabout theory dealing with the length of a Babylonian mile.

**In addition to θ (theta), other Greek letters such as α (alpha) and β (beta) are sometimes used to name angles.

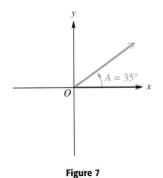

Acute angle
$0° < \theta < 90°$

Right angle
$\theta = 90°$

Obtuse angle
$90° < \theta < 180°$

Straight angle
$\theta = 180°$

Figure 5

If the sum of the measures of two positive angles is $90°$, the angles are called **complementary.** Two positive angles with measures whose sum is $180°$ are **supplementary.**

EXAMPLE 1   Finding Measures of Complementary and Supplementary Angles

Find the measure of each angle in Figure 6.

Solution

(a) In Figure 6(a), since the two angles form a right angle (as indicated by the ⌐ symbol), they are complementary angles. Thus,

$$6m + 3m = 90$$
$$9m = 90 \qquad \text{Combine terms.}$$
$$m = 10. \qquad \text{Divide by 9. (Appendix A)}$$

The two angles have measures of $6(10) = 60°$ and $3(10) = 30°$.

(b) The angles in Figure 6(b) are supplementary, so

$$4k + 6k = 180$$
$$10k = 180$$
$$k = 18.$$

These angle measures are $4(18) = 72°$ and $6(18) = 108°$.

Now try Exercises 13 and 15.

Do not confuse an angle with its measure. Angle A of Figure 7 is a rotation; the measure of the rotation is $35°$. This measure is often expressed by saying that $m(\text{angle } A)$ is $35°$, where $m(\text{angle } A)$ is read "the measure of angle A." It is convenient, however, to abbreviate $m(\text{angle } A) = 35°$ as $A = 35°$.

Traditionally, portions of a degree have been measured with minutes and seconds. One **minute,** written $1'$, is $\frac{1}{60}$ of a degree.

$$1' = \frac{1}{60}^{\circ} \qquad \text{or} \qquad 60' = 1°$$

One **second,** $1''$, is $\frac{1}{60}$ of a minute.

$$1'' = \frac{1}{60}' = \frac{1}{3600}^{\circ} \qquad \text{or} \qquad 60'' = 1'$$

The measure $12° \, 42' \, 38''$ represents 12 degrees, 42 minutes, 38 seconds.

EXAMPLE 2 Calculating with Degrees, Minutes, and Seconds

Perform each calculation.

(a) $51° 29' + 32° 46'$ **(b)** $90° - 73° 12'$

Solution

(a) Add the degrees and the minutes separately.

$$
\begin{array}{r}
51° \; 29' \\
+ \; 32° \; 46' \\
\hline
83° \; 75'
\end{array}
$$

Since $75' = 60' + 15' = 1° 15'$, the sum is written

$$
\begin{array}{r}
83° \\
+ \quad 1° \; 15' \\
\hline
84° \; 15'.
\end{array}
$$

(b) $89° 60'$ Write $90°$ as $89° 60'$.

$$
\begin{array}{r}
89° \; 60' \\
- \; 73° \; 12' \\
\hline
16° \; 48'
\end{array}
$$

Now try Exercises 23 and 27.

Because calculators are now so prevalent, angles are commonly measured in decimal degrees. For example, $12.4238°$ represents

$$12.4238° = 12\frac{4238}{10,000}°.$$

EXAMPLE 3 Converting Between Decimal Degrees and Degrees, Minutes, and Seconds

(a) Convert $74° 8' 14''$ to decimal degrees.

(b) Convert $34.817°$ to degrees, minutes, and seconds.

Solution

```
74°8'14"
         74.13722222
74°8'14"
              74.137
34.817▶DMS
         34°49'1.2"
```

A graphing calculator performs the conversions in Example 3 as shown above.

(a) $74° 8' 14'' = 74° + \dfrac{8}{60}° + \dfrac{14}{3600}°$ $1' = \frac{1}{60}°$ and $1'' = \frac{1}{3600}°$

          $\approx 74° + .1333° + .0039°$

          $\approx 74.137°$ Add; round to the nearest thousandth.

(b) $34.817° = 34° + .817°$

          $= 34° + .817(60')$ $1° = 60'$

          $= 34° + 49.02'$

          $= 34° + 49' + .02'$

          $= 34° + 49' + .02(60'')$ $1' = 60''$

          $= 34° + 49' + 1.2''$

          $= 34° 49' 1.2''$

Now try Exercises 33 and 37.

Standard Position An angle is in **standard position** if its vertex is at the origin and its initial side is along the positive *x*-axis. The angles in Figures 8(a) and 8(b) are in standard position. An angle in standard position is said to lie in the quadrant in which its terminal side lies. An acute angle is in quadrant I (Figure 8(a)) and an obtuse angle is in quadrant II (Figure 8(b)). Figure 8(c) shows ranges of angle measures for each quadrant when $0° < \theta < 360°$. Angles in standard position having their terminal sides along the *x*-axis or *y*-axis, such as angles with measures 90°, 180°, 270°, and so on, are called **quadrantal angles.**

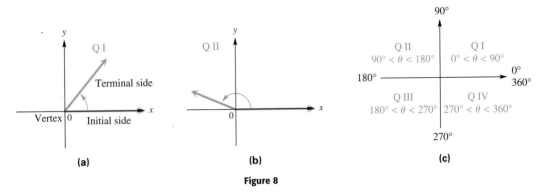

(a) (b) (c)

Figure 8

Coterminal Angles A complete rotation of a ray results in an angle measuring 360°. By continuing the rotation, angles of measure larger than 360° can be produced. The angles in Figure 9 with measures 60° and 420° have the same initial side and the same terminal side, but different amounts of rotation. Such angles are called **coterminal angles;** their measures differ by a multiple of 360°. As shown in Figure 10, angles with measures 110° and 830° are coterminal.

Figure 9 **Figure 10**

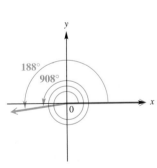

Figure 11

EXAMPLE 4 Finding Measures of Coterminal Angles

Find the angles of smallest possible positive measure coterminal with each angle.

(a) 908° **(b)** −75°

Solution

(a) Add or subtract 360° as many times as needed to obtain an angle with measure greater than 0° but less than 360°. Since $908° - 2 \cdot 360° = 188°$, an angle of 188° is coterminal with an angle of 908°. See Figure 11.

(b) Use a rotation of $360° + (-75°) = 285°$. See Figure 12.

Now try Exercises 45 and 49.

Figure 12

Sometimes it is necessary to find an expression that will generate all angles coterminal with a given angle. For example, we can obtain any angle coterminal with 60° by adding an appropriate integer multiple of 360° to 60°. Let n represent any integer; then the expression

$$60° + n \cdot 360°$$

represents all such coterminal angles. The table shows a few possibilities.

Value of n	Angle Coterminal with 60°
2	$60° + 2 \cdot 360° = 780°$
1	$60° + 1 \cdot 360° = 420°$
0	$60° + 0 \cdot 360° = 60°$ (the angle itself)
−1	$60° + (−1) \cdot 360° = −300°$

EXAMPLE 5 Analyzing the Revolutions of a CD Player

CAV (Constant Angular Velocity) CD players always spin at the same speed. Suppose a CAV player makes 480 revolutions per min. Through how many degrees will a point on the edge of a CD move in 2 sec?

Solution The player revolves 480 times in 1 min or $\frac{480}{60}$ times = 8 times per sec (since 60 sec = 1 min). In 2 sec, the player will revolve $2 \cdot 8 = 16$ times. Each revolution is 360°, so a point on the edge of the CD will revolve $16 \cdot 360° = 5760°$ in 2 sec.

Now try Exercise 75.

1.1 Exercises

 1. Explain the difference between a segment and a ray.

2. What part of a complete revolution is an angle of 45°?

3. *Concept Check* What angle is its own complement?

4. *Concept Check* What angle is its own supplement?

Find (a) the complement and (b) the supplement of each angle.

5. 30°   **6.** 60°   **7.** 45°   **8.** 18°   **9.** 54°   **10.** 89°

Find the measure of the smaller angle formed by the hands of a clock at the following times.

11.

12.

Find the measure of each angle in Exercises 13–18. See Example 1.

13.   **14.**   **15.**

16. supplementary angles with measures $10m + 7$ and $7m + 3$ degrees

17. supplementary angles with measures $6x - 4$ and $8x - 12$ degrees

18. complementary angles with measures $9z + 6$ and $3z$ degrees

Concept Check   Answer each question.

19. If an angle measures $x°$, how can we represent its complement?

20. If an angle measures $x°$, how can we represent its supplement?

21. If a positive angle has measure $x°$ between $0°$ and $60°$, how can we represent the first negative angle coterminal with it?

22. If a negative angle has measure $x°$ between $0°$ and $-60°$, how can we represent the first positive angle coterminal with it?

Perform each calculation. See Example 2.

23. $62° \, 18' + 21° \, 41'$ **24.** $75° \, 15' + 83° \, 32'$ **25.** $71° \, 18' - 47° \, 29'$

26. $47° \, 23' - 73° \, 48'$ **27.** $90° - 51° \, 28'$ **28.** $180° - 124° \, 51'$

29. $90° - 72° \, 58' \, 11''$ **30.** $90° - 36° \, 18' \, 47''$

Convert each angle measure to decimal degrees. Round to the nearest thousandth of a degree. See Example 3.

31. $20° \, 54'$ **32.** $38° \, 42'$ **33.** $91° \, 35' \, 54''$

34. $34° \, 51' \, 35''$ **35.** $274° \, 18' \, 59''$ **36.** $165° \, 51' \, 9''$

Convert each angle measure to degrees, minutes, and seconds. See Example 3.

37. $31.4296°$ **38.** $59.0854°$ **39.** $89.9004°$

40. $102.3771°$ **41.** $178.5994°$ **42.** $122.6853°$

43. Read about the degree symbol (°) in the manual for your graphing calculator. How is it used?

44. Show that 1.21 hr is the same as 1 hr, 12 min, 36 sec. Discuss the similarity between converting hours, minutes, and seconds to decimal hours and converting degrees, minutes, and seconds to decimal degrees.

Find the angle of smallest positive measure coterminal with each angle. See Example 4.

45. $-40°$ **46.** $-98°$ **47.** $-125°$ **48.** $-203°$

49. $539°$ **50.** $699°$ **51.** $850°$ **52.** $1000°$

Give an expression that generates all angles coterminal with each angle. Let n represent any integer.

53. $30°$ **54.** $45°$ **55.** $135°$ **56.** $270°$ **57.** $-90°$ **58.** $-135°$

59. Explain why the answers to Exercises 56 and 57 give the same set of angles.

60. *Concept Check*   Which two of the following are not coterminal with $r°$?

A. $360° + r°$ **B.** $r° - 360°$ **C.** $360° - r°$ **D.** $r° + 180°$

Concept Check   Sketch each angle in standard position. Draw an arrow representing the correct amount of rotation. Find the measure of two other angles, one positive and one negative, that are coterminal with the given angle. Give the quadrant of each angle.

61. 75°   **62.** 89°   **63.** 174°   **64.** 234°

65. 300°   **66.** 512°   **67.** −61°   **68.** −159°

Concept Check   Locate each point in a coordinate system. Draw a ray from the origin through the given point. Indicate with an arrow the angle in standard position having smallest positive measure. Then find the distance r from the origin to the point, using the distance formula of Appendix B.

69. $(-3, -3)$ **70.** $(-5, 2)$ **71.** $(-3, -5)$

72. $(\sqrt{3}, 1)$ **73.** $(-2, 2\sqrt{3})$ **74.** $(4\sqrt{3}, -4)$

Solve each problem. See Example 5.

75. *Revolutions of a Turntable*   A turntable in a shop makes 45 revolutions per min. How many revolutions does it make per second?

76. *Revolutions of a Windmill*   A windmill makes 90 revolutions per min. How many revolutions does it make per second?

77. *Rotating Tire*   A tire is rotating 600 times per min. Through how many degrees does a point on the edge of the tire move in $\frac{1}{2}$ sec?

78. *Rotating Airplane Propeller*   An airplane propeller rotates 1000 times per min. Find the number of degrees that a point on the edge of the propeller will rotate in 1 sec.

79. *Rotating Pulley*   A pulley rotates through 75° in 1 min. How many rotations does the pulley make in an hour?

80. *Surveying*   One student in a surveying class measures an angle as 74.25°, while another student measures the same angle as 74° 20′. Find the difference between these measurements, both to the nearest minute and to the nearest hundredth of a degree.

81. *Viewing Field of a Telescope*   Due to Earth's rotation, celestial objects like the moon and the stars appear to move across the sky, rising in the east and setting in the west. As a result, if a telescope on Earth remains stationary while viewing a celestial object, the object will slowly move outside the viewing field of the telescope. For this reason, a motor is often attached to telescopes so that the telescope rotates at the same rate as Earth. Determine how long it should take the motor to turn the telescope through an angle of 1 min in a direction perpendicular to Earth's axis.

82. *Angle Measure of a Star on the American Flag* Determine the measure of the angle in each point of the five-pointed star appearing on the American flag. (*Hint:* Inscribe the star in a circle, and use the following theorem from geometry: *An angle whose vertex lies on the circumference of a circle is equal to half the central angle that cuts off the same arc.* See the figure.)

1.2 | Angle Relationships and Similar Triangles

Geometric Properties ▪ **Triangles**

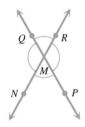

Vertical angles

Figure 13

Geometric Properties   In Figure 13, we extended the sides of angle *NMP* to form another angle, *RMQ*. The pair of angles *NMP* and *RMQ* are called **vertical angles.** Another pair of vertical angles, *NMQ* and *PMR*, are formed at the same time. Vertical angles have the following important property.

> ## Vertical Angles
>
> Vertical angles have equal measures.

Parallel lines are lines that lie in the same plane and do not intersect. Figure 14 shows parallel lines *m* and *n*. When a line *q* intersects two parallel lines, *q* is called a **transversal.** In Figure 14, the transversal intersecting the parallel lines forms eight angles, indicated by numbers.

We learn in geometry that the degree measures of angles 1 through 8 in Figure 14 possess some special properties. The following chart gives the names of these angles and rules about their measures.

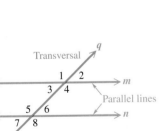

Figure 14

Name	Sketch	Rule
Alternate interior angles	*(sketch: 5 and 4, also 3 and 6)*	Angle measures are equal.
Alternate exterior angles	*(sketch: 1 and 8, also 2 and 7)*	Angle measures are equal.

(continued)

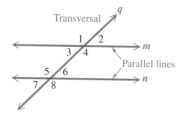

Figure 14 (repeated)

Name	Sketch	Rule
Interior angles on same side of transversal	 (also 3 and 5)	Angle measures add to 180°.
Corresponding angles	 (also 1 and 5, 3 and 7, 4 and 8)	Angle measures are equal.

Figure 15

EXAMPLE 1 Finding Angle Measures

Find the measure of each marked angle in Figure 15, given that lines m and n are parallel.

Solution The marked angles are alternate exterior angles, which are equal. Thus,

$$3x + 2 = 5x - 40$$
$$42 = 2x \qquad \text{Subtract } 3x; \text{ add } 40. \text{ (Appendix A)}$$
$$21 = x. \qquad \text{Divide by } 2.$$

One angle has measure

$$3x + 2 = 3 \cdot 21 + 2 = 65°, \qquad \text{Substitute } 21 \text{ for } x.$$

and the other has measure

$$5x - 40 = 5 \cdot 21 - 40 = 65°. \qquad \text{Substitute } 21 \text{ for } x.$$

Now try Exercise 11.

Triangles An important property of triangles, first proved by Greek geometers, deals with the sum of the measures of the angles of any triangle.

Angle Sum of a Triangle

The sum of the measures of the angles of any triangle is 180°.

(a)

(b)

Figure 16

While it is not an actual proof, we give a rather convincing argument for the truth of this statement, using any size triangle cut from a piece of paper. Tear each corner from the triangle, as suggested in Figure 16(a). You should be able to rearrange the pieces so that the three angles form a straight angle, which has measure 180°, as shown in Figure 16(b).

EXAMPLE 2 Applying the Angle Sum of a Triangle Property

The measures of two of the angles of a triangle are 48° and 61°. (See the figure.) Find the measure of the third angle, x.

Solution $48° + 61° + x = 180°$ The sum of the angles is 180°.

$$109° + x = 180°$$

$$x = 71°$$ Subtract 109°.

The third angle of the triangle measures 71°.

Now try Exercise 15.

We classify triangles according to angles and sides, as shown in the following chart.

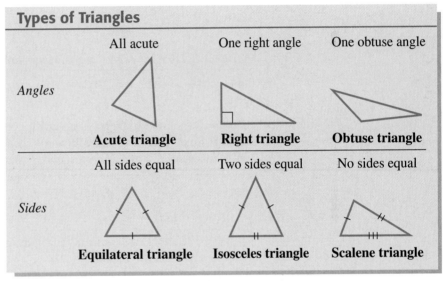

Types of Triangles

	All acute	One right angle	One obtuse angle
Angles	**Acute triangle**	**Right triangle**	**Obtuse triangle**
	All sides equal	Two sides equal	No sides equal
Sides	**Equilateral triangle**	**Isosceles triangle**	**Scalene triangle**

Now try Exercises 25, 27, and 31.

(a)

(b)

(c)

Figure 17

Many key ideas of trigonometry depend on **similar triangles,** triangles of exactly the same shape but not necessarily the same size. Figure 17 shows three pairs of similar triangles.

The two triangles in Figure 17(c) not only have the same shape but also the same size. Triangles that are both the same size and the same shape are called **congruent triangles.** If two triangles are congruent, then it is possible to pick one of them up and place it on top of the other so that they coincide. If two triangles are congruent, then they must be similar. However, two similar triangles need not be congruent.

The triangular supports for a child's swing are congruent (and thus similar) triangles, machine-produced with exactly the same dimensions each time. These supports are just one example of similar triangles. The supports of a long bridge, all the same shape but decreasing in size toward the center of the bridge, are examples of similar (but not congruent) triangles.

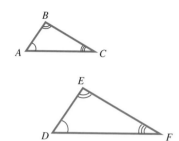

Figure 18

Suppose a correspondence between two triangles ABC and DEF is set up as shown in Figure 18.

Angle A corresponds to angle D.   Side AB corresponds to side DE.
Angle B corresponds to angle E.   Side BC corresponds to side EF.
Angle C corresponds to angle F.   Side AC corresponds to side DF.

For triangle ABC to be similar to triangle DEF, the following conditions must hold.

Conditions for Similar Triangles

1. Corresponding angles must have the same measure.

2. Corresponding sides must be proportional. (That is, their ratios must be equal.)

Now try Exercise 41.

EXAMPLE 3 Finding Angle Measures in Similar Triangles

Figure 19

In Figure 19, triangles ABC and NMP are similar. Find the measures of angles B and C.

Solution   Since the triangles are similar, corresponding angles have the same measure. Since C corresponds to P and P measures $104°$, angle C also measures $104°$. Since angles B and M correspond, B measures $31°$.

Now try Exercise 47.

N O T E   The small arcs found at the angles in Figures 17–19 denote the corresponding angles in the triangles.

EXAMPLE 4 Finding Side Lengths in Similar Triangles

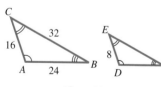

Figure 20

Given that triangle ABC and triangle DFE in Figure 20 are similar, find the lengths of the unknown sides of triangle DFE.

Solution   As mentioned before, similar triangles have corresponding sides in proportion. Use this fact to find the unknown side lengths in triangle DFE. Side DF of triangle DFE corresponds to side AB of triangle ABC, and sides DE and AC correspond. This leads to the proportion

$$\frac{8}{16} = \frac{DF}{24}.$$

Recall the *cross-multiplication property* of proportions from algebra.

$$\text{If} \quad \frac{a}{b} = \frac{c}{d}, \quad \text{then} \quad ad = bc.$$

We use this property to solve the equation for *DF*.

$$\frac{8}{16} = \frac{DF}{24}$$

$$8 \cdot 24 = 16 \cdot DF \quad \text{Cross multiply.}$$

$$192 = 16 \cdot DF \quad \text{Multiply.}$$

$$12 = DF \quad \text{Divide by 16.}$$

Side *DF* has length 12.

Side *EF* corresponds to *CB*. This leads to another proportion.

$$\frac{8}{16} = \frac{EF}{32}$$

$$8 \cdot 32 = 16 \cdot EF \quad \text{Cross multiply.}$$

$$16 = EF \quad \text{Solve for } EF.$$

Side *EF* has length 16.

Now try Exercise 53.

EXAMPLE 5 Finding the Height of a Flagpole

Firefighters at the Arcade Fire Station need to measure the height of the station flagpole. They find that at the instant when the shadow of the station is 18 ft long, the shadow of the flagpole is 99 ft long. The station is 10 ft high. Find the height of the flagpole.

Solution Figure 21 shows the information given in the problem. The two triangles are similar, so corresponding sides are in proportion.

$$\frac{MN}{10} = \frac{99}{18}$$

$$\frac{MN}{10} = \frac{11}{2} \quad \text{Lowest terms}$$

$$2 \cdot MN = 110 \quad \text{Cross multiply.}$$

$$MN = 55 \quad \text{Divide by 2.}$$

Figure 21

The flagpole is 55 ft high.

Now try Exercise 57.

EXAMPLE 6 Determining When a Solar Eclipse Can Occur

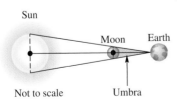

Not to scale

Figure 22

The sun has a diameter of about 865,000 mi with a maximum distance from Earth's surface of about 94,500,000 mi. The moon has a smaller diameter of 2159 mi. For a total solar eclipse to occur, the moon must pass between Earth and the sun. The moon must also be close enough to Earth for the moon's *umbra* (shadow) to reach the surface of Earth. See Figure 22. (*Source:* Karttunen, H., P. Kröger, H. Oja, M. Putannen, and K. Donners, Editors, *Fundamental Astronomy,* Fourth Edition, Springer-Verlag, 2003.)

(a) Calculate the maximum distance that the moon can be from Earth and still have a total solar eclipse occur.

(b) In a recent year, the closest approach of the moon to Earth's surface was 225,745 mi and the farthest was 251,978 mi. (*Source: The World Almanac and Book of Facts.*) Can a total solar eclipse occur every time the moon is between Earth and the sun? Explain.

Solution

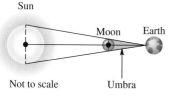

Figure 22 (repeated)

(a) Let D_s be the Earth-sun distance, d_s the diameter of the sun, D_m the Earth-moon distance, and d_m the diameter of the moon.

$$\frac{D_s}{D_m} = \frac{d_s}{d_m} \quad \text{Similar triangles}$$

$$D_m = \frac{D_s d_m}{d_s} = \frac{94{,}500{,}000 \times 2159}{865{,}000} \approx 236{,}000 \text{ mi}$$

(b) No, a total solar eclipse cannot occur every time. The moon must be less than 236,000 mi away from Earth for an eclipse to occur, and sometimes it is farther than this.

Now try Exercise 73.

1.2 Exercises

1. A geometry book states, "When two straight lines intersect, the opposite angles are equal." What term do we use for "opposite angles"?

2. Consider Figure 14. If the measure of one of the angles is known, can the measures of the remaining seven angles be determined? Explain.

Find the measure of each marked angle. In Exercises 11–14, m and n are parallel. See Examples 1 and 2.

3.
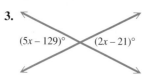
$(5x - 129)°$ $(2x - 21)°$

4.

$(11x - 37)°$ $(7x + 27)°$

5.
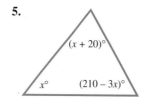
$(x + 20)°$
$x°$ $(210 - 3x)°$

6.
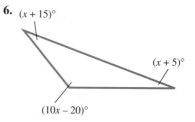
$(x + 15)°$
$(x + 5)°$
$(10x - 20)°$

7.

8.

9.

10.

11.

12.

13.

14.

The measures of two angles of a triangle are given. Find the measure of the third angle. See Example 2.

15. $37°, 52°$

16. $29°, 104°$

17. $147° \, 12', 30° \, 19'$

18. $136° \, 50', 41° \, 38'$

19. $74.2°, 80.4°$

20. $29.6°, 49.7°$

21. Can a triangle have angles of measures $85°$ and $100°$? Explain.

22. Can a triangle have two obtuse angles? Explain.

23. *Concept Check* Use the given figure to find the measures of the numbered angles, given that lines m and n are parallel.

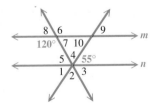

24. *Concept Check* Find the measures of the marked angles, given that $x + y = 40$. (*Hint:* You must solve a system of equations.)

Concept Check *Classify each triangle in Exercises 25–36 as* acute, right, *or* obtuse.
Also classify each as equilateral, isosceles, *or* scalene.

25.

26.

120°

27.

60°

8   8

60°   60°

8

28.

9   9

6

29.

3

90°

5   4

30.

8

130°

8

31.

4

4

32.

60°

33.

10

96°

14

34.

60°

r   r

60°   60°

r

35.

50°   50°

36.

13

5

90°

12

37. Write a definition of *isosceles right triangle.*

38. Explain why the sum of the lengths of any two sides of a triangle must be greater than the length of the third side.

39. Must all equilateral triangles be similar? Explain.

40. *Carpentry Technique*   The following technique is used by carpenters to draw a 60° angle with a straightedge and compass. Explain why this technique works. (*Source:* Hamilton, J. E. and M. S. Hamilton, *Math to Build On,* Construction Trades Press, 1993.)

Step 1   Draw a straight line segment, and mark a point near the center of the line. See the figure at the top of the next page.

Step 2   Place the compass tip on the marked point and draw a semicircle.

Step 3   Without changing the setting of the compass, place the tip of the compass at the right intersection of the line and the semicircle and then mark a small arc across the semicircle.

Step 4   Draw a line segment from the marked point on the line to the point where the arc crosses the semicircle. This line will make a 60° angle with the original line.

Point marked
on line in Step 1

Compass tip placed
here in Step 3

Concept Check   *Name the corresponding angles and the corresponding sides of each pair of similar triangles.*

41.

42.
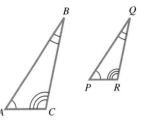

43. (*EA* is parallel to *CD*.)
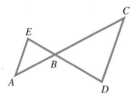

44. (*HK* is parallel to *EF*.)

Find all unknown angle measures in each pair of similar triangles. See Example 3.

45.

46.

47.

48.

49. **50.**

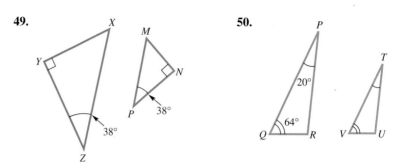

Find the unknown side lengths labeled with a variable in each pair of similar triangles. See Example 4.

51. **52.**

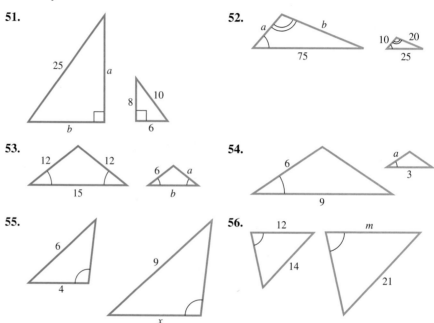

53. **54.**

55. **56.**

Solve each problem. (Hint: Draw a sketch.) See Example 5.

57. *Height of a Tree* A tree casts a shadow 45 m long. At the same time, the shadow cast by a vertical 2-m stick is 3 m long. Find the height of the tree.

58. *Height of a Lookout Tower* A forest fire lookout tower casts a shadow 180 ft long at the same time that the shadow of a 9-ft truck is 15 ft long. Find the height of the tower.

59. *Lengths of Sides of a Triangle* On a photograph of a triangular piece of land, the lengths of the three sides are 4 cm, 5 cm, and 7 cm, respectively. The shortest side of the actual piece of land is 400 m long. Find the lengths of the other two sides.

60. *Height of a Lighthouse* The Santa Cruz lighthouse in the figure casts a shadow 28 m long at 7 P.M. At the same time, the shadow of the lighthouse keeper, who is 1.75 m tall, is 3.5 m long. How tall is the lighthouse?

Not to scale

61. *Height of a Building*   A house is 15 ft tall. Its shadow is 40 ft long at the same time the shadow of a nearby building is 300 ft long. Find the height of the building.

62. *Distance Between Two Cities*   By drawing lines on a map, a triangle can be formed by the cities of Phoenix, Tucson, and Yuma. On the map, the distance between Phoenix and Tucson is 8 cm, the distance between Phoenix and Yuma is 12 cm, and the distance between Tucson and Yuma is 17 cm. The actual straight-line distance from Phoenix to Yuma is 230 km. Find the distances between the other pairs of cities.

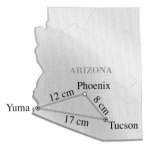

In each diagram, there are two similar triangles. Find the unknown measurement. (Hint: In the sketch for Exercise 63, the side of length 100 in the small triangle corresponds to the side of the length $100 + 120 = 220$ in the large triangle.)

63.

64.

65.

66.

Work each problem.

67. *Lengths of Sides of a Quadrilateral*   Two quadrilaterals (four-sided figures) are similar. The lengths of the three shortest sides of the first quadrilateral are 18 cm, 24 cm, and 32 cm. The lengths of the two longest sides of the second quadrilateral are 48 cm and 60 cm. Find the unknown lengths of the sides of these two figures.

68. *Height of a Carving of Lincoln*   Assume that Lincoln was $6\frac{1}{3}$ ft tall and his head $\frac{3}{4}$ ft long. Knowing that the carved head of Lincoln at Mt. Rushmore is 60 ft tall, find how tall his entire body would be if it were carved into the mountain.

In each figure, two similar triangles are present. Find the value of each variable.

69.

70.

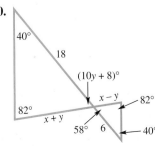

Solve each problem.

71. *Sizes and Distances in the Sky* Astron- omers use degrees, minutes, and seconds to measure sizes and distances in the sky along an arc from the horizon to the zenith point directly overhead. An adult observer on Earth can judge distances in the sky using his or her hand at arm's length. An outstretched hand will be about 20 arc degrees wide from the tip of the thumb to the tip of the little finger. A clenched fist at arm's length measures about 10 arc degrees, and a thumb corresponds to about 2 arc degrees. (*Source:* Levy, D. H., *Skywatching,* The Nature Company, 1994.)

(a) The apparent size of the moon is about 31 arc minutes. What part of your thumb would cover the moon?

(b) If an outstretched hand plus a fist cover the distance between two bright stars, about how far apart in arc degrees are the stars?

72. *Eclipse on Mars* Refer to Example 6. The sun's distance from the surface of Mars is approximately 142,000,000 mi. One of Mars' two moons, Phobos, has a max- imum diameter of 17.4 mi. (*Source:* Zeilik, M., S. Gregory, and E. Smith, *Introductory Astronomy and Astrophysics,* Second Edition, Saunders College Publishers, 1998.)

(a) Calculate the maximum distance that the moon Phobos can be from Mars for a total eclipse of the sun to occur on Mars.

(b) Phobos is approximately 5800 mi from Mars. Is it possible for Phobos to cause a total eclipse on Mars?

73. *Eclipse on Jupiter* Refer to Example 6. The sun's distance from the surface of Jupiter is approximately 484,000,000 mi. One of Jupiter's moons, Ganymede, is the largest moon in the solar system with a diameter of 3270 mi. (*Source:* Wright, J. W., General Editor, *The Universal Almanac,* Andrews and McMeel, 1997.)

(a) Calculate the maximum distance that the moon Ganymede can be from Jupiter for a total eclipse of the sun to occur on Jupiter.

(b) Ganymede is approximately 665,000 mi from Jupiter. Is it possible for Ganymede to cause a total eclipse on Jupiter?

1.3 Trigonometric Functions

Trigonometric Functions ▪ **Quadrantal Angles**

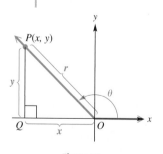

Figure 23

Trigonometric Functions The study of trigonometry covers the six *trigonometric functions* defined here. To define these functions, we start with an angle θ in standard position, and choose any point P having coordinates (x, y) on the terminal side of angle θ. (The point P must not be the vertex of the angle.) See Figure 23. A perpendicular from P to the x-axis at point Q determines a right triangle, having vertices at O, P, and Q. We find the distance r from $P(x, y)$ to the origin, $(0, 0)$, using the distance formula.

$$r = \sqrt{(x - 0)^2 + (y - 0)^2} = \sqrt{x^2 + y^2} \quad \text{(Appendix B)}$$

Notice that $r > 0$ since distance is never negative.

The six trigonometric functions of angle θ are **sine, cosine, tangent, cotangent, secant,** and **cosecant.** In the following definitions, we use the customary abbreviations for the names of these functions.

Trigonometric Functions

Let (x, y) be a point other than the origin on the terminal side of an angle θ in standard position. The distance from the point to the origin is $r = \sqrt{x^2 + y^2}$. The six trigonometric functions of θ are defined as follows.

$$\sin \theta = \frac{y}{r} \qquad \cos \theta = \frac{x}{r} \qquad \tan \theta = \frac{y}{x} \;\; (x \neq 0)$$

$$\csc \theta = \frac{r}{y} \;\; (y \neq 0) \qquad \sec \theta = \frac{r}{x} \;\; (x \neq 0) \qquad \cot \theta = \frac{x}{y} \;\; (y \neq 0)$$

EXAMPLE 1 Finding Function Values of an Angle

The terminal side of an angle θ in standard position passes through the point $(8, 15)$. Find the values of the six trigonometric functions of angle θ.

Solution   Figure 24 shows angle θ and the triangle formed by dropping a perpendicular from the point $(8, 15)$ to the x-axis. The point $(8, 15)$ is 8 units to the right of the y-axis and 15 units above the x-axis, so $x = 8$ and $y = 15$. Since $r = \sqrt{x^2 + y^2}$,

$$r = \sqrt{8^2 + 15^2} = \sqrt{64 + 225} = \sqrt{289} = 17.$$

We can now find the values of the six trigonometric functions of angle θ.

$$\sin \theta = \frac{y}{r} = \frac{15}{17} \qquad \cos \theta = \frac{x}{r} = \frac{8}{17} \qquad \tan \theta = \frac{y}{x} = \frac{15}{8}$$

$$\csc \theta = \frac{r}{y} = \frac{17}{15} \qquad \sec \theta = \frac{r}{x} = \frac{17}{8} \qquad \cot \theta = \frac{x}{y} = \frac{8}{15}$$

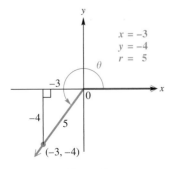

Figure 24

Now try Exercise 9.

EXAMPLE 2 Finding Function Values of an Angle

The terminal side of an angle θ in standard position passes through the point $(-3, -4)$. Find the values of the six trigonometric functions of angle θ.

Solution   As shown in Figure 25, $x = -3$ and $y = -4$. The value of r is

$$r = \sqrt{(-3)^2 + (-4)^2} = \sqrt{25} = 5. \quad \text{Remember that } r > 0.$$

Then use the definitions of the trigonometric functions.

$$\sin \theta = \frac{-4}{5} = -\frac{4}{5} \qquad \cos \theta = \frac{-3}{5} = -\frac{3}{5} \qquad \tan \theta = \frac{-4}{-3} = \frac{4}{3}$$

$$\csc \theta = \frac{5}{-4} = -\frac{5}{4} \qquad \sec \theta = \frac{5}{-3} = -\frac{5}{3} \qquad \cot \theta = \frac{-3}{-4} = \frac{3}{4}$$

Figure 25

Now try Exercise 5.

Figure 26

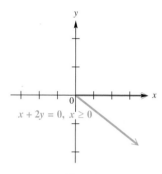

Figure 27

We can find the six trigonometric functions using *any* point other than the origin on the terminal side of an angle. To see why any point may be used, refer to Figure 26, which shows an angle θ and two distinct points on its terminal side. Point P has coordinates (x, y), and point P' (read "P-prime") has coordinates (x', y'). Let r be the length of the hypotenuse of triangle OPQ, and let r' be the length of the hypotenuse of triangle $OP'Q'$. Since corresponding sides of similar triangles are proportional,

$$\frac{y}{r} = \frac{y'}{r'}, \quad \text{(Section 1.2)}$$

so $\sin \theta = \frac{y}{r}$ is the same no matter which point is used to find it. A similar result holds for the other five trigonometric functions.

We can also find the trigonometric function values of an angle if we know the equation of the line coinciding with the terminal ray. Recall from algebra that the graph of the equation

$$Ax + By = 0 \quad \text{(Appendix B)}$$

is a line that passes through the origin. If we restrict x to have only nonpositive or only nonnegative values, we obtain as the graph a ray with endpoint at the origin. For example, the graph of $x + 2y = 0$, $x \geq 0$, shown in Figure 27, is a ray that can serve as the terminal side of an angle in standard position. By choosing a point on the ray, we can find the trigonometric function values of the angle.

EXAMPLE 3 Finding Function Values of an Angle

Find the six trigonometric function values of the angle θ in standard position, if the terminal side of θ is defined by $x + 2y = 0$, $x \geq 0$.

Figure 28

Solution   The angle is shown in Figure 28. We can use *any* point except $(0, 0)$ on the terminal side of θ to find the trigonometric function values. We choose $x = 2$ and find the corresponding y-value.

$$x + 2y = 0, \ x \geq 0$$

$$2 + 2y = 0 \qquad \text{Let } x = 2.$$

$$2y = -2 \qquad \text{Subtract 2. (Appendix A)}$$

$$y = -1 \qquad \text{Divide by 2.}$$

The point $(2, -1)$ lies on the terminal side, and the corresponding value of r is $r = \sqrt{2^2 + (-1)^2} = \sqrt{5}$. Now we use the definitions of the trigonometric functions.

$$\sin \theta = \frac{y}{r} = \frac{-1}{\sqrt{5}} = \frac{-1}{\sqrt{5}} \cdot \frac{\sqrt{5}}{\sqrt{5}} = -\frac{\sqrt{5}}{5} \qquad \text{Multiply the numerators and denominators by } \sqrt{5} \text{ to rationalize the denominators.}$$

$$\cos \theta = \frac{x}{r} = \frac{2}{\sqrt{5}} = \frac{2}{\sqrt{5}} \cdot \frac{\sqrt{5}}{\sqrt{5}} = \frac{2\sqrt{5}}{5}$$

$$\tan \theta = \frac{y}{x} = \frac{-1}{2} = -\frac{1}{2}$$

$$\csc \theta = \frac{r}{y} = \frac{\sqrt{5}}{-1} = -\sqrt{5} \qquad \sec \theta = \frac{r}{x} = \frac{\sqrt{5}}{2} \qquad \cot \theta = \frac{x}{y} = \frac{2}{-1} = -2$$

Now try Exercise 25.

Recall that when the equation of a line is written in the form $y = mx + b$, the coefficient of x is the slope of the line. In Example 3, $x + 2y = 0$ can be written as $y = -\frac{1}{2}x$, so the slope is $-\frac{1}{2}$. Notice that $\tan \theta = -\frac{1}{2}$. *In general, it is true that $m = \tan \theta$.*

N O T E The trigonometric function values we found in Examples 1–3 are *exact*. If we were to use a calculator to approximate these values, the decimal results would not be acceptable if exact values were required.

Quadrantal Angles If the terminal side of an angle in standard position lies along the y-axis, any point on this terminal side has x-coordinate 0. Similarly, an angle with terminal side on the x-axis has y-coordinate 0 for any point on the terminal side. Since the values of x and y appear in the denominators of some trigonometric functions, and since a fraction is undefined if its denominator is 0, some trigonometric function values of quadrantal angles (i.e., those with terminal side on an axis) are undefined.

EXAMPLE 4 Finding Function Values of Quadrantal Angles

Find the values of the six trigonometric functions for each angle.

(a) an angle of 90°

(b) an angle θ in standard position with terminal side through $(-3, 0)$

Solution

A calculator in degree mode returns the correct values for sin 90° and cos 90°. The second screen shows an ERROR message for tan 90°, because 90° is not in the domain of the tangent function.

(a) First, we select any point on the terminal side of a 90° angle. We choose the point $(0, 1)$, as shown in Figure 29. Here $x = 0$ and $y = 1$, so $r = 1$. Then,

$$\sin 90° = \frac{1}{1} = 1 \qquad \cos 90° = \frac{0}{1} = 0 \qquad \tan 90° = \frac{1}{0} \ \text{(undefined)}$$

$$\csc 90° = \frac{1}{1} = 1 \qquad \sec 90° = \frac{1}{0} \ \text{(undefined)} \qquad \cot 90° = \frac{0}{1} = 0.$$

Figure 29   **Figure 30**

(b) Figure 30 shows the angle. Here, $x = -3$, $y = 0$, and $r = 3$, so the trigonometric functions have the following values.

$$\sin \theta = \frac{0}{3} = 0 \qquad\qquad \cos \theta = \frac{-3}{3} = -1 \qquad \tan \theta = \frac{0}{-3} = 0$$

$$\csc \theta = \frac{3}{0} \ \text{(undefined)} \qquad \sec \theta = \frac{3}{-3} = -1 \qquad \cot \theta = \frac{-3}{0} \ \text{(undefined)}$$

Now try Exercises 7 and 11.

The conditions under which the trigonometric function values of quadrantal angles are undefined are summarized here.

Undefined Function Values

If the terminal side of a quadrantal angle lies along the *y*-axis, then the tangent and secant functions are undefined. If it lies along the *x*-axis, then the cotangent and cosecant functions are undefined.

The function values of the most commonly used quadrantal angles, 0°, 90°, 180°, 270°, and 360°, are summarized in the following table.

θ	$\sin \theta$	$\cos \theta$	$\tan \theta$	$\cot \theta$	$\sec \theta$	$\csc \theta$
0°	0	1	0	Undefined	1	Undefined
90°	1	0	Undefined	0	Undefined	1
180°	0	−1	0	Undefined	−1	Undefined
270°	−1	0	Undefined	0	Undefined	−1
360°	0	1	0	Undefined	1	Undefined

The values given in this table can be found with a calculator that has trigonometric function keys. Make sure the calculator is set in *degree mode.*

CAUTION One of the most common errors involving calculators in trigonometry occurs when the calculator is set for *radian measure,* rather than *degree measure.* (Radian measure of angles is discussed in Chapter 3.) Be sure you know how to set your calculator in *degree mode.*

1.3 Exercises

Concept Check   Sketch an angle θ in standard position such that θ has the smallest possible positive measure, and the given point is on the terminal side of θ.

1. $(5, -12)$ **2.** $(-12, -5)$ **3.** $(-3, 4)$ **4.** $(-4, -3)$

Find the values of the six trigonometric functions for each angle in standard position having the given point on its terminal side. Rationalize denominators when applicable. See Examples 1, 2, and 4.

5. $(-3, 4)$ **6.** $(-4, -3)$ **7.** $(0, 2)$ **8.** $(-4, 0)$
9. $(1, \sqrt{3})$ **10.** $(-2\sqrt{3}, -2)$ **11.** $(-2, 0)$ **12.** $(3, -4)$

13. For any nonquadrantal angle θ, sin θ and csc θ will have the same sign. Explain why.

14. How is the value of *r* interpreted geometrically in the definitions of the sine, cosine, secant, and cosecant functions?

15. *Concept Check*   If cot θ is undefined, then what is the value of tan θ?

16. *Concept Check*   If the terminal side of an angle θ is in quadrant III, then what is the sign of each of the trigonometric function values of θ?

Concept Check   Suppose that the point (x, y) *is in the indicated quadrant. Decide whether the given ratio is positive or negative. (Hint: Drawing a sketch may help.)*

17. II, $\dfrac{x}{r}$ **18.** III, $\dfrac{y}{r}$ **19.** IV, $\dfrac{y}{x}$ **20.** IV, $\dfrac{x}{y}$

21. II, $\dfrac{y}{r}$ **22.** III, $\dfrac{x}{r}$ **23.** IV, $\dfrac{x}{r}$ **24.** IV, $\dfrac{y}{r}$

In Exercises 25–30, an equation of the terminal side of an angle θ *in standard position is given with a restriction on x. Sketch the smallest positive such angle* θ*, and find the values of the six trigonometric functions of* θ*. See Example 3.*

25. $2x + y = 0, x \geq 0$ **26.** $3x + 5y = 0, x \geq 0$

27. $-6x - y = 0, x \leq 0$ **28.** $-5x - 3y = 0, x \leq 0$

29. $-4x + 7y = 0, x \leq 0$ **30.** $6x - 5y = 0, x \geq 0$

31. Find the six trigonometric function values of the quadrantal angle $450°$.

32. Rework Example 3 using the values of x and y for which the point (x, y) is on the circle of radius 1 having center at the origin, and then show that the six trigonometric function values you obtain are the same as those obtained in Example 3.

Use the trigonometric function values of quadrantal angles given in this section to evaluate each expression. An expression such as $\cot^2 90°$ *means* $(\cot 90°)^2$*, which is equal to* $0^2 = 0$*.*

33. $\cos 90° + 3 \sin 270°$ **34.** $\tan 0° - 6 \sin 90°$

35. $3 \sec 180° - 5 \tan 360°$ **36.** $4 \csc 270° + 3 \cos 180°$

37. $\tan 360° + 4 \sin 180° + 5 \cos^2 180°$ **38.** $2 \sec 0° + 4 \cot^2 90° + \cos 360°$

39. $\sin^2 180° + \cos^2 180°$ **40.** $\sin^2 360° + \cos^2 360°$

41. $\sec^2 180° - 3 \sin^2 360° + 2 \cos 180°$

42. $5 \sin^2 90° + 2 \cos^2 270° - 7 \tan 360°$

If n is an integer, $n \cdot 180°$ *represents an integer multiple of* $180°$*, and* $(2n + 1) \cdot 90°$ *represents an odd integer multiple of* $90°$*. Decide whether each expression is equal to 0, 1, -1, or is* undefined.

43. $\cos[(2n + 1) \cdot 90°]$ **44.** $\sin[n \cdot 180°]$

45. $\tan[n \cdot 180°]$ **46.** $\tan[(2n + 1) \cdot 90°]$

Provide conjectures in Exercises 47–50.

47. The angles $15°$ and $75°$ are complementary. With your calculator determine $\sin 15°$ and $\cos 75°$. Make a conjecture about the sines and cosines of complementary angles, and test your hypothesis with other pairs of complementary angles. (*Note:* This relationship will be discussed in detail in Section 2.1.)

48. The angles $25°$ and $65°$ are complementary. With your calculator determine $\tan 25°$ and $\cot 65°$. Make a conjecture about the tangents and cotangents of complementary angles, and test your hypothesis with other pairs of complementary angles. (*Note:* This relationship will be discussed in detail in Section 2.1.)

49. With your calculator determine $\sin 10°$ and $\sin(-10°)$. Make a conjecture about the sines of an angle and its negative, and test your hypothesis with other angles. Also, use a geometry argument with the definition of $\sin \theta$ to justify your hypothesis. (*Note:* This relationship will be discussed in detail in Section 5.1.)

50. With your calculator determine cos 20° and cos(−20°). Make a conjecture about the cosines of an angle and its negative, and test your hypothesis with other angles. Also, use a geometry argument with the definition of cos θ to justify your hypothesis. (*Note:* This relationship will be discussed in detail in Section 5.1.)

Concept Check *Work each problem.*

51. Give two values of $\theta, 0° \leq \theta < 360°$, for which sin $\theta = \cos \theta$.

52. Give two values of $\theta, 0° \leq \theta < 360°$, for which tan $\theta = 1$.

53. The equation cos $\theta = .8$ has the approximate solution $\theta = 36.87°$. Find an approximate solution for cos $\theta = −.8$.

54. The equation sin $\theta = .8$ has the approximate solution $\theta = 53.13°$. Find an approximate solution for sin $\theta = −.8$.

In Exercises 55–60, set your graphing calculator in parametric and degree modes. Set the window and functions (see the third screen) as shown here, and graph. A circle of radius 1 will appear on the screen. Trace to move a short distance around the circle. In the screen, the point on the circle corresponds to an angle T = 25°. Since r = 1, cos 25° is X = .90630779, and sin 25° is Y = .42261826.

  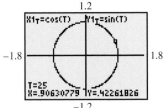

This screen is a continuation of the previous one.

55. Use the right- and left-arrow keys to move to the point corresponding to 20°. What are cos 20° and sin 20°?

56. For what angle T, $0° \leq T \leq 90°$, is cos T $\approx .766$?

57. For what angle T, $0° \leq T \leq 90°$, is sin T $\approx .574$?

58. For what angle T, $0° \leq T \leq 90°$, does cos T $=$ sin T?

59. As T increases from 0° to 90°, does the cosine increase or decrease? What about the sine?

60. As T increases from 90° to 180°, does the cosine increase or decrease? What about the sine?

61. *Area of a Solar Cell* A solar cell converts the energy of sunlight directly into electrical energy. The amount of energy a cell produces depends on its area. Suppose a solar cell is hexagonal, as shown in the figure. Express its area in terms of sin θ and any side x. (*Hint:* Consider one of the six equilateral triangles from the hexagon. See the figure.) (*Source:* Kastner, B., *Space Mathematics,* NASA, 1985.)

62. *(Modeling) Distance Between the Sun and a Star*
Suppose that a star forms an angle θ with respect to Earth and the sun. Let the coordinates of Earth be (x, y), those of the star $(0, 0)$, and those of the sun $(x, 0)$. See the figure. Find an equation for x, the distance between the sun and the star, as follows.

Not to scale

(a) Write an equation involving a trigonometric function that relates x, y, and θ.

(b) Solve your equation for x.

1.4 Using the Definitions of the Trigonometric Functions

Reciprocal Identities ▪ **Signs and Ranges of Function Values** ▪ **Pythagorean Identities** ▪ **Quotient Identities**

Identities are equations that are true for all values of the variables for which all expressions are defined. Identities are studied in more detail in Chapter 5.

$$(x + y)^2 = x^2 + 2xy + y^2 \qquad 2(x + 3) = 2x + 6 \quad \text{Identities (Appendix A)}$$

Reciprocal Identities Recall the definition of a reciprocal: the *reciprocal* of the nonzero number x is $\frac{1}{x}$. For example, the reciprocal of 2 is $\frac{1}{2}$, and the reciprocal of $\frac{8}{11}$ is $\frac{11}{8}$. There is no reciprocal for 0. Scientific calculators have a reciprocal key, usually labeled $\boxed{1/x}$ or $\boxed{x^{-1}}$. Using this key gives the reciprocal of any nonzero number entered in the display.

The definitions of the trigonometric functions in the previous section were written so that functions in the same column are reciprocals of each other. Since $\sin \theta = \frac{y}{r}$ and $\csc \theta = \frac{r}{y}$,

$$\sin \theta = \frac{1}{\csc \theta} \qquad \text{and} \qquad \csc \theta = \frac{1}{\sin \theta},$$

provided $\sin \theta \neq 0$. Also, $\cos \theta$ and $\sec \theta$ are reciprocals, as are $\tan \theta$ and $\cot \theta$. In summary, we have the **reciprocal identities** that hold for any angle θ that does not lead to a 0 denominator.

```
1/sin(90)
              1
1/cos(180)
             -1
(sin(-270))⁻¹
              1
1/cos(90)
```

(a)

Reciprocal Identities

$$\sin \theta = \frac{1}{\csc \theta} \qquad \cos \theta = \frac{1}{\sec \theta} \qquad \tan \theta = \frac{1}{\cot \theta}$$

$$\csc \theta = \frac{1}{\sin \theta} \qquad \sec \theta = \frac{1}{\cos \theta} \qquad \cot \theta = \frac{1}{\tan \theta}$$

```
ERR:DIVIDE BY 0
1 Quit
2:Goto
```

(b)

Figure 31

The screen in Figure 31(a) shows how to find csc 90°, sec 180°, and csc(−270°), using the appropriate reciprocal identities and the reciprocal key of a graphing calculator in degree mode. Be sure *not* to use the *inverse trigonometric function* keys to find the reciprocal function values. Attempting to find sec 90° by entering $\frac{1}{\cos 90°}$ produces an ERROR message, indicating the reciprocal is undefined. See Figure 31(b). Compare these results with the ones found in the table of quadrantal angle function values in Section 1.3. ■

NOTE   Identities can be written in different forms. For example,

$$\sin \theta = \frac{1}{\csc \theta} \quad \text{can be written} \quad \csc \theta = \frac{1}{\sin \theta} \quad \text{or} \quad (\sin \theta)(\csc \theta) = 1.$$

EXAMPLE 1   Using the Reciprocal Identities

Find each function value.

(a) $\cos \theta$, if $\sec \theta = \dfrac{5}{3}$

(b) $\sin \theta$, if $\csc \theta = -\dfrac{\sqrt{12}}{2}$

Solution

(a) Since $\cos \theta$ is the reciprocal of $\sec \theta$,

$$\cos \theta = \frac{1}{\sec \theta} = \frac{1}{\frac{5}{3}} = 1 \div \frac{5}{3} = 1 \cdot \frac{3}{5} = \frac{3}{5}.$$

Simplify the complex fraction.

(b) $\sin \theta = \dfrac{1}{-\dfrac{\sqrt{12}}{2}}$ $\sin \theta = \frac{1}{\csc \theta}$

$$= -\frac{2}{\sqrt{12}}$$ Simplify the complex fraction as in part (a).

$$= -\frac{2}{2\sqrt{3}}$$ $\sqrt{12} = \sqrt{4 \cdot 3} = 2\sqrt{3}$

$$= -\frac{1}{\sqrt{3}}$$ Simplify.

$$= -\frac{1}{\sqrt{3}} \cdot \frac{\sqrt{3}}{\sqrt{3}} = -\frac{\sqrt{3}}{3}$$ Rationalize the denominator.

Now try Exercises 5 and 7.

Signs and Ranges of Function Values   In the definitions of the trigonometric functions, r is the distance from the origin to the point (x, y). Distance is never negative, so $r > 0$. If we choose a point (x, y) in quadrant I, then both x and y will be positive, and the values of all six functions will be positive.

A point (x, y) in quadrant II has $x < 0$ and $y > 0$. This makes the values of sine and cosecant positive for quadrant II angles, while the other four functions take on negative values. Similar results can be obtained for the other quadrants, as summarized here.

Signs of Function Values

θ in Quadrant	$\sin \theta$	$\cos \theta$	$\tan \theta$	$\cot \theta$	$\sec \theta$	$\csc \theta$
I	+	+	+	+	+	+
II	+	−	−	−	−	+
III	−	−	+	+	−	−
IV	−	+	−	−	+	−

$x < 0, y > 0, r > 0$	$x > 0, y > 0, r > 0$
II	I
Sine and cosecant positive	All functions positive
$x < 0, y < 0, r > 0$	$x > 0, y < 0, r > 0$
III	IV
Tangent and cotangent positive	Cosine and secant positive

EXAMPLE 2 Identifying the Quadrant of an Angle

Identify the quadrant (or quadrants) of any angle θ that satisfies $\sin \theta > 0$, $\tan \theta < 0$.

Solution Since $\sin \theta > 0$ in quadrants I and II, while $\tan \theta < 0$ in quadrants II and IV, both conditions are met only in quadrant II.

Now try Exercise 27.

Figure 32

Figure 32 shows an angle θ as it increases in measure from near $0°$ toward $90°$. In each case, the value of r is the same. As the measure of the angle increases, y increases but never exceeds r, so $y \leq r$. Dividing both sides by the positive number r gives $\frac{y}{r} \leq 1$.

In a similar way, angles in quadrant IV suggest that

$$-1 \leq \frac{y}{r},$$

so

$$-1 \leq \frac{y}{r} \leq 1$$

and

$$-1 \leq \sin \theta \leq 1. \quad \tfrac{y}{r} = \sin \theta \text{ for any angle } \theta. \text{ (Section 1.3)}$$

Similarly, $\qquad -1 \leq \cos \theta \leq 1.$

The tangent of an angle is defined as $\frac{y}{x}$. It is possible that $x < y$, $x = y$, or $x > y$. Thus, $\frac{y}{x}$ can take any value, so $\tan \theta$ can be any real number, as can $\cot \theta$.

The functions $\sec \theta$ and $\csc \theta$ are reciprocals of the functions $\cos \theta$ and $\sin \theta$, respectively, making

$$\sec \theta \leq -1 \text{ or } \sec \theta \geq 1 \qquad \text{and} \qquad \csc \theta \leq -1 \text{ or } \csc \theta \geq 1.$$

In summary, the ranges of the trigonometric functions are as follows.

> **Ranges of Trigonometric Functions**
>
>
>
> For any angle θ for which the indicated functions exist:
> **1.** $-1 \leq \sin \theta \leq 1$ and $-1 \leq \cos \theta \leq 1$;
> **2.** $\tan \theta$ and $\cot \theta$ can equal any real number;
> **3.** $\sec \theta \leq -1$ or $\sec \theta \geq 1$ and $\csc \theta \leq -1$ or $\csc \theta \geq 1$.
> (Notice that $\sec \theta$ and $\csc \theta$ are *never* between -1 and 1.)

EXAMPLE 3 Deciding Whether a Value Is in the Range of a Trigonometric Function

Decide whether each statement is *possible* or *impossible*.

(a) $\sin \theta = \sqrt{8}$ **(b)** $\tan \theta = 110.47$ **(c)** $\sec \theta = .6$

Solution

(a) For any value of θ, $-1 \leq \sin \theta \leq 1$. Since $\sqrt{8} > 1$, it is impossible to find a value of θ with $\sin \theta = \sqrt{8}$.

(b) Tangent can equal any value. Thus, $\tan \theta = 110.47$ is possible.

(c) Since $\sec \theta \leq -1$ or $\sec \theta \geq 1$, the statement $\sec \theta = .6$ is impossible.

Now try Exercises 47 and 49.

The six trigonometric functions are defined in terms of x, y, and r, where the Pythagorean theorem shows that $r^2 = x^2 + y^2$ and $r > 0$. With these relationships, knowing the value of only one function and the quadrant in which the angle lies makes it possible to find the values of the other trigonometric functions.

EXAMPLE 4 Finding All Function Values Given One Value and the Quadrant

Suppose that angle θ is in quadrant II and $\sin \theta = \frac{2}{3}$. Find the values of the other five functions.

Solution Choose any point on the terminal side of angle θ. For simplicity, since $\sin \theta = \frac{y}{r}$, choose the point with $r = 3$.

$$\sin \theta = \frac{2}{3} \qquad \text{Given value}$$

$$\frac{y}{r} = \frac{2}{3} \qquad \text{Substitute } \tfrac{y}{r} \text{ for } \sin \theta.$$

If $r = 3$, then $y = 2$. To find x, use the result $x^2 + y^2 = r^2$.

$$x^2 + y^2 = r^2$$
$$x^2 + 2^2 = 3^2 \qquad \text{Substitute.}$$
$$x^2 + 4 = 9$$
$$x^2 = 5 \qquad \text{Subtract 4. (Appendix A)}$$
$$x = \sqrt{5} \quad \text{or} \quad x = -\sqrt{5} \qquad \text{Square root property (Appendix A)}$$

Since θ is in quadrant II, x must be negative, as shown in Figure 33, so $x = -\sqrt{5}$, and the point $\left(-\sqrt{5}, 2\right)$ is on the terminal side of θ. Now we can find the values of the remaining trigonometric functions.

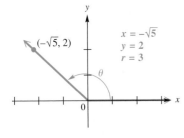

Figure 33

$$\cos \theta = \frac{x}{r} = \frac{-\sqrt{5}}{3} = -\frac{\sqrt{5}}{3}$$

$$\sec \theta = \frac{r}{x} = \frac{3}{-\sqrt{5}} = -\frac{3}{\sqrt{5}} \cdot \frac{\sqrt{5}}{\sqrt{5}} = -\frac{3\sqrt{5}}{5}$$

$$\tan \theta = \frac{y}{x} = \frac{2}{-\sqrt{5}} = -\frac{2}{\sqrt{5}} \cdot \frac{\sqrt{5}}{\sqrt{5}} = -\frac{2\sqrt{5}}{5}$$

$$\cot \theta = \frac{x}{y} = \frac{-\sqrt{5}}{2} = -\frac{\sqrt{5}}{2}$$

$$\csc \theta = \frac{r}{y} = \frac{3}{2}$$

Now try Exercise 67.

Pythagorean Identities   We derive three new identities from the relationship $x^2 + y^2 = r^2$.

$$\frac{x^2}{r^2} + \frac{y^2}{r^2} = \frac{r^2}{r^2} \qquad \text{Divide by } r^2.$$

$$\left(\frac{x}{r}\right)^2 + \left(\frac{y}{r}\right)^2 = 1 \qquad \text{Power rule for exponents; } \frac{a^m}{b^m} = \left(\frac{a}{b}\right)^m$$

$$(\cos\theta)^2 + (\sin\theta)^2 = 1 \qquad \cos\theta = \frac{x}{r}, \sin\theta = \frac{y}{r} \text{ (Section 1.3)}$$

or $\qquad \sin^2\theta + \cos^2\theta = 1$

Starting again with $x^2 + y^2 = r^2$ and dividing through by x^2 gives

$$\frac{x^2}{x^2} + \frac{y^2}{x^2} = \frac{r^2}{x^2} \qquad \text{Divide by } x^2.$$

$$1 + \left(\frac{y}{x}\right)^2 = \left(\frac{r}{x}\right)^2 \qquad \text{Power rule for exponents}$$

$$1 + (\tan\theta)^2 = (\sec\theta)^2 \qquad \tan\theta = \frac{y}{x}, \sec\theta = \frac{r}{x} \text{ (Section 1.3)}$$

or $\qquad \tan^2\theta + 1 = \sec^2\theta.$

On the other hand, dividing through by y^2 leads to

$$1 + \cot^2\theta = \csc^2\theta.$$

These three identities are called the **Pythagorean identities** since the original equation that led to them, $x^2 + y^2 = r^2$, comes from the Pythagorean theorem.

Pythagorean Identities

$$\sin^2\theta + \cos^2\theta = 1 \qquad \tan^2\theta + 1 = \sec^2\theta \qquad 1 + \cot^2\theta = \csc^2\theta$$

Although we usually write $\sin^2\theta$, for example, it should be entered as $(\sin\theta)^2$ in your calculator. To test this, verify that in degree mode, $\sin^2 30° = .25 = \frac{1}{4}$. ∎

As before, we have given only one form of each identity. However, algebraic transformations produce equivalent identities. For example, by subtracting $\sin^2\theta$ from both sides of $\sin^2\theta + \cos^2\theta = 1$, we get the equivalent identity

$$\cos^2\theta = 1 - \sin^2\theta.$$

You should be able to transform these identities quickly and also recognize their equivalent forms.

Looking Ahead to Calculus

The reciprocal, Pythagorean, and quotient identities are used in calculus to find derivatives and integrals of trigonometric functions. A standard technique of integration called *trigonometric substitution* relies on the Pythagorean identities.

Quotient Identities   Consider the quotient of $\sin\theta$ and $\cos\theta$, for $\cos\theta \neq 0$.

$$\frac{\sin\theta}{\cos\theta} = \frac{\frac{y}{r}}{\frac{x}{r}} = \frac{y}{r} \div \frac{x}{r} = \frac{y}{r} \cdot \frac{r}{x} = \frac{y}{x} = \tan\theta$$

Similarly, $\frac{\cos\theta}{\sin\theta} = \cot\theta$, for $\sin\theta \neq 0$. Thus, we have the **quotient identities.**

> ## Quotient Identities
>
> $$\frac{\sin \theta}{\cos \theta} = \tan \theta \qquad\qquad \frac{\cos \theta}{\sin \theta} = \cot \theta$$

EXAMPLE 5 Finding Other Function Values Given One Value and the Quadrant

Find $\sin \theta$ and $\tan \theta$, if $\cos \theta = -\frac{\sqrt{3}}{4}$ and θ is in quadrant II.

Solution   Start with $\sin^2 \theta + \cos^2 \theta = 1$.

$$\sin^2 \theta + \left(-\frac{\sqrt{3}}{4}\right)^2 = 1 \qquad \text{Replace } \cos \theta \text{ with } -\frac{\sqrt{3}}{4}.$$

$$\sin^2 \theta + \frac{3}{16} = 1$$

$$\sin^2 \theta = \frac{13}{16} \qquad \text{Subtract } \tfrac{3}{16}.$$

$$\sin \theta = \pm\frac{\sqrt{13}}{4} \qquad \text{Take square roots.}$$

$$\sin \theta = \frac{\sqrt{13}}{4} \qquad \begin{array}{l}\text{Choose the positive square root since}\\ \sin \theta \text{ is positive when } \theta \text{ is in quadrant II.}\end{array}$$

To find $\tan \theta$, use the quotient identity $\tan \theta = \frac{\sin \theta}{\cos \theta}$.

$$\tan \theta = \frac{\sin \theta}{\cos \theta} = \frac{\frac{\sqrt{13}}{4}}{-\frac{\sqrt{3}}{4}} = \frac{\sqrt{13}}{4}\left(-\frac{4}{\sqrt{3}}\right) = -\frac{\sqrt{13}}{\sqrt{3}}$$

$$= -\frac{\sqrt{13}}{\sqrt{3}} \cdot \frac{\sqrt{3}}{\sqrt{3}} = -\frac{\sqrt{39}}{3} \qquad \text{Rationalize the denominator.}$$

Now try Exercise 55.

C A U T I O N   In problems like those in Examples 4 and 5, be careful to choose the correct sign when square roots are taken.

EXAMPLE 6 Finding Other Function Values Given One Value and the Quadrant

Find $\sin \theta$ and $\cos \theta$, if $\tan \theta = \frac{4}{3}$ and θ is in quadrant III.

Solution   Since θ is in quadrant III, $\sin \theta$ and $\cos \theta$ will both be negative. It is tempting to say that since $\tan \theta = \frac{\sin \theta}{\cos \theta}$ and $\tan \theta = \frac{4}{3}$, then $\sin \theta = -4$ and $\cos \theta = -3$. This is *incorrect*, however, since both $\sin \theta$ and $\cos \theta$ must be in the interval $[-1, 1]$.

We use the Pythagorean identity $\tan^2 \theta + 1 = \sec^2 \theta$ to find $\sec \theta$, and then the reciprocal identity $\cos \theta = \frac{1}{\sec \theta}$ to find $\cos \theta$.

$$\tan^2 \theta + 1 = \sec^2 \theta$$

$$\left(\frac{4}{3}\right)^2 + 1 = \sec^2 \theta \qquad \tan \theta = \frac{4}{3}$$

$$\frac{16}{9} + 1 = \sec^2 \theta$$

$$\frac{25}{9} = \sec^2 \theta$$

$$-\frac{5}{3} = \sec \theta \qquad \text{Choose the negative square root since } \sec \theta \text{ is negative when } \theta \text{ is in quadrant III.}$$

$$-\frac{3}{5} = \cos \theta \qquad \text{Secant and cosine are reciprocals.}$$

Since $\sin^2 \theta = 1 - \cos^2 \theta$,

$$\sin^2 \theta = 1 - \left(-\frac{3}{5}\right)^2 \qquad \cos \theta = -\frac{3}{5}$$

$$\sin^2 \theta = 1 - \frac{9}{25}$$

$$\sin^2 \theta = \frac{16}{25}$$

$$\sin \theta = -\frac{4}{5}. \qquad \text{Choose the negative square root.}$$

Now try Exercise 59.

N O T E Example 6 can also be worked by drawing θ in standard position in quadrant III, finding r to be 5, and then using the definitions of $\sin \theta$ and $\cos \theta$ in terms of x, y, and r.

1.4 Exercises

1. *Concept Check* What positive number a is its own reciprocal? Find a value of θ for which $\sin \theta = \csc \theta = a$.

2. *Concept Check* What negative number a is its own reciprocal? Find a value of θ for which $\cos \theta = \sec \theta = a$.

Use the appropriate reciprocal identity to find each function value. Rationalize denominators when applicable. In Exercises 9 and 10, use a calculator. See Example 1.

3. $\cos \theta$, if $\sec \theta = -2.5$

4. $\cot \theta$, if $\tan \theta = -\dfrac{1}{5}$

5. $\sin \theta$, if $\csc \theta = 3$

6. $\sec \theta$, if $\cos \theta = -\dfrac{\sqrt{7}}{7}$

7. $\sin \theta$, if $\csc \theta = \sqrt{15}$

8. $\tan \theta$, if $\cot \theta = -\dfrac{\sqrt{5}}{3}$

9. $\sin \theta$, if $\csc \theta = 1.42716321$

10. $\cos \theta$, if $\sec \theta = 9.80425133$

11. Can a given angle θ satisfy both $\sin \theta > 0$ and $\csc \theta < 0$? Explain.

12. Explain what is wrong with the following item that appears on a trigonometry test:

$$\text{Find } \sec \theta, \text{ given that } \cos \theta = \frac{3}{2}.$$

13. What is wrong with the following statement? $\tan 90° = \frac{1}{\cot 90°}$.

14. *Concept Check*   One form of a particular reciprocal identity is $\tan \theta = \frac{1}{\cot \theta}$. Give two other equivalent forms of this identity.

Find the tangent of each angle. See Example 1.

15. $\cot \theta = -3$

16. $\cot \theta = 2$

17. $\cot \theta = \dfrac{\sqrt{6}}{12}$

18. $\cot \theta = \dfrac{\sqrt{3}}{3}$

19. $\cot \theta = .4$

20. $\cot \theta = -.01$

Find a value of each variable.

21. $\tan(3\theta - 4°) = \dfrac{1}{\cot(5\theta - 8°)}$

22. $\sec(2\theta + 6°) \cos(5\theta + 3°) = 1$

23. $\sin(4\theta + 2°) \csc(3\theta + 5°) = 1$

24. $\cos(6A + 5°) = \dfrac{1}{\sec(4A + 15°)}$

25. The screen was obtained with the calculator in degree mode. How can we use it to justify that an angle of 14,879° is a quadrant II angle?

```
cos(14879)
       -.4848096202
sin(14879)
        .8746197071
```

26. The screen was obtained with the calculator in degree mode. An angle of 1294° lies in which quadrant? Explain.

```
tan(1294)
        .6745085168
sin(1294)
       -.5591929035
```

Identify the quadrant or quadrants for the angle satisfying the given conditions. See Example 2.

27. $\sin \theta > 0$, $\cos \theta < 0$

28. $\cos \theta > 0$, $\tan \theta > 0$

29. $\tan \theta > 0$, $\cot \theta > 0$

30. $\tan \theta < 0$, $\cot \theta < 0$

31. $\cos \theta < 0$

32. $\tan \theta > 0$

Concept Check   *Give the signs of the sine, cosine, and tangent functions for each angle.*

33. 74°   **34.** 298°   **35.** 129°   **36.** 183°

37. 406°   **38.** 412°   **39.** −82°   **40.** −121°

Concept Check   *Without using a calculator, decide which is greater.*

41. sin 30° or tan 30°   **42.** sin 20° or sin 21°   **43.** sin 33° or sec 33°

44. cos 5° or cos² 5°   **45.** cos 26° or cos 27°   **46.** cos 2° or cot 2°

Decide whether each statement is possible *or* impossible *for an angle θ. See Example 3.*

47. sin θ = 2   **48.** cos θ = −1.001   **49.** tan θ = .92

50. cot θ = −12.1   **51.** sec θ = 1   **52.** tan θ = 1

53. sin θ = $\dfrac{1}{2}$ and csc θ = 2   **54.** tan θ = 2 and cot θ = −2

Use identities to find each function value. Use a calculator in Exercises 61 and 62. See Examples 4–6.

55. tan θ, if sec θ = 3, with θ in quadrant IV

56. sin θ, if cos θ = −$\dfrac{1}{4}$, with θ in quadrant II

57. csc θ, if cot θ = −$\dfrac{1}{2}$, with θ in quadrant IV

58. sec θ, if tan θ = $\dfrac{\sqrt{7}}{3}$, with θ in quadrant III

59. cos θ, if csc θ = −4, with θ in quadrant III

60. sin θ, if sec θ = 2, with θ in quadrant IV

61. cot θ, if csc θ = −3.5891420, with θ in quadrant III

62. tan θ, if sin θ = .49268329, with θ in quadrant II

63. *Concept Check*   Does there exist an angle θ with cos θ = −.6 and sin θ = .8?

Find all trigonometric function values for each angle. Use a calculator in Exercises 70 and 71. See Example 4.

64. cos θ = −$\dfrac{3}{5}$, with θ in quadrant III   **65.** tan θ = −$\dfrac{15}{8}$, with θ in quadrant II

66. tan θ = $\sqrt{3}$, with θ in quadrant III   **67.** sin θ = $\dfrac{\sqrt{5}}{7}$, with θ in quadrant I

68. csc θ = 2, with θ in quadrant II   **69.** cot θ = $\dfrac{\sqrt{3}}{8}$, with θ in quadrant I

70. cot θ = −1.49586, with θ in quadrant IV

71. sin θ = .164215, with θ in quadrant II

Work each problem.

72. Derive the identity 1 + cot² θ = csc² θ by dividing $x^2 + y^2 = r^2$ by y^2.

73. Using a method similar to the one given in this section showing that $\frac{\sin θ}{\cos θ}$ = tan θ, show that $\frac{\cos θ}{\sin θ}$ = cot θ.

74. *Concept Check* True or false: For all angles θ, $\sin \theta + \cos \theta = 1$. If false, give an example showing why it is false.

75. *Concept Check* True or false: Since $\cot \theta = \frac{\cos \theta}{\sin \theta}$, if $\cot \theta = \frac{1}{2}$ with θ in quadrant I, then $\cos \theta = 1$ and $\sin \theta = 2$. If false, explain why.

Use a trigonometric function ratio to solve each problem. (Source for Exercises 76 and 77: Parker, M., Editor, She Does Math, Mathematical Association of America, 1995.)

76. *Height of a Tree* A civil engineer must determine the height of the tree shown in the figure. The given angle was measured with a *clinometer*. She knows that $\sin 70° \approx .9397$, $\cos 70° \approx .3420$, and $\tan 70° \approx 2.747$. Use the pertinent trigonometric function and the measurement given in the figure to find the height of the tree to the nearest whole number.

This is a picture of one type of clinometer, called an Abney hand level and clinometer. The picture is courtesy of Keuffel & Esser Co.

77. *(Modeling) Double Vision* To correct mild double vision, a small amount of prism is added to a patient's eyeglasses. The amount of light shift this causes is measured in *prism diopters*. A patient needs 12 prism diopters horizontally and 5 prism diopters vertically. A prism that corrects for both requirements should have length r and be set at angle θ. See the figure.

(a) Use the Pythagorean theorem to find r.

(b) Write an equation involving a trigonometric function of θ and the known prism measurements 5 and 12.

Concept Check Suppose that $90° < \theta < 180°$. Find the sign of each function value.

78. $\sin 2\theta$ **79.** $\tan \dfrac{\theta}{2}$ **80.** $\cot(\theta + 180°)$ **81.** $\cos(-\theta)$

82. The straight line in the figure determines both angle α (alpha) and angle β (beta) with the positive x-axis. Explain why $\tan \alpha = \tan \beta$.

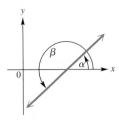

Chapter 1 Summary

KEY TERMS

1.1 line
line segment (or
  segment)
ray
angle
initial side
terminal side
vertex of an angle
positive angle

negative angle
degree
acute angle
right angle
obtuse angle
straight angle
complementary angles
supplementary angles
minute

second
angle in standard
  position
quadrantal angle
coterminal angles
1.2 vertical angles
parallel lines
transversal
similar triangles

congruent triangles
1.3 sine
cosine
tangent
cotangent
secant
cosecant

NEW SYMBOLS

θ Greek letter theta
°   degree

′   minute
″   second

QUICK REVIEW

CONCEPTS	EXAMPLES

1.1 Angles

Types of Angles

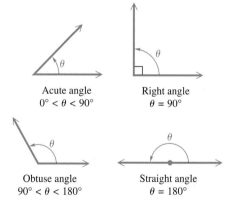

Acute angle
$0° < \theta < 90°$

Right angle
$\theta = 90°$

Obtuse angle
$90° < \theta < 180°$

Straight angle
$\theta = 180°$

If $\theta = 32°$, then angle θ is an acute angle.

If $\theta = 90°$, then angle θ is a right angle.

If $\theta = 148°$, then angle θ is an obtuse angle.

If $\theta = 180°$, then angle θ is a straight angle.

1.2 Angle Relationships and Similar Triangles

Vertical angles have equal measures.

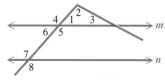

m and n are parallel lines.

Vertical angles 4 and 5 are equal.

(continued)

CONCEPTS	EXAMPLES

The sum of the measures of the angles of any triangle is 180°.

m and *n* are parallel lines.

The sum of angles 1, 2, and 3 is 180°.

When a transversal intersects parallel lines, the following angles formed have equal measure: alternate interior, alternate exterior, and corresponding. Interior angles on the same side of the transversal are supplementary.

Angles 5 and 7 are alternate interior angles, so they are equal. Angles 4 and 8 are alternate exterior angles, so they are equal. Angles 4 and 7 are corresponding angles, so they are equal. Angles 6 and 7 are interior angles on the same side of the transversal, so they are supplementary.

Similar triangles have corresponding angles with the same measures, and corresponding sides proportional.

Pairs of corresponding angles as marked in triangles *ABC* and *DEF* are equal.

Also, $\dfrac{AB}{DE} = \dfrac{BC}{EF} = \dfrac{AC}{DF}$.

Congruent triangles are the same size and the same shape.

Corresponding angles are equal, and corresponding sides are equal.

1.3 Trigonometric Functions

Definitions of the Trigonometric Functions
Let (x, y) be a point other than the origin on the terminal side of an angle θ in standard position. Let $r = \sqrt{x^2 + y^2}$, the distance from the origin to (x, y). Then

$$\sin \theta = \frac{y}{r} \qquad \cos \theta = \frac{x}{r} \qquad \tan \theta = \frac{y}{x} \ (x \neq 0)$$

$$\csc \theta = \frac{r}{y} \ (y \neq 0) \ \sec \theta = \frac{r}{x} \ (x \neq 0) \ \cot \theta = \frac{x}{y} \ (y \neq 0).$$

See the summary table of trigonometric function values for quadrantal angles on page 24.

If the point $(-2, 3)$ is on the terminal side of angle θ in standard position, then $x = -2, y = 3, r = \sqrt{4 + 9} = \sqrt{13}$, and

$$\sin \theta = \frac{3\sqrt{13}}{13}, \qquad \cos \theta = -\frac{2\sqrt{13}}{13}, \qquad \tan \theta = -\frac{3}{2},$$

$$\csc \theta = \frac{\sqrt{13}}{3}, \qquad \sec \theta = -\frac{\sqrt{13}}{2}, \qquad \cot \theta = -\frac{2}{3}.$$

1.4 Using the Definitions of the Trigonometric Functions

Reciprocal Identities

$$\sin \theta = \frac{1}{\csc \theta} \qquad \cos \theta = \frac{1}{\sec \theta} \qquad \tan \theta = \frac{1}{\cot \theta}$$

$$\csc \theta = \frac{1}{\sin \theta} \qquad \sec \theta = \frac{1}{\cos \theta} \qquad \cot \theta = \frac{1}{\tan \theta}$$

The preceding examples given for Section 1.3 satisfy these identities. For example,

$$\tan \theta = \frac{1}{\cot \theta} = \frac{1}{-\frac{2}{3}} = -\frac{3}{2}.$$

CONCEPTS	EXAMPLES

Pythagorean Identities

$$\sin^2 \theta + \cos^2 \theta = 1 \qquad \tan^2 \theta + 1 = \sec^2 \theta$$

$$1 + \cot^2 \theta = \csc^2 \theta$$

Using the preceding trigonometric values for Section 1.3,

$$\sin^2 \theta + \cos^2 \theta = \left(\frac{3\sqrt{13}}{13}\right)^2 + \left(-\frac{2\sqrt{13}}{13}\right)^2$$

$$= \frac{9}{13} + \frac{4}{13} = 1,$$

$$\tan^2 \theta + 1 = \left(-\frac{3}{2}\right)^2 + 1 = \frac{13}{4} = \left(-\frac{\sqrt{13}}{2}\right)^2 = \sec^2 \theta,$$

$$1 + \cot^2 \theta = 1 + \left(-\frac{2}{3}\right)^2 = \frac{13}{9} = \left(\frac{\sqrt{13}}{3}\right)^2 = \csc^2 \theta.$$

Quotient Identities

$$\frac{\sin \theta}{\cos \theta} = \tan \theta \qquad \frac{\cos \theta}{\sin \theta} = \cot \theta$$

From the preceding trigonometric values,

$$\frac{\sin \theta}{\cos \theta} = \frac{\frac{3\sqrt{13}}{13}}{-\frac{2\sqrt{13}}{13}} = \frac{3\sqrt{13}}{13}\left(-\frac{13}{2\sqrt{13}}\right) = -\frac{3}{2} = \tan \theta.$$

Signs of the Trigonometric Functions

$x < 0, y > 0, r > 0$ **II** Sine and cosecant positive	$x > 0, y > 0, r > 0$ **I** All functions positive
$x < 0, y < 0, r > 0$ **III** Tangent and cotangent positive	$x > 0, y < 0, r > 0$ **IV** Cosine and secant positive

Identify the quadrant(s) of any angle θ that satisfies $\sin \theta < 0$, $\tan \theta > 0$.

Since $\sin \theta < 0$ in quadrants III and IV, while $\tan \theta > 0$ in quadrants I and III, both conditions are met only in quadrant III.

Chapter 1 Review Exercises

1. Give the complement and the supplement of an angle of 35°.

Find the angle of smallest possible positive measure coterminal with each angle.

2. −51°   **3.** −174°   **4.** 792°

5. Let n represent any integer, and write an expression for all angles coterminal with an angle of 270°.

Work each problem.

6. *Rotating Pulley* A pulley is rotating 320 times per min. Through how many degrees does a point on the edge of the pulley move in $\frac{2}{3}$ sec?

7. *Rotating Propeller* The propeller of a speedboat rotates 650 times per min. Through how many degrees will a point on the edge of the propeller rotate in 2.4 sec?

Convert decimal degrees to degrees, minutes, seconds, and convert degrees, minutes, seconds to decimal degrees. Round to the nearest second or the nearest thousandth of a degree, as appropriate. Use a calculator as necessary.

8. 47° 25′ 11″   **9.** 119° 8′ 3″   **10.** −61.5034°   **11.** 275.1005°

Find the measure of each marked angle.

12.

13.

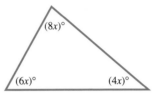

14. Express θ in terms of α and β.

15. *Length of a Road* The flight path CP of a satellite carrying a camera with its lens at C is shown in the figure. Length PC represents the distance from the lens to the film PQ, and BA represents a straight road on the ground. Use the measurements given in the figure to find the length of the road. (*Source:* Kastner, B., *Space Mathematics*, NASA, 1985.)

Find all unknown angle measures in each pair of similar triangles.

16.

17.

Find the unknown side lengths in each pair of similar triangles.

18.

19.

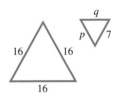

Find the unknown measurement. There are two similar triangles in each figure.

20.

21.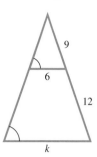

22. *Concept Check* Complete the following statement: If two triangles are similar, their corresponding sides are _____ and their corresponding angles are _____ .

23. *Length of a Shadow* If a tree 20 ft tall casts a shadow 8 ft long, how long would the shadow of a 30-ft tree be at the same time and place?

Find the six trigonometric function values for each angle. If a value is undefined, say so.

24.

25.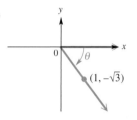

26. $180°$

Find the values of the six trigonometric functions for an angle in standard position having each point on its terminal side.

27. $(3, -4)$ **28.** $(9, -2)$ **29.** $(-8, 15)$

30. $(1, -5)$ **31.** $\left(6\sqrt{3}, -6\right)$ **32.** $\left(-2\sqrt{2}, 2\sqrt{2}\right)$

33. *Concept Check* If the terminal side of a quadrantal angle lies along the y-axis, which of its trigonometric functions are undefined?

34. Find the values of all six trigonometric functions for an angle in standard position having its terminal side defined by the equation $5x - 3y = 0, x \geq 0$.

In Exercises 35 and 36, consider an angle θ in standard position whose terminal side has the equation $y = -5x$, with $x \leq 0$.

35. Sketch θ and use an arrow to show the rotation if $0° \leq \theta < 360°$.

36. Find the exact values of $\sin \theta$, $\cos \theta$, and $\tan \theta$.

Evaluate each expression.

37. $4 \sec 180° - 2 \sin^2 270°$ **38.** $-\cot^2 90° + 4 \sin 270° - 3 \tan 180°$

39. Decide whether each statement is *possible* or *impossible*.

  (a) $\sec \theta = -\dfrac{2}{3}$ **(b)** $\tan \theta = 1.4$ **(c)** $\csc \theta = 5$

Find all six trigonometric function values for each angle. Rationalize denominators when applicable.

40. $\sin \theta = \dfrac{\sqrt{3}}{5}$ and $\cos \theta < 0$ **41.** $\cos \theta = -\dfrac{5}{8}$, with θ in quadrant III

42. $\tan \alpha = 2$, with α in quadrant III   **43.** $\sec \beta = -\sqrt{5}$, with β in quadrant II

44. $\sin \theta = -\dfrac{2}{5}$, with θ in quadrant III   **45.** $\sec \alpha = \dfrac{5}{4}$, with α in quadrant IV

46. *Concept Check*   If, for some particular angle θ, $\sin \theta < 0$ and $\cos \theta > 0$, in what quadrant must θ lie? What is the sign of $\tan \theta$?

47. Explain how you would find the cotangent of an angle θ whose tangent is 1.6778490 using a calculator. Then find $\cot \theta$.

Solve each problem.

48. *Angle the Celestial North Pole Moves*   At present, the north star Polaris is located very near the celestial north pole. However, because Earth is inclined 23.5°, the moon's gravitational pull on Earth is uneven. As a result, Earth slowly precesses (moves in) like a spinning top and the direction of the celestial north pole traces out a circular path once every 26,000 yr. See the figure. For example, in approximately A.D. 14,000 the star Vega will be located at the celestial north pole—and not the star Polaris. As viewed from the center C of this circular path, calculate the angle in seconds that the celestial north pole moves each year. (*Source:* Zeilik, M., S. Gregory, and E. Smith, *Introductory Astronomy and Astrophysics,* Second Edition, Saunders College Publishers, 1998.)

49. *Depth of a Crater on the Moon*   The depths of unknown craters on the moon can be approximated by comparing the lengths of their shadows to the shadows of nearby craters with known depths. The crater Aristillus is 11,000 ft deep, and its shadow was measured as 1.5 mm on a photograph. Its companion crater, Autolycus, had a shadow of 1.3 mm on the same photograph. Use similar triangles to determine the depth of the crater Autolycus. (*Source:* Webb, T., *Celestial Objects for Common Telescopes,* Dover Publications, 1962.)

50. *Height of a Lunar Peak*   The lunar mountain peak Huygens has a height of 21,000 ft. The shadow of Huygens on a photograph was 2.8 mm, while the nearby mountain Bradley had a shadow of 1.8 mm on the same photograph. Calculate the height of Bradley. (*Source:* Webb, T., *Celestial Objects for Common Telescopes,* Dover Publications, 1962.)

Chapter 1 Test

1. Convert 74° 17′ 54″ to decimal degrees.

2. Find the angle of smallest positive measure coterminal with −157°.

3. *Rotating Tire*   A tire rotates 450 times per min. Through how many degrees does a point on the edge of the tire move in 1 sec?

4. Find the measure of each marked angle.

5. *Swimmer in Distress*   A lifeguard located 20 yd from the water spots a swimmer in distress. The swimmer is 30 yd from shore and 100 yd east of the lifeguard. Suppose the lifeguard runs, then swims to the swimmer in a direct line, as shown in the figure. How far east from his original position will he enter the water? (*Hint:* Find the value of x in the sketch.)

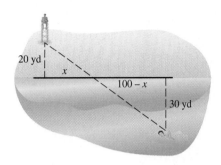

6. If $(2, -5)$ is on the terminal side of an angle θ in standard position, find $\sin \theta$, $\cos \theta$, and $\tan \theta$.

7. If $\cos \theta < 0$ and $\cot \theta > 0$, in what quadrant does θ terminate?

8. If $\cos \theta = \frac{4}{5}$ and θ is in quadrant IV, find the values of the other trigonometric functions of θ.

9. Is it possible for $\sin \theta$, $\cos \theta$, and $\tan \theta$ to all be negative for the same value of θ? Explain.

10. Suppose θ is in the interval $(90°, 180°)$. Find the sign of each function value.

(a) $\cos \dfrac{\theta}{2}$
   (b) $\cot(\theta - 180°)$

11. Find $\cos \theta$ if $\sec \theta = -10$.

12. Use the reciprocal, quotient, and Pythagorean identities to give three expressions that complete the following statement:

$$\cot \theta = \underline{\hspace{2cm}}.$$

Chapter 1 Quantitative Reasoning

Have you ever gazed up at a redwood tree, a skyscraper, a public monument, or perhaps a dinosaur in a museum and wondered how tall it was?

There is a relatively simple way to make a reasonable estimate. All you need is a 1-ft ruler. Hold the ruler vertically at arm's length as you approach the object to be measured. Stop when one end of the ruler lines up with the top of the object and the other end with its base. Now pace off the distance to the object, taking normal strides. The number of paces will be the approximate height of the object in feet.

The reason this method works depends on a concept presented in this chapter. Furnish the reasons for the following steps, which refer to the figure. (Assume that the length of one pace is *EF*.) Then answer the question.

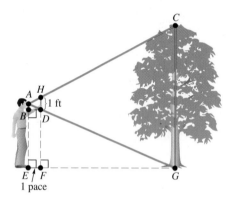

Reasons

Step 1 $CG = \dfrac{CG}{1} = \dfrac{AG}{AD}$ _____

Step 2 $\dfrac{AG}{AD} = \dfrac{EG}{EF}$ _____

Step 3 $\dfrac{EG}{EF} = \dfrac{EG}{BD} = \dfrac{EG}{1}$ _____

Step 4 CG ft $= EG$ paces _____

What is the height of the tree?

2

Acute Angles and Right Triangles

Highway transportation is critical to the economy of the United States. In 1970 there were 1150 billion miles traveled, and by the year 2000 this increased to approximately 2500 billion miles. When an automobile travels around a curve, objects like trees, buildings, and fences situated on the curve may obstruct a driver's vision. Trigonometry is used to determine how far inside the curve land must be cleared to provide visibility for a safe stopping distance. (*Source:* Mannering, F. and W. Kilareski, *Principles of Highway Engineering and Traffic Analysis,* Second Edition, John Wiley & Sons, 1998.)

A problem like this is presented in Exercise 35 of Section 2.5.

2.1 Trigonometric Functions of Acute Angles

Right-Triangle-Based Definitions of the Trigonometric Functions ▪ **Cofunctions** ▪ **Trigonometric Function Values of Special Angles**

Figure 1

Right-Triangle-Based Definitions of the Trigonometric Functions

In Section 1.3 we used angles in standard position to define the trigonometric functions. There is another way to approach them: as ratios of the lengths of the sides of right triangles. Figure 1 shows an acute angle A in standard position. The definitions of the trigonometric function values of angle A require x, y, and r. As drawn in Figure 1, x and y are the lengths of the two legs of the right triangle ABC, and r is the length of the hypotenuse.

The side of length y is called the **side opposite** angle A, and the side of length x is called the **side adjacent** to angle A. We use the lengths of these sides to replace x and y in the definitions of the trigonometric functions, and the length of the hypotenuse to replace r, to get the following right-triangle-based definitions.

Right-Triangle-Based Definitions of Trigonometric Functions

For any acute angle A in standard position,

$$\sin A = \frac{y}{r} = \frac{\text{side opposite}}{\text{hypotenuse}} \qquad \csc A = \frac{r}{y} = \frac{\text{hypotenuse}}{\text{side opposite}}$$

$$\cos A = \frac{x}{r} = \frac{\text{side adjacent}}{\text{hypotenuse}} \qquad \sec A = \frac{r}{x} = \frac{\text{hypotenuse}}{\text{side adjacent}}$$

$$\tan A = \frac{y}{x} = \frac{\text{side opposite}}{\text{side adjacent}} \qquad \cot A = \frac{x}{y} = \frac{\text{side adjacent}}{\text{side opposite}}.$$

EXAMPLE 1 Finding Trigonometric Function Values of an Acute Angle

Figure 2

Find the sine, cosine, and tangent values for angles A and B in the right triangle in Figure 2.

Solution The length of the side opposite angle A is 7, the length of the side adjacent to angle A is 24, and the length of the hypotenuse is 25. Use the relationships given in the box.

$$\sin A = \frac{\text{side opposite}}{\text{hypotenuse}} = \frac{7}{25} \qquad \cos A = \frac{\text{side adjacent}}{\text{hypotenuse}} = \frac{24}{25} \qquad \tan A = \frac{\text{side opposite}}{\text{side adjacent}} = \frac{7}{24}$$

The length of the side opposite angle B is 24, while the length of the side adjacent to B is 7, so

$$\sin B = \frac{24}{25} \qquad \cos B = \frac{7}{25} \qquad \tan B = \frac{24}{7}.$$

Now try Exercise 1.

N O T E Because the cosecant, secant, and cotangent ratios are the reciprocals of the sine, cosine, and tangent values, respectively, in Example 1 we can conclude that $\csc A = \frac{25}{7}$, $\sec A = \frac{25}{24}$, $\cot A = \frac{24}{7}$, $\csc B = \frac{25}{24}$, $\sec B = \frac{25}{7}$, and $\cot B = \frac{7}{24}$.

Figure 3

Cofunctions   In Example 1, you may have noticed that $\sin A = \cos B$ and $\cos A = \sin B$. Such relationships are always true for the two acute angles of a right triangle. Figure 3 shows a right triangle with acute angles A and B and a right angle at C. (Whenever we use A, B, and C to name angles in a right triangle, C will be the right angle.) The length of the side opposite A is a, and the length of the side opposite angle B is b. The length of the hypotenuse is c.

By the preceding definitions, $\sin A = \frac{a}{c}$. Since $\cos B$ is also equal to $\frac{a}{c}$,

$$\sin A = \frac{a}{c} = \cos B.$$

Similarly, $\tan A = \dfrac{a}{b} = \cot B$ and $\sec A = \dfrac{c}{b} = \csc B.$

Since the sum of the three angles in any triangle is $180°$ and angle C equals $90°$, angles A and B must have a sum of $180° - 90° = 90°$. As mentioned in Chapter 1, angles with a sum of $90°$ are complementary angles. Since angles A and B are complementary and $\sin A = \cos B$, the functions sine and cosine are called **cofunctions.** Tangent and cotangent are also cofunctions, as are secant and cosecant. And since the angles A and B are complementary, $A + B = 90°$, or $B = 90° - A$, giving

$$\sin A = \cos B = \cos(90° - A).$$

Similar results, called the **cofunction identities,** are true for the other trigonometric functions.

Cofunction Identities

For any acute angle A,

$\sin A = \cos(90° - A)$ $\csc A = \sec(90° - A)$ $\tan A = \cot(90° - A)$

$\cos A = \sin(90° - A)$ $\sec A = \csc(90° - A)$ $\cot A = \tan(90° - A).$

EXAMPLE 2   Writing Functions in Terms of Cofunctions

Write each function in terms of its cofunction.

(a) $\cos 52°$ **(b)** $\tan 71°$ **(c)** $\sec 24°$

Solution

(a) Since $\cos A = \sin(90° - A)$,

$$\cos 52° = \sin(90° - 52°) = \sin 38°.$$

(b) $\tan 71° = \cot(90° - 71°) = \cot 19°$ **(c)** $\sec 24° = \csc 66°$

Now try Exercises **17** and **21.**

EXAMPLE 3 Solving Equations Using the Cofunction Identities

Find one solution for each equation. Assume all angles involved are acute angles.

(a) $\cos(\theta + 4°) = \sin(3\theta + 2°)$ **(b)** $\tan(2\theta - 18°) = \cot(\theta + 18°)$

Solution

(a) Since sine and cosine are cofunctions, $\cos(\theta + 4°) = \sin(3\theta + 2°)$ is true if the sum of the angles is 90°.

$$(\theta + 4°) + (3\theta + 2°) = 90°$$
$$4\theta + 6° = 90° \quad \text{Combine terms.}$$
$$4\theta = 84° \quad \text{Subtract 6°. (Appendix A)}$$
$$\theta = 21° \quad \text{Divide by 4.}$$

(b) Tangent and cotangent are cofunctions, so

$$(2\theta - 18°) + (\theta + 18°) = 90°$$
$$3\theta = 90°$$
$$\theta = 30°.$$

Now try Exercises 23 and 25.

Figure 4 shows three right triangles. From left to right, the length of each hypotenuse is the same, but angle A increases in measure. As angle A increases in measure from 0° to 90°, the length of the side opposite angle A also increases.

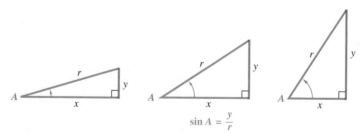

$$\sin A = \frac{y}{r}$$

As A increases, y increases. Since r is fixed, $\sin A$ increases.

Figure 4

Since
$$\sin A = \frac{\text{side opposite}}{\text{hypotenuse}},$$

as angle A increases, the numerator of this fraction also increases, while the denominator is fixed. This means that $\sin A$ *increases* as A increases from 0° to 90°.

As angle A increases from 0° to 90°, the length of the side adjacent to A decreases. Since r is fixed, the ratio $\frac{x}{r}$ will decrease. This ratio gives $\cos A$, showing that the values of cosine *decrease* as the angle measure changes from 0° to 90°. Finally, increasing A from 0° to 90° causes y to increase and x to decrease, making the values of $\frac{y}{x} = \tan A$ increase.

A similar discussion shows that as A increases from 0° to 90°, the values of $\sec A$ increase, while the values of $\cot A$ and $\csc A$ decrease.

EXAMPLE 4 Comparing Function Values of Acute Angles

Tell whether each statement is *true* or *false*.

(a) $\sin 21° > \sin 18°$ **(b)** $\cos 49° \leq \cos 56°$

Solution

(a) In the interval from $0°$ to $90°$, as the angle increases, so does the sine of the angle, which makes $\sin 21° > \sin 18°$ a true statement.

(b) As the angle increases, the cosine of the angle decreases. The given statement $\cos 49° \leq \cos 56°$ is false.

Now try Exercise 29.

Trigonometric Function Values of Special Angles Certain special angles, such as $30°$, $45°$, and $60°$, occur so often in trigonometry and in more advanced mathematics that they deserve special study. We start with an equilateral triangle, a triangle with all sides of equal length. Each angle of such a triangle measures $60°$. While the results we will obtain are independent of the length, for convenience we choose the length of each side to be 2 units. See Figure 5(a).

Bisecting one angle of this equilateral triangle leads to two right triangles, each of which has angles of $30°$, $60°$, and $90°$, as shown in Figure 5(b). Since the hypotenuse of each of these right triangles has length 2, the shortest sides will have length 1. (Why?) If x represents the length of the medium side, then,

$$2^2 = 1^2 + x^2 \quad \text{Pythagorean theorem (Appendix B)}$$
$$4 = 1 + x^2$$
$$3 = x^2 \quad \text{Subtract 1.}$$
$$\sqrt{3} = x. \quad \text{Square root property (Appendix A);}$$
choose the positive square root.

Figure 6 summarizes our results using a $30°$–$60°$ right triangle. As shown in the figure, the side opposite the $30°$ angle has length 1; that is, for the $30°$ angle,

$$\text{hypotenuse} = 2, \quad \text{side opposite} = 1, \quad \text{side adjacent} = \sqrt{3}.$$

Now we use the definitions of the trigonometric functions.

$$\sin 30° = \frac{\text{side opposite}}{\text{hypotenuse}} = \frac{1}{2}$$

$$\cos 30° = \frac{\text{side adjacent}}{\text{hypotenuse}} = \frac{\sqrt{3}}{2}$$

$$\tan 30° = \frac{\text{side opposite}}{\text{side adjacent}} = \frac{1}{\sqrt{3}} = \frac{1}{\sqrt{3}} \cdot \frac{\sqrt{3}}{\sqrt{3}} = \frac{\sqrt{3}}{3}$$

$$\csc 30° = \frac{2}{1} = 2$$

$$\sec 30° = \frac{2}{\sqrt{3}} = \frac{2}{\sqrt{3}} \cdot \frac{\sqrt{3}}{\sqrt{3}} = \frac{2\sqrt{3}}{3}$$

$$\cot 30° = \frac{\sqrt{3}}{1} = \sqrt{3}$$

Equilateral triangle

(a)

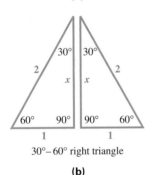

$30°$–$60°$ right triangle

(b)

Figure 5

Figure 6

Figure 6 (repeated)

EXAMPLE 5 Finding Trigonometric Function Values for 60°

Find the six trigonometric function values for a 60° angle.

Solution Refer to Figure 6 to find the following ratios.

$$\sin 60° = \frac{\sqrt{3}}{2} \qquad \cos 60° = \frac{1}{2} \qquad \tan 60° = \frac{\sqrt{3}}{1} = \sqrt{3}$$

$$\csc 60° = \frac{2}{\sqrt{3}} = \frac{2\sqrt{3}}{3} \qquad \sec 60° = \frac{2}{1} = 2 \qquad \cot 60° = \frac{1}{\sqrt{3}} = \frac{\sqrt{3}}{3}$$

Now try Exercises 35, 37, and 39.

NOTE The results in Example 5 can also be found using the fact that cofunctions of complementary angles are equal.

We find the values of the trigonometric functions for 45° by starting with a 45°–45° right triangle, as shown in Figure 7. This triangle is isosceles; we choose the lengths of the equal sides to be 1 unit. (As before, the results are independent of the length of the equal sides.) Since the shorter sides each have length 1, if r represents the length of the hypotenuse, then

$$1^2 + 1^2 = r^2 \quad \text{Pythagorean theorem}$$
$$2 = r^2$$
$$\sqrt{2} = r. \quad \text{Choose the positive square root.}$$

45°–45° right triangle

Figure 7

Now we use the measures indicated on the 45°–45° right triangle in Figure 7.

$$\sin 45° = \frac{1}{\sqrt{2}} = \frac{\sqrt{2}}{2} \qquad \cos 45° = \frac{1}{\sqrt{2}} = \frac{\sqrt{2}}{2} \qquad \tan 45° = \frac{1}{1} = 1$$

$$\csc 45° = \frac{\sqrt{2}}{1} = \sqrt{2} \qquad \sec 45° = \frac{\sqrt{2}}{1} = \sqrt{2} \qquad \cot 45° = \frac{1}{1} = 1$$

Function values for 30°, 45°, and 60° are summarized in the table that follows.

Function Values of Special Angles

θ	$\sin \theta$	$\cos \theta$	$\tan \theta$	$\cot \theta$	$\sec \theta$	$\csc \theta$
30°	$\dfrac{1}{2}$	$\dfrac{\sqrt{3}}{2}$	$\dfrac{\sqrt{3}}{3}$	$\sqrt{3}$	$\dfrac{2\sqrt{3}}{3}$	2
45°	$\dfrac{\sqrt{2}}{2}$	$\dfrac{\sqrt{2}}{2}$	1	1	$\sqrt{2}$	$\sqrt{2}$
60°	$\dfrac{\sqrt{3}}{2}$	$\dfrac{1}{2}$	$\sqrt{3}$	$\dfrac{\sqrt{3}}{3}$	2	$\dfrac{2\sqrt{3}}{3}$

Now try Exercises 5, 7, and 9.

NOTE You will be able to reproduce this table quickly if you learn the values of sin 30°, sin 45°, and sin 60°. Then you can complete the rest of the table using the reciprocal, cofunction, and quotient identities.

Since a calculator finds trigonometric function values at the touch of a key, you may wonder why we spend so much time finding values for special angles. We do this because a calculator gives only *approximate* values in most cases, while we often need *exact* values. For example, tan 30° can be found on a scientific calculator by first setting it in *degree mode,* then entering 30 and pressing the $\boxed{\tan}$ key to get

$$\tan 30° \approx .57735027. \quad \approx \text{ means "is approximately equal to."}$$

Earlier, however, we found the exact value:

$$\tan 30° = \frac{\sqrt{3}}{3}.$$

 To use a graphing calculator to approximate sine, cosine, or tangent function values, press the appropriate function key *first,* and then enter the angle measure. (The calculator must be in degree mode to enter the angle measure in degrees.) See Figure 8. ■

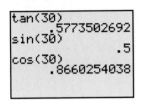

Figure 8

2.1 Exercises

Find exact values or expressions for sin *A,* cos *A, and* tan *A. See Example 1.*

1.

2.

3.

4.

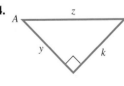

Concept Check   For each trigonometric function in Column I, choose its value from Column II.

I		II		
5. sin 30°	**6.** cos 45°	**A.** $\sqrt{3}$	**B.** 1	**C.** $\dfrac{1}{2}$
7. tan 45°	**8.** sec 60°	**D.** $\dfrac{\sqrt{3}}{2}$	**E.** $\dfrac{2\sqrt{3}}{3}$	**F.** $\dfrac{\sqrt{3}}{3}$
9. csc 60°	**10.** cot 30°	**G.** 2	**H.** $\dfrac{\sqrt{2}}{2}$	**I.** $\sqrt{2}$

Suppose ABC is a right triangle with sides of lengths a, b, and c and right angle at C. (See Figure 3.) Find the unknown side length using the Pythagorean theorem (Appendix B), and then find the values of the six trigonometric functions for angle B. Rationalize denominators when applicable.

11. $a = 5, b = 12$

12. $a = 3, b = 5$

13. $a = 6, c = 7$

14. $b = 7, c = 12$

15. *Concept Check*   Give a summary of the six cofunction relationships.

Write each function in terms of its cofunction. Assume that all angles in which an unknown appears are acute angles. See Example 2.

16. $\cot 73°$

17. $\sec 39°$

18. $\cos(\alpha + 20°)$

19. $\cot(\theta - 10°)$

20. $\tan 25.4°$

21. $\sin 38.7°$

22. With a calculator, evaluate $\sin(90° - A)$ and $\cos A$ for various values of A. (Include values greater than 90° and less than 0°.) What do you find?

Find one solution for each equation. Assume that all angles in which an unknown appears are acute angles. See Example 3.

23. $\tan \alpha = \cot(\alpha + 10°)$

24. $\cos \theta = \sin 2\theta$

25. $\sin(2\theta + 10°) = \cos(3\theta - 20°)$

26. $\sec(\beta + 10°) = \csc(2\beta + 20°)$

27. $\tan(3B + 4°) = \cot(5B - 10°)$

28. $\cot(5\theta + 2°) = \tan(2\theta + 4°)$

Tell whether each statement is true *or* false. *See Example 4.*

29. $\sin 50° > \sin 40°$

30. $\tan 28° \le \tan 40°$

31. $\sin 46° < \cos 46°$
   (*Hint:* $\cos 46° = \sin 44°$)

32. $\cos 28° < \sin 28°$
   (*Hint:* $\sin 28° = \cos 62°$)

33. $\tan 41° < \cot 41°$

34. $\cot 30° < \tan 40°$

For each expression, give the exact value. See Example 5.

35. $\tan 30°$

36. $\cot 30°$

37. $\sin 30°$

38. $\cos 30°$

39. $\sec 30°$

40. $\csc 30°$

41. $\csc 45°$

42. $\sec 45°$

43. $\cos 45°$

44. $\cot 45°$

Relating Concepts

For individual or collaborative investigation
(Exercises 45–48)

The figure shows a 45° central angle in a circle with radius 4 units. To find the coordinates of point P on the circle, **work Exercises 45–48 in order.**

45. Sketch a line segment from P perpendicular to the x-axis.

46. Use the trigonometric ratios for a 45° angle to label the sides of the right triangle you sketched in Exercise 45.

47. Which sides of the right triangle give the coordinates of point P? What are the coordinates of P?

48. Follow the same procedure as in Exercises 45–47 to find the coordinates of P in the figure given here.

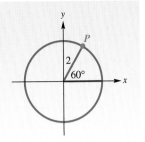

49. *Concept Check* Refer to the table. What trigonometric functions are y_1 and y_2?

$x°$	y_1	y_2
0	0	0
15	.25882	.26795
30	.5	.57735
45	.70711	1
60	.86603	1.7321
75	.96593	3.7321
90	1	undefined

50. *Concept Check* Refer to the table. What trigonometric functions are y_1 and y_2?

$x°$	y_1	y_2
0	1	undefined
15	.96593	3.8637
30	.86603	2
45	.70711	1.4142
60	.5	1.1547
75	.25882	1.0353
90	0	1

51. *Concept Check* What value of A between 0° and 90° will produce the output shown on the graphing calculator screen?

```
√(3)/2
          .8660254038
sin(A)
          .8660254038
```

52. A student was asked to give the exact value of sin 45°. Using a calculator, he gave the answer .7071067812. The teacher did not give him credit. What was the teacher's reason for this?

53. With a graphing calculator, find the coordinates of the point of intersection of $y = x$ and $y = \sqrt{1 - x^2}$. These coordinates are the cosine and sine of what angle between 0° and 90°?

Concept Check Work each problem.

54. Find the equation of the line passing through the origin and making a 60° angle with the x-axis.

55. Find the equation of the line passing through the origin and making a 30° angle with the x-axis.

56. What angle does the line $y = \frac{\sqrt{3}}{3}x$ make with the positive x-axis?

57. What angle does the line $y = \sqrt{3}x$ make with the positive x-axis?

58. Construct a square with each side of length k.

   (a) Draw a diagonal of the square. What is the measure of each angle formed by a side of the square and this diagonal?

   (b) What is the length of the diagonal?

   (c) From the results of parts (a) and (b), complete the following statement: In a $45°$–$45°$ right triangle, the hypotenuse has a length that is _____ times as long as either leg.

59. Construct an equilateral triangle with each side having length $2k$.

   (a) What is the measure of each angle?

   (b) Label one angle A. Drop a perpendicular from A to the side opposite A. Two $30°$ angles are formed at A, and two right triangles are formed. What is the length of the sides opposite the $30°$ angles?

   (c) What is the length of the perpendicular constructed in part (b)?

   (d) From the results of parts (a)–(c), complete the following statement: In a $30°$–$60°$ right triangle, the hypotenuse is always _____ times as long as the shorter leg, and the longer leg has a length that is _____ times as long as that of the shorter leg. Also, the shorter leg is opposite the _____ angle, and the longer leg is opposite the _____ angle.

Find the exact value of each part labeled with a variable in each figure.

60.

61.

62.

63.

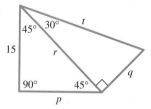

Find a formula for the area of each figure in terms of s.

64.

65.

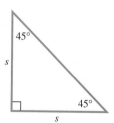

66. Suppose you know the length of one side and one acute angle of a right triangle. Can you determine the measures of all the sides and angles of the triangle?

67. Why is it important to be able to find trigonometric values for the special angles without using a calculator?

2.2 | Trigonometric Functions of Non-Acute Angles

Reference Angles ▪ Special Angles as Reference Angles ▪ Finding Angle Measures with Special Angles

Reference Angles   Associated with every nonquadrantal angle in standard position is a positive acute angle called its *reference angle*. A **reference angle** for an angle θ, written θ', is the positive acute angle made by the terminal side of angle θ and the x-axis. Figure 9 shows several angles θ (each less than one complete counterclockwise revolution) in quadrants II, III, and IV, respectively, with the reference angle θ' also shown. In quadrant I, θ and θ' are the same. If an angle θ is negative or has measure greater than $360°$, its reference angle is found by first finding its coterminal angle that is between $0°$ and $360°$, and then using the diagrams in Figure 9.

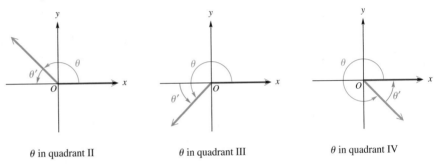

θ in quadrant II θ in quadrant III θ in quadrant IV

Figure 9

CAUTION   A common error is to find the reference angle by using the terminal side of θ and the y-axis. *The reference angle is always found with reference to the x-axis.*

$218° - 180° = 38°$

Figure 10

EXAMPLE 1   Finding Reference Angles

Find the reference angle for each angle.

(a) $218°$ **(b)** $1387°$

Solution

(a) As shown in Figure 10, the positive acute angle made by the terminal side of this angle and the x-axis is $218° - 180° = 38°$. For $\theta = 218°$, the reference angle $\theta' = 38°$.

(b) First find a coterminal angle between $0°$ and $360°$. Divide $1387°$ by $360°$ to get a quotient of about 3.9. Begin by subtracting $360°$ three times (because of the whole number 3 in 3.9):

$$1387° - 3 \cdot 360° = 1387° - 1080° = 307°. \quad \text{(Section 1.1)}$$

The reference angle for $307°$ (and thus for $1387°$) is $360° - 307° = 53°$. See Figure 11.

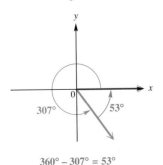

$360° - 307° = 53°$

Figure 11

Now try Exercises 1 and 5.

The preceding example suggests the following table for finding the reference angle θ' for any angle θ between 0° and 360°.

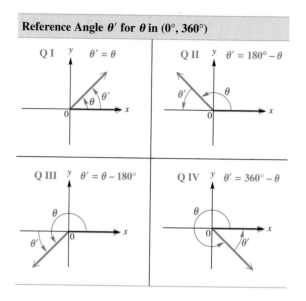

| Reference Angle θ' for θ in (0°, 360°) | |

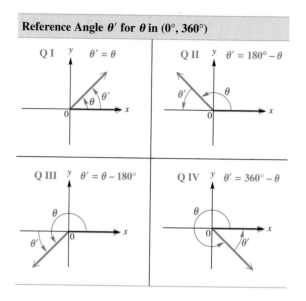

Special Angles as Reference Angles

We can now find exact trigonometric function values of angles with reference angles of 30°, 45°, or 60°.

EXAMPLE 2 Finding Trigonometric Function Values of a Quadrant III Angle

Find the values of the trigonometric functions for 210°.

Solution An angle of 210° is shown in Figure 12. The reference angle is $210° - 180° = 30°$. To find the trigonometric function values of 210°, choose point P on the terminal side of the angle so that the distance from the origin O to P is 2. By the results from 30°–60° right triangles, the coordinates of point P become $\left(-\sqrt{3}, -1\right)$, with $x = -\sqrt{3}$, $y = -1$, and $r = 2$. Then, by the definitions of the trigonometric functions in Section 2.1,

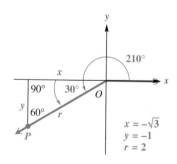

$x = -\sqrt{3}$
$y = -1$
$r = 2$

Figure 12

$$\sin 210° = \frac{-1}{2} = -\frac{1}{2} \qquad\qquad \csc 210° = \frac{2}{-1} = -2$$

$$\cos 210° = \frac{-\sqrt{3}}{2} = -\frac{\sqrt{3}}{2} \qquad\qquad \sec 210° = \frac{2}{-\sqrt{3}} = -\frac{2\sqrt{3}}{3}$$

$$\tan 210° = \frac{-1}{-\sqrt{3}} = \frac{\sqrt{3}}{3} \qquad\qquad \cot 210° = \frac{-\sqrt{3}}{-1} = \sqrt{3}.$$

Now try Exercise 19.

Notice in Example 2 that the trigonometric function values of 210° correspond in absolute value to those of its reference angle 30°. The signs are different for the sine, cosine, secant, and cosecant functions because 210° is a quadrant III angle. These results suggest a shortcut for finding the trigonometric function values of a non-acute angle, using the reference angle. In Example 2, the reference angle for 210° is 30°. Using the trigonometric function values of 30°, and choosing the correct signs for a quadrant III angle, we obtain the results found in Example 2.

Similarly, we determine the values of the trigonometric functions for any nonquadrantal angle θ by finding the function values for its reference angle between 0° and 90°, and choosing the appropriate signs.

Finding Trigonometric Function Values for Any Nonquadrantal Angle θ

Step 1   If $\theta > 360°$, or if $\theta < 0°$, then find a coterminal angle by adding or subtracting 360° as many times as needed to get an angle greater than 0° but less than 360°.

Step 2   Find the reference angle θ'.

Step 3   Find the trigonometric function values for reference angle θ'.

Step 4   Determine the correct signs for the values found in Step 3. (Use the table of signs in Section 1.4, if necessary.) This gives the values of the trigonometric functions for angle θ.

EXAMPLE 3   Finding Trigonometric Function Values Using Reference Angles

Find the exact value of each expression.

(a) $\cos(-240°)$ **(b)** $\tan 675°$

Solution

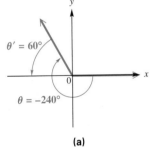

(a) Since an angle of $-240°$ is coterminal with an angle of

$$-240° + 360° = 120°, \quad \text{(Section 1.1)}$$

the reference angle is $180° - 120° = 60°$, as shown in Figure 13(a). Since the cosine is negative in quadrant II,

$$\cos(-240°) = \cos 120° = -\cos 60° = -\frac{1}{2}.$$

$$\underset{\substack{\uparrow \\ \text{Coterminal} \\ \text{angle}}}{} \qquad \underset{\substack{\uparrow \\ \text{Reference} \\ \text{angle}}}{}$$

(b) Begin by subtracting 360° to get a coterminal angle between 0° and 360°.

$$675° - 360° = 315°$$

As shown in Figure 13(b), the reference angle is $360° - 315° = 45°$. An angle of 315° is in quadrant IV, so the tangent will be negative, and

$$\tan 675° = \tan 315° = -\tan 45° = -1.$$

Figure 13

Now try Exercises 35 and 37.

EXAMPLE 4 Evaluating an Expression with Function Values of Special Angles

Evaluate $\cos 120° + 2 \sin^2 60° - \tan^2 30°$.

Solution Since $\cos 120° = -\frac{1}{2}$, $\sin 60° = \frac{\sqrt{3}}{2}$, and $\tan 30° = \frac{\sqrt{3}}{3}$,

$$\cos 120° + 2 \sin^2 60° - \tan^2 30° = -\frac{1}{2} + 2\left(\frac{\sqrt{3}}{2}\right)^2 - \left(\frac{\sqrt{3}}{3}\right)^2$$

$$= -\frac{1}{2} + 2\left(\frac{3}{4}\right) - \frac{3}{9} = \frac{2}{3}.$$

Now try Exercise 39.

Earlier we saw that trigonometric function values of coterminal angles are the same.

EXAMPLE 5 Using Coterminal Angles to Find Function Values

Evaluate each function by first expressing the function in terms of an angle between 0° and 360°.

(a) $\cos 780°$ **(b)** $\tan(-405°)$

Solution

(a) Add or subtract 360° as many times as necessary to get an angle between 0° and 360°. Subtracting 720°, which is $2 \cdot 360°$, gives

$$\cos 780° = \cos(780° - 2 \cdot 360°)$$

$$= \cos 60° = \frac{1}{2}.$$

(b) We add 360° to get $-405° + 360° = -45°$. We find that $\tan(-45°) = -\tan 45°$ because the reference angle is 45°, and $-45°$ is in quadrant IV, where the tangent function is negative. Thus,

$$\tan(-405°) = \tan(-45°) = -\tan 45° = -1.$$

Now try Exercises 27 and 31.

Finding Angle Measures with Special Angles The ideas discussed in this section can also be used to find the measures of certain angles, given a trigonometric function value and an interval in which the angle must lie. We are most often interested in the interval $[0°, 360°)$.

EXAMPLE 6 Finding Angle Measures Given an Interval and a Function Value

Find all values of θ, if θ is in the interval $[0°, 360°)$ and $\cos \theta = -\frac{\sqrt{2}}{2}$.

Solution Since $\cos \theta$ is negative, θ must lie in quadrant II or III. Since the absolute value of $\cos \theta$ is $\frac{\sqrt{2}}{2}$, the reference angle θ' must be 45°. The two possible angles θ are sketched in Figure 14. The quadrant II angle θ equals $180° - 45° = 135°$, and the quadrant III angle θ equals $180° + 45° = 225°$.

Figure 14

Now try Exercise 61.

2.2 Exercises

Match each angle in Column I with its reference angle in Column II. Choices may be used once, more than once, or not at all. See Example 1.

I		II	
1. 98°	**2.** 212°	**A.** 45°	**B.** 60°
3. −135°	**4.** −60°	**C.** 82°	**D.** 30°
5. 750°	**6.** 480°	**E.** 38°	**F.** 32°

Give a short explanation in Exercises 7–9.

7. In Example 2, why was 2 a good choice for r? Could any other positive number have been used?

8. Explain how the reference angle is used to find values of the trigonometric functions for an angle in quadrant III.

9. Explain why two coterminal angles have the same values for their trigonometric functions.

Complete the table with exact trigonometric function values. Do not use a calculator. See Examples 2 and 3.

	θ	$\sin \theta$	$\cos \theta$	$\tan \theta$	$\cot \theta$	$\sec \theta$	$\csc \theta$
10.	30°	$\dfrac{1}{2}$	$\dfrac{\sqrt{3}}{2}$			$\dfrac{2\sqrt{3}}{3}$	2
11.	45°			1	1		
12.	60°		$\dfrac{1}{2}$	$\sqrt{3}$		2	
13.	120°	$\dfrac{\sqrt{3}}{2}$		$-\sqrt{3}$			$\dfrac{2\sqrt{3}}{3}$

(continued)

θ	$\sin \theta$	$\cos \theta$	$\tan \theta$	$\cot \theta$	$\sec \theta$	$\csc \theta$
14. 135°	$\dfrac{\sqrt{2}}{2}$	$-\dfrac{\sqrt{2}}{2}$			$-\sqrt{2}$	$\sqrt{2}$
15. 150°		$-\dfrac{\sqrt{3}}{2}$	$-\dfrac{\sqrt{3}}{3}$			2
16. 210°	$-\dfrac{1}{2}$		$\dfrac{\sqrt{3}}{3}$	$\sqrt{3}$		-2
17. 240°	$-\dfrac{\sqrt{3}}{2}$	$-\dfrac{1}{2}$			-2	$-\dfrac{2\sqrt{3}}{3}$

Find exact values of the six trigonometric functions for each angle. Rationalize denominators when applicable. See Examples 2, 3, and 5.

18. 300°   **19.** 315°   **20.** 405°   **21.** −300°   **22.** 420°   **23.** 480°

24. 495°   **25.** 570°   **26.** 750°   **27.** 1305°   **28.** 1500°   **29.** 2670°

30. −390°   **31.** −510°   **32.** −1020°   **33.** −1290°

Find the exact value of each expression. See Example 3.

34. $\sin 1305°$ **35.** $\cos(-510°)$ **36.** $\tan(-1020°)$ **37.** $\sin 1500°$

Tell whether each statement is true *or* false. *If false, tell why. See Example 4.*

38. $\sin 30° + \sin 60° = \sin(30° + 60°)$

39. $\sin(30° + 60°) = \sin 30° \cdot \cos 60° + \sin 60° \cdot \cos 30°$

40. $\cos 60° = 2\cos^2 30° - 1$

41. $\cos 60° = 2\cos 30°$

42. $\sin 120° = \sin 150° - \sin 30°$

43. $\sin 120° = \sin 180° \cdot \cos 60° - \sin 60° \cdot \cos 180°$

Concept Check   Find the coordinates of the point P on the circumference of each circle. (Hint: Add x- and y-axes, assuming that the angle is in standard position.)

44.

45.

46. *Concept Check*   Does there exist an angle θ with the function values $\cos \theta = .6$ and $\sin \theta = -.8$?

47. *Concept Check*   Does there exist an angle θ with the function values $\cos \theta = \frac{2}{3}$ and $\sin \theta = \frac{3}{4}$?

Suppose θ is in the interval $(90°, 180°)$. *Find the sign of each of the following.*

48. $\sin \dfrac{\theta}{2}$ **49.** $\cos \dfrac{\theta}{2}$ **50.** $\cot(\theta + 180°)$

51. $\sec(\theta + 180°)$ **52.** $\cos(-\theta)$ **53.** $\sin(-\theta)$

54. Explain why $\sin \theta = \sin(\theta + n \cdot 360°)$ for any angle θ and any integer n.

55. Explain why $\cos \theta = \cos(\theta + n \cdot 360°)$ for any angle θ and any integer n.

Concept Check   *Work Exercises 56–59.*

56. Without using a calculator, determine which of the following numbers is closest to $\cos 115°$: $.4, .6, 0, -.4,$ or $-.6$.

57. Without using a calculator, determine which of the following numbers is closest to $\sin 115°$: $.9, .1, 0, -.9,$ or $-.1$.

58. For what angles θ between $0°$ and $360°$ does $\cos \theta = -\sin \theta$?

59. For what angles θ between $0°$ and $360°$ does $\cos \theta = \sin \theta$?

60. *(Modeling) Length of a Sag Curve*   When a highway goes downhill and then uphill, it is said to have a *sag curve*. Sag curves are designed so that at night, headlights shine sufficiently far down the road to allow a safe stopping distance. See the figure.

The minimum length L of a sag curve is determined by the height h of the car's headlights above the pavement, the downhill grade $\theta_1 < 0°$, the uphill grade $\theta_2 > 0°$, and the safe stopping distance S for a given speed limit. In addition, L is dependent on the vertical alignment of the headlights. Headlights are usually pointed upward at a slight angle α above the horizontal of the car. Using these quantities, for a 55 mph speed limit, L can be modeled by the formula

$$L = \frac{(\theta_2 - \theta_1)S^2}{200(h + S \tan \alpha)},$$

where $S < L$. (*Source:* Mannering, F. and W. Kilareski, *Principles of Highway Engineering and Traffic Analysis,* Second Edition, John Wiley & Sons, 1998.)

(a) Compute L if $h = 1.9$ ft, $\alpha = .9°$, $\theta_1 = -3°$, $\theta_2 = 4°$, and $S = 336$ ft.

(b) Repeat part (a) with $\alpha = 1.5°$.

(c) How does the alignment of the headlights affect the value of L?

Find all values of θ, if θ is in the interval $[0°, 360°)$ *and has the given function value. See Example 6.*

61. $\sin \theta = \dfrac{1}{2}$ **62.** $\cos \theta = \dfrac{\sqrt{3}}{2}$ **63.** $\tan \theta = -\sqrt{3}$

64. $\sec \theta = -\sqrt{2}$ **65.** $\cos \theta = \dfrac{\sqrt{2}}{2}$ **66.** $\cot \theta = -\dfrac{\sqrt{3}}{3}$

2.3 Finding Trigonometric Function Values Using a Calculator

Finding Function Values Using a Calculator ▪ Finding Angle Measures Using a Calculator

Degree mode

Figure 15

Finding Function Values Using a Calculator Calculators are capable of finding trigonometric function values. For example, the values of cos(−240°) and tan 675° in Example 3 of the previous section are found with a calculator as shown in Figure 15.

C A U T I O N We have studied only degree measure of angles; radian measure will be introduced in Chapter 3. When evaluating trigonometric functions of angles given in degrees, remember that the calculator must be set in *degree mode*. Get in the habit of always starting work by entering sin 90. If the displayed answer is 1, then the calculator is set for degree measure. Remember that most calculator values of trigonometric functions are *approximations*.

EXAMPLE 1 Finding Function Values with a Calculator

Approximate the value of each expression.

(a) sin 49° 12′   **(b)** sec 97.977°   **(c)** cot 51.4283°   **(d)** sin(−246°)

Solution

(a) $49° \, 12' = 49 \dfrac{12}{60}^{°} = 49.2°$ Convert 49° 12′ to decimal degrees. (Section 1.1)

   $\sin 49° \, 12' = \sin 49.2° \approx .75699506$ To eight decimal places

(b) Calculators do not have secant keys. However, $\sec \theta = \dfrac{1}{\cos \theta}$ for all angles θ where $\cos \theta \neq 0$. First find cos 97.977°, and then take the reciprocal to get

$$\sec 97.977° \approx -7.20587921.$$

(c) $\cot 51.4283° \approx .79748114$ Use the identity $\cot \theta = \dfrac{1}{\tan \theta}$. (Section 1.4)

(d) $\sin(-246°) \approx .91354546$

These screens support the results of Example 1. We entered the angle measure in degrees and minutes for part (a). In the fifth line of the first screen, Ans⁻¹ tells the calculator to find the reciprocal of the answer given in the previous line.

Now try Exercises 5, 9, 11, and 15.

Finding Angle Measures Using a Calculator Sometimes we need to find the measure of an angle having a certain trigonometric function value. Graphing calculators have three *inverse functions* (denoted $\sin^{-1}$, $\cos^{-1}$, and $\tan^{-1}$) that do just that. If x is an appropriate number, then $\sin^{-1}x$, $\cos^{-1}x$, or $\tan^{-1}x$ gives the measure of an angle whose sine, cosine, or tangent is x. For the applications in this chapter, these functions will return values of x in quadrant I.

EXAMPLE 2 Using Inverse Trigonometric Functions to Find Angles

Use a calculator to find an angle θ in the interval $[0°, 90°]$ that satisfies each condition.

(a) $\sin \theta \approx .9677091705$ **(b)** $\sec \theta \approx 1.0545829$

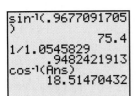

Degree mode

Figure 16

Solution

(a) Using degree mode and the inverse sine function, we find that an angle θ having sine value .9677091705 is 75.4°. (While there are infinitely many such angles, the calculator gives only this one.) We write this result as

$$\sin^{-1} .9677091705 \approx 75.4°.$$

See Figure 16.

(b) Use the identity $\cos \theta = \frac{1}{\sec \theta}$. Find the reciprocal of 1.0545829 to get $\cos \theta \approx .9482421913$. Now find θ as shown in part (a), using the inverse cosine function. The result is $\theta \approx 18.514704°$. See Figure 16.

Now try Exercise 23.

CAUTION Compare Examples 1(b) and 2(b). Note that the reciprocal is used *before* the inverse cosine key when finding the angle, but *after* the cosine key when finding the trigonometric function value.

EXAMPLE 3 Finding Grade Resistance

Figure 17

When an automobile travels uphill or downhill on a highway, it experiences a force due to gravity. This force F in pounds is called *grade resistance* and is modeled by the equation $F = W \sin \theta$, where θ is the grade and W is the weight of the automobile. If the automobile is moving uphill, then $\theta > 0°$; if downhill, then $\theta < 0°$. See Figure 17. (*Source:* Mannering, F. and W. Kilareski, *Principles of Highway Engineering and Traffic Analysis,* Second Edition, John Wiley & Sons, 1998.)

(a) Calculate F to the nearest 10 lb for a 2500-lb car traveling an uphill grade with $\theta = 2.5°$.

(b) Calculate F to the nearest 10 lb for a 5000-lb truck traveling a downhill grade with $\theta = -6.1°$.

(c) Calculate F for $\theta = 0°$ and $\theta = 90°$. Do these answers agree with your intuition?

Solution

(a) $F = W \sin \theta = 2500 \sin 2.5° \approx 110$ lb

(b) $F = W \sin \theta = 5000 \sin(-6.1°) \approx -530$ lb

F is negative because the truck is moving downhill.

(c) $F = W \sin \theta = W \sin 0° = W(0) = 0$ lb

$F = W \sin \theta = W \sin 90° = W(1) = W$ lb

This agrees with intuition because if $\theta = 0°$, then there is level ground and gravity does not cause the vehicle to roll. If $\theta = 90°$, the road would be vertical and the full weight of the vehicle would be pulled downward by gravity, so $F = W$.

Now try Exercises 49 and 51.

2.3 Exercises

Concept Check   *Fill in the blanks to complete each statement.*

1. The CAUTION at the beginning of this section suggests verifying that a calculator is in degree mode by finding _____ 90°. If the calculator is in degree
(sin/cos/tan)
 mode, the display should be _____.

2. When a scientific or graphing calculator is used to find a trigonometric function value, in most cases the result is a(n) _____ value.
(exact/approximate)

3. To find values of the cotangent, secant, and cosecant functions with a calculator, it is necessary to find the _____ of the _____ function value.

4. The reciprocal is used _____ the inverse function key when finding the
(before/after)
 angle, but _____ the function key when finding the trigonometric function
(before/after)
 value.

Use a calculator to find a decimal approximation for each value. Give as many digits as your calculator displays. In Exercises 16–21, simplify the expression before using the calculator. See Example 1.

5. sin 38° 42′
6. cot 41° 24′
7. sec 13° 15′

8. csc 145° 45′
9. cot 183° 48′
10. cos 421° 30′

11. sec 312° 12′
12. tan(−80° 6′)
13. sin(−317° 36′)

14. cot(−512° 20′)
15. cos(−15′)
16. $\dfrac{1}{\sec 14.8°}$

17. $\dfrac{1}{\cot 23.4°}$
18. $\dfrac{\sin 33°}{\cos 33°}$
19. $\dfrac{\cos 77°}{\sin 77°}$

20. cos(90° − 3.69°)
21. cot(90° − 4.72°)

Find a value of θ in the interval [0°, 90°] that satisfies each statement. Leave answers in decimal degrees. See Example 2.

22. sin θ = .84802194
23. tan θ = 1.4739716
24. tan θ = 6.4358841

25. sin θ = .27843196
26. sec θ = 1.1606249
27. cot θ = 1.2575516

28. csc θ = 1.3861147
29. sec θ = 2.7496222

30. A student, wishing to use a calculator to verify the value of sin 30°, enters the information correctly but gets a display of −.98803162. He knows that the display should be .5, and he also knows that his calculator is in good working order. What do you think is the problem?

31. At one time, a certain make of calculator did not allow the input of angles outside of a particular interval when finding trigonometric function values. For example, trying to find cos 2000° using the methods of this section gave an error message, despite the fact that cos 2000° can be evaluated. Explain how you would find cos 2000° using this calculator.

32. What value of A between $0°$ and $90°$ will produce the output in the graphing calculator screen?

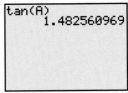

33. What value of A will produce the output (in degrees) in the graphing calculator screen?

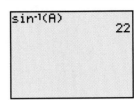

Use a calculator to evaluate each expression.

34. $\sin 35° \cos 55° + \cos 35° \sin 55°$ **35.** $\cos 100° \cos 80° - \sin 100° \sin 80°$

36. $\cos 75° 29' \cos 14° 31' - \sin 75° 29' \sin 14° 31'$

37. $\sin 28° 14' \cos 61° 46' + \cos 28° 14' \sin 61° 46'$

Work each problem.

38. *Measuring Speed by Radar*   Any offset between a stationary radar gun and a moving target creates a "cosine effect" that reduces the radar mileage reading by the cosine of the angle between the gun and the vehicle. That is, the radar speed reading is the product of the actual reading and the cosine of the angle. Find the radar readings for Auto A and Auto B shown in the figure. (*Source:* Fischetti, M., "Working Knowledge," *Scientific American,* March 2001.)

39. *Measuring Speed by Radar*   In Exercise 38, we saw that the mileage reported by a radar gun is reduced by the cosine of angle θ, shown in the figure. In the figure, r represents reduced speed and a represents the actual speed. Use the figure to show why this "cosine effect" occurs.

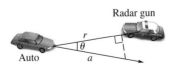

Use a calculator to decide whether each statement is true *or* false. *It may be that a true statement will lead to results that differ in the last decimal place due to rounding error.*

40. $\cos 40° = 2 \cos 20°$ **41.** $\sin 10° + \sin 10° = \sin 20°$

42. $\cos 70° = 2 \cos^2 35° - 1$ **43.** $\sin 50° = 2 \sin 25° \cos 25°$

44. $2 \cos 38° 22' = \cos 76° 44'$ **45.** $\cos 40° = 1 - 2 \sin^2 80°$

46. $\dfrac{1}{2} \sin 40° = \sin \dfrac{1}{2}(40°)$

47. $\sin 39° \, 48' + \cos 39° \, 48' = 1$

(Modeling) Grade Resistance   *See Example 3 to work Exercises 48–53.*

48. What is the grade resistance of a 2400-lb car traveling on a $-2.4°$ downhill grade?

49. What is the grade resistance of a 2100-lb car traveling on a $1.8°$ uphill grade?

50. A car traveling on a $-3°$ downhill grade has a grade resistance of -145 lb. What is the weight of the car?

51. A 2600-lb car traveling downhill has a grade resistance of -130 lb. What is the angle of the grade?

52. Which has the greater grade resistance: a 2200-lb car on a $2°$ uphill grade or a 2000-lb car on a $2.2°$ uphill grade?

53. A 3000-lb car traveling uphill has a grade resistance of 150 lb. What is the angle of the grade?

54. *Highway Grades*   Complete the table for the values of $\sin \theta$, $\tan \theta$, and $\dfrac{\pi\theta}{180}$ to four decimal places.

θ	$\sin \theta$	$\tan \theta$	$\dfrac{\pi\theta}{180}$
0°			
.5°			
1°			
1.5°			
2°			
2.5°			
3°			
3.5°			
4°			

(a) How do $\sin \theta$, $\tan \theta$, and $\dfrac{\pi\theta}{180}$ compare for small grades θ?

(b) Highway grades are usually small. Give two approximations of the grade resistance $F = W \sin \theta$ that do not use the sine function.

(c) A stretch of highway has a 4-ft vertical rise for every 100 ft of horizontal run. Use an approximation from part (a) to estimate the grade resistance for a 2000-lb car on this stretch of highway.

(d) A stretch of highway has a $3.75°$ grade. Without evaluating a trigonometric function, estimate the grade resistance for an 1800-lb car on this stretch of highway.

(Modeling)   *Solve each problem.*

55. *Design of Highway Curves*   When highway curves are designed, the outside of the curve is often slightly elevated or inclined above the inside of the curve. See the figure. This inclination is called *superelevation*. For safe-

ty reasons, it is important that both the curve's radius and superelevation are correct for a given speed limit. If an automobile is traveling at velocity V (in feet per second), the safe radius R for a curve with superelevation θ is modeled by the formula

$$R = \frac{V^2}{g(f + \tan \theta)},$$

where f and g are constants. (*Source:* Mannering, F. and W. Kilareski, *Principles of Highway Engineering and Traffic Analysis,* Second Edition, John Wiley & Sons, 1998.)

(a) A roadway is being designed for automobiles traveling at 45 mph. If $\theta = 3°$, $g = 32.2$, and $f = .14$, calculate R. (*Hint:* 45 mph = 66 ft per sec)

(b) What should the radius of the curve be if the speed in part (a) is increased to 70 mph?

(c) How would increasing the angle θ affect the results? Verify your answer by repeating parts (a) and (b) with $\theta = 4°$.

56. *Speed Limit on a Curve* Refer to Exercise 55 and use the same values for f and g. A highway curve has radius $R = 1150$ ft and a superelevation of $\theta = 2.1°$. What should the speed limit (in miles per hour) be for this curve?

(Modeling) Speed of Light When a light ray travels from one medium, such as air, to another medium, such as water or glass, the speed of the light changes, and the direction in which the ray is traveling changes. (This is why a fish under water is in a different position than it appears to be.) These changes are given by Snell's law

Medium 1

θ_1

If this medium is less dense, light travels at a faster speed, c_1.

If this medium is more dense, light travels at a slower speed, c_2.

Medium 2 θ_2

$$\frac{c_1}{c_2} = \frac{\sin \theta_1}{\sin \theta_2},$$

where c_1 is the speed of light in the first medium, c_2 is the speed of light in the second medium, and θ_1 and θ_2 are the angles shown in the figure. (*Source: The Physics Classroom,* www.glenbrook.k12.il.us) *In Exercises 57 and 58, assume that $c_1 = 3 \times 10^8$ m per sec.*

57. Find the speed of light in the second medium for each of the following.

(a) $\theta_1 = 46°, \theta_2 = 31°$ **(b)** $\theta_1 = 39°, \theta_2 = 28°$

58. Find θ_2 for each of the following values of θ_1 and c_2. Round to the nearest degree.

(a) $\theta_1 = 40°, c_2 = 1.5 \times 10^8$ m per sec **(b)** $\theta_1 = 62°, c_2 = 2.6 \times 10^8$ m per sec

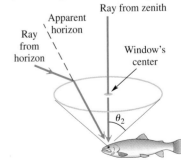

Ray from zenith

Apparent horizon

Ray from horizon

Window's center

θ_2

(Modeling) Fish's View of the World The figure shows a fish's view of the world above the surface of the water. (*Source:* Walker, J., "The Amateur Scientist," *Scientific American,* March 1984.) *Suppose that a light ray comes from the horizon, enters the water, and strikes the fish's eye.*

59. Assume that this ray gives a value of 90° for angle θ_1 in the formula for Snell's law. (In a practical situation, this angle would probably be a little less than 90°.) The speed of light in water is about 2.254×10^8 m per sec. Find angle θ_2.

60. Refer to Exercise 59. Suppose an object is located at a true angle of 29.6° above the horizon. Find the apparent angle above the horizon to a fish.

61. *(Modeling) Braking Distance* If aerodynamic resistance is ignored, the braking distance D (in feet) for an automobile to change its velocity from V_1 to V_2 (feet per second) can be modeled using the equation

$$D = \frac{1.05(V_1^2 - V_2^2)}{64.4(K_1 + K_2 + \sin\theta)}.$$

K_1 is a constant determined by the efficiency of the brakes and tires, K_2 is a constant determined by the rolling resistance of the automobile, and θ is the grade of the high-way. (*Source:* Mannering, F. and W. Kilareski, *Principles of Highway Engineering and Traffic Analysis,* Second Edition, John Wiley & Sons, 1998.)

(a) Compute the number of feet required to slow a car from 55 mph to 30 mph while traveling uphill with a grade of $\theta = 3.5°$. Let $K_1 = .4$ and $K_2 = .02$. (*Hint:* Change miles per hour to feet per second.)

(b) Repeat part (a) with $\theta = -2°$.

(c) How is braking distance affected by grade θ? Does this agree with your driving experience?

62. *(Modeling) Car's Speed at Collision* Refer to Exercise 61. An automobile is traveling at 90 mph on a highway with a downhill grade of $\theta = -3.5°$. The driver sees a stalled truck in the road 200 ft away and immediately applies the brakes. Assuming that a collision cannot be avoided, how fast (in miles per hour) is the car traveling when it hits the truck? (Use the same values for K_1 and K_2 as in Exercise 61.)

2.4 Solving Right Triangles

Significant Digits ▪ **Solving Triangles** ▪ **Angles of Elevation or Depression**

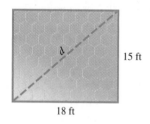

Figure 18

Significant Digits Suppose we quickly measure a room as 15 ft by 18 ft. See Figure 18. To calculate the length of a diagonal of the room, we can use the Pythagorean theorem.

$$d^2 = 15^2 + 18^2 \quad \text{(Appendix B)}$$
$$d^2 = 549$$
$$d = \sqrt{549} \approx 23.430749$$

Should this answer be given as the length of the diagonal of the room? Of course not. The number 23.430749 contains six decimal places, while the original data of 15 ft and 18 ft are only accurate to the nearest foot. Since the results of a problem can be no more accurate than the least accurate number in any cal-culation, we really should say that the diagonal of the 15-by-18-ft room is about 23 ft.

If a wall measured to the nearest foot is 18 ft long, this actually means that the wall has length between 17.5 ft and 18.5 ft. If the wall is measured more ac-curately as 18.3 ft long, then its length is really between 18.25 ft and 18.35 ft. A measurement of 18.00 ft would indicate that the length of the wall is between 17.995 ft and 18.005 ft. The measurement 18 ft is said to have two *significant digits* of accuracy; 18.0 has three significant digits, and 18.00 has four.

A **significant digit** is a digit obtained by actual measurement. A number that represents the result of counting, or a number that results from theoretical work and is not the result of a measurement, is an **exact number.** There are 50 states in the United States, so in that statement, 50 is an exact number.

Most values obtained for trigonometric functions are approximations, and virtually all measurements are approximations. When performing calculations involving approximate numbers, start by determining the number that has the least number of significant digits. Round your final answer to the same number of significant digits as this number. Remember that *your answer is no more accurate than the least accurate number in your calculation.*

Use the following table to determine the significant digits in angle measure.

Significant Digits for Angles

Number of Significant Digits	Angle Measure to Nearest:
2	Degree
3	Ten minutes, or nearest tenth of a degree
4	Minute, or nearest hundredth of a degree
5	Tenth of a minute, or nearest thousandth of a degree

For example, an angle measuring $52° \, 30'$ has three significant digits (assuming that $30'$ is measured to the nearest ten minutes).

Solving Triangles To *solve a triangle* means to find the measures of all the angles and sides of the triangle. As shown in Figure 19, we use a to represent the length of the side opposite angle A, b for the length of the side opposite angle B, and so on. In a right triangle, the letter c is reserved for the hypotenuse.

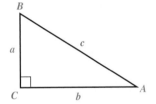

When solving triangles, a labeled sketch is an important aid.

Figure 19

EXAMPLE 1 Solving a Right Triangle Given an Angle and a Side

Solve right triangle ABC, if $A = 34° \, 30'$ and $c = 12.7$ in. See Figure 20.

Solution To solve the triangle, find the measures of the remaining sides and angles. To find the value of a, use a trigonometric function involving the known values of angle A and side c. Since the sine of angle A is given by the quotient of the side opposite A and the hypotenuse, use $\sin A$.

Figure 20

$$\sin A = \frac{a}{c} \qquad \sin A = \frac{\text{side opposite}}{\text{hypotenuse}} \text{ (Section 2.1)}$$

$$\sin 34° \, 30' = \frac{a}{12.7} \qquad A = 34° \, 30', c = 12.7$$

$$a = 12.7 \sin 34° \, 30' \qquad \text{Multiply by 12.7; rewrite.}$$

$$a = 12.7 \sin 34.5° \qquad \text{Convert to decimal degrees. (Section 1.1)}$$

$$a = 12.7(.56640624) \qquad \text{Use a calculator.}$$

$$a \approx 7.19 \text{ in.} \qquad \text{Three significant digits}$$

To find the value of b, we could substitute the value of a just calculated and the given value of c in the Pythagorean theorem. It is better, however, to use the

Looking Ahead to Calculus

The derivatives of the *parametric equations* $x = f(t)$ and $y = g(t)$ often represent the rate of change of physical quantities, such as velocities. When x and y are related by an equation, the derivatives are called *related rates* because a change in one causes a related change in the other. Determining these rates in calculus often requires solving a right triangle. Some problems that ask for the maximum or minimum value of a quantity also involve solving a right triangle.

information given in the problem rather than a result just calculated. If a mistake is made in finding a, then b also would be incorrect. Also, rounding more than once may cause the result to be less accurate. Use cos A.

$$\cos A = \frac{b}{c} \qquad \cos A = \frac{\text{side adjacent}}{\text{hypotenuse}} \text{ (Section 2.1)}$$

$$\cos 34° 30' = \frac{b}{12.7}$$

$$b = 12.7 \cos 34° 30'$$

$$b \approx 10.5 \text{ in.}$$

Once b is found, the Pythagorean theorem can be used as a check. All that remains to solve triangle ABC is to find the measure of angle B. Since $A + B = 90°$,

$$B = 90° - A \qquad \text{(Section 1.1)}$$

$$B = 89° 60' - 34° 30' \quad A = 34° 30'$$

$$B = 55° 30'.$$

Now try Exercise 19.

NOTE In Example 1, we could have found the measure of angle B first, and then used the trigonometric function values of B to find the unknown sides. The process of solving a right triangle can usually be done in several ways, each producing the correct answer. To maintain accuracy, always use given information as much as possible, and avoid rounding off in intermediate steps.

EXAMPLE 2 Solving a Right Triangle Given Two Sides

Solve right triangle ABC, if $a = 29.43$ cm and $c = 53.58$ cm.

Solution We draw a sketch showing the given information, as in Figure 21. One way to begin is to find angle A by using the sine function.

Figure 21

$$\sin A = \frac{\text{side opposite}}{\text{hypotenuse}} = \frac{29.43}{53.58} \approx .5492721165 \quad \text{(Section 2.1)}$$

$$A = \sin^{-1} .5492721165 \approx 33.32° \qquad \text{(Section 2.3)}$$

The measure of B is approximately

$$90° - 33.32° = 56.68°.$$

We now find b from the Pythagorean theorem, using $a^2 + b^2 = c^2$, or $b^2 = c^2 - a^2$. Since $c = 53.58$ and $a = 29.43$,

$$b^2 = c^2 - a^2 \qquad \text{Pythagorean theorem (Appendix B)}$$

$$b^2 = 53.58^2 - 29.43^2$$

$$b \approx 44.77 \text{ cm.}$$

Now try Exercise 23.

Figure 22

Angles of Elevation or Depression Many applications of right triangles involve angles of elevation or depression. The **angle of elevation** from point X to point Y (above X) is the acute angle formed by ray XY and a horizontal ray with endpoint at X. See Figure 22(a). The **angle of depression** from point X to point Y (below X) is the acute angle formed by ray XY and a horizontal ray with endpoint X. See Figure 22(b).

C A U T I O N   Be careful when interpreting the angle of depression. Both the angle of elevation and the angle of depression are measured between the line of sight and a *horizontal* line.

To solve applied trigonometry problems, follow the same procedure as solving a triangle.

Solving an Applied Trigonometry Problem

Step 1 Draw a sketch, and label it with the given information. Label the quantity to be found with a variable.

Step 2 Use the sketch to write an equation relating the given quantities to the variable.

Step 3 Solve the equation, and check that your answer makes sense.

Drawing a triangle and labeling it correctly in Step 1 is crucial.

EXAMPLE 3   Finding a Length When the Angle of Elevation Is Known

Shelly McCarthy knows that when she stands 123 ft from the base of a flagpole, the angle of elevation to the top of the flagpole is $26° \, 40'$. If her eyes are 5.30 ft above the ground, find the height of the flagpole.

Solution

Step 1 The length of the side adjacent to Shelly is known, and the length of the side opposite her must be found. See Figure 23.

Figure 23

Step 2 The tangent ratio involves the given values. Write an equation.

$$\tan 26° \, 40' = \frac{a}{123} \qquad \tan A = \frac{\text{side opposite}}{\text{side adjacent}} \text{ (Section 2.1)}$$

Step 3
$$a = 123 \tan 26° \, 40' \qquad \text{Multiply by 123; rewrite.}$$
$$a \approx 61.8 \text{ ft} \qquad \text{Use a calculator.}$$

Since Shelly's eyes are 5.30 ft above the ground, the height of the flagpole is

$$61.8 + 5.30 = 67.1 \text{ ft.}$$

Now try Exercise 45.

34.09 m

37.62 m   *B*

Figure 24

EXAMPLE 4 Finding the Angle of Elevation When Lengths Are Known

The length of the shadow of a building 34.09 m tall is 37.62 m. Find the angle of elevation of the sun.

Solution   As shown in Figure 24, the angle of elevation of the sun is angle *B*. Since the side opposite *B* and the side adjacent to *B* are known, use the tangent ratio to find *B*.

$$\tan B = \frac{34.09}{37.62}, \quad \text{so} \quad B = \tan^{-1}\frac{34.09}{37.62} \approx 42.18°. \quad \text{(Section 2.3)}$$

The angle of elevation of the sun is 42.18°.

Now try Exercise 35.

George Polya (1887–1985)

CONNECTIONS   George Polya proposed an excellent general outline for solving applied problems in his classic book *How to Solve It.*

1. Understand the problem.   **2.** Devise a plan.

3. Carry out the plan.   **4.** Look back.

Polya, a native of Budapest, Hungary, wrote more than 250 papers in many languages, as well as a number of books. He was a brilliant lecturer and teacher, and numerous mathematical properties and theorems bear his name. He once was asked why so many good mathematicians came out of Hungary at the turn of the century. He theorized that it was because mathematics was the cheapest science, requiring no expensive equipment, only pencil and paper.

For Discussion or Writing

Look back at the problem-solving steps given in this section, and compare them to Polya's four steps.

2.4 Exercises

Concept Check   Refer to the discussion of accuracy and significant digits in this section to work Exercises 1–8.

1. *Leading NFL Receiver*   At the end of the 2001 National Football League season, Oakland Raider Jerry Rice was the leading career receiver with 20,386 yd. State the range represented by this number. (*Source: The World Almanac and Book of Facts,* 2003.)

2. *Height of Mt. Everest*   When Mt. Everest was first surveyed, the surveyors obtained a height of 29,000 ft to the nearest foot. State the range represented by this number. (The surveyors thought no one would believe a measurement of 29,000 ft, so they reported it as 29,002.) (*Source:* Dunham, W., *The Mathematical Universe,* John Wiley & Sons, 1994.)

3. *Longest Vehicular Tunnel* The E. Johnson Memorial Tunnel in Colorado, which measures 8959 ft, is one of the longest land vehicular tunnels in the United States. What is the range of this number? (*Source: The World Almanac and Book of Facts,* 2003.)

4. *Top WNBA Scorer* Women's National Basketball Association player Tamika Catchings of the Indiana Fever received the 2002 award for most points scored, 594. Is it appropriate to consider this number as between 593.5 and 594.5? Why or why not? (*Source: The World Almanac and Book of Facts,* 2003.)

5. *Circumference of a Circle* The formula for the circumference of a circle is $C = 2\pi r$. Suppose you use the $\boxed{\pi}$ key on your calculator to find the circumference of a circle with radius 54.98 cm, getting 345.44953. Since 2 has only one significant digit, the answer should be given as 3×10^2, or 300 cm. Is this conclusion correct? If not, explain how the answer should be given.

6. Explain the difference between a measurement of 23.0 ft and a measurement of 23.00 ft.

7. If h is the actual height of a building and the height is measured as 58.6 ft, then $|h - 58.6| \le$ _____.

8. If w is the actual weight of a car and the weight is measured as 15.00×10^2 lb, then $|w - 1500| \le$ _____.

Solve each right triangle. See Example 1.

9.

10.

11.

12.

13.

14.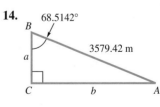

15. Can a right triangle be solved if we are given measures of its two acute angles and no side lengths? Explain.

16. *Concept Check* If we are given an acute angle and a side in a right triangle, what unknown part of the triangle requires the least work to find?

17. Explain why you can always solve a right triangle if you know the measures of one side and one acute angle.

18. Explain why you can always solve a right triangle if you know the lengths of two sides.

Solve each right triangle. In each case, C = 90°. If angle information is given in degrees and minutes, give answers in the same way. If given in decimal degrees, do likewise in answers. When two sides are given, give angles in degrees and minutes. See Examples 1 and 2.

19. $A = 28.00°, c = 17.4$ ft

20. $B = 46.00°, c = 29.7$ m

21. $B = 73.00°, b = 128$ in.

22. $A = 61° 00', b = 39.2$ cm

23. $a = 76.4$ yd, $b = 39.3$ yd

24. $a = 958$ m, $b = 489$ m

25. $a = 18.9$ cm, $c = 46.3$ cm

26. $b = 219$ m, $c = 647$ m

27. $A = 53° 24', c = 387.1$ ft

28. $A = 13° 47', c = 1285$ m

29. $B = 39° 9', c = .6231$ m

30. $B = 82° 51', c = 4.825$ cm

31. *Concept Check* When is an angle of elevation equal to 90°?

32. *Concept Check* Can an angle of elevation be more than 90°?

33. Explain why the angle of depression *DAB* has the same measure as the angle of elevation *ABC* in the figure.

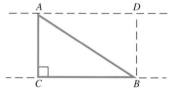

AD is parallel to BC.

34. Why is angle *CAB not* an angle of depression in the figure for Exercise 33?

Solve each problem involving triangles. See Examples 1–4.

35. *Height of a Ladder on a Wall* A 13.5-m fire truck ladder is leaning against a wall. Find the distance *d* the ladder goes up the wall (above the top of the fire truck) if the ladder makes an angle of 43° 50′ with the horizontal.

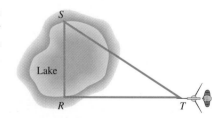

36. *Distance Across a Lake* To find the distance *RS* across a lake, a surveyor lays off *RT* = 53.1 m, with angle *T* = 32° 10′ and angle *S* = 57° 50′. Find length *RS*.

37. *Height of a Building* From a window 30 ft above the street, the angle of elevation to the top of the building across the street is 50.0° and the angle of depression to the base of this building is 20.0°. Find the height of the building across the street.

38. *Diameter of the Sun*   To determine the diameter of the sun, an astronomer might sight with a *transit* (a device used by surveyors for measuring angles) first to one edge of the sun and then to the other, finding that the included angle equals 32′. Assuming that the distance *d* from Earth to the sun is 92,919,800 mi, calculate the diameter of the sun.

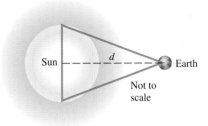

39. *Side Lengths of a Triangle*   The length of the base of an isosceles triangle is 42.36 in. Each base angle is 38.12°. Find the length of each of the two equal sides of the triangle. (*Hint:* Divide the triangle into two right triangles.)

40. *Altitude of a Triangle*   Find the altitude of an isosceles triangle having base 184.2 cm if the angle opposite the base is 68° 44′.

Solve each problem involving an angle of elevation or depression. See Examples 3 and 4.

41. *Angle of Elevation of Pyramid of the Sun*   The Pyramid of the Sun in the ancient Mexican city of Teotihuacan was the largest and most important structure in the city. The base is a square with sides 700 ft long, and the height of the pyramid is 200 ft. Find the angle of elevation of the edge indicated in the figure to two significant digits. (*Hint:* The base of the triangle in the figure is half the diagonal of the square base of the pyramid.)

42. *Cloud Ceiling*   The U.S. Weather Bureau defines a *cloud ceiling* as the altitude of the lowest clouds that cover more than half the sky. To determine a cloud ceiling, a powerful searchlight projects a circle of light vertically on the bottom of the cloud. An observer sights the circle of light in the crosshairs of a tube called a *clinometer*. A pendant hanging vertically from the tube and resting on a protractor gives the angle of elevation. Find the cloud ceiling if the searchlight is located 1000 ft from the observer and the angle of elevation is 30.0° as measured with a clinometer at eye-height 6 ft. (Assume three significant digits.)

43. *Height of a Tower* The shadow of a vertical tower is 40.6 m long when the angle of elevation of the sun is 34.6°. Find the height of the tower.

44. *Distance from the Ground to the Top of a Building* The angle of depression from the top of a building to a point on the ground is 32° 30′. How far is the point on the ground from the top of the building if the building is 252 m high?

45. *Length of a Shadow* Suppose that the angle of elevation of the sun is 23.4°. Find the length of the shadow cast by Diane Carr, who is 5.75 ft tall.

46. *Airplane Distance* An airplane is flying 10,500 ft above the level ground. The angle of depression from the plane to the base of a tree is 13° 50′. How far horizontally must the plane fly to be directly over the tree?

47. *Height of a Building* The angle of elevation from the top of a small building to the top of a nearby taller building is 46° 40′, while the angle of depression to the bottom is 14° 10′. If the smaller building is 28.0 m high, find the height of the taller building.

48. *Angle of Depression of a Light* A company safety committee has recommended that a floodlight be mounted in a parking lot so as to illuminate the employee exit. Find the angle of depression of the light.

49. *Height of Mt. Everest* The highest mountain peak in the world is Mt. Everest, located in the Himalayas. The height of this enormous mountain was determined in 1856 by surveyors using trigonometry long before it was first climbed in 1953. This difficult measurement had to be done from a great distance. At an altitude of 14,545 ft on a different mountain, the straight line distance to the peak of Mt. Everest is 27.0134 mi and its angle of elevation is $\theta = 5.82°$. (*Source:* Dunham, W., *The Mathematical Universe,* John Wiley & Sons, 1994.)

(a) Approximate the height (in feet) of Mt. Everest.
(b) In the actual measurement, Mt. Everest was over 100 mi away and the curvature of Earth had to be taken into account. Would the curvature of Earth make the peak appear taller or shorter than it actually is?

50. *Error in Measurement*   A degree may seem like a very small unit, but an error of one degree in measuring an angle may be very significant. For example, suppose a laser beam directed toward the visible center of the moon misses its assigned target by 30 sec. How far is it (in miles) from its assigned target? Take the distance from the surface of Earth to that of the moon to be 234,000 mi. (*Source: A Sourcebook of Applications of School Mathematics* by Donald Bushaw et al. Copyright © 1980 by The Mathematical Association of America.)

2.5 | **Further Applications of Right Triangles**

Bearing ▪ Further Applications

Bearing   Other applications of right triangles involve **bearing,** an important idea in navigation. There are two methods for expressing bearing. When a single angle is given, such as 164°, it is understood that the bearing is measured in a clockwise direction from due north. Several sample bearings using this first method are shown in Figure 25.

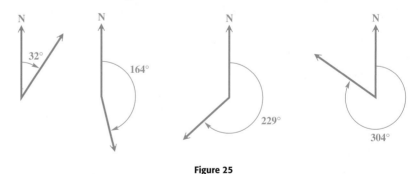

Figure 25

EXAMPLE 1   Solving a Problem Involving Bearing (First Method)

Radar stations *A* and *B* are on an east-west line, 3.7 km apart. Station *A* detects a plane at *C*, on a bearing of 61°. Station *B* simultaneously detects the same plane, on a bearing of 331°. Find the distance from *A* to *C*.

Solution   Draw a sketch showing the given information, as in Figure 26. Since a line drawn due north is perpendicular to an east-west line, right angles are formed at *A* and *B*, so angles *CAB* and *CBA* can be found as shown in Figure 26. Angle *C* is a right angle because angles *CAB* and *CBA* are complementary. Find distance *b* by using the cosine function for angle *A*.

Figure 26

Figure 26 (repeated)

$$\cos 29° = \frac{b}{3.7} \qquad \text{(Section 2.1)}$$

$$3.7 \cos 29° = b \qquad \text{Multiply by 3.7.}$$

$$b \approx 3.2 \text{ km} \qquad \text{Use a calculator; round to the nearest tenth.}$$

Now try Exercise 11.

C A U T I O N *A correctly labeled sketch is crucial* when solving applications like that in Example 1. Some of the necessary information is often not directly stated in the problem and can be determined only from the sketch.

The second method for expressing bearing starts with a north-south line and uses an acute angle to show the direction, either east or west, from this line. Figure 27 shows several sample bearings using this system. Either N or S always comes first, followed by an acute angle, and then E or W.

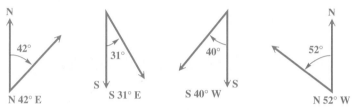

Figure 27

EXAMPLE 2 Solving a Problem Involving Bearing (Second Method)

The bearing from A to C is S 52° E. The bearing from A to B is N 84° E. The bearing from B to C is S 38° W. A plane flying at 250 mph takes 2.4 hr to go from A to B. Find the distance from A to C.

Solution Make a sketch. First draw the two bearings from point A. Choose a point B on the bearing N 84° E from A, and draw the bearing to C. Point C will be located where the bearing lines from A and B intersect, as shown in Figure 28.

Since the bearing from A to B is N 84° E, angle ABD is $180° - 84° = 96°$. Thus, angle ABC is 46°. Also, angle BAC is $180° - (84° + 52°) = 44°$. Angle C is $180° - (44° + 46°) = 90°$. Since a plane flying at 250 mph takes 2.4 hr to go from A to B, the distance from A to B is

$$c = \text{rate} \times \text{time} = 250(2.4) = 600 \text{ mi.}$$

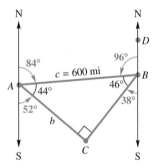

Figure 28

To find b, the distance from A to C, use the sine. (The cosine could also be used.)

$$\sin 46° = \frac{b}{c} \qquad \text{(Section 2.1)}$$

$$\sin 46° = \frac{b}{600}$$

$$600 \sin 46° = b$$

$$b \approx 430 \text{ mi}$$

Now try Exercise 13.

Further Applications

EXAMPLE 3   Using Trigonometry to Measure a Distance

Figure 29

A method that surveyors use to determine a small distance d between two points P and Q is called the *subtense bar method.* The subtense bar with length b is centered at Q and situated perpendicular to the line of sight between P and Q. See Figure 29. Angle θ is measured, and then the distance d can be determined.

(a) Find d when $\theta = 1° \ 23' \ 12''$ and $b = 2.0000$ cm.

(b) Angle θ usually cannot be measured more accurately than to the nearest $1''$. How much change would there be in the value of d if θ were measured $1''$ larger?

Solution

(a) From Figure 29, we see that

$$\cot \frac{\theta}{2} = \frac{d}{\frac{b}{2}}$$

$$d = \frac{b}{2} \cot \frac{\theta}{2}. \quad \text{Multiply; rewrite.}$$

Let $b = 2$. To evaluate $\frac{\theta}{2}$, we change θ to decimal degrees.

$$1° \ 23' \ 12'' \approx 1.386667° \quad \text{(Section 1.1)}$$

Then

$$d = \frac{2}{2} \cot \frac{1.386667°}{2} \approx 82.6341 \text{ cm.}$$

(b) Since θ is $1''$ larger, use $\theta = 1° \ 23' \ 13'' \approx 1.386944°$.

$$d = \frac{2}{2} \cot \frac{1.386944°}{2} \approx 82.6176 \text{ cm}$$

The difference is $82.6341 - 82.6176 \approx .0170$ cm.

Now try Exercise 29.

EXAMPLE 4   Solving a Problem Involving Angles of Elevation

Francisco needs to know the height of a tree. From a given point on the ground, he finds that the angle of elevation to the top of the tree is $36.7°$. He then moves back 50 ft. From the second point, the angle of elevation to the top of the tree is $22.2°$. See Figure 30. Find the height of the tree.

Figure 30

Algebraic Solution

Figure 30 shows two unknowns: x, the distance from the center of the trunk of the tree to the point where the first observation was made, and h, the height of the tree. See Figure 31 in the Graphing Calculator Solution. Since nothing is given about the length of the hypotenuse of either triangle ABC or triangle BCD, use a ratio that does not involve the hypotenuse—namely, the tangent.

In triangle ABC, $\quad \tan 36.7° = \dfrac{h}{x} \quad$ or $\quad h = x \tan 36.7°$.

In triangle BCD, $\quad \tan 22.2° = \dfrac{h}{50 + x} \quad$ or $\quad h = (50 + x) \tan 22.2°$.

Each expression equals h, so the expressions must be equal.

$$x \tan 36.7° = (50 + x) \tan 22.2°$$
Solve for x.

$$x \tan 36.7° = 50 \tan 22.2° + x \tan 22.2°$$
Distributive property

$$x \tan 36.7° - x \tan 22.2° = 50 \tan 22.2°$$
Get x-terms on one side.

$$x(\tan 36.7° - \tan 22.2°) = 50 \tan 22.2°$$
Factor out x.

$$x = \frac{50 \tan 22.2°}{\tan 36.7° - \tan 22.2°}$$
Divide by the coefficient of x.

We saw above that $h = x \tan 36.7°$. Substituting for x,

$$h = \left(\frac{50 \tan 22.2°}{\tan 36.7° - \tan 22.2°} \right) \tan 36.7°.$$

Using a calculator,

$$\tan 36.7° = .74537703 \quad \text{and} \quad \tan 22.2° = .40809244,$$

so $\quad \tan 36.7° - \tan 22.2° = .74537703 - .40809244 = .33728459$

and $\qquad h = \left(\dfrac{50(.40809244)}{.33728459} \right).74537703 \approx 45.$

The height of the tree is approximately 45 ft.

Graphing Calculator Solution*

In Figure 31, we superimposed Figure 30 on coordinate axes with the origin at D. By definition, the tangent of the angle between the x-axis and the graph of a line with equation $y = mx + b$ is the slope of the line, m. For line DB, $m = \tan 22.2°$. Since b equals 0, the equation of line DB is $y_1 = (\tan 22.2°)x$. The equation of line AB is $y_2 = (\tan 36.7°)x + b$. Since $b \neq 0$ here, we use the point $A(50, 0)$ and the point-slope form to find the equation.

$$y_2 - y_1 = m(x - x_1)$$
$$y_2 - 0 = m(x - 50) \quad x_1 = 50, y_1 = 0$$
$$y_2 = \tan 36.7°(x - 50)$$

Lines y_1 and y_2 are graphed in Figure 32. The y-coordinate of the point of intersection of the graphs gives the length of BC, or h. Thus, $h \approx 45$.

Figure 31

Figure 32

Now try Exercise 23.

NOTE In practice, we usually do not write down intermediate calculator approximation steps. We did in Example 4 so you could follow the steps more easily.

---

*Source: Adapted with permission from "Letter to the Editor," by Robert Ruzich (*Mathematics Teacher*, Volume 88, Number 1). Copyright © 1995 by the National Council of Teachers of Mathematics.

2.5 Exercises

Concept Check   *Give a short written answer to each question.*

1. When bearing is given as a single angle measure, how is the angle represented in a sketch?

2. When bearing is given as N (or S), then the angle measure, and then E (or W), how is the angle represented in a sketch?

3. Why is it important to draw a sketch before solving trigonometric problems like those in the last two sections of this chapter?

4. How should the angle of elevation (or depression) from a point X to a point Y be represented?

Concept Check   *An observer for a radar station is located at the origin of a coordinate system. For each of the points in Exercises 5–8, find the bearing of an airplane located at that point. Express the bearing using both methods.*

5. $(-4, 0)$ 6. $(-3, -3)$ 7. $(-5, 5)$ 8. $(0, -2)$

9. The ray $y = x$, $x \geq 0$, contains the origin and all points in the coordinate system whose bearing from the origin is $45°$. Determine the equation of a ray consisting of the origin and all points whose bearing from the origin is $240°$.

10. Repeat Exercise 9 for a bearing of $150°$.

Work each problem. In these exercises, assume the course of a plane or ship is on the indicated bearing. See Examples 1 and 2.

11. *Distance Flown by a Plane*   A plane flies 1.3 hr at 110 mph on a bearing of $40°$. It then turns and flies 1.5 hr at the same speed on a bearing of $130°$. How far is the plane from its starting point?

12. *Distance Traveled by a Ship*   A ship travels 50 km on a bearing of $27°$, then travels on a bearing of $117°$ for 140 km. Find the distance traveled from the starting point to the ending point.

13. *Distance Between Two Ships* Two ships leave a port at the same time. The first ship sails on a bearing of 40° at 18 knots (nautical miles per hour) and the second at a bearing of 130° at 26 knots. How far apart are they after 1.5 hr?

14. *Distance Between Two Lighthouses* Two lighthouses are located on a north-south line. From lighthouse *A*, the bearing of a ship 3742 m away is 129° 43′. From lighthouse *B*, the bearing of the ship is 39° 43′. Find the distance between the lighthouses.

15. *Distance Between Two Cities* The bearing from Winston-Salem, North Carolina, to Danville, Virginia, is N 42° E. The bearing from Danville to Goldsboro, North Carolina, is S 48° E. A car driven by Mark Ferrari, traveling at 60 mph, takes 1 hr to go from Winston-Salem to Danville and 1.8 hr to go from Danville to Goldsboro. Find the distance from Winston-Salem to Goldsboro.

16. *Distance Between Two Cities* The bearing from Atlanta to Macon is S 27° E, and the bearing from Macon to Augusta is N 63° E. An automobile traveling at 60 mph needs $1\frac{1}{4}$ hr to go from Atlanta to Macon and $1\frac{3}{4}$ hr to go from Macon to Augusta. Find the distance from Atlanta to Augusta.

17. *Distance Between Two Ships* A ship leaves its home port and sails on a bearing of N 28° 10′ E. Another ship leaves the same port at the same time and sails on a bearing of S 61° 50′ E. If the first ship sails at 24.0 mph and the second sails at 28.0 mph, find the distance between the two ships after 4 hr.

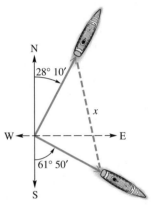

18. *Distance Between Transmitters* Radio direction finders are set up at two points *A* and *B*, which are 2.50 mi apart on an east-west line. From *A*, it is found that the bearing of a signal from a radio transmitter is N 36° 20′ E, while from *B* the bearing of the same signal is N 53° 40′ W. Find the distance of the transmitter from *B*.

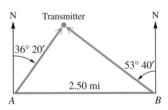

19. Solve the equation $ax = b + cx$ for *x* in terms of *a*, *b*, and *c*. (*Note:* This is in essence the calculation carried out in Example 4.)

20. Explain why the line $y = (\tan \theta)(x - a)$ passes through the point $(a, 0)$ and makes an angle θ with the *x*-axis.

21. Find the equation of the line passing through the point $(25, 0)$ that makes an angle of 35° with the *x*-axis.

22. Find the equation of the line passing through the point $(5, 0)$ that makes an angle of 15° with the *x*-axis.

In Exercises 23–28, use the method of Example 4.

23. Find *h* as indicated in the figure.

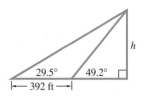

24. Find *h* as indicated in the figure.

25. *Height of a Pyramid* The angle of elevation from a point on the ground to the top of a pyramid is 35° 30′. The angle of elevation from a point 135 ft farther back to the top of the pyramid is 21° 10′. Find the height of the pyramid.

26. *Distance Between a Whale and a Lighthouse* Debbie Glockner-Ferrari, a whale researcher, is watching a whale approach directly toward a lighthouse as she observes from the top of this lighthouse. When she first begins watching the whale, the angle of depression to the whale is 15° 50′. Just as the whale turns away from the lighthouse, the angle of depression is 35° 40′. If the height of the lighthouse is 68.7 m, find the distance traveled by the whale as it approaches the lighthouse.

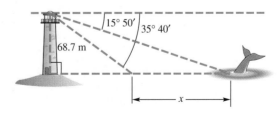

27. *Height of an Antenna* A scanner antenna is on top of the center of a house. The angle of elevation from a point 28.0 m from the center of the house to the top of the antenna is 27° 10′, and the angle of elevation to the bottom of the antenna is 18° 10′. Find the height of the antenna.

28. *Height of Mt. Whitney* The angle of elevation from Lone Pine to the top of Mt. Whitney is 10° 50′. Van Dong Le, traveling 7.00 km from Lone Pine along a straight, level road toward Mt. Whitney, finds the angle of elevation to be 22° 40′. Find the height of the top of Mt. Whitney above the level of the road.

Solve each problem.

29. *(Modeling) Distance Between Two Points*   Refer to Example 3. A variation of the subtense bar method that surveyors use to determine larger distances *d* between two points *P* and *Q* is shown in the figure. In this case the subtense bar with length *b* is placed between the points *P* and *Q* so that the bar is centered on and perpendicular to the line of sight connecting *P* and *Q*. The angles α and β are measured from points *P* and *Q*, respectively. (*Source:* Mueller, I. and K. Ramsayer, *Introduction to Surveying,* Frederick Ungar Publishing Co., 1979.)

(a) Find a formula for *d* involving α, β, and *b*.

(b) Use your formula to determine *d* if $\alpha = 37'\ 48''$, $\beta = 42'\ 3''$, and $b = 2.000$ cm.

30. *Height of a Plane Above Earth*   Find the minimum height *h* above the surface of Earth so that a pilot at point *A* in the figure can see an object on the horizon at *C*, 125 mi away. Assume that the radius of Earth is 4.00×10^3 mi.

Not to scale

31. *Distance of a Plant from a Fence*   In one area, the lowest angle of elevation of the sun in winter is $23°\ 20'$. Find the minimum distance *x* that a plant needing full sun can be placed from a fence 4.65 ft high.

4.65 ft

$23°\ 20'$

x

Plant

32. *Distance Through a Tunnel*   A tunnel is to be dug from *A* to *B*. Both *A* and *B* are visible from *C*. If *AC* is 1.4923 mi and *BC* is 1.0837 mi, and if *C* is 90°, find the measures of angles *A* and *B*.

Tunnel

A

B

1.4923 mi

1.0837 mi

C

33. *(Modeling) Highway Curves*   A basic highway curve connecting two straight sections of road is often circular. In the figure, the points P and S mark the beginning and end of the curve. Let Q be the point of intersection where the two straight sections of highway leading into the curve would meet if extended. The radius of the curve is R, and the central angle θ denotes how many degrees the curve turns. (*Source:* Mannering, F. and W. Kilareski, *Principles of Highway Engineering and Traffic Analysis,* Second Edition, John Wiley & Sons, 1998.)

(a) If $R = 965$ ft and $\theta = 37°$, find the distance d between P and Q.

(b) Find an expression in terms of R and θ for the distance between points M and N.

34. *Length of a Side of a Piece of Land*   A piece of land has the shape shown in the figure. Find x.

198.4 m

$52°\ 20'$

$30°\ 50'$

x

35. *(Modeling) Stopping Distance on a Curve*   Refer to Exercise 33. When an automobile travels along a circular curve, objects like trees and buildings situated on the inside of the curve can obstruct the driver's vision. These obstructions prevent the driver from seeing sufficiently far down the highway to ensure a safe stopping distance. In the figure, the *minimum* distance d that should be cleared on the inside of the highway is modeled by the equation

$$d = R\left(1 - \cos\frac{\theta}{2}\right).$$

(*Source:* Mannering, F. and W. Kilareski, *Principles of Highway Engineering and Traffic Analysis,* Second Edition, John Wiley & Sons, 1998.)

Not to scale

(a) It can be shown that if θ is measured in degrees, then $\theta \approx \frac{57.3S}{R}$, where S is safe stopping distance for the given speed limit. Compute d for a 55 mph speed limit if $S = 336$ ft and $R = 600$ ft.

(b) Compute d for a 65 mph speed limit if $S = 485$ ft and $R = 600$ ft.

(c) How does the speed limit affect the amount of land that should be cleared on the inside of the curve?

Chapter 2 Summary

KEY TERMS

2.1 side opposite
side adjacent
cofunctions

2.2 reference angle
2.4 significant digit

exact number
angle of elevation
angle of depression

2.5 bearing

QUICK REVIEW

CONCEPTS

EXAMPLES

2.1 Trigonometric Functions of Acute Angles

Right-Triangle-Based Definitions of the Trigonometric Functions

For any acute angle A in standard position,

$$\sin A = \frac{y}{r} = \frac{\text{side opposite}}{\text{hypotenuse}} \qquad \csc A = \frac{r}{y} = \frac{\text{hypotenuse}}{\text{side opposite}}$$

$$\cos A = \frac{x}{r} = \frac{\text{side adjacent}}{\text{hypotenuse}} \qquad \sec A = \frac{r}{x} = \frac{\text{hypotenuse}}{\text{side adjacent}}$$

$$\tan A = \frac{y}{x} = \frac{\text{side opposite}}{\text{side adjacent}} \qquad \cot A = \frac{x}{y} = \frac{\text{side adjacent}}{\text{side opposite}}.$$

$$\sin A = \frac{7}{25} \qquad \cos A = \frac{24}{25} \qquad \tan A = \frac{7}{24}$$

$$\csc A = \frac{25}{7} \qquad \sec A = \frac{25}{24} \qquad \cot A = \frac{24}{7}$$

Cofunction Identities

For any acute angle A,

$$\sin A = \cos(90° - A) \quad \cos A = \sin(90° - A)$$

$$\csc A = \sec(90° - A) \quad \sec A = \csc(90° - A)$$

$$\tan A = \cot(90° - A) \quad \cot A = \tan(90° - A).$$

$$\sin 55° = \cos(90° - 55°) = \cos 35°$$

$$\csc 48° = \sec(90° - 48°) = \sec 42°$$

$$\tan 72° = \cot(90° - 72°) = \cot 18°$$

Function Values of Special Angles

θ	$\sin \theta$	$\cos \theta$	$\tan \theta$	$\cot \theta$	$\sec \theta$	$\csc \theta$
30°	$\dfrac{1}{2}$	$\dfrac{\sqrt{3}}{2}$	$\dfrac{\sqrt{3}}{3}$	$\sqrt{3}$	$\dfrac{2\sqrt{3}}{3}$	2
45°	$\dfrac{\sqrt{2}}{2}$	$\dfrac{\sqrt{2}}{2}$	1	1	$\sqrt{2}$	$\sqrt{2}$
60°	$\dfrac{\sqrt{3}}{2}$	$\dfrac{1}{2}$	$\sqrt{3}$	$\dfrac{\sqrt{3}}{3}$	2	$\dfrac{2\sqrt{3}}{3}$

CONCEPTS	EXAMPLES

2.2 Trigonometric Functions of Non-Acute Angles

Reference Angle θ' for θ in (0°, 360°)

θ in Quadrant	I	II	III	IV
θ' is	θ	$180° - \theta$	$\theta - 180°$	$360° - \theta$

See the figure on page 56 for illustrations of reference angles.

Quadrant I: For $\theta = 25°$, $\theta' = 25°$
Quadrant II: For $\theta = 152°$, $\theta' = 28°$
Quadrant III: For $\theta = 200°$, $\theta' = 20°$
Quadrant IV: For $\theta = 320°$, $\theta' = 40°$

Finding Trigonometric Function Values for Any Nonquadrantal Angle θ

Step 1   Add or subtract 360° as many times as needed to get an angle greater than 0° but less than 360°.
Step 2   Find the reference angle θ'.
Step 3   Find the trigonometric function values for θ'.
Step 4   Determine the correct signs for the values found in Step 3.

Find sin 1050°.

$$1050° - 2(360°) = 330°$$

Thus, $\theta' = 30°$.

$$\sin 1050° = -\sin 30° = -\frac{1}{2}$$

2.4 Solving Right Triangles

Solving an Applied Trigonometry Problem

Step 1   Draw a sketch, and label it with the given information. Label the quantity to be found with a variable.

Find the angle of elevation of the sun if a 48.6-ft flagpole casts a shadow 63.1 ft long.

Step 1   See the sketch. We must find θ.

Step 2   Use the sketch to write an equation relating the given quantities to the variable.
Step 3   Solve the equation, and check that your answer makes sense.

Step 2 $\tan \theta = \dfrac{48.6}{63.1} \approx .770206$

Step 3 $\theta = \tan^{-1} .770206 \approx 37.6°$

The angle of elevation is 37.6°.

2.5 Further Applications of Right Triangles

Figures 25 and 27 on pages 77 and 78 illustrate the two methods for expressing bearing.

The bearing represented by S 40° W is also represented by $180° + 40° = 220°$.

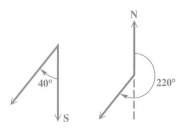

Chapter 2 Review Exercises

Find the values of the six trigonometric functions for each angle A.

1.

2.

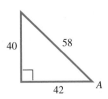

Find one solution for each equation. Assume that all angles are acute angles.

3. $\sin 4\beta = \cos 5\beta$

4. $\sec(2\theta + 10°) = \csc(4\theta + 20°)$

5. $\tan(5x + 11°) = \cot(6x + 2°)$

6. $\cos\left(\dfrac{3\theta}{5} + 11°\right) = \sin\left(\dfrac{7\theta}{10} + 40°\right)$

Tell whether each statement is true *or* false. *If false, tell why.*

7. $\sin 46° < \sin 58°$

8. $\cos 47° < \cos 58°$

9. $\sec 48° \geq \cos 42°$

10. $\sin 22° \geq \csc 68°$

11. Explain why, in the figure, the cosine of angle A is equal to the sine of angle B.

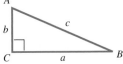

Find exact values of the six trigonometric functions for each angle. Do not use a calculator. Rationalize denominators when applicable.

12. $120°$

13. $300°$

14. $-225°$

15. $-390°$

Find all values of θ, if θ is in the interval $[0°, 360°)$ and θ has the given function value.

16. $\sin \theta = -\dfrac{1}{2}$

17. $\cos \theta = -\dfrac{1}{2}$

18. $\cot \theta = -1$

19. $\sec \theta = -\dfrac{2\sqrt{3}}{3}$

Evaluate each expression. Give exact values.

20. $\cos 60° + 2 \sin^2 30°$

21. $\tan^2 120° - 2 \cot 240°$

22. $\sec^2 300° - 2 \cos^2 150° + \tan 45°$

23. Find the sine, cosine, and tangent function values for each angle.

(a)

(b)

Use a calculator to find each value.

24. sin 72° 30′　　　　　　**25.** sec 222° 30′　　　　　　**26.** cot 305.6°

27. csc 78° 21′　　　　　　**28.** sec 58.9041°　　　　　　**29.** tan 11.7689°

30. *Concept Check* Which one of the following cannot be *exactly* determined using the methods of this chapter?

　　A. cos 135°　　　**B.** cot(−45°)　　　**C.** sin 300°　　　**D.** tan 140°

Use a calculator to find each value of θ, where θ is in the interval [0°, 90°). *Give answers in decimal degrees.*

31. sin θ = .82584121　　**32.** cot θ = 1.1249386　　**33.** cos θ = .97540415

34. sec θ = 1.2637891　　**35.** tan θ = 1.9633124　　**36.** csc θ = 9.5670466

Find two angles in the interval [0°, 360°) *that satisfy each of the following. Leave answers in decimal degrees.*

37. sin θ = .73254290　　　　　　　　　　**38.** tan θ = 1.3865342

Tell whether each statement is true *or* false. *If false, tell why. Use a calculator for Exercises 39 and 42.*

39. sin 50° + sin 40° = sin 90°

40. cos 210° = cos 180° · cos 30° − sin 180° · sin 30°

41. sin 240° = 2 sin 120° · cos 120°

42. sin 42° + sin 42° = sin 84°

 43. A student wants to use a calculator to find the value of cot 25°. However, instead of entering $\frac{1}{\tan 25}$, he enters $\tan^{-1} 25$. Assuming the calculator is in degree mode, will this produce the correct answer? Explain.

For each angle θ, use a calculator to find cos θ *and* sin θ. *Use your results to decide in which quadrant the angle lies.*

44. θ = 2976°　　　　　　**45.** θ = 1997°　　　　　　**46.** θ = 4000°

Solve each right triangle. In Exercise 48, give angles to the nearest minute. In Exercises 49 and 50, label the triangle ABC as in Exercises 47 and 48.

47.

48.

49. A = 39.72°, b = 38.97 m　　　　**50.** B = 47° 53′, b = 298.6 m

Solve each problem.

51. *Height of a Tower*   The angle of elevation from a point
93.2 ft from the base of a tower to the top of the tower is
38° 20′. Find the height of the tower.

52. *Height of a Tower*   The angle of depression of a tele-
vision tower to a point on the ground 36.0 m from the
bottom of the tower is 29.5°. Find the height of the
tower.

53. *Length of a Diagonal*   One side of a rectangle measures 15.24 cm. The angle
between the diagonal and that side is 35.65°. Find the length of the diagonal.

54. *Length of Sides of an Isosceles Triangle*   An isosceles triangle has a base of length
49.28 m. The angle opposite the base is 58.746°. Find the length of each of the two
equal sides.

55. *Distance Between Two Points*   The bearing of point B from point C is 254°. The
bearing of point A from point C is 344°. The bearing of point A from point B is 32°.
If the distance from A to C is 780 m, find the distance from A to B.

56. *Distance a Ship Sails*   The bearing from point A to point B is S 55° E and from
point B to point C is N 35° E. If a ship sails from A to B, a distance of 80 km, and
then from B to C, a distance of 74 km, how far is it from A to C?

57. *Distance Between Two Points*   Two cars leave an intersection at the same time.
One heads due south at 55 mph. The other travels due west. After 2 hr, the bearing
of the car headed west from the car headed south is 324°. How far apart are they at
that time?

58. Find a formula for h in terms of k, A, and B.
Assume $A < B$.

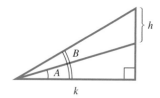

59. Make up a right triangle problem whose solution is 3 tan 25°.

60. Make up a right triangle problem whose solution is found from $\sin \theta = \frac{3}{4}$.

61. *(Modeling) Height of a Satellite*   Artificial satellites that orbit Earth often use VHF
signals to communicate with the ground. VHF signals travel in straight lines. The
height h of the satellite above Earth and the time T that the satellite can communicate with a fixed location on the ground are related by the model

$$h = R\left(\frac{1}{\cos \frac{180T}{P}} - 1\right),$$

where $R = 3955$ mi is the radius of Earth and P is the period for the satellite to orbit
Earth. (*Source:* Schlosser, W., T. Schmidt-Kaler, and E. Milone, *Challenges of
Astronomy,* Springer-Verlag, 1991.)

(a) Find h when $T = 25$ min and $P = 140$ min. (Evaluate the cosine function in degree mode.)

(b) What is the value of h if T is increased to 30 min?

62. *(Modeling) Fundamental Surveying Problem* The first fundamental problem of surveying is to determine the coordinates of a point Q given the coordinates of a point P, the distance between P and Q, and the bearing θ from P to Q. See the figure. (*Source:* Mueller, I. and K. Ramsayer, *Introduction to Surveying,* Frederick Ungar Publishing Co., 1979.)

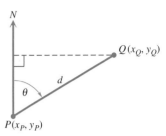

(a) Find a formula for the coordinates (x_Q, y_Q) of the point Q given θ, the coordinates (x_P, y_P) of P, and the distance d between P and Q.

(b) Use your formula to determine (x_Q, y_Q) if $(x_P, y_P) = (123.62, 337.95)$, $\theta = 17° \, 19' \, 22''$, and $d = 193.86$ ft.

Chapter 2 Test

1. Give the six trigonometric function values of angle A.

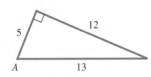

2. Find the exact values of each part labeled with a letter.

3. Find a solution for $\sin(B + 15°) = \cos(2B + 30°)$.

4. Find a value of θ in $[0°, 90°)$ in decimal degrees, if $\sin \theta = .27843196$.

5. Give two angles in $[0°, 360°)$ that satisfy $\cos \theta = -\frac{\sqrt{2}}{2}$.

6. How would you find $\cot \theta$ using a calculator, if $\tan \theta = 1.6778490$? Give $\cot \theta$.

7. Tell whether each statement is *true* or *false*. If false, tell why.

(a) $\sin 24° < \sin 48°$ (b) $\cos 24° < \cos 48°$ (c) $\tan 24° < \tan 48°$

8. Find the exact value of $\cot(-750°)$.

9. Use a calculator to approximate each value.

(a) $\sin 78° \, 21'$ (b) $\tan 117.689°$ (c) $\sec 58.9041°$

10. Solve the triangle.

11. *Antenna Mast Guy Wire* A guy wire 77.4 m long is attached to the top of an antenna mast that is 71.3 m high. Find the angle that the wire makes with the ground.

12. *Height of a Flagpole*   To measure the height of a flagpole, Amado Carillo found that the angle of elevation from a point 24.7 ft from the base to the top is 32° 10′. What is the height of the flagpole?

13. *Distance of a Ship from a Pier*   A ship leaves a pier on a bearing of S 62° E and travels for 75 km. It then turns and continues on a bearing of N 28° E for 53 km. How far is the ship from the pier?

14. Find *h* as indicated in the figure.

Chapter 2 Quantitative Reasoning

Can trigonometry be used to win an Olympic medal?

A shot-putter trying to improve performance may wonder: Is there an optimal angle to aim for, or is the velocity (speed) at which the ball is thrown more important? The figure shows the path of a steel ball thrown by a shot-putter. The distance *D* depends on initial velocity *v*, height *h*, and angle *θ* when the ball is released.

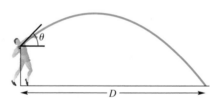

One model developed for this situation gives *D* as

$$D = \frac{v^2 \sin \theta \cos \theta + v \cos \theta \sqrt{(v \sin \theta)^2 + 64h}}{32}.$$

Typical ranges for the variables are *v*: 33–46 ft per sec; *h*: 6–8 ft; and *θ*: 40°–45°. (*Source:* Kreighbaum, E. and K. Barthels, *Biomechanics*, Allyn & Bacon, 1996.)

1. To see how angle *θ* affects distance *D*, let *v* = 44 ft per sec and *h* = 7 ft. Calculate *D* for *θ* = 40°, 42°, and 45°. How does distance *D* change as *θ* increases?

2. To see how velocity *v* affects distance *D*, let *h* = 7 and *θ* = 42°. Calculate *D* for *v* = 43, 44, and 45 ft per sec. How does distance *D* change as *v* increases?

3. Which affects distance *D* more, *v* or *θ*? What should the shot-putter do to improve performance?

3

Radian Measure and Circular Functions

In August 2003, the planet Mars passed closer to Earth than it had in almost 60,000 years. Like Earth, Mars rotates on its axis and thus has days (also called *sols*) and nights. The photos here were taken by the Hubble telescope and show two nearly opposite sides of Mars. In early 2004, the rovers *Spirit* and *Opportunity* landed on Mars and have provided scientists a wealth of information about the "Red Planet." (*Source:* www.hubblesite.org)

In Exercise 36 of Section 3.4, we examine the length of a Martian *sol* using *radian measure,* an alternative to measuring with degrees.

3.1 | Radian Measure

Radian Measure **·** Converting Between Degrees and Radians **·** Finding Function Values for Angles
in Radians

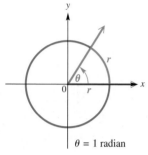

$\theta = 1$ radian

Figure 1

Radian Measure   In most applications of trigonometry, angles are measured
in degrees. In more advanced work in mathematics, *radian measure* of angles is
preferred. Radian measure allows us to treat the trigonometric functions as functions with domains of *real numbers,* rather than angles.

Figure 1 shows an angle θ in standard position along with a circle of
radius r. The vertex of θ is at the center of the circle. Because angle θ intercepts
an arc on the circle equal in length to the radius of the circle, we say that angle θ
has a measure of 1 radian.

Radian

An angle with its vertex at the center of a circle that intercepts an arc on the
circle equal in length to the radius of the circle has a measure of **1 radian.**

It follows that an angle of measure 2 radians intercepts an arc equal in
length to twice the radius of the circle, an angle of measure $\frac{1}{2}$ radian intercepts
an arc equal in length to half the radius of the circle, and so on. In general, if θ is
a central angle of a circle of radius r and θ intercepts an arc of length s, then the
radian measure of θ is $\frac{s}{r}$.

Converting Between Degrees and Radians   The circumference of a
circle—the distance around the circle—is given by $C = 2\pi r$, where r is the radius of the circle. The formula $C = 2\pi r$ shows that the radius can be laid off 2π
times around a circle. Therefore, an angle of 360°, which corresponds to a complete circle, intercepts an arc equal in length to 2π times the radius of the circle.
Thus, an angle of 360° has a measure of 2π radians:

$$360° = 2\pi \text{ radians.}$$

An angle of 180° is half the size of an angle of 360°, so an angle of 180° has
half the radian measure of an angle of 360°.

$$180° = \frac{1}{2}(2\pi) \text{ radians} = \pi \text{ radians} \quad \text{Degree/radian relationship}$$

We can use the relationship $180° = \pi$ radians to develop a method for converting between degrees and radians as follows.

$$180° = \pi \text{ radians}$$

$$1° = \frac{\pi}{180} \text{ radian} \quad \text{Divide by 180.} \qquad \text{or} \qquad 1 \text{ radian} = \frac{180°}{\pi} \quad \text{Divide by } \pi.$$

Converting Between Degrees and Radians

1. Multiply a degree measure by $\frac{\pi}{180}$ radian and simplify to convert to radians.

2. Multiply a radian measure by $\frac{180°}{\pi}$ and simplify to convert to degrees.

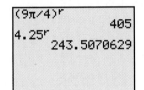

Some calculators (in radian mode) have the capability to convert directly between decimal degrees and radians. This screen shows the conversions for Example 1. Note that when *exact* values involving π are required, such as $\frac{\pi}{4}$ in part (a), calculator approximations are not acceptable.

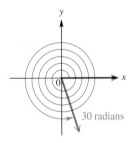

This screen shows how a TI-83 Plus calculator in degree mode converts the radian measures in Example 2 to degree measures.

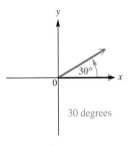

Figure 2

EXAMPLE 1 Converting Degrees to Radians

Convert each degree measure to radians.

(a) 45°   **(b)** 249.8°

Solution

(a) $45° = 45\left(\frac{\pi}{180} \text{ radian}\right) = \frac{\pi}{4} \text{ radian}$ Multiply by $\frac{\pi}{180}$ radian.

(b) $249.8° = 249.8\left(\frac{\pi}{180} \text{ radian}\right) \approx 4.360 \text{ radians}$ Nearest thousandth

Now try Exercises 5 and 37.

EXAMPLE 2 Converting Radians to Degrees

Convert each radian measure to degrees.

(a) $\frac{9\pi}{4}$ **(b)** 4.25

Solution

(a) $\frac{9\pi}{4} = \frac{9\pi}{4}\left(\frac{180°}{\pi}\right) = 405°$ **(b)** $4.25 = 4.25\left(\frac{180°}{\pi}\right) \approx 243.5° = 243° \, 30'$

Multiply by $\frac{180°}{\pi}$.   Use a calculator.

Now try Exercises 23 and 49.

If no unit of angle measure is specified, then radian measure is understood.

C A U T I O N   Figure 2 shows angles measuring 30 radians and 30°. Be careful not to confuse them.

The following table and Figure 3 on the next page give some equivalent angle measures in degrees and radians. Keep in mind that **180° = π radians.**

Degrees	Radians		Degrees	Radians	
	Exact	Approximate		Exact	Approximate
0°	0	0	90°	$\frac{\pi}{2}$	1.57
30°	$\frac{\pi}{6}$	.52	180°	π	3.14
45°	$\frac{\pi}{4}$	.79	270°	$\frac{3\pi}{2}$	4.71
60°	$\frac{\pi}{3}$	1.05	360°	2π	6.28

Looking Ahead to Calculus

In calculus, radian measure is much easier to work with than degree measure. If x is measured in radians, then the derivative of $f(x) = \sin x$ is

$$f'(x) = \cos x.$$

However, if x is measured in degrees, then the derivative of $f(x) = \sin x$ is

$$f'(x) = \frac{\pi}{180} \cos x.$$

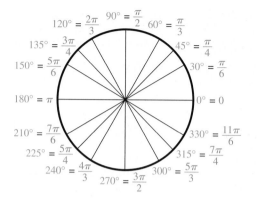

Figure 3

Finding Function Values for Angles in Radians
Trigonometric function values for angles measured in radians can be found by first converting radian measure to degrees. (You should try to skip this intermediate step as soon as possible, and find the function values directly from radian measure.)

EXAMPLE 3 Finding Function Values of Angles in Radian Measure

Find each function value.

(a) $\tan \dfrac{2\pi}{3}$ **(b)** $\sin \dfrac{3\pi}{2}$ **(c)** $\cos\left(-\dfrac{4\pi}{3}\right)$

Solution

(a) First convert $\frac{2\pi}{3}$ radians to degrees.

$$\tan \frac{2\pi}{3} = \tan\left(\frac{2}{3} \cdot 180°\right) \quad \text{Substitute } 180° \text{ for } \pi.$$

$$= \tan 120°$$

$$= -\sqrt{3} \qquad \text{(Section 2.2)}$$

(b) From the table on the preceding page and Figure 3, $\frac{3\pi}{2}$ radians $= 270°$, so

$$\sin \frac{3\pi}{2} = \sin 270° = -1.$$

(c) $\cos\left(-\dfrac{4\pi}{3}\right) = \cos(-240°) \quad \text{Substitute } 180° \text{ for } \pi.$

$$= -\cos 60° \qquad \text{(Section 2.2)}$$

$$= -\frac{1}{2}$$

Now try Exercises 57, 69, and 73.

3.1 Exercises

Concept Check In Exercises 1–4, each angle θ is an integer when measured in radians. Give the radian measure of the angle.

1.

2.

3.

4.
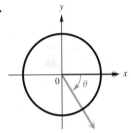

Convert each degree measure to radians. Leave answers as multiples of π. See Example 1(a).

5. 60°	**6.** 30°	**7.** 90°	**8.** 120°
9. 150°	**10.** 270°	**11.** 300°	**12.** 315°
13. 450°	**14.** 480°		

Give a short explanation in Exercises 15–20.

15. In your own words, explain how to convert degree measure to radian measure.

16. In your own words, explain how to convert radian measure to degree measure.

17. In your own words, explain the meaning of radian measure.

18. Explain the difference between degree measure and radian measure.

19. Use an example to show that you can convert from radian measure to degree measure by multiplying by $\frac{180°}{\pi}$.

20. Explain why an angle of radian measure t in standard position intercepts an arc of length t on a circle of radius 1.

Convert each radian measure to degrees. See Example 2(a).

21. $\dfrac{\pi}{3}$	**22.** $\dfrac{8\pi}{3}$	**23.** $\dfrac{7\pi}{4}$	**24.** $\dfrac{2\pi}{3}$
25. $\dfrac{11\pi}{6}$	**26.** $\dfrac{15\pi}{4}$	**27.** $-\dfrac{\pi}{6}$	**28.** $\dfrac{8\pi}{5}$
29. $\dfrac{7\pi}{10}$	**30.** $\dfrac{11\pi}{15}$	**31.** $\dfrac{4\pi}{15}$	**32.** $\dfrac{7\pi}{20}$
33. $\dfrac{17\pi}{20}$	**34.** $\dfrac{11\pi}{30}$		

Convert each degree measure to radians. See Example 1(b).

35. 39°   **36.** 74°   **37.** 42.5°   **38.** 264.9°

39. 139° 10′   **40.** 174° 50′   **41.** 64.29°   **42.** 85.04°

43. 56° 25′   **44.** 122° 37′   **45.** 47.6925°   **46.** 23.0143°

Convert each radian measure to degrees. Write answers to the nearest minute. See Example 2(b).

47. 2   **48.** 5   **49.** 1.74   **50.** 3.06

51. .3417   **52.** 9.84763   **53.** 5.01095   **54.** −3.47189

55. The value of sin 30 is not $\frac{1}{2}$. Explain why.

56. Explain in your own words what is meant by an angle of one radian.

Find the exact value of each expression without using a calculator. See Example 3.

57. $\sin \dfrac{\pi}{3}$ **58.** $\cos \dfrac{\pi}{6}$ **59.** $\tan \dfrac{\pi}{4}$ **60.** $\cot \dfrac{\pi}{3}$

61. $\sec \dfrac{\pi}{6}$ **62.** $\csc \dfrac{\pi}{4}$ **63.** $\sin \dfrac{\pi}{2}$ **64.** $\csc \dfrac{\pi}{2}$

65. $\tan \dfrac{5\pi}{3}$ **66.** $\cot \dfrac{2\pi}{3}$ **67.** $\sin \dfrac{5\pi}{6}$ **68.** $\tan \dfrac{5\pi}{6}$

69. $\cos 3\pi$ **70.** $\sec \pi$ **71.** $\sin\left(-\dfrac{8\pi}{3}\right)$ **72.** $\cot\left(-\dfrac{2\pi}{3}\right)$

73. $\sin\left(-\dfrac{7\pi}{6}\right)$ **74.** $\cos\left(-\dfrac{\pi}{6}\right)$

75. *Concept Check*   The figure shows the same angles measured in both degrees and radians. Complete the missing measures.

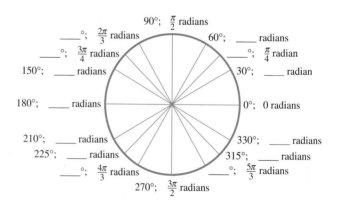

Solve each problem.

76. *Railroad Engineering*   The term *grade* has several different meanings in construction work. Some engineers use the term *grade* to represent $\frac{1}{100}$ of a right angle and express grade as a percent. For instance, an angle of .9° would be referred to as a 1% grade. (*Source:* Hay, W., *Railroad Engineering*, John Wiley & Sons, 1982.)

   (a) By what number should you multiply a grade (disregarding the % symbol) to convert it to radians?

   (b) In a rapid-transit rail system, the maximum grade allowed between two stations is 3.5%. Express this angle in degrees and radians.

77. *Rotating Hour Hand on a Clock*   Through how many radians will the hour hand on a clock rotate in **(a)** 24 hr and **(b)** 4 hr?

78. *Rotating Pulley*   A circular pulley is rotating about its center. Through how many radians would it turn in **(a)** 8 rotations and **(b)** 30 rotations?

79. *Orbits of a Space Vehicle*   A space vehicle is orbiting Earth in a circular orbit. What radian measure corresponds to **(a)** 2.5 orbits and **(b)** $\frac{4}{3}$ orbit?

3.2 Applications of Radian Measure

Arc Length on a Circle   ▪   Area of a Sector of a Circle

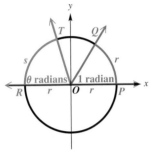

Figure 4

Arc Length on a Circle   We use radian measure in the formula to find the length of an arc of a circle. This formula is derived from the fact (proven in geometry) that the length of an arc is proportional to the measure of its central angle. In Figure 4, angle *QOP* has measure 1 radian and intercepts an arc of length *r* on the circle. Angle *ROT* has measure θ radians and intercepts an arc of length *s* on the circle. Since the lengths of the arcs are proportional to the measures of their central angles,

$$\frac{s}{r} = \frac{\theta}{1}.$$

Multiplying both sides by *r* gives the following result.

Arc Length

The length *s* of the arc intercepted on a circle of radius *r* by a central angle of measure θ radians is given by the product of the radius and the radian measure of the angle, or

$$s = r\theta, \qquad \theta \text{ in radians.}$$

CAUTION   When applying the formula $s = r\theta$, the value of θ *must be expressed in radians.*

EXAMPLE 1   Finding Arc Length Using $s = r\theta$

A circle has radius 18.2 cm. Find the length of the arc intercepted by a central angle having each of the following measures.

(a) $\frac{3\pi}{8}$ radians   **(b)** 144°

Solution

(a) As shown in Figure 5, $r = 18.2$ cm and $\theta = \frac{3\pi}{8}$.

$$s = r\theta \qquad \text{Arc length formula}$$

$$s = 18.2\left(\frac{3\pi}{8}\right) \text{ cm} \qquad \text{Substitute for } r \text{ and } \theta.$$

$$s = \frac{54.6\pi}{8} \text{ cm} \approx 21.4 \text{ cm}$$

Figure 5

(b) The formula $s = r\theta$ requires that θ be measured in radians. First, convert θ to radians by multiplying $144°$ by $\frac{\pi}{180}$ radian.

$$144° = 144\left(\frac{\pi}{180}\right) = \frac{4\pi}{5} \text{ radians} \quad \begin{array}{l}\text{Convert from degrees to radians.}\\\text{(Section 3.1)}\end{array}$$

The length s is given by

$$s = r\theta = 18.2\left(\frac{4\pi}{5}\right) = \frac{72.8\pi}{5} \approx 45.7 \text{ cm.}$$

Now try Exercises 7 and 9.

EXAMPLE 2   Using Latitudes to Find the Distance Between Two Cities

Reno, Nevada, is approximately due north of Los Angeles. The latitude of Reno is $40°$ N, while that of Los Angeles is $34°$ N. (The N in $34°$ N means *north* of the equator.) The radius of Earth is 6400 km. Find the north-south distance between the two cities.

Solution   Latitude gives the measure of a central angle with vertex at Earth's center whose initial side goes through the equator and whose terminal side goes through the given location. As shown in Figure 6, the central angle between Reno and Los Angeles is $40° - 34° = 6°$. The distance between the two cities can be found by the formula $s = r\theta$, after $6°$ is first converted to radians.

$$6° = 6\left(\frac{\pi}{180}\right) = \frac{\pi}{30} \text{ radian}$$

The distance between the two cities is

$$s = r\theta = 6400\left(\frac{\pi}{30}\right) \approx 670 \text{ km.} \quad \text{Let } r = 6400 \text{ and } \theta = \frac{\pi}{30}.$$

Now try Exercise 13.

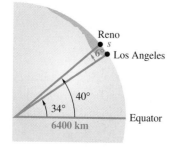

Reno
s
Los Angeles
$40°$
$34°$
6400 km
Equator

Figure 6

EXAMPLE 3   Finding a Length Using $s = r\theta$

A rope is being wound around a drum with radius .8725 ft. (See Figure 7.) How much rope will be wound around the drum if the drum is rotated through an angle of $39.72°$?

Solution   The length of rope wound around the drum is the arc length for a circle of radius .8725 ft and a central angle of $39.72°$. Use the formula $s = r\theta$, with the angle converted to radian measure. The length of the rope wound around the drum is approximately

$$s = r\theta = .8725\left[39.72\left(\frac{\pi}{180}\right)\right] \approx .6049 \text{ ft.}$$

Now try Exercise 19(a).

$39.72°$
.8725 ft

Figure 7

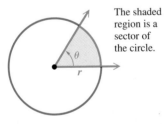

Figure 8

EXAMPLE 4 Finding an Angle Measure Using $s = r\theta$

Two gears are adjusted so that the smaller gear drives the larger one, as shown in Figure 8. If the smaller gear rotates through 225°, through how many degrees will the larger gear rotate?

Solution First find the radian measure of the angle, and then find the arc length on the smaller gear that determines the motion of the larger gear. Since $225° = \frac{5\pi}{4}$ radians, for the smaller gear,

$$s = r\theta = 2.5\left(\frac{5\pi}{4}\right) = \frac{12.5\pi}{4} = \frac{25\pi}{8} \text{ cm.}$$

An arc with this length on the larger gear corresponds to an angle measure θ, in radians, where

$$s = r\theta$$

$$\frac{25\pi}{8} = 4.8\theta \qquad \text{Substitute } \tfrac{25\pi}{8} \text{ for } s \text{ and } 4.8 \text{ for } r.$$

$$\frac{125\pi}{192} \approx \theta. \qquad \begin{array}{l}4.8 = \tfrac{48}{10} = \tfrac{24}{5}; \text{ multiply by } \tfrac{5}{24} \\ \text{to solve for } \theta.\end{array}$$

Converting θ back to degrees shows that the larger gear rotates through

$$\frac{125\pi}{192}\left(\frac{180°}{\pi}\right) \approx 117°. \qquad \text{Convert } \theta = \tfrac{125\pi}{192} \text{ to degrees.}$$

Now try Exercise 21.

The shaded region is a sector of the circle.

Figure 9

Area of a Sector of a Circle

A **sector of a circle** is the portion of the interior of a circle intercepted by a central angle. Think of it as a "piece of pie." See Figure 9. A complete circle can be thought of as an angle with measure 2π radians. If a central angle for a sector has measure θ radians, then the sector makes up the fraction $\frac{\theta}{2\pi}$ of a complete circle. The area of a complete circle with radius r is $A = \pi r^2$. Therefore,

$$\text{area of the sector} = \frac{\theta}{2\pi}(\pi r^2) = \frac{1}{2}r^2\theta, \qquad \theta \text{ in radians.}$$

This discussion is summarized as follows.

Area of a Sector

The area of a sector of a circle of radius r and central angle θ is given by

$$A = \frac{1}{2}r^2\theta, \qquad \theta \text{ in radians.}$$

CAUTION   As in the formula for arc length, the value of **θ *must be in radians*** when using this formula for the area of a sector.

Figure 10

EXAMPLE 5 Finding the Area of a Sector-Shaped Field

Find the area of the sector-shaped field shown in Figure 10.

Solution First, convert 15° to radians.

$$15° = 15\left(\frac{\pi}{180}\right) = \frac{\pi}{12} \text{ radian}$$

Now use the formula to find the area of a sector of a circle with radius $r = 321$.

$$A = \frac{1}{2}r^2\theta = \frac{1}{2}(321)^2\left(\frac{\pi}{12}\right) \approx 13{,}500 \text{ m}^2$$

Now try Exercise 35.

CONNECTIONS *Longitude* is the angular distance (expressed in degrees) East or West of the prime meridian, which goes from the North Pole to the South Pole through Greenwich, England. Arcs of longitude are 110 km apart at the equator. As the figure shows, these sections are similar to those of an orange.

Because Earth revolves 15° per hr, longitude is found by taking the difference between time zones multiplied by 15°. For example, if it is 12 noon where you are (in the United States) and 5 P.M. in Greenwich, you are located at longitude 5(15°) = longitude 75° W. Thus, determining longitude requires only an accurate measure of time. Before 1772, sailors were unable to determine their position at sea because there were no clocks capable of precise measure of time at sea. In 1772, a clock invented by John Harrison, a carpenter's son with no formal education, solved the problem. (*Source:* Ola, P. and E. D'Aulaire, "Taking the Measure of Time," *Smithsonian*, December 1999.)

For Discussion or Writing

Use time zones to determine the longitude where you live. What would the longitude be at Greenwich, England? Visit the Internet to learn more about longitude.

3.2 Exercises

Concept Check   *Find the exact length of each arc intercepted by the given central angle.*

1.

2.
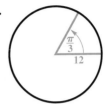

Concept Check   *Find the radius of each circle.*

3.

4.

Concept Check   *Find the measure of each central angle (in radians).*

5.

6.

Unless otherwise directed, give calculator approximations in your answers in the rest of this exercise set.

Find the length of each arc intercepted by a central angle θ in a circle of radius r. See Example 1.

7. $r = 12.3$ cm, $\theta = \dfrac{2\pi}{3}$ radians

8. $r = .892$ cm, $\theta = \dfrac{11\pi}{10}$ radians

9. $r = 4.82$ m, $\theta = 60°$

10. $r = 71.9$ cm, $\theta = 135°$

11. *Concept Check*   If the radius of a circle is doubled, how is the length of the arc intercepted by a fixed central angle changed?

12. *Concept Check*   Radian measure simplifies many formulas, such as the formula for arc length, $s = r\theta$. Give the corresponding formula when θ is measured in degrees instead of radians.

Distance Between Cities   *Find the distance in kilometers between each pair of cities, assuming they lie on the same north-south line. See Example 2.*

13. Panama City, Panama, 9° N, and Pittsburgh, Pennsylvania, 40° N

14. Farmersville, California, 36° N, and Penticton, British Columbia, 49° N

15. New York City, New York, 41° N, and Lima, Peru, 12° S

16. Halifax, Nova Scotia, 45° N, and Buenos Aires, Argentina, 34° S

17. *Latitude of Madison*   Madison, South Dakota, and Dallas, Texas, are 1200 km apart and lie on the same north-south line. The latitude of Dallas is 33° N. What is the latitude of Madison?

18. *Latitude of Toronto*   Charleston, South Carolina, and Toronto, Canada, are 1100 km apart and lie on the same north-south line. The latitude of Charleston is 33° N. What is the latitude of Toronto?

Work each problem. See Examples 3 and 4.

19. *Pulley Raising a Weight*

  (a) How many inches will the weight in the figure rise if the pulley is rotated through an angle of 71° 50′?

  (b) Through what angle, to the nearest minute, must the pulley be rotated to raise the weight 6 in.?

20. *Pulley Raising a Weight*   Find the radius of the pulley in the figure if a rotation of 51.6° raises the weight 11.4 cm.

21. *Rotating Wheels*   The rotation of the smaller wheel in the figure causes the larger wheel to rotate. Through how many degrees will the larger wheel rotate if the smaller one rotates through 60.0°?

22. *Rotating Wheels*   Find the radius of the larger wheel in the figure if the smaller wheel rotates 80.0° when the larger wheel rotates 50.0°.

23. *Bicycle Chain Drive*   The figure shows the chain drive of a bicycle. How far will the bicycle move if the pedals are rotated through 180°? Assume the radius of the bicycle wheel is 13.6 in.

24. *Pickup Truck Speedometer*   The speedometer of Terry's small pickup truck is designed to be accurate with tires of radius 14 in.

  (a) Find the number of rotations of a tire in 1 hr if the truck is driven at 55 mph.

(b) Suppose that oversize tires of radius 16 in. are placed on the truck. If the truck is now driven for 1 hr with the speedometer reading 55 mph, how far has the truck gone? If the speed limit is 55 mph, does Terry deserve a speeding ticket?

If a central angle is very small, there is little difference in length between an arc and the inscribed chord. See the figure. Approximate each of the following lengths by finding the necessary arc length. (Note: When a central angle intercepts an arc, the arc is said to **subtend** *the angle.)*

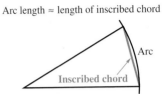

Arc length ≈ length of inscribed chord

25. *Length of a Train* A railroad track in the desert is 3.5 km away. A train on the track subtends (horizontally) an angle of 3° 20′. Find the length of the train.

26. *Distance to a Boat* The mast of Brent Simon's boat is 32 ft high. If it subtends an angle of 2° 10′, how far away is it?

Concept Check *Find the area of each sector.*

27.

28.

Concept Check *Find the measure (in radians) of each central angle. The number inside the sector is the area.*

29.

30.

Find the area of a sector of a circle having radius r and central angle θ. See Example 5.

31. $r = 29.2$ m, $\theta = \dfrac{5\pi}{6}$ radians

32. $r = 59.8$ km, $\theta = \dfrac{2\pi}{3}$ radians

33. $r = 30.0$ ft, $\theta = \dfrac{\pi}{2}$ radians

34. $r = 90.0$ yd, $\theta = \dfrac{5\pi}{6}$ radians

35. $r = 12.7$ cm, $\theta = 81°$

36. $r = 18.3$ m, $\theta = 125°$

37. $r = 40.0$ mi, $\theta = 135°$

38. $r = 90.0$ km, $\theta = 270°$

Work each problem.

39. Find the measure (in radians) of a central angle of a sector of area 16 in.^{2} in a circle of radius 3.0 in.

40. Find the radius of a circle in which a central angle of $\frac{\pi}{6}$ radian determines a sector of area 64 m^2.

41. *Measures of a Structure* The figure shows Medicine Wheel, a Native American structure in northern Wyoming. This circular structure is perhaps 2500 yr old. There are 27 aboriginal spokes in the wheel, all equally spaced.

(a) Find the measure of each central angle in degrees and in radians.
(b) If the radius of the wheel is 76.0 ft, find the circumference.
(c) Find the length of each arc intercepted by consecutive pairs of spokes.
(d) Find the area of each sector formed by consecutive spokes.

42. *Area Cleaned by a Windshield Wiper* The Ford Model A, built from 1928 to 1931, had a single windshield wiper on the driver's side. The total arm and blade was 10 in. long and rotated back and forth through an angle of 95°. The shaded region in the figure is the portion of the windshield cleaned by the 7-in. wiper blade. What is the area of the region cleaned?

43. *Circular Railroad Curves* In the United States, circular railroad curves are designated by the *degree of curvature,* the central angle subtended by a chord of 100 ft. Suppose a portion of track has curvature 42°. (*Source:* Hay, W., *Railroad Engineering,* John Wiley & Sons, 1982.)

(a) What is the radius of the curve?
(b) What is the length of the arc determined by the 100-ft chord?
(c) What is the area of the portion of the circle bounded by the arc and the 100-ft chord?

44. *Land Required for a Solar-Power Plant* A 300-megawatt solar-power plant requires approximately 950,000 m² of land area in order to collect the required amount of energy from sunlight.

(a) If this land area is circular, what is its radius?
(b) If this land area is a 35° sector of a circle, what is its radius?

45. *Area of a Lot* A frequent problem in surveying city lots and rural lands adjacent to curves of highways and railways is that of finding the area when one or more of the boundary lines is the arc of a circle. Find the area of the lot shown in the figure. (*Source:* Anderson, J. and E. Michael, *Introduction to Surveying,* McGraw-Hill, 1985.)

46. *Nautical Miles* *Nautical miles* are used by ships and airplanes. They are different from *statute miles,* which equal 5280 ft. A nautical mile is defined to be the arc length along the equator intercepted by a central angle *AOB* of 1 min, as illustrated in the figure. If the equatorial radius of Earth is 3963 mi, use the arc length formula to approximate the number of statute miles in 1 nautical mile. Round your answer to two decimal places.

47. *Circumference of Earth* The first accurate estimate of the distance around Earth was done by the Greek astronomer Eratosthenes (276–195 B.C.), who noted that the noontime position of the sun at the summer solstice differed by 7° 12′ from the city of Syene to the city of Alexandria. (See the figure.) The distance between these two cities is 496 mi. Use the arc length formula to estimate the radius of Earth. Then find the circumference of Earth. (*Source:* Zeilik, M., *Introductory Astronomy and Astrophysics,* Third Edition, Saunders College Publishers, 1992.)

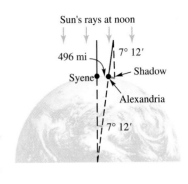

Sun's rays at noon

496 mi

Syene

7° 12′

Shadow

Alexandria

7° 12′

48. *Diameter of the Moon* The distance to the moon is approximately 238,900 mi. Use the arc length formula to estimate the diameter *d* of the moon if angle θ in the figure is measured to be .517°.

θ

d

Not to scale

Relating Concepts

For individual or collaborative investigation

(Exercises 49–52)

(Modeling) Measuring Paper Curl Manufacturers of paper determine its quality by its curl. The curl of a sheet of paper is measured by holding it at the center of one edge and comparing the arc formed by the free end to arcs on a chart lying flat on a table. Each arc in the chart corresponds to a number d that gives the depth of the arc. See the figure. (Source: Tabakovic, H., J. Paullet, and R. Bertram, "Measuring the Curl of Paper," The College Mathematics Journal, Vol. 30 No. 4, September 1999.)

Hand

Paper

L Chart

θ r

h

d

L

To produce the chart, it is necessary to find a function that relates d to the length of arc L. **Work Exercises 49–52 in order,** to determine that function. Refer to the figure on the right.

49. Express *L* in terms of *r* and θ, and then solve for *r*.

50. Use a right triangle to relate *r*, *h*, and θ. Solve for *h*.

51. Express *d* in terms of *r* and *h*, then substitute your answer from Exercise 50 for *h*. Factor out *r*.

52. Use your answer from Exercise 49 to substitute for *r* in the result from Exercise 51. This result is a formula that gives *d* for specific values of θ.

53. *Concept Check* If the radius of a circle is doubled and the central angle of a sector is unchanged, how is the area of the sector changed?

54. *Concept Check*   Give the corresponding formula for the area of a sector when the angle is measured in degrees.

Volume of a Solid   *Multiply the area of the base by the height to find the volume of each solid.*

55.

56.

Outside radius is r_1, inside radius is r_2.

3.3 | The Unit Circle and Circular Functions

> Circular Functions ▪ Finding Values of Circular Functions ▪ Determining a Number with a Given Circular Function Value ▪ Applying Circular Functions

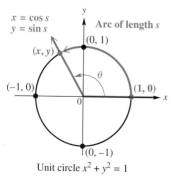

$x = \cos s$
$y = \sin s$

Arc of length s

$(0, 1)$

(x, y)

$(-1, 0)$

θ

$(1, 0)$

0

$(0, -1)$

Unit circle $x^2 + y^2 = 1$

Figure 11

In Section 1.3, we defined the six trigonometric functions in such a way that the domain of each function was a set of *angles* in standard position. These angles can be measured in degrees or in radians. In advanced courses, such as calculus, it is necessary to modify the trigonometric functions so that their domains consist of *real numbers* rather than angles. We do this by using the relationship between an angle θ and an arc of length s on a circle.

Circular Functions   In Figure 11, we start at the point $(1, 0)$ and measure an arc of length s along the circle. If $s > 0$, then the arc is measured in a counterclockwise direction, and if $s < 0$, then the direction is clockwise. (If $s = 0$, then no arc is measured.) Let the endpoint of this arc be at the point (x, y). The circle in Figure 11 is a **unit circle**—it has center at the origin and radius 1 unit (hence the name *unit circle*). Recall from algebra that the equation of this circle is

$$x^2 + y^2 = 1. \quad \text{(Appendix B)}$$

We saw in the previous section that the radian measure of θ is related to the arc length s. In fact, for θ measured in radians, we know that $s = r\theta$. Here, $r = 1$, so s, which is measured in linear units such as inches or centimeters, is numerically equal to θ, measured in radians. Thus, the trigonometric functions of angle θ in radians found by choosing a point (x, y) on the unit circle can be rewritten as functions of the arc length s, a real number. When interpreted this way, they are called **circular functions.**

Looking Ahead to Calculus

If you plan to study calculus, you must become very familiar with radian measure. In calculus, the trigonometric or circular functions are always understood to have real number domains.

Circular Functions		
$\sin s = y$	$\cos s = x$	$\tan s = \dfrac{y}{x} \quad (x \neq 0)$
$\csc s = \dfrac{1}{y} \quad (y \neq 0)$	$\sec s = \dfrac{1}{x} \quad (x \neq 0)$	$\cot s = \dfrac{x}{y} \quad (y \neq 0)$

Since x represents the cosine of s and y represents the sine of s, and because of the discussion in Section 3.1 on converting between degrees and radians, we can summarize a great deal of information in a concise manner, as seen in Figure 12.*

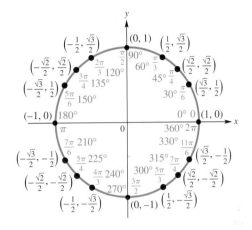

Unit circle $x^2 + y^2 = 1$

Figure 12

NOTE   Since $\sin s = y$ and $\cos s = x$, we can replace x and y in the equation $x^2 + y^2 = 1$ and obtain the Pythagorean identity

$$\cos^2 s + \sin^2 s = 1. \quad \text{(Section 1.4)}$$

The ordered pair (x, y) represents a point on the unit circle, and therefore

$$-1 \le x \le 1 \qquad \text{and} \qquad -1 \le y \le 1,$$

so

$$-1 \le \cos s \le 1 \qquad \text{and} \qquad -1 \le \sin s \le 1.$$

For any value of s, both $\sin s$ and $\cos s$ exist, so the domain of these functions is the set of all real numbers. For $\tan s$, defined as $\frac{y}{x}$, x must not equal 0. The only way x can equal 0 is when the arc length s is $\frac{\pi}{2}$, $-\frac{\pi}{2}$, $\frac{3\pi}{2}$, $-\frac{3\pi}{2}$, and so on. To avoid a 0 denominator, the domain of the tangent function must be restricted to those values of s satisfying

$$s \neq (2n + 1)\frac{\pi}{2}, \qquad n \text{ any integer.}$$

The definition of secant also has x in the denominator, so the domain of secant is the same as the domain of tangent. Both cotangent and cosecant are defined with a denominator of y. To guarantee that $y \neq 0$, the domain of these functions must be the set of all values of s satisfying

$$s \neq n\pi, \qquad n \text{ any integer.}$$

---

*The authors thank Professor Marvel Townsend of the University of Florida for her suggestion to include this figure.

In summary, the domains of the circular functions are as follows.

Domains of the Circular Functions

Assume that n is any integer and s is a real number.

Sine and Cosine Functions: $(-\infty, \infty)$

Tangent and Secant Functions: $\left\{ s \mid s \neq (2n + 1)\dfrac{\pi}{2} \right\}$

Cotangent and Cosecant Functions: $\{s \mid s \neq n\pi\}$

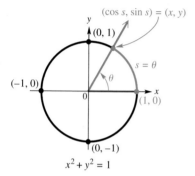

$x^2 + y^2 = 1$

Figure 13

Finding Values of Circular Functions The circular functions of real numbers are closely related to the trigonometric functions of angles measured in radians. Let us assume that angle θ is in standard position, superimposed on the unit circle. See Figure 13. Suppose that θ is the *radian* measure of this angle. Using the arc length formula $s = r\theta$ with $r = 1$, we have $s = \theta$. Thus, the length of the intercepted arc is the real number that corresponds to the radian measure of θ. Using the trigonometric function definitions from Section 1.3,

$$\sin \theta = \frac{y}{r} = \frac{y}{1} = y = \sin s, \qquad \text{and} \qquad \cos \theta = \frac{x}{r} = \frac{x}{1} = x = \cos s,$$

and so on. As shown here, the trigonometric functions and the circular functions lead to the same function values, provided we think of the angles as being in radian measure. This leads to the following important result.

Evaluating a Circular Function

Circular function values of real numbers are obtained in the same manner as trigonometric function values of angles measured in radians. This applies both to methods of finding exact values (such as reference angle analysis) and to calculator approximations. *Calculators must be in radian mode when finding circular function values.*

EXAMPLE 1 Finding Exact Circular Function Values

Find the exact values of $\sin \frac{3\pi}{2}$, $\cos \frac{3\pi}{2}$, and $\tan \frac{3\pi}{2}$.

Solution Evaluating a circular function at the real number $\frac{3\pi}{2}$ is equivalent to evaluating it at $\frac{3\pi}{2}$ radians. An angle of $\frac{3\pi}{2}$ radians intersects the unit circle at the point $(0, -1)$, as shown in Figure 14. Since

$$\sin s = y, \qquad \cos s = x, \qquad \text{and} \qquad \tan s = \frac{y}{x},$$

it follows that

$$\sin \frac{3\pi}{2} = -1, \qquad \cos \frac{3\pi}{2} = 0, \qquad \text{and} \qquad \tan \frac{3\pi}{2} \text{ is undefined.}$$

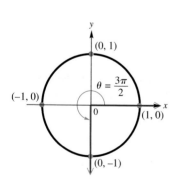

Figure 14

Now try Exercise 1.

EXAMPLE 2 Finding Exact Circular Function Values

(a) Use Figure 12 to find the exact values of $\cos \frac{7\pi}{4}$ and $\sin \frac{7\pi}{4}$.

(b) Use Figure 12 to find the exact value of $\tan\left(-\frac{5\pi}{3}\right)$.

(c) Use reference angles and degree/radian conversion to find the exact value of $\cos \frac{2\pi}{3}$.

Solution

(a) In Figure 12, we see that the terminal side of $\frac{7\pi}{4}$ radians intersects the unit circle at $\left(\frac{\sqrt{2}}{2}, -\frac{\sqrt{2}}{2}\right)$. Thus,

$$\cos \frac{7\pi}{4} = \frac{\sqrt{2}}{2} \qquad \text{and} \qquad \sin \frac{7\pi}{4} = -\frac{\sqrt{2}}{2}.$$

(b) Angles of $-\frac{5\pi}{3}$ radians and $\frac{\pi}{3}$ radians are coterminal. Their terminal sides intersect the unit circle at $\left(\frac{1}{2}, \frac{\sqrt{3}}{2}\right)$, so

$$\tan\left(-\frac{5\pi}{3}\right) = \tan \frac{\pi}{3} = \frac{\frac{\sqrt{3}}{2}}{\frac{1}{2}} = \frac{\sqrt{3}}{2} \div \frac{1}{2} = \frac{\sqrt{3}}{2} \cdot \frac{2}{1} = \sqrt{3}.$$

(c) An angle of $\frac{2\pi}{3}$ radians corresponds to an angle of $120°$. In standard position, $120°$ lies in quadrant II with a reference angle of $60°$, so

Cosine is negative in quadrant II.

$$\cos \frac{2\pi}{3} = \cos 120° = -\cos 60° = -\frac{1}{2}.$$

Reference angle (Section 2.2)

Now try Exercises 7, 17, and 21.

EXAMPLE 3 Approximating Circular Function Values

Find a calculator approximation for each circular function value.

(a) $\cos 1.85$ (b) $\cos .5149$ (c) $\cot 1.3209$ (d) $\sec(-2.9234)$

Solution

(a) With a calculator in radian mode, we find $\cos 1.85 \approx -.2756$.

(b) $\cos .5149 \approx .8703$ Use a calculator in radian mode.

(c) As before, to find cotangent, secant, and cosecant function values, we must use the appropriate reciprocal functions. To find $\cot 1.3209$, first find $\tan 1.3209$ and then find the reciprocal.

$$\cot 1.3209 = \frac{1}{\tan 1.3209} \approx .2552$$

(d) $\sec(-2.9234) = \dfrac{1}{\cos(-2.9234)} \approx -1.0243$

Now try Exercises 23, 29, and 33.

cos(1.85)
         -.2756

Radian mode

This is how a calculator displays the result of Example 3(a), correct to four decimal digits.

CAUTION A common error in trigonometry is using calculators in degree mode when radian mode should be used. Remember, *if you are finding a circular function value of a real number, the calculator must be in radian mode.*

Determining a Number with a Given Circular Function Value
Recall from Section 2.3 how we used a calculator to determine an angle measure, given a trigonometric function value of the angle.

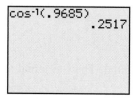

The calculator is set to show four decimal digits in the answer.

Figure 15

EXAMPLE 4 Finding a Number Given Its Circular Function Value

(a) Approximate the value of s in the interval $\left[0, \frac{\pi}{2}\right]$, if $\cos s = .9685$.

(b) Find the exact value of s in the interval $\left[\pi, \frac{3\pi}{2}\right]$, if $\tan s = 1$.

Solution

(a) Since we are given a cosine value and want to determine the real number in $\left[0, \frac{\pi}{2}\right]$ having this cosine value, we use the *inverse cosine* function of a calculator. With the calculator in radian mode, we find

$$\cos^{-1}(.9685) \approx .2517. \quad \text{(Section 2.3)}$$

See Figure 15. (Refer to your owner's manual to determine how to evaluate the $\sin^{-1}$, $\cos^{-1}$, and $\tan^{-1}$ functions with your calculator.)

(b) Recall that $\tan \frac{\pi}{4} = 1$, and in quadrant III $\tan s$ is positive. Therefore,

$$\tan\left(\pi + \frac{\pi}{4}\right) = \tan \frac{5\pi}{4} = 1,$$

and $s = \frac{5\pi}{4}$. Figure 16 supports this result.

This screen supports the result of Example 4(b). The calculator is in radian mode.

Figure 16

Now try Exercises 49 and 55.

Applying Circular Functions

EXAMPLE 5 Modeling the Angle of Elevation of the Sun

The angle of elevation θ of the sun in the sky at any latitude L is calculated with the formula

$$\sin \theta = \cos D \cos L \cos \omega + \sin D \sin L,$$

where $\theta = 0$ corresponds to sunrise and $\theta = \frac{\pi}{2}$ occurs if the sun is directly overhead. ω (the Greek letter *omega*) is the number of radians that Earth has rotated through since noon, when $\omega = 0$. D is the declination of the sun, which varies because Earth is tilted on its axis. (*Source:* Winter, C., R. Sizmann, and Vant-Hunt, Editors, *Solar Power Plants,* Springer-Verlag, 1991.)

Sacramento, California, has latitude $L = 38.5°$ or .6720 radian. Find the angle of elevation θ of the sun at 3 P.M. on February 29, 2000, where at that time $D \approx -.1425$ and $\omega \approx .7854$.

Solution   Use the given formula for $\sin \theta$.

$$\sin \theta = \cos D \cos L \cos \omega + \sin D \sin L$$
$$= \cos(-.1425) \cos(.6720) \cos(.7854) + \sin(-.1425) \sin(.6720)$$
$$\approx .4593$$

Thus, $\theta \approx .4773$ radian or $27.3°$.

Now try Exercise 71.

CONNECTIONS   A convenient way to see the sine, cosine, and tangent trigonometric ratios geometrically is shown in Figure 17 for θ in quadrants I and II. The circle shown is the unit circle, which has radius 1. By remembering this figure and the segments that represent the sine, cosine, and tangent functions, you can quickly recall properties of the trigonometric functions. Horizontal line segments to the left of the origin and vertical line segments below the x-axis represent negative values. Note that the tangent line must be tangent to the circle at $(1, 0)$, for any quadrant in which θ lies.

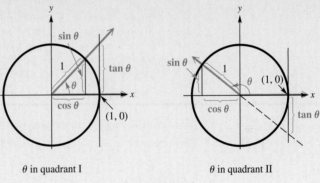

θ in quadrant I θ in quadrant II

Figure 17

For Discussion or Writing

See Figure 18. Use the definition of the trigonometric functions and similar triangles to show that $PQ = \sin \theta$, $OQ = \cos \theta$, and $AB = \tan \theta$.

Figure 18

3.3 Exercises

For each value of θ, find **(a)** $\sin \theta$, **(b)** $\cos \theta$, *and* **(c)** $\tan \theta$. *See Example 1.*

1. $\theta = \dfrac{\pi}{2}$ **2.** $\theta = \pi$ **3.** $\theta = 2\pi$

4. $\theta = 3\pi$ **5.** $\theta = -\pi$ **6.** $\theta = -\dfrac{3\pi}{2}$

Find the exact circular function value for each of the following. See Example 2.

7. $\sin \dfrac{7\pi}{6}$ **8.** $\cos \dfrac{5\pi}{3}$ **9.** $\tan \dfrac{3\pi}{4}$ **10.** $\sec \dfrac{2\pi}{3}$

11. $\csc \dfrac{11\pi}{6}$ **12.** $\cot \dfrac{5\pi}{6}$ **13.** $\cos\left(-\dfrac{4\pi}{3}\right)$ **14.** $\tan \dfrac{17\pi}{3}$

15. $\cos \dfrac{7\pi}{4}$ **16.** $\sec \dfrac{5\pi}{4}$ **17.** $\sin\left(-\dfrac{4\pi}{3}\right)$ **18.** $\sin\left(-\dfrac{5\pi}{6}\right)$

19. $\sec \dfrac{23\pi}{6}$ **20.** $\csc \dfrac{13\pi}{3}$ **21.** $\tan \dfrac{5\pi}{6}$ **22.** $\cos \dfrac{3\pi}{4}$

Find a calculator approximation for each circular function value. See Example 3.

23. $\sin .6109$ **24.** $\sin .8203$ **25.** $\cos(-1.1519)$

26. $\cos(-5.2825)$ **27.** $\tan 4.0203$ **28.** $\tan 6.4752$

29. $\csc(-9.4946)$ **30.** $\csc 1.3875$ **31.** $\sec 2.8440$

32. $\sec(-8.3429)$ **33.** $\cot 6.0301$ **34.** $\cot 3.8426$

Concept Check The figure displays a unit circle and an angle of 1 radian. The tick marks on the circle are spaced at every two-tenths radian. Use the figure to estimate each value.

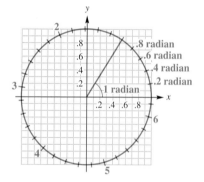

35. $\cos .8$

36. $\sin 4$

37. a positive angle whose cosine is $-.65$

38. a positive angle whose sine is $-.95$

Concept Check Without using a calculator, decide whether each function value is positive or negative. (Hint: Consider the radian measures of the quadrantal angles.)

39. $\cos 2$ **40.** $\sin(-1)$ **41.** $\sin 5$

42. $\cos 6$ **43.** $\tan 6.29$ **44.** $\tan(-6.29)$

Concept Check Each figure in Exercises 45–48 shows an angle θ in standard position with its terminal side intersecting the unit circle. Evaluate the six circular function values of θ.

45.

46.

47.

$\left(\frac{5}{13}, -\frac{12}{13}\right)$

48.

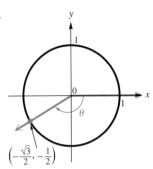

$\left(-\frac{\sqrt{3}}{2}, -\frac{1}{2}\right)$

Find the value of s in the interval $\left[0, \frac{\pi}{2}\right]$ *that makes each statement true. See Example 4(a).*

49. $\tan s = .2126$ **50.** $\cos s = .7826$ **51.** $\sin s = .9918$

52. $\cot s = .2994$ **53.** $\sec s = 1.0806$ **54.** $\csc s = 1.0219$

Find the exact value of s in the given interval that has the given circular function value. Do not use a calculator. See Example 4(b).

55. $\left[\frac{\pi}{2}, \pi\right]$; $\sin s = \frac{1}{2}$ **56.** $\left[\frac{\pi}{2}, \pi\right]$; $\cos s = -\frac{1}{2}$

57. $\left[\pi, \frac{3\pi}{2}\right]$; $\tan s = \sqrt{3}$ **58.** $\left[\pi, \frac{3\pi}{2}\right]$; $\sin s = -\frac{1}{2}$

59. $\left[\frac{3\pi}{2}, 2\pi\right]$; $\tan s = -1$ **60.** $\left[\frac{3\pi}{2}, 2\pi\right]$; $\cos s = \frac{\sqrt{3}}{2}$

Suppose an arc of length s lies on the unit circle $x^2 + y^2 = 1$, *starting at the point* $(1, 0)$ *and terminating at the point* (x, y). *(See Figure 11.) Use a calculator to find the approximate coordinates for* (x, y). *(Hint:* $x = \cos s$ *and* $y = \sin s$.*)*

61. $s = 2.5$ **62.** $s = 3.4$ **63.** $s = -7.4$ **64.** $s = -3.9$

Concept Check   *For each value of s, use a calculator to find* $\sin s$ *and* $\cos s$ *and then use the results to decide in which quadrant an angle of s radians lies.*

65. $s = 51$ **66.** $s = 49$ **67.** $s = 65$ **68.** $s = 79$

Concept Check   *In Exercises 69 and 70, each graphing calculator screen shows a point on the unit circle. What is the length of the shortest arc of the circle from* $(1, 0)$ *to the point?*

69.

$x^2 + y^2 = 1$

1.55

1

−2.35   2.35

X=.55319149   Y=.83305413

−1.55

70.

$x^2 + y^2 = 1$

1.55

1

−2.35   2.35

X=-.9361702   Y=.35154709

−1.55

(Modeling) Solve each problem. See Example 5.

71. *Length of a Day* The ability to calculate the number of daylight hours H at any location is important in estimating potential solar energy production. H can be calculated using the formula

$$\cos(.1309H) = -\tan D \tan L,$$

where D and L are defined in Example 5. Use this trigonometric equation to calculate the shortest and longest days in Minneapolis, Minnesota, if its latitude $L = 44.88°$, the shortest day occurs when $D = -23.44°$, and the longest day occurs when $D = 23.44°$. Remember to convert degrees to radians. (*Source:* Winter, C., R. Sizmann, and Vant-Hunt, Editors, *Solar Power Plants,* Springer-Verlag, 1991.)

72. *Elevation of the Sun* Refer to Example 5.

   (a) Repeat the example for New Orleans, which has latitude $L = 30°$.

   (b) Compare your answers. Do they agree with your intuition?

73. *Maximum Temperatures* Because the values of the circular functions repeat every 2π, they are used to describe things that repeat periodically. For example, the maximum afternoon temperature in a given city might be modeled by

$$t = 60 - 30 \cos \frac{x\pi}{6},$$

where t represents the maximum afternoon temperature in month x, with $x = 0$ representing January, $x = 1$ representing February, and so on. Find the maximum afternoon temperature for each of the following months.

   (a) January   **(b)** April   **(c)** May

   (d) June   **(e)** August   **(f)** October

74. *Temperature in Fairbanks* The temperature in Fairbanks is modeled by

$$T(x) = 37 \sin \left[\frac{2\pi}{365}(x - 101) \right] + 25,$$

where $T(x)$ is the temperature in degrees Fahrenheit on day x, with $x = 1$ corresponding to January 1 and $x = 365$ corresponding to December 31. Use a calculator to estimate the temperature on the following days. (*Source:* Lando, B. and C. Lando, "Is the Graph of Temperature Variation a Sine Curve?" *The Mathematics Teacher,* 70, September 1977.)

   (a) March 1 (day 60)   **(b)** April 1 (day 91)   **(c)** Day 150

   (d) June 15   **(e)** September 1   **(f)** October 31

3.4 Linear and Angular Speed

Linear Speed ▪ Angular Speed

Linear Speed In many situations we need to know how fast a point on a circular disk is moving or how fast the central angle of such a disk is changing. Some examples occur with machinery involving gears or pulleys or the speed of a car around a curved portion of highway.

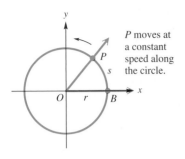

y

P moves at a constant speed along the circle.

O   r   B

x

s

P

Figure 19

Suppose that point P moves at a constant speed along a circle of radius r and center O. See Figure 19. The measure of how fast the position of P is changing is called **linear speed.** If v represents linear speed, then

$$\text{speed} = \frac{\text{distance}}{\text{time}} \quad \text{or} \quad v = \frac{s}{t},$$

where s is the length of the arc traced by point P at time t. (This formula is just a restatement of $d = rt$ with s as distance, v as rate (speed), and t as time.)

Angular Speed As point P in Figure 19 moves along the circle, ray OP rotates around the origin. Since ray OP is the terminal side of angle POB, the measure of the angle changes as P moves along the circle. The measure of how fast angle POB is changing is called **angular speed.** Angular speed, symbolized ω, is given as

$$\omega = \frac{\theta}{t}, \qquad \theta \text{ in radians,}$$

where θ is the measure of angle POB at time t. As with earlier formulas in this chapter, θ must be measured in radians, with ω expressed as radians per unit of time. Angular speed is used in physics and engineering, among other applications.

In Section 3.2, the length s of the arc intercepted on a circle of radius r by a central angle of measure θ radians was found to be $s = r\theta$. Using this formula, the formula for linear speed, $v = \frac{s}{t}$, becomes

$$v = \frac{s}{t} = \frac{r\theta}{t} = r \cdot \frac{\theta}{t} = r\omega. \quad s = r\theta;\ \omega = \frac{\theta}{t}$$

The formula $v = r\omega$ relates linear and angular speeds.

As an example of linear and angular speeds, consider the following. The human joint that can be flexed the fastest is the wrist, which can rotate through $90°$, or $\frac{\pi}{2}$ radians, in .045 sec while holding a tennis racket. The angular speed of a human wrist swinging a tennis racket is

$$\omega = \frac{\theta}{t} = \frac{\frac{\pi}{2}}{.045} \approx 35 \text{ radians per sec.}$$

If the radius (distance) from the tip of the racket to the wrist joint is 2 ft, then the speed at the tip of the racket is

$$v = r\omega \approx 2(35) = 70 \text{ ft per sec,} \qquad \text{or about 48 mph.}$$

In a tennis serve the arm rotates at the shoulder, so the final speed of the racket is considerably faster. (*Source:* Cooper, J. and R. Glassow, *Kinesiology,* Second Edition, C.V. Mosby, 1968.)

The formulas for angular and linear speed are summarized in the table.

Angular Speed	Linear Speed
$\omega = \dfrac{\theta}{t}$	$v = \dfrac{s}{t}$
	$v = \dfrac{r\theta}{t}$
(ω in radians per unit time, θ in radians)	$v = r\omega$

EXAMPLE 1 Using Linear and Angular Speed Formulas

Suppose that point P is on a circle with radius 10 cm, and ray OP is rotating with angular speed $\frac{\pi}{18}$ radian per sec.

(a) Find the angle generated by P in 6 sec.

(b) Find the distance traveled by P along the circle in 6 sec.

(c) Find the linear speed of P.

Solution

(a) The speed of ray OP is $\omega = \frac{\pi}{18}$ radian per sec. Since $\omega = \frac{\theta}{t}$, then in 6 sec

$$\frac{\pi}{18} = \frac{\theta}{6} \qquad \text{Let } \omega = \frac{\pi}{18} \text{ and } t = 6 \text{ in the angular speed formula.}$$

$$\theta = \frac{6\pi}{18} = \frac{\pi}{3} \text{ radians.} \quad \text{Solve for } \theta.$$

(b) From part (a), P generates an angle of $\frac{\pi}{3}$ radians in 6 sec. The distance traveled by P along the circle is

$$s = r\theta = 10\left(\frac{\pi}{3}\right) = \frac{10\pi}{3} \text{ cm.} \quad \text{(Section 3.2)}$$

(c) $v = \dfrac{s}{t} = \dfrac{\frac{10\pi}{3}}{6} = \dfrac{10\pi}{3} \div 6 = \dfrac{10\pi}{3} \cdot \dfrac{1}{6} = \dfrac{5\pi}{9}$ cm per sec

Now try Exercise 3.

EXAMPLE 2 Finding Angular Speed of a Pulley and Linear Speed of a Belt

A belt runs a pulley of radius 6 cm at 80 revolutions per min.

(a) Find the angular speed of the pulley in radians per second.

(b) Find the linear speed of the belt in centimeters per second.

Solution

(a) In 1 min, the pulley makes 80 revolutions. Each revolution is 2π radians, so

$$80(2\pi) = 160\pi \text{ radians per min.}$$

Since there are 60 sec in 1 min, we find ω, the angular speed in radians per second, by dividing 160π by 60.

$$\omega = \frac{160\pi}{60} = \frac{8\pi}{3} \text{ radians per sec}$$

(b) The linear speed of the belt will be the same as that of a point on the circumference of the pulley. Thus,

$$v = r\omega = 6\left(\frac{8\pi}{3}\right) = 16\pi \approx 50.3 \text{ cm per sec.}$$

Now try Exercise 39.

EXAMPLE 3 Finding Linear Speed and Distance Traveled by a Satellite

A satellite traveling in a circular orbit 1600 km above the surface of Earth takes 2 hr to make an orbit. The radius of Earth is 6400 km. See Figure 20.

(a) Find the linear speed of the satellite.

(b) Find the distance the satellite travels in 4.5 hr.

Solution

(a) The distance of the satellite from the center of Earth is

$$r = 1600 + 6400 = 8000 \text{ km.}$$

For one orbit, $\theta = 2\pi$, and

$$s = r\theta = 8000(2\pi) \text{ km.}$$

Since it takes 2 hr to complete an orbit, the linear speed is

$$v = \frac{s}{t} = \frac{8000(2\pi)}{2} = 8000\pi \approx 25{,}000 \text{ km per hr.}$$

(b) $s = vt = 8000\pi(4.5) = 36{,}000\pi \approx 110{,}000 \text{ km}$

1600 km

6400 km

Satellite

Earth

Not to scale

Figure 20

Now try Exercise 37.

3.4 Exercises

1. *Concept Check* If a point moves around the circumference of the unit circle at an angular velocity of 1 radian per sec, how long will it take for the point to move around the entire circle?

2. *Concept Check* If a point moves around the circumference of the unit circle at a speed of 1 unit per sec, how long will it take for the point to move around the entire circle?

Suppose that point P is on a circle with radius r, and ray OP is rotating with angular speed ω. For the given values of r, ω, and t, find each of the following.

(a) *the angle generated by P in time t*
(b) *the distance traveled by P along the circle in time t*
(c) *the linear speed of P*

See Example 1.

3. $r = 20$ cm, $\omega = \dfrac{\pi}{12}$ radian per sec, $t = 6$ sec

4. $r = 30$ cm, $\omega = \dfrac{\pi}{10}$ radian per sec, $t = 4$ sec

Use the formula $\omega = \frac{\theta}{t}$ to find the value of the missing variable.

5. $\omega = \dfrac{2\pi}{3}$ radians per sec, $t = 3$ sec

6. $\omega = \dfrac{\pi}{4}$ radian per min, $t = 5$ min

7. $\theta = \dfrac{3\pi}{4}$ radians, $t = 8$ sec

8. $\theta = \dfrac{2\pi}{5}$ radians, $t = 10$ sec

9. $\theta = \dfrac{2\pi}{9}$ radian, $\omega = \dfrac{5\pi}{27}$ radian per min

10. $\theta = \dfrac{3\pi}{8}$ radians, $\omega = \dfrac{\pi}{24}$ radian per min

11. $\theta = 3.871142$ radians, $t = 21.4693$ sec

12. $\omega = .90674$ radian per min, $t = 11.876$ min

Use the formula $v = r\omega$ to find the value of the missing variable.

13. $r = 12$ m, $\omega = \dfrac{2\pi}{3}$ radians per sec   **14.** $r = 8$ cm, $\omega = \dfrac{9\pi}{5}$ radians per sec

15. $v = 9$ m per sec, $r = 5$ m   **16.** $v = 18$ ft per sec, $r = 3$ ft

17. $v = 107.692$ m per sec, $r = 58.7413$ m

18. $r = 24.93215$ cm, $\omega = .372914$ radian per sec

The formula $\omega = \frac{\theta}{t}$ can be rewritten as $\theta = \omega t$. Using ωt for θ changes $s = r\theta$ to $s = r\omega t$. Use the formula $s = r\omega t$ to find the value of the missing variable.

19. $r = 6$ cm, $\omega = \dfrac{\pi}{3}$ radians per sec, $t = 9$ sec

20. $r = 9$ yd, $\omega = \dfrac{2\pi}{5}$ radians per sec, $t = 12$ sec

21. $s = 6\pi$ cm, $r = 2$ cm, $\omega = \dfrac{\pi}{4}$ radian per sec

22. $s = \dfrac{12\pi}{5}$ m, $r = \dfrac{3}{2}$ m, $\omega = \dfrac{2\pi}{5}$ radians per sec

23. $s = \dfrac{3\pi}{4}$ km, $r = 2$ km, $t = 4$ sec   **24.** $s = \dfrac{8\pi}{9}$ m, $r = \dfrac{4}{3}$ m, $t = 12$ sec

Find ω for each of the following.

25. the hour hand of a clock

26. a line from the center to the edge of a CD revolving 300 times per min

27. the minute hand of a clock   **28.** the second hand of a clock

Find v for each of the following.

29. the tip of the minute hand of a clock, if the hand is 7 cm long

30. the tip of the second hand of a clock, if the hand is 28 mm long

31. a point on the edge of a flywheel of radius 2 m, rotating 42 times per min

32. a point on the tread of a tire of radius 18 cm, rotating 35 times per min

33. the tip of an airplane propeller 3 m long, rotating 500 times per min (*Hint:* $r = 1.5$ m)

34. a point on the edge of a gyroscope of radius 83 cm, rotating 680 times per min

Solve each problem. See Examples 1–3.

35. *Speed of a Bicycle* The tires of a bicycle have radius 13 in. and are turning at the rate of 200 revolutions per min. See the figure. How fast is the bicycle traveling in miles per hour? (*Hint:* 5280 ft = 1 mi)

13 in.

36. *Hours in a Martian Day* Mars rotates on its axis at the rate of about .2552 radian per hr. Approximately how many hours are in a Martian day (or *sol*)? (*Source:* Wright, J. W., General Editor, *The Universal Almanac,* Andrews and McMeel, 1997.)

37. *Angular and Linear Speeds of Earth* Earth travels about the sun in an orbit that is almost circular. Assume that the orbit is a circle with radius 93,000,000 mi. Its angular and linear speeds are used in designing solar-power facilities.

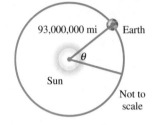

(a) Assume that a year is 365 days, and find the angle formed by Earth's movement in one day.

(b) Give the angular speed in radians per hour.

(c) Find the linear speed of Earth in miles per hour.

38. *Angular and Linear Speeds of Earth* Earth revolves on its axis once every 24 hr. Assuming that Earth's radius is 6400 km, find the following.

(a) angular speed of Earth in radians per day and radians per hour

(b) linear speed at the North Pole or South Pole

(c) linear speed at Quito, Ecuador, a city on the equator

(d) linear speed at Salem, Oregon (halfway from the equator to the North Pole)

39. *Speeds of a Pulley and a Belt* The pulley shown has a radius of 12.96 cm. Suppose it takes 18 sec for 56 cm of belt to go around the pulley.

(a) Find the angular speed of the pulley in radians per second.

(b) Find the linear speed of the belt in centimeters per second.

40. *Angular Speeds of Pulleys* The two pulleys in the figure have radii of 15 cm and 8 cm, respectively. The larger pulley rotates 25 times in 36 sec. Find the angular speed of each pulley in radians per second.

41. *Radius of a Spool of Thread* A thread is being pulled off a spool at the rate of 59.4 cm per sec. Find the radius of the spool if it makes 152 revolutions per min.

42. *Time to Move Along a Railroad Track* A railroad track is laid along the arc of a circle of radius 1800 ft. The circular part of the track subtends a central angle of 40°. How long (in seconds) will it take a point on the front of a train traveling 30 mph to go around this portion of the track?

43. *Angular Speed of a Motor Propeller* A 90-horsepower outboard motor at full throttle will rotate its propeller at 5000 revolutions per min. Find the angular speed of the propeller in radians per second.

44. *Linear Speed of a Golf Club* The shoulder joint can rotate at about 25 radians per sec. If a golfer's arm is straight and the distance from the shoulder to the club head is 5 ft, estimate the linear speed of the club head from shoulder rotation. (*Source:* Cooper, J. and R. Glassow, *Kinesiology,* Second Edition, C.V. Mosby, 1968.)

Chapter 3 Summary

KEY TERMS

3.1 radian
3.2 sector of a circle

3.3 unit circle
circular functions

3.4 linear speed v
angular speed ω

QUICK REVIEW

CONCEPTS

EXAMPLES

3.1 Radian Measure

An angle with its vertex at the center of a circle that intercepts an arc on the circle equal in length to the radius of the circle has a measure of **1 radian.**

Degree/Radian Relationship $180° = \pi$ **radians**

Converting Between Degrees and Radians

1. Multiply a degree measure by $\frac{\pi}{180}$ radian and simplify to convert to radians.

2. Multiply a radian measure by $\frac{180°}{\pi}$ and simplify to convert to degrees.

Convert $135°$ to radians.

$$135° = 135\left(\frac{\pi}{180} \text{ radian}\right) = \frac{3\pi}{4} \text{ radians}$$

Convert $-\frac{5\pi}{3}$ radians to degrees.

$$-\frac{5\pi}{3} \text{ radians} = -\frac{5\pi}{3}\left(\frac{180°}{\pi}\right) = -300°$$

3.2 Applications of Radian Measure

Arc Length

The length s of the arc intercepted on a circle of radius r by a central angle of measure θ radians is given by the product of the radius and the radian measure of the angle, or

$$s = r\theta, \qquad \theta \text{ in radians.}$$

In the figure, $s = r\theta$ so

$$\theta = \frac{s}{r} = \frac{3}{4} \text{ radian.}$$

Area of a Sector

The area of a sector of a circle of radius r and central angle θ is given by

$$A = \frac{1}{2}r^2\theta, \qquad \theta \text{ in radians.}$$

The area of the sector in the figure is

$$A = \frac{1}{2}(4)^2\left(\frac{3}{4}\right) = 6 \text{ sq units.}$$

CONCEPTS	EXAMPLES

3.3 The Unit Circle and Circular Functions

Circular Functions

Start at the point $(1, 0)$ on the unit circle $x^2 + y^2 = 1$ and lay off an arc of length $|s|$ along the circle, going counterclockwise if s is positive, and clockwise if s is negative. Let the endpoint of the arc be at the point (x, y). The six circular functions of s are defined as follows. (Assume that no denominators are 0.)

$$\sin s = y \qquad \cos s = x \qquad \tan s = \frac{y}{x}$$

$$\csc s = \frac{1}{y} \qquad \sec s = \frac{1}{x} \qquad \cot s = \frac{x}{y}$$

The Unit Circle

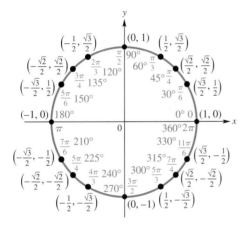

Unit circle $x^2 + y^2 = 1$

Use the unit circle to find each value.

$$\sin \frac{5\pi}{6} = \frac{1}{2}$$

$$\cos \frac{3\pi}{2} = 0$$

$$\tan \frac{\pi}{4} = \frac{\frac{\sqrt{2}}{2}}{\frac{\sqrt{2}}{2}} = 1$$

$$\csc \frac{7\pi}{4} = \frac{1}{-\frac{\sqrt{2}}{2}} = -\sqrt{2}$$

$$\sec \frac{7\pi}{6} = \frac{1}{-\frac{\sqrt{3}}{2}} = -\frac{2\sqrt{3}}{3}$$

$$\cot \frac{\pi}{3} = \frac{\frac{1}{2}}{\frac{\sqrt{3}}{2}} = \frac{\sqrt{3}}{3}$$

3.4 Linear and Angular Speed

Formulas for Angular and Linear Speed

Angular Speed	Linear Speed
$\omega = \dfrac{\theta}{t}$	$v = \dfrac{s}{t}$
(ω in radians per unit time, θ in radians)	$v = \dfrac{r\theta}{t}$
	$v = r\omega$

A belt runs a pulley of radius 8 in. at 60 revolutions per min. Find the angular speed ω in radians per minute, and the linear speed v of the belt in inches per minute.

$$\omega = 60(2\pi) = 120\pi \text{ radians per min}$$
$$v = r\omega = 8(120\pi) = 960\pi \text{ in. per min}$$

Chapter 3 Review Exercises

 1. Which is larger—an angle of 1° or an angle of 1 radian? Discuss and justify your answer.

2. Consider each angle in standard position having the given radian measure. In what quadrant does the terminal side lie?

    (a) 3   **(b)** 4   **(c)** −2   **(d)** 7

3. Find three angles coterminal to an angle of 1 radian.

4. Give an expression that generates all angles coterminal with an angle of $\frac{\pi}{6}$ radian. Let *n* represent any integer.

Convert each degree measure to radians. Leave answers as multiples of π.

    5. 45°   **6.** 120°   **7.** 175°

    8. 330°   **9.** 800°   **10.** 1020°

Convert each radian measure to degrees.

    11. $\dfrac{5\pi}{4}$ **12.** $\dfrac{9\pi}{10}$ **13.** $\dfrac{8\pi}{3}$

    14. $-\dfrac{6\pi}{5}$ **15.** $-\dfrac{11\pi}{18}$ **16.** $\dfrac{21\pi}{5}$

Suppose the tip of the minute hand of a clock is 2 in. from the center of the clock. For each duration, determine the distance traveled by the tip of the minute hand.

    17. 15 min   **18.** 20 min

    19. 3 hr   **20.** $10\frac{1}{2}$ hr

Solve each problem. Use a calculator as necessary.

21. *Arc Length*   The radius of a circle is 15.2 cm. Find the length of an arc of the circle intercepted by a central angle of $\frac{3\pi}{4}$ radians.

22. *Arc Length*   Find the length of an arc intercepted by a central angle of .769 radian on a circle with radius 11.4 cm.

23. *Arc Length*   A circle has radius 8.973 cm. Find the length of an arc on this circle intercepted by a central angle of 49.06°.

24. *Area of a Sector*   A central angle of $\frac{7\pi}{4}$ radians forms a sector of a circle. Find the area of the sector if the radius of the circle is 28.69 in.

25. *Area of a Sector*   Find the area of a sector of a circle having a central angle of 21° 40′ in a circle of radius 38.0 m.

26. *Height of a Tree*   A tree 2000 yd away subtends an angle of 1° 10′. Find the height of the tree to two significant digits.

Distance Between Cities   *Assume that the radius of Earth is 6400 km in Exercises 27 and 28.*

27. Find the distance in kilometers between cities on a north-south line that are on latitudes 28° N and 12° S, respectively.

28. Two cities on the equator have longitudes of 72° E and 35° W, respectively. Find the distance between the cities.

Concept Check   *In Exercises 29 and 30, find the measure of the central angle θ (in radians) and the area of the sector.*

29.

30.

31. *Concept Check*   The hour hand of a wall clock measures 6 in. from its tip to the center of the clock.

 (a) Through what angle (in radians) does the hour hand pass between 1 o'clock and 3 o'clock?

 (b) What distance does the tip of the hour hand travel during the time period from 1 o'clock to 3 o'clock?

32. Describe what would happen to the central angle for a given arc length of a circle if the circle's radius were doubled. (Assume everything else is unchanged.)

Find each exact function value. Do not use a calculator.

33. $\tan \dfrac{\pi}{3}$
34. $\cos \dfrac{2\pi}{3}$
35. $\sin\left(-\dfrac{5\pi}{6}\right)$

36. $\tan\left(-\dfrac{7\pi}{3}\right)$
37. $\csc\left(-\dfrac{11\pi}{6}\right)$
38. $\cot\left(-\dfrac{17\pi}{3}\right)$

Without using a calculator, determine which of the following is greater.

39. tan 1 or tan 2
40. sin 1 or tan 1

41. cos 2 or sin 2
42. cos(sin 0) or sin(cos 0)

Use a calculator to find an approximation for each circular function value. Be sure your calculator is set in radian mode.

43. sin 1.0472
44. tan 1.2275
45. cos(−.2443)

46. cot 3.0543
47. sec 7.3159
48. csc 4.8386

Find the value of s in the interval $\left[0, \frac{\pi}{2}\right]$ that makes each statement true.

49. cos s = .9250
50. tan s = 4.0112
51. sin s = .4924

52. csc s = 1.2361
53. cot s = .5022
54. sec s = 4.5600

Find the exact value of s in the given interval that has the given circular function value. Do not use a calculator.

55. $\left[0, \dfrac{\pi}{2}\right];$ $\cos s = \dfrac{\sqrt{2}}{2}$

56. $\left[\dfrac{\pi}{2}, \pi\right];$ $\tan s = -\sqrt{3}$

57. $\left[\pi, \dfrac{3\pi}{2}\right]$; $\sec s = -\dfrac{2\sqrt{3}}{3}$ **58.** $\left[\dfrac{3\pi}{2}, 2\pi\right]$; $\sin s = -\dfrac{1}{2}$

Solve each problem, where t, ω, θ, and s are as defined in Section 3.4.

59. Find t if $\theta = \dfrac{5\pi}{12}$ radians and $\omega = \dfrac{8\pi}{9}$ radians per sec.

60. Find θ if $t = 12$ sec and $\omega = 9$ radians per sec.

61. Find ω if $t = 8$ sec and $\theta = \dfrac{2\pi}{5}$ radians.

62. Find ω if $s = \dfrac{12\pi}{25}$ ft, $r = \dfrac{3}{5}$ ft, and $t = 15$ sec.

63. Find s if $r = 11.46$ cm, $\omega = 4.283$ radians per sec, and $t = 5.813$ sec.

64. *Linear Speed of a Flywheel*   Find the linear speed of a point on the edge of a flywheel of radius 7 cm if the flywheel is rotating 90 times per sec.

65. *Angular Speed of a Ferris Wheel*   A Ferris wheel has radius 25 ft. If it takes 30 sec for the wheel to turn $\dfrac{5\pi}{6}$ radians, what is the angular speed of the wheel?

(Modeling)   Solve each problem.

66. *Phase Angle of the Moon*   Because the moon orbits Earth, we observe different phases of the moon during the period of a month. In the figure, t is called the *phase angle*.

The *phase F* of the moon is modeled by

$$F(t) = \frac{1}{2}(1 - \cos t),$$

and gives the fraction of the moon's face that is illuminated by the sun. (*Source:* Duffet-Smith, P., *Practical Astronomy with Your Calculator,* Cambridge University Press, 1988.) Evaluate each expression and interpret the result.

(a) $F(0)$ **(b)** $F\left(\dfrac{\pi}{2}\right)$ **(c)** $F(\pi)$ **(d)** $F\left(\dfrac{3\pi}{2}\right)$

67. *Atmospheric Effect on Sunlight*   The shortest path for the sun's rays through Earth's atmosphere occurs when the sun is directly overhead. Disregarding the curvature of Earth, as the sun moves lower on the horizon, the distance that sunlight passes through the atmosphere increases by a factor of csc θ, where θ is the angle of elevation of the sun. This increased distance reduces both the intensity of the sun and the amount of ultraviolet light that reaches Earth's surface. See the figure below. (*Source:* Winter, C., R. Sizmann, and Vant-Hunt, Editors, *Solar Power Plants,* Springer-Verlag, 1991.)

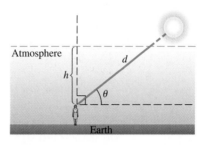

(a) Verify that $d = h \csc \theta$.

(b) Determine θ when $d = 2h$.

(c) The atmosphere filters out the ultraviolet light that causes skin to burn. Compare the difference between sunbathing when $\theta = \frac{\pi}{2}$ and when $\theta = \frac{\pi}{3}$. Which measure gives less ultraviolet light?

68. *Engine Specifications*   The figure shows a rotating wheel (crankshaft) with radius r and a connecting rod AP with length L. The piston slides back and forth along the x-axis as the wheel rotates counterclockwise at a rate of R revolutions per minute (RPM). The 1937 John Deere B engine, used in all models from 1935–1938, had the following specifications.

Maximum RPM: 1340 (no load) $\approx$ 505,168.1 radians per hr
Connecting rod: 10.500000 in.
Stroke: 5.250000 in.

(*Source:* Drost, J. P. and R. H. Kunferman, "Related Rates Challenge Problem: Calculate the Velocity of a Piston," *The AMATYC Review,* Vol. 21 No. 1, Fall 1999.)

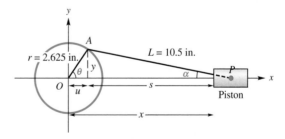

Find an expression for each measure.

(a) y (b) u (c) s (d) x

(e) The velocity v of the piston at the maximum RPM is given by

$$v = -2.625 \sin \theta \left(1 + \frac{\cos \theta}{\sqrt{15 + \cos^2 \theta}} \right) (505,168.1) \left(\frac{1}{12 \cdot 5280} \right),$$

where the last quantity changes inches to miles, so the velocity will be in miles per hour. Use the maximum-finding feature of a graphing calculator, with the window $[0, 6]$ by $[-25, 25]$, to find the value of θ that maximizes velocity. Give this maximum velocity.

Chapter 3 Test

Convert each degree measure to radians.

1. 120°

2. −45°

3. 5° (to the nearest hundredth)

Convert each radian measure to degrees.

4. $\dfrac{3\pi}{4}$

5. $-\dfrac{7\pi}{6}$

6. 4 (to the nearest hundredth)

7. A central angle of a circle with radius 150 cm cuts off an arc of 200 cm. Find each measure.

    (a) the radian measure of the angle

    (b) the area of a sector with that central angle

8. *Rotation of Gas Gauge Arrow*   The arrow on a car's gasoline gauge is $\frac{1}{2}$ in. long. See the figure. Through what angle does the arrow rotate when it moves 1 in. on the gauge?

Empty   Full

Find each circular function value.

9. $\sin \dfrac{3\pi}{4}$

10. $\cos\left(-\dfrac{7\pi}{6}\right)$

11. $\tan \dfrac{3\pi}{2}$

12. $\sec \dfrac{8\pi}{3}$

13. $\tan \pi$

14. $\cos \dfrac{3\pi}{2}$

15. Give the sine, cosine, and tangent of *s*.

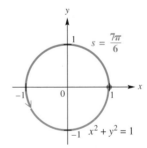

16. What are the domains of the tangent and secant circular functions?

17. (a) Use a calculator to approximate *s* in the interval $\left[0, \frac{\pi}{2}\right]$, if $\sin s = .8258$.

    (b) Find the exact value of *s* in the interval $\left[0, \frac{\pi}{2}\right]$, if $\cos s = \frac{1}{2}$.

18. *Angular and Linear Speed of a Point*   Suppose that point *P* is on a circle with radius 60 cm, and ray *OP* is rotating with angular speed $\frac{\pi}{12}$ radian per sec.

    (a) Find the angle generated by *P* in 8 sec.

    (b) Find the distance traveled by *P* along the circle in 8 sec.

    (c) Find the linear speed of *P*.

19. *Orbital Speed of Jupiter*   It takes Jupiter 11.64 yr to complete one orbit around the sun. See the figure. If Jupiter's average distance from the sun is 483,600,000 mi, find its orbital speed (speed along its orbital path) in miles per second. (*Source:* Wright, J. W., General Editor, *The Universal Almanac,* Andrews and McMeel, 1997.)

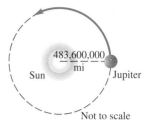

Not to scale

20. *Ferris Wheel*   A Ferris wheel has radius 50 ft. A person takes a seat and then the wheel turns $\frac{2\pi}{3}$ radians.

(a) How far is the person above the ground?

(b) If it takes 30 sec for the wheel to turn $\frac{2\pi}{3}$ radians, what is the angular speed of the wheel?

Chapter 3 Quantitative Reasoning

Can you find the radius of an Indian artifact given an arc of a circle?

Suppose you find an interesting Indian pottery fragment. The archaeology professor at your college believes it is a piece of the edge of a ceremonial plate and shows you a formula that will give the radius of the original plate using measurements from your fragment, shown in Figure A. Measurements are in inches.

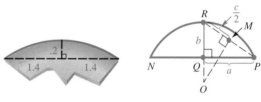

| Figure A | Figure B |

In Figure B, a is $\frac{1}{2}$ the length of chord *NP* and b is the distance from the midpoint of chord *NP* to the circle. According to the professor's formula, the radius of the circle, *OR*, is given by

$$r = \frac{a^2 + b^2}{2b}.$$

Why does this formula work? See if you can use the trigonometry you've studied so far to explain it by answering the following.

1. Extend line segment *RQ* to a radius of the arc, and draw line segment *RP* and line segment *MO*, which is the perpendicular bisector of line segment *RP*. See Figure B. Find a pair of similar triangles in the figure. How do you know they are similar?

2. Find two pairs of corresponding sides, and set their ratios equal. Solve the equation for *r*.

3. How are a, b, and c related? Use this relationship to substitute for c in terms of a and b.

4. What is the radius of the original plate from which your fragment came?

4

Graphs of the Circular Functions

Phenomena that repeat in a regular pattern, such as rotation of a planet on its axis, high and low tides, and average monthly temperature, can be modeled by *periodic functions*. In Example 6 of Section 4.2, we model the average monthly temperature in New Orleans, a city famous for its Mardi Gras celebration, using a graph of the sine function. Since the 18th century, residents of New Orleans have hosted this annual carnival, featuring festive parades and pageants, elaborate costumes, and dancing in the streets, to mark the final days before the Christian season of Lent. (*Source: Microsoft Encarta Encyclopedia 2002.*)

4.1 | Graphs of the Sine and Cosine Functions

Periodic Functions ▪ **Graph of the Sine Function** ▪ **Graph of the Cosine Function** ▪ **Graphing Techniques, Amplitude, and Period** ▪ **Using a Trigonometric Model**

Periodic Functions Many things in daily life repeat with a predictable pattern: in warm areas electricity use goes up in summer and down in winter, the price of fresh fruit goes down in summer and up in winter, and attendance at amusement parks increases in spring and declines in autumn. Because the sine and cosine functions repeat their values in a regular pattern, they are **periodic functions.** Figure 1 shows a periodic graph that represents a normal heartbeat.

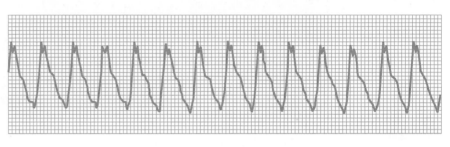

Figure 1

Looking Ahead to Calculus

Periodic functions are used throughout calculus, so you will need to know their characteristics. One use of these functions is to describe the location of a point in the plane using *polar coordinates,* an alternative to rectangular coordinates. (See Chapter 8).

Periodic Function

A **periodic function** is a function f such that

$$f(x) = f(x + np),$$

for every real number x in the domain of f, every integer n, and some positive real number p. The smallest possible positive value of p is the **period** of the function.

The circumference of the unit circle is 2π, so the smallest value of p for which the sine and cosine functions repeat is 2π. Therefore, the sine and cosine functions are periodic functions with period 2π.

Graph of the Sine Function In Section 3.3 we saw that for a real number s, the point on the unit circle corresponding to s has coordinates $(\cos s, \sin s)$. See Figure 2. Trace along the circle to verify the results shown in the table.

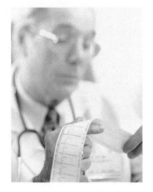

Figure 2

As s Increases from	sin s	cos s
0 to $\frac{\pi}{2}$	Increases from 0 to 1	Decreases from 1 to 0
$\frac{\pi}{2}$ to π	Decreases from 1 to 0	Decreases from 0 to -1
π to $\frac{3\pi}{2}$	Decreases from 0 to -1	Increases from -1 to 0
$\frac{3\pi}{2}$ to 2π	Increases from -1 to 0	Increases from 0 to 1

To avoid confusion when graphing the sine function, we use x rather than s; this corresponds to the letters in the xy-coordinate system. Selecting key values of x and finding the corresponding values of $\sin x$ leads to the table in Figure 3. To obtain the traditional graph in Figure 3, we plot the points from the table, use symmetry, and join them with a smooth curve. Since $y = \sin x$ is periodic with period 2π and has domain $(-\infty, \infty)$, the graph continues in the same pattern in both directions. This graph is called a **sine wave** or **sinusoid.**

SINE FUNCTION $f(x) = \sin x$

Domain: $(-\infty, \infty)$ Range: $[-1, 1]$

x	y
0	0
$\frac{\pi}{6}$	$\frac{1}{2}$
$\frac{\pi}{4}$	$\frac{\sqrt{2}}{2}$
$\frac{\pi}{3}$	$\frac{\sqrt{3}}{2}$
$\frac{\pi}{2}$	1
π	0
$\frac{3\pi}{2}$	-1
2π	0

$f(x) = \sin x, \ -2\pi \le x \le 2\pi$

$f(x) = \sin x$

Figure 3

- The graph is continuous over its entire domain, $(-\infty, \infty)$.
- Its x-intercepts are of the form $n\pi$, where n is an integer.
- Its period is 2π.
- The graph is symmetric with respect to the origin, so the function is an odd function. For all x in the domain, $\sin(-x) = -\sin x$.

NOTE   In algebra, we say that a function f is *odd* if for all x in the domain of f, $f(-x) = -f(x)$. The graph of an odd function is symmetric with respect to the origin (Appendix D); that is, if (x, y) belongs to the function, then $(-x, -y)$ also belongs to the function. For example, $\left(\frac{\pi}{2}, 1\right)$ and $\left(-\frac{\pi}{2}, -1\right)$ are points on the graph of $y = \sin x$, illustrating the property $\sin(-x) = -\sin x$.

Sine graphs occur in many practical applications. For example, look back at Figure 2 and assume that the line from the origin to the point (p, q) is part of the pedal of a bicycle, with a foot placed at (p, q). Since $q = \sin x$, the height of the pedal from the horizontal axis in Figure 2 is given by $\sin x$. By choosing various angles for the pedal and calculating q for each angle, the height of the pedal leads to the sine curve shown in Figure 4 on the next page.

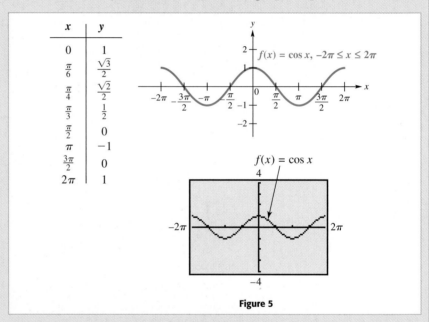

Figure 4

Graph of the Cosine Function   The graph of $y = \cos x$ in Figure 5 has the same shape as the graph of $y = \sin x$. It is, in fact, the graph of the sine function shifted, or translated, $\frac{\pi}{2}$ units to the left.

COSINE FUNCTION $f(x) = \cos x$

Domain: $(-\infty, \infty)$ Range: $[-1, 1]$

x	y
0	1
$\frac{\pi}{6}$	$\frac{\sqrt{3}}{2}$
$\frac{\pi}{4}$	$\frac{\sqrt{2}}{2}$
$\frac{\pi}{3}$	$\frac{1}{2}$
$\frac{\pi}{2}$	0
π	-1
$\frac{3\pi}{2}$	0
2π	1

Figure 5

- The graph is continuous over its entire domain, $(-\infty, \infty)$.
- Its x-intercepts are of the form $(2n + 1)\frac{\pi}{2}$, where n is an integer.
- Its period is 2π.
- The graph is symmetric with respect to the y-axis, so the function is an even function. For all x in the domain, $\cos(-x) = \cos x$.

NOTE   A function f is *even* if for all x in the domain of f, $f(-x) = f(x)$. The graph of an even function is symmetric with respect to the y-axis (Appendix D); that is, if (x, y) belongs to the function, then $(-x, y)$ also belongs to the function. For example, $\left(\frac{\pi}{2}, 0\right)$ and $\left(-\frac{\pi}{2}, 0\right)$ are points on the graph of $y = \cos x$, illustrating the property $\cos(-x) = \cos x$.

Notice that the calculator graphs of $f(x) = \sin x$ in Figure 3 and $f(x) = \cos x$ in Figure 5 are graphed in the window $[-2\pi, 2\pi]$ by $[-4, 4]$, with $\text{Xscl} = \frac{\pi}{2}$ and $\text{Yscl} = 1$. This is called the **trig viewing window.** (Your model may use a different "standard" trigonometric viewing window. Consult your owner's manual.) ■

Graphing Techniques, Amplitude, and Period The examples that follow show graphs that are "stretched" or "compressed" either vertically, horizontally, or both when compared with the graphs of $y = \sin x$ or $y = \cos x$.

EXAMPLE 1 Graphing $y = a \sin x$

Graph $y = 2 \sin x$, and compare to the graph of $y = \sin x$.

Solution For a given value of x, the value of y is twice as large as it would be for $y = \sin x$, as shown in the table of values. The only change in the graph is the range, which becomes $[-2, 2]$. See Figure 6, which includes a graph of $y = \sin x$ for comparison.

x	0	$\frac{\pi}{2}$	π	$\frac{3\pi}{2}$	2π
$\sin x$	0	1	0	-1	0
$2 \sin x$	0	2	0	-2	0

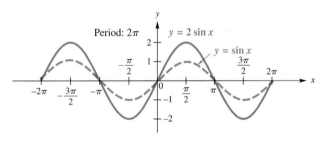

Figure 6

The thick graph style represents the function in Example 1.

The **amplitude** of a periodic function is half the difference between the maximum and minimum values. Thus, for both the basic sine and cosine functions, the amplitude is

$$\frac{1}{2}[1 - (-1)] = \frac{1}{2}(2) = 1.$$

Generalizing from Example 1 gives the following.

Amplitude

The graph of **$y = a \sin x$ or $y = a \cos x$,** with $a \neq 0$, will have the same shape as the graph of $y = \sin x$ or $y = \cos x$, respectively, except with range $[-|a|, |a|]$. The amplitude is $|a|$.

Now try Exercise 11.

No matter what the value of the amplitude, the periods of $y = a \sin x$ and $y = a \cos x$ are still 2π. Consider $y = \sin 2x$. We can complete a table of values for the interval $[0, 2\pi]$.

x	0	$\frac{\pi}{4}$	$\frac{\pi}{2}$	$\frac{3\pi}{4}$	π	$\frac{5\pi}{4}$	$\frac{3\pi}{2}$	$\frac{7\pi}{4}$	2π
$\sin 2x$	0	1	0	-1	0	1	0	-1	0

Note that one complete cycle occurs in π units, not 2π units. Therefore, the period here is π, which equals $\frac{2\pi}{2}$. Now consider $y = \sin 4x$. Look at the next table.

x	0	$\frac{\pi}{8}$	$\frac{\pi}{4}$	$\frac{3\pi}{8}$	$\frac{\pi}{2}$	$\frac{5\pi}{8}$	$\frac{3\pi}{4}$	$\frac{7\pi}{8}$	π
$\sin 4x$	0	1	0	-1	0	1	0	-1	0

These values suggest that one complete cycle is achieved in $\frac{\pi}{2}$ or $\frac{2\pi}{4}$ units, which is reasonable since

$$\sin\left(4 \cdot \frac{\pi}{2}\right) = \sin 2\pi = 0.$$

In general, the graph of a function of the form $y = \sin bx$ or $y = \cos bx$, for $b > 0$, will have a period different from 2π when $b \neq 1$. To see why this is so, remember that the values of $\sin bx$ or $\cos bx$ will take on all possible values as bx ranges from 0 to 2π. Therefore, to find the period of either of these functions, we must solve the three-part inequality

$$0 \leq bx \leq 2\pi \quad \text{(Appendix A)}$$

$$0 \leq x \leq \frac{2\pi}{b}. \quad \text{Divide by the positive number } b.$$

Thus, the period is $\frac{2\pi}{b}$. By dividing the interval $\left[0, \frac{2\pi}{b}\right]$ into four equal parts, we obtain the values for which $\sin bx$ or $\cos bx$ is -1, 0, or 1. These values will give minimum points, x-intercepts, and maximum points on the graph. Once these points are determined, we can sketch the graph by joining the points with a smooth sinusoidal curve. (If a function has $b < 0$, then the identities of the next chapter can be used to rewrite the function so that $b > 0$.)

N O T E   One method to divide an interval into four equal parts is as follows.

Step 1 Find the midpoint of the interval by adding the x-values of the endpoints and dividing by 2.   (Appendix B)

Step 2 Find the midpoints of the two intervals found in Step 1, using the same procedure.

EXAMPLE 2   Graphing $y = \sin bx$

Graph $y = \sin 2x$, and compare to the graph of $y = \sin x$.

Solution   In this function the coefficient of x is 2, so the period is $\frac{2\pi}{2} = \pi$. Therefore, the graph will complete one period over the interval $[0, \pi]$.

The endpoints are 0 and π, and the three middle points are

$$\frac{1}{4}(0 + \pi), \qquad \frac{1}{2}(0 + \pi), \qquad \text{and} \qquad \frac{3}{4}(0 + \pi),$$

which give the following *x*-values.

0,	$\dfrac{\pi}{4}$,	$\dfrac{\pi}{2}$,	$\dfrac{3\pi}{4}$,	π
↑	↑	↑	↑	↑
Left endpoint	First-quarter point	Midpoint	Third-quarter point	Right endpoint

We plot the points from the table of values given on page 136, and join them with a smooth sinusoidal curve. More of the graph can be sketched by repeating this cycle, as shown in Figure 7. The amplitude is not changed.

Figure 7

Now try Exercise 21.

Generalizing from Example 2 leads to the following result.

Period

For $b > 0$, the graph of $y = \sin bx$ will resemble that of $y = \sin x$, but with period $\frac{2\pi}{b}$. Also, the graph of $y = \cos bx$ will resemble that of $y = \cos x$, but with period $\frac{2\pi}{b}$.

EXAMPLE 3 Graphing $y = \cos bx$

Graph $y = \cos \dfrac{2}{3}x$ over one period.

Solution The period is $\dfrac{2\pi}{\frac{2}{3}} = 2\pi \div \frac{2}{3} = 2\pi \cdot \frac{3}{2} = 3\pi$. We divide the interval $[0, 3\pi]$ into four equal parts to get the *x*-values 0, and 3π that yield minimum points, maximum points, and *x*-intercepts. We use these values to obtain a table of key points for one period.

x	0	$\frac{3\pi}{4}$	$\frac{3\pi}{2}$	$\frac{9\pi}{4}$	3π
$\frac{2}{3}x$	0	$\frac{\pi}{2}$	π	$\frac{3\pi}{2}$	2π
$\cos \frac{2}{3}x$	1	0	-1	0	1

The amplitude is 1 because the maximum value is 1, the minimum value is -1, and half of $1 - (-1)$ is $\frac{1}{2}(2) = 1$. We plot these points and join them with a smooth curve. The graph is shown in Figure 8.

Figure 8

This screen shows a graph of the function in Example 3. By choosing $Xscl = \dfrac{3\pi}{4}$, the *x*-intercepts, maxima, and minima coincide with tick marks on the *x*-axis.

Now try Exercise 19.

N O T E Look back at the middle row of the table in Example 3. Dividing the interval $\left[0, \frac{2\pi}{b}\right]$ into four equal parts will always give the values 0, $\frac{\pi}{2}$, π, $\frac{3\pi}{2}$, and 2π for this row, resulting in values of -1, 0, or 1 for the circular function. These lead to key points on the graph, which can then be easily sketched.

The method used in Examples 1–3 is summarized as follows.

Guidelines for Sketching Graphs of Sine and Cosine Functions

To graph $y = a \sin bx$ or $y = a \cos bx$, with $b > 0$, follow these steps.

Step 1 Find the period, $\frac{2\pi}{b}$. Start at 0 on the x-axis, and lay off a distance of $\frac{2\pi}{b}$.

Step 2 Divide the interval into four equal parts. (See the Note preceding Example 2.)

Step 3 Evaluate the function for each of the five x-values resulting from Step 2. The points will be maximum points, minimum points, and x-intercepts.

Step 4 Plot the points found in Step 3, and join them with a sinusoidal curve having amplitude $|a|$.

Step 5 Draw the graph over additional periods, to the right and to the left, as needed.

The function in Example 4 has both amplitude and period affected by the values of a and b.

EXAMPLE 4 Graphing $y = a \sin bx$

Graph $y = -2 \sin 3x$ over one period using the preceding guidelines.

Solution

Step 1 For this function, $b = 3$, so the period is $\frac{2\pi}{3}$. The function will be graphed over the interval $\left[0, \frac{2\pi}{3}\right]$.

Step 2 Divide the interval $\left[0, \frac{2\pi}{3}\right]$ into four equal parts to get the x-values 0, $\frac{\pi}{6}$, $\frac{\pi}{3}$, $\frac{\pi}{2}$, and $\frac{2\pi}{3}$.

Step 3 Make a table of values determined by the x-values from Step 2.

x	0	$\frac{\pi}{6}$	$\frac{\pi}{3}$	$\frac{\pi}{2}$	$\frac{2\pi}{3}$
$3x$	0	$\frac{\pi}{2}$	π	$\frac{3\pi}{2}$	2π
$\sin 3x$	0	1	0	-1	0
$-2 \sin 3x$	0	-2	0	2	0

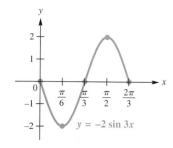

$y = -2 \sin 3x$

Figure 9

Step 4 Plot the points $(0, 0)$, $\left(\frac{\pi}{6}, -2\right)$, $\left(\frac{\pi}{3}, 0\right)$, $\left(\frac{\pi}{2}, 2\right)$, and $\left(\frac{2\pi}{3}, 0\right)$, and join them with a sinusoidal curve with amplitude 2. See Figure 9.

Step 5 The graph can be extended by repeating the cycle.

Notice that when *a* is negative, the graph of $y = a \sin bx$ is the reflection across the *x*-axis of the graph of $y = |a| \sin bx$.

Now try Exercise 25.

EXAMPLE 5 Graphing $y = a \cos bx$ for *b* Equal to a Multiple of π

Graph $y = -3 \cos \pi x$ over one period.

Solution

Step 1 Since $b = \pi$, the period is $\frac{2\pi}{\pi} = 2$, so we will graph the function over the interval $[0, 2]$.

Step 2 Dividing $[0, 2]$ into four equal parts yields the *x*-values $0, \frac{1}{2}, 1, \frac{3}{2}$, and 2.

Step 3 Make a table using these *x*-values.

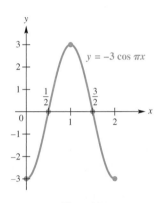

x	0	$\frac{1}{2}$	1	$\frac{3}{2}$	2
πx	0	$\frac{\pi}{2}$	π	$\frac{3\pi}{2}$	2π
$\cos \pi x$	1	0	-1	0	1
$-3 \cos \pi x$	-3	0	3	0	-3

Step 4 Plot the points $(0, -3)$, $\left(\frac{1}{2}, 0\right)$, $(1, 3)$, $\left(\frac{3}{2}, 0\right)$, and $(2, -3)$, and join them with a sinusoidal curve having amplitude $|-3| = 3$. See Figure 10.

Step 5 The graph can be extended by repeating the cycle.

Notice that when *b* is a multiple of π, the *x*-intercepts of the graph are rational numbers.

Now try Exercise 29.

Figure 10

Using a Trigonometric Model

EXAMPLE 6 Interpreting a Sine Function Model

The average temperature (in °F) at Mould Bay, Canada, can be approximated by the function defined by

$$f(x) = 34 \sin\left[\frac{\pi}{6}(x - 4.3)\right],$$

where *x* is the month and $x = 1$ corresponds to January.

(a) Graph *f* in the window $[0, 25]$ by $[-45, 45]$.

(b) What is the average temperature during the month of May?

(c) What would be an approximation for the average *yearly* temperature at Mould Bay?

Figure 11

Solution

(a) The graph of $f(x) = 34 \sin\left[\frac{\pi}{6}(x - 4.3)\right]$ is shown in Figure 11. Its amplitude is 34, and the period is $\dfrac{2\pi}{\frac{\pi}{6}} = 2\pi \div \frac{\pi}{6} = 2\pi \cdot \frac{6}{\pi} = 12$. The function f has a period of 12 months or 1 year, which agrees with the changing of the seasons.

(b) May is the fifth month, so the average temperature during May is

$$f(5) = 34 \sin\left[\frac{\pi}{6}(5 - 4.3)\right] \approx 12°F. \quad \text{Let } x = 5. \text{ (Appendix C)}$$

See the display at the bottom of the screen in Figure 11.

(c) From the graph, it appears that the average yearly temperature is about 0°F since the graph is centered vertically about the line $y = 0$.

Now try Exercise 41.

CONNECTIONS   Using a TI-83 Plus calculator, adjust the settings to correspond to the following screens.

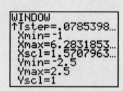

In the final two screens, Tmax is 2π, Tstep is $\frac{\pi}{40}$, Xmax is 2π, and Xscl is $\frac{\pi}{2}$. Now graph these two equations (which are in *parametric form*), and watch as the unit circle and the sine function are graphed simultaneously. Press the TRACE key once to get the screen shown on the left below, and then press the up-arrow key to get the screen shown on the right below. The screen on the left gives a unit circle interpretation of cos 0 = 1 and sin 0 = 0. The screen on the right gives a rectangular coordinate graph interpretation of sin 0 = 0.

Now go back and redefine Y_{2T} as cos(T). Graph both equations again; the second screen will look like the one in the margin, after the TRACE and up-arrow keys are pressed. This screen indicates that cos 0 = 1.

For Discussion or Writing

1. On the unit circle, let T = 2. What values of X and Y are displayed? Interpret these values.

2. On the sine graph, trace to the point where T = 1.9. What values of X and Y are displayed? Interpret these values with an equation in X and Y.

3. Repeat Exercise 2, but use the cosine graph.

4.1 Exercises

Concept Check   *In Exercises 1–8, match each function with its graph.*

1. $y = \sin x$ **2.** $y = \cos x$ **3.** $y = -\sin x$ **4.** $y = -\cos x$

5. $y = \sin 2x$ **6.** $y = \cos 2x$ **7.** $y = 2 \sin x$ **8.** $y = 2 \cos x$

A.   **B.**   **C.**

D.   **E.**   **F.**

G.   **H.**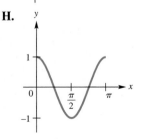

Graph each function over the interval $[-2\pi, 2\pi]$. Give the amplitude. See Example 1.

9. $y = 2 \cos x$ **10.** $y = 3 \sin x$ **11.** $y = \dfrac{2}{3} \sin x$

12. $y = \dfrac{3}{4} \cos x$ **13.** $y = -\cos x$ **14.** $y = -\sin x$

15. $y = -2 \sin x$ **16.** $y = -3 \cos x$

Graph each function over a two-period interval. Give the period and amplitude. See Examples 2–5.

17. $y = \sin \dfrac{1}{2}x$

18. $y = \sin \dfrac{2}{3}x$

19. $y = \cos \dfrac{3}{4}x$

20. $y = \cos \dfrac{1}{3}x$

21. $y = \sin 3x$

22. $y = \cos 2x$

23. $y = 2 \sin \dfrac{1}{4}x$

24. $y = 3 \sin 2x$

25. $y = -2 \cos 3x$

26. $y = -5 \cos 2x$

27. $y = \cos \pi x$

28. $y = -\sin \pi x$

29. $y = -2 \sin 2\pi x$

30. $y = 3 \cos 2\pi x$

31. $y = \dfrac{1}{2} \cos \dfrac{\pi}{2}x$

32. $y = -\dfrac{2}{3} \sin \dfrac{\pi}{4}x$

(Modeling)   *Solve each problem.*

33. *Average Annual Temperature*   Scientists believe that the average annual temperature in a given location is periodic. The average temperature at a given place during a given season fluctuates as time goes on, from colder to warmer, and back to colder. The graph shows an idealized description of the temperature (in °F) for the last few thousand years of a location at the same latitude as Anchorage, Alaska.

Average Annual Temperature (Idealized)

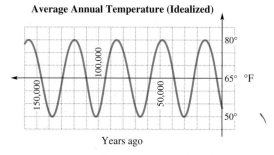

Years ago

(a) Find the highest and lowest temperatures recorded.
(b) Use these two numbers to find the amplitude.
(c) Find the period of the function.
(d) What is the trend of the temperature now?

34. *Blood Pressure Variation*   The graph gives the variation in blood pressure for a typical person. Systolic and diastolic pressures are the upper and lower limits of the periodic changes in pressure that produce the pulse. The length of time between peaks is called the period of the pulse.

Blood Pressure Variation

Time (in seconds)

(a) Find the amplitude of the graph.
(b) Find the pulse rate (the number of pulse beats in 1 min) for this person.

35. *Activity of a Nocturnal Animal*   Many of the activities of living organisms are periodic. For example, the graph at the right shows the time that a certain nocturnal animal begins its evening activity.

(a) Find the amplitude of this graph.

(b) Find the period.

Activity of a Nocturnal Animal

36. *Position of a Moving Arm*   The figure shows schematic diagrams of a rhythmically moving arm. The upper arm *RO* rotates back and forth about the point *R*; the position of the arm is measured by the angle *y* between the actual position and the downward vertical position. (*Source:* De Sapio, R., *Calculus for the Life Sciences*. Copyright © 1978 by W. H. Freeman and Company. Reprinted by permission.)

This graph shows the relationship between angle *y* and time *t* in seconds.

(a) Find an equation of the form $y = a \sin kt$ for the graph shown.

(b) How long does it take for a complete movement of the arm?

(Modeling)   *Solve each problem.*

37. *Voltage of an Electrical Circuit*   The voltage *E* in an electrical circuit is modeled by

$$E = 5 \cos 120\pi t,$$

where *t* is time measured in seconds.

(a) Find the amplitude and the period.

(b) How many cycles are completed in 1 sec? (The number of cycles (periods) completed in 1 sec is the *frequency* of the function.)

(c) Find *E* when $t = 0, .03, .06, .09, .12$.

(d) Graph *E* for $0 \le t \le \frac{1}{30}$.

38. *Voltage of an Electrical Circuit*   For another electrical circuit, the voltage *E* is modeled by

$$E = 3.8 \cos 40\pi t,$$

where *t* is time measured in seconds.

(a) Find the amplitude and the period.

(b) Find the frequency. See Exercise 37(b).

(c) Find *E* when $t = .02, .04, .08, .12, .14$.

(d) Graph one period of *E*.

39. *Atmospheric Carbon Dioxide* At Mauna Loa, Hawaii, atmospheric carbon dioxide levels in parts per million (ppm) have been measured regularly since 1958. The function defined by

$$L(x) = .022x^2 + .55x + 316 + 3.5 \sin 2\pi x$$

can be used to model these levels, where x is in years and $x = 0$ corresponds to 1960. (*Source:* Nilsson, A., *Greenhouse Earth,* John Wiley & Sons, 1992.)

(a) Graph L in the window $[15, 35]$ by $[325, 365]$.

(b) When do the seasonal maximum and minimum carbon dioxide levels occur?

(c) L is the sum of a quadratic function and a sine function. What is the significance of each of these functions? Discuss what physical phenomena may be responsible for each function.

40. *Atmospheric Carbon Dioxide* Refer to Exercise 39. The carbon dioxide content in the atmosphere at Barrow, Alaska, in parts per million (ppm) can be modeled using the function defined by

$$C(x) = .04x^2 + .6x + 330 + 7.5 \sin 2\pi x,$$

where $x = 0$ corresponds to 1970. (*Source:* Zeilik, M. and S. Gregory, *Introductory Astronomy and Astrophysics,* Brooks/Cole, 1998.)

(a) Graph C in the window $[5, 25]$ by $[320, 380]$.

(b) Discuss possible reasons why the amplitude of the oscillations in the graph of C is larger than the amplitude of the oscillations in the graph of L in Exercise 39, which models Hawaii.

(c) Define a new function C that is valid if x represents the actual year, where $1970 \le x \le 1995$. (See horizontal translations in Appendix D.)

41. *Temperature in Fairbanks* The temperature in Fairbanks is modeled by

$$T(x) = 37 \sin\left[\frac{2\pi}{365}(x - 101)\right] + 25,$$

where $T(x)$ is the temperature in degrees Fahrenheit on day x, with $x = 1$ corresponding to January 1 and $x = 365$ corresponding to December 31. Use a calculator to estimate the temperature on the following days. (*Source:* Lando, B. and C. Lando, "Is the Graph of Temperature Variation a Sine Curve?", *The Mathematics Teacher,* 70, September 1977.)

(a) March 1 (day 60)   **(b)** April 1 (day 91)   **(c)** Day 150

(d) June 15   **(e)** September 1   **(f)** October 31

42. *Fluctuation in the Solar Constant* The *solar constant S* is the amount of energy per unit area that reaches Earth's atmosphere from the sun. It is equal to 1367 watts per sq m but varies slightly throughout the seasons. This fluctuation ΔS in S can be calculated using the formula

$$\Delta S = .034S \sin\left[\frac{2\pi(82.5 - N)}{365.25}\right].$$

In this formula, N is the day number covering a four-year period, where $N = 1$ corresponds to January 1 of a leap year and $N = 1461$ corresponds to December 31 of the fourth year. (*Source:* Winter, C., R. Sizmann, and Vant-Hunt, Editors, *Solar Power Plants,* Springer-Verlag, 1991.)

(a) Calculate ΔS for $N = 80$, which is the spring equinox in the first year.

(b) Calculate ΔS for $N = 1268$, which is the summer solstice in the fourth year.

(c) What is the maximum value of ΔS?

(d) Find a value for N where ΔS is equal to 0.

Tides for Kahului Harbor   The chart shows the tides for Kahului Harbor (on the island of Maui, Hawaii). To identify high and low tides and times for other Maui areas, the following adjustments must be made.

Hana: High, +40 min, +.1 ft;
          Low, +18 min, −.2 ft

Makena: High, +1:21, −.5 ft;
             Low, +1:09, −.2 ft

Maalaea: High, +1:52, −.1 ft;
             Low, +1:19, −.2 ft

Lahaina: High, +1:18, −.2 ft;
             Low, +1:01, −.1 ft

JANUARY

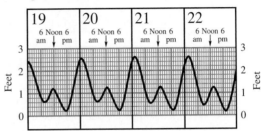

Source: Maui News. Original chart prepared by Edward K. Noda and Associates.

Use the graph to work Exercises 43–48.

43. The graph is an example of a periodic function. What is the period (in hours)?

44. What is the amplitude?

45. At what time on January 20 was low tide at Kahului? What was the height?

46. Repeat Exercise 45 for Maalaea.

47. At what time on January 22 was high tide at Kahului? What was the height?

48. Repeat Exercise 47 for Lahaina.

Musical Sound Waves   Pure sounds produce single sine waves on an oscilloscope. Find the amplitude and period of each sine wave graph in Exercises 49 and 50. On the vertical scale, each square represents .5; on the horizontal scale, each square represents $30°$ or $\frac{\pi}{6}$.

49.

50.

51. Compare the graphs of $y = \sin 2x$ and $y = 2 \sin x$ over the interval $[0, 2\pi]$. Can we say that, in general, $\sin bx = b \sin x$? Explain.

52. Compare the graphs of $y = \cos 3x$ and $y = 3 \cos x$ over the interval $[0, 2\pi]$. Can we say that, in general, $\cos bx = b \cos x$? Explain.

53. Refer to the graph of $y = \sin x$ in Figure 3. The graph completes one cycle between $x = 0$ and $x = 2\pi$. Consider the statement, "The function $y = \sin bx$ completes b cycles between 0 and 2π." Use your graphing calculator to confirm the statement for some positive integer values of b, such as 3, 4, and 5. Interpret and confirm the statement for $b = \frac{1}{2}$ and $b = \frac{3}{2}$.

4.2 | Translations of the Graphs of the Sine and Cosine Functions

Horizontal Translations ▪ **Vertical Translations** ▪ **Combinations of Translations** ▪ **Determining a Trigonometric Model Using Curve Fitting**

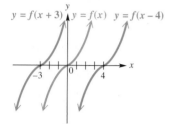

Horizontal translations of $y = f(x)$
(Appendix D)

Figure 12

Horizontal Translations The graph of the function defined by $y = f(x - d)$ is translated *horizontally* compared to the graph of $y = f(x)$. The translation is d units to the right if $d > 0$ and $|d|$ units to the left if $d < 0$. See Figure 12. With circular functions, a horizontal translation is called a **phase shift**. In the function $y = f(x - d)$, the expression $x - d$ is called the **argument.** In Examples 1–3, we give two methods that can be used to sketch the graph of a circular function involving a phase shift.

EXAMPLE 1 Graphing $y = \sin(x - d)$

Graph $y = \sin\left(x - \dfrac{\pi}{3}\right)$ over one period.

Solution *Method 1* For the argument $x - \frac{\pi}{3}$ to result in all possible values throughout one period, it must take on all values between 0 and 2π, inclusive. Therefore, to find an interval of one period, we solve the three-part inequality

$$0 \le x - \frac{\pi}{3} \le 2\pi \quad \text{(Appendix A)}$$

$$\frac{\pi}{3} \le x \le \frac{7\pi}{3}. \qquad \text{Add } \tfrac{\pi}{3} \text{ to each part.}$$

Divide the interval $\left[\frac{\pi}{3}, \frac{7\pi}{3}\right]$ into four equal parts to get the following x-values.

$$\frac{\pi}{3}, \quad \frac{5\pi}{6}, \quad \frac{4\pi}{3}, \quad \frac{11\pi}{6}, \quad \frac{7\pi}{3}$$

A table of values using these x-values follows.

x	$\frac{\pi}{3}$	$\frac{5\pi}{6}$	$\frac{4\pi}{3}$	$\frac{11\pi}{6}$	$\frac{7\pi}{3}$
$x - \frac{\pi}{3}$	0	$\frac{\pi}{2}$	π	$\frac{3\pi}{2}$	2π
$\sin\left(x - \frac{\pi}{3}\right)$	0	1	0	-1	0

Figure 13

We join the corresponding points with a smooth curve to get the solid blue graph shown in Figure 13. The period is 2π, and the amplitude is 1.

Method 2 We can also graph $y = \sin\left(x - \frac{\pi}{3}\right)$ using a horizontal translation of the graph of $y = \sin x$. The argument $x - \frac{\pi}{3}$ indicates that the graph will be translated $\frac{\pi}{3}$ units to the *right* (the phase shift) as compared to the graph of $y = \sin x$. In Figure 13 we show the graph of $y = \sin x$ as a dashed red curve, and the graph of $y = \sin\left(x - \frac{\pi}{3}\right)$ as a solid blue curve. Therefore, to graph a function using this method, first graph the basic circular function, and then graph the desired function by using the appropriate translation.

Now try Exercise 25.

NOTE The graph in Figure 13 of Example 1 can be extended through additional periods by repeating the given portion of the graph over and over, as necessary.

EXAMPLE 2 Graphing $y = a \cos(x - d)$

Graph $y = 3 \cos\left(x + \dfrac{\pi}{4}\right)$ over one period.

Solution *Method 1* The graph can be sketched over one period by first solving the three-part inequality

$$0 \leq x + \frac{\pi}{4} \leq 2\pi$$

$$-\frac{\pi}{4} \leq x \leq \frac{7\pi}{4}. \qquad \text{Subtract } \tfrac{\pi}{4} \text{ from each part.}$$

Dividing this interval into four equal parts gives x-values of

$$-\frac{\pi}{4}, \quad \frac{\pi}{4}, \quad \frac{3\pi}{4}, \quad \frac{5\pi}{4}, \quad \frac{7\pi}{4}.$$

A table of points for these x-values leads to maximum points, minimum points, and x-intercepts.

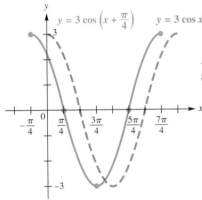

$y = 3 \cos\left(x + \frac{\pi}{4}\right)$ $y = 3 \cos x$

Figure 14

x	$-\frac{\pi}{4}$	$\frac{\pi}{4}$	$\frac{3\pi}{4}$	$\frac{5\pi}{4}$	$\frac{7\pi}{4}$
$x + \frac{\pi}{4}$	0	$\frac{\pi}{2}$	π	$\frac{3\pi}{2}$	2π
$\cos\left(x + \frac{\pi}{4}\right)$	1	0	-1	0	1
$3 \cos\left(x + \frac{\pi}{4}\right)$	3	0	-3	0	3

We join the corresponding points with a smooth curve to get the solid blue graph shown in Figure 14. The period is 2π, and the amplitude is 3.

Method 2 Write $3 \cos\left(x + \frac{\pi}{4}\right)$ in the form $a \cos(x - d)$.

$$y = 3 \cos\left(x + \frac{\pi}{4}\right) = 3 \cos\left[x - \left(-\frac{\pi}{4}\right)\right]$$

This result shows that $d = -\frac{\pi}{4}$. Since $-\frac{\pi}{4}$ is negative, the phase shift is $\left|-\frac{\pi}{4}\right| = \frac{\pi}{4}$ unit to the left. The graph is the same as that of $y = 3 \cos x$ (the dashed red graph in Figure 14), except that it is translated $\frac{\pi}{4}$ unit to the left (the solid blue graph in Figure 14).

This screen shows the graph of

$$Y_1 = 3 \cos\left(X + \frac{\pi}{4}\right)$$

in Example 2 as a thick line. The graph of

$$Y_2 = 3 \cos X$$

is shown as a thin line for comparison.

Now try Exercise 27.

EXAMPLE 3 Graphing $y = a \cos b(x - d)$

Graph $y = -2 \cos(3x + \pi)$ over two periods.

Solution *Method 1* The function can be sketched over one period by solving the three-part inequality

$$0 \le 3x + \pi \le 2\pi$$

to get the interval $\left[-\frac{\pi}{3}, \frac{\pi}{3}\right]$. Divide this interval into four equal parts to get the following points.

$$\left(-\frac{\pi}{3}, -2\right), \quad \left(-\frac{\pi}{6}, 0\right), \quad (0, 2), \quad \left(\frac{\pi}{6}, 0\right), \quad \left(\frac{\pi}{3}, -2\right)$$

Plot these points and join them with a smooth curve. By graphing an additional half period to the left and to the right, we obtain the graph shown in Figure 15.

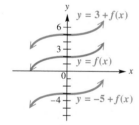

$y = -2 \cos(3x + \pi)$

Figure 15

Method 2 First write the expression in the form $a \cos b(x - d)$.

$$y = -2 \cos(3x + \pi) = -2 \cos 3\left(x + \frac{\pi}{3}\right) \quad \text{Rewrite } 3x + \pi \text{ as } 3\left(x + \frac{\pi}{3}\right).$$

Then $a = -2$, $b = 3$, and $d = -\frac{\pi}{3}$. The amplitude is $|-2| = 2$, and the period is $\frac{2\pi}{3}$ (since the value of b is 3). The phase shift is $\left|-\frac{\pi}{3}\right| = \frac{\pi}{3}$ units to the left as compared to the graph of $y = -2 \cos 3x$. Again, see Figure 15.

Now try Exercise 31.

Vertical Translations The graph of a function of the form $y = c + f(x)$ is translated *vertically* as compared with the graph of $y = f(x)$. See Figure 16. The translation is c units up if $c > 0$ and $|c|$ units down if $c < 0$.

Vertical translations of $y = f(x)$
(Appendix D)

Figure 16

EXAMPLE 4 Graphing $y = c + a \cos bx$

Graph $y = 3 - 2 \cos 3x$ over two periods.

Solution The values of y will be 3 greater than the corresponding values of y in $y = -2 \cos 3x$. This means that the graph of $y = 3 - 2 \cos 3x$ is the same as the graph of $y = -2 \cos 3x$, vertically translated 3 units up. Since the period of $y = -2 \cos 3x$ is $\frac{2\pi}{3}$, the key points have x-values

$$0, \quad \frac{\pi}{6}, \quad \frac{\pi}{3}, \quad \frac{\pi}{2}, \quad \frac{2\pi}{3}.$$

Use these x-values to make a table of points.

x	0	$\frac{\pi}{6}$	$\frac{\pi}{3}$	$\frac{\pi}{2}$	$\frac{2\pi}{3}$
$\cos 3x$	1	0	-1	0	1
$2 \cos 3x$	2	0	-2	0	2
$3 - 2 \cos 3x$	1	3	5	3	1

The key points are shown on the graph in Figure 17, along with more of the graph, sketched using the fact that the function is periodic.

The function in Example 4 is shown using the thick graph style. Notice also the thin graph style for $y = -2 \cos 3x$.

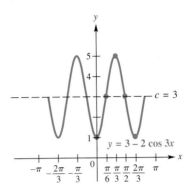

Figure 17

Now try Exercise 37.

Combinations of Translations A function of the form

$$y = c + a \sin b(x - d) \qquad \text{or} \qquad y = c + a \cos b(x - d), \qquad b > 0,$$

which involves stretching, compressing, and translating, can be graphed according to the following guidelines.

Further Guidelines for Sketching Graphs of Sine and Cosine Functions

Method 1 Follow these steps.

Step 1 Find an interval whose length is one period $\frac{2\pi}{b}$ by solving the three-part inequality

$$0 \le b(x - d) \le 2\pi. \quad \text{(Appendix A)}$$

Step 2 Divide the interval into four equal parts.

Step 3 Evaluate the function for each of the five x-values resulting from Step 2. The points will be maximum points, minimum points, and points that intersect the line $y = c$ ("middle" points of the wave).

Step 4 Plot the points found in Step 3, and join them with a sinusoidal curve having amplitude $|a|$.

Step 5 Draw the graph over additional periods, to the right and to the left, as needed.

Method 2 First graph the basic circular function. The amplitude of the function is $|a|$, and the period is $\frac{2\pi}{b}$. Then use translations to graph the desired function. The vertical translation is c units up if $c > 0$ and $|c|$ units down if $c < 0$. The horizontal translation (phase shift) is d units to the right if $d > 0$ and $|d|$ units to the left if $d < 0$.

EXAMPLE 5 Graphing $y = c + a \sin b(x - d)$

Graph $y = -1 + 2 \sin(4x + \pi)$ over two periods.

Solution We use Method 1. First write the expression on the right side in the form $c + a \sin b(x - d)$.

$$y = -1 + 2 \sin(4x + \pi) = -1 + 2 \sin\left[4\left(x + \frac{\pi}{4}\right)\right]$$

Step 1 Find an interval whose length is one period.

$$0 \le 4\left(x + \frac{\pi}{4}\right) \le 2\pi$$

$$0 \le x + \frac{\pi}{4} \le \frac{\pi}{2} \qquad \text{Divide each part by 4.}$$

$$-\frac{\pi}{4} \le x \le \frac{\pi}{4} \qquad \text{Subtract } \tfrac{\pi}{4} \text{ from each part.}$$

Step 2 Divide the interval $\left[-\frac{\pi}{4}, \frac{\pi}{4}\right]$ into four equal parts to get the x-values

$$-\frac{\pi}{4}, \quad -\frac{\pi}{8}, \quad 0, \quad \frac{\pi}{8}, \quad \frac{\pi}{4}.$$

Step 3 Make a table of values.

x	$-\frac{\pi}{4}$	$-\frac{\pi}{8}$	0	$\frac{\pi}{8}$	$\frac{\pi}{4}$
$x + \frac{\pi}{4}$	0	$\frac{\pi}{8}$	$\frac{\pi}{4}$	$\frac{3\pi}{8}$	$\frac{\pi}{2}$
$4\left(x + \frac{\pi}{4}\right)$	0	$\frac{\pi}{2}$	π	$\frac{3\pi}{2}$	2π
$\sin 4\left(x + \frac{\pi}{4}\right)$	0	1	0	-1	0
$2 \sin 4\left(x + \frac{\pi}{4}\right)$	0	2	0	-2	0
$-1 + 2 \sin(4x + \pi)$	-1	1	-1	-3	-1

Steps 4 and 5 Plot the points found in the table and join them with a sinusoidal curve. Figure 18 shows the graph, extended to the right and left to include two full periods.

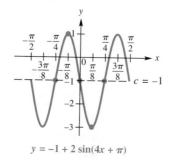

$$y = -1 + 2 \sin(4x + \pi)$$

Figure 18

Now try Exercise 43.

Determining a Trigonometric Model Using Curve Fitting A sinusoidal function is often a good approximation of a set of real data points.

EXAMPLE 6 Modeling Temperature with a Sine Function

The maximum average monthly temperature in New Orleans is 82°F and the minimum is 54°F. The table shows the average monthly temperatures. The scatter diagram for a two-year interval in Figure 19 strongly suggests that the temperatures can be modeled with a sine curve.

Month	°F	Month	°F
Jan	54	July	82
Feb	55	Aug	81
Mar	61	Sept	77
Apr	69	Oct	71
May	73	Nov	59
June	79	Dec	55

Source: Miller, A., J. Thompson, and R. Peterson, *Elements of Meteorology, Fourth Edition,* Charles E. Merrill Publishing Co., 1983.

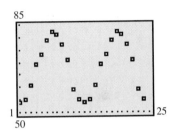

Figure 19

(a) Using only the maximum and minimum temperatures, determine a function of the form $f(x) = a \sin[b(x - d)] + c$, where a, b, c, and d are constants, that models the average monthly temperature in New Orleans. Let x represent the month, with January corresponding to $x = 1$.

(b) On the same coordinate axes, graph f for a two-year period together with the actual data values found in the table.

(c) Use the *sine regression* feature of a graphing calculator to determine a second model for these data.

Solution

(a) We use the maximum and minimum average monthly temperatures to find the amplitude a.

$$a = \frac{82 - 54}{2} = 14$$

The average of the maximum and minimum temperatures is a good choice for c. The average is

$$\frac{82 + 54}{2} = 68.$$

Since the coldest month is January, when $x = 1$, and the hottest month is July, when $x = 7$, we should choose d to be about 4. We experiment with values just greater than 4 to find d. Trial and error using a calculator leads to $d = 4.2$. Since temperatures repeat every 12 months, b is $\frac{2\pi}{12} = \frac{\pi}{6}$. Thus,

$$f(x) = a \sin[b(x - d)] + c = 14 \sin\left[\frac{\pi}{6}(x - 4.2)\right] + 68.$$

(b) Figure 20 shows the data points from the table, the graph of $y = 14 \sin\left[\frac{\pi}{6}(x - 4.2)\right] + 68$, and the graph of $y = 14 \sin \frac{\pi}{6}x + 68$ for comparison. The horizontal translation of the model is fairly obvious here.

Values are rounded to the
nearest hundredth.

(a)

Figure 20

(b)

Figure 21

(c) We used the given data for a two-year period to produce the model described in Figure 21(a). Figure 21(b) shows its graph along with the data points.

Now try Exercise 47.

4.2 Exercises

Concept Check   *In Exercises 1–8, match each function with its graph in A–H.*

1. $y = \sin\left(x - \dfrac{\pi}{4}\right)$ **2.** $y = \sin\left(x + \dfrac{\pi}{4}\right)$ **3.** $y = \cos\left(x - \dfrac{\pi}{4}\right)$

4. $y = \cos\left(x + \dfrac{\pi}{4}\right)$ **5.** $y = 1 + \sin x$ **6.** $y = -1 + \sin x$

7. $y = 1 + \cos x$ **8.** $y = -1 + \cos x$

A.

B.

C.

D.

E.

F.

G.

H.

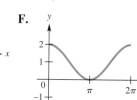

Concept Check   *Match each function in Column I with the appropriate description in Column II.*

I	**II**
9. $y = 3 \sin(2x - 4)$	**A.** amplitude $= 2$, period $= \dfrac{\pi}{2}$, phase shift $= \dfrac{3}{4}$
10. $y = 2 \sin(3x - 4)$	**B.** amplitude $= 3$, period $= \pi$, phase shift $= 2$
11. $y = 4 \sin(3x - 2)$	**C.** amplitude $= 4$, period $= \dfrac{2\pi}{3}$, phase shift $= \dfrac{2}{3}$
12. $y = 2 \sin(4x - 3)$	**D.** amplitude $= 2$, period $= \dfrac{2\pi}{3}$, phase shift $= \dfrac{4}{3}$

Concept Check   *In Exercises 13 and 14, fill in the blanks with the word* right *or the word* left.

13. If the graph of $y = \cos x$ is translated $\frac{\pi}{2}$ units horizontally to the _____ , it will coincide with the graph of $y = \sin x$.

14. If the graph of $y = \sin x$ is translated $\frac{\pi}{2}$ units horizontally to the _____ , it will coincide with the graph of $y = \cos x$.

Find the amplitude, the period, any vertical translation, and any phase shift of the graph of each function. See Examples 1–5.

15. $y = 2 \sin(x - \pi)$

16. $y = \dfrac{2}{3} \sin\left(x + \dfrac{\pi}{2}\right)$

17. $y = 4 \cos\left(\dfrac{x}{2} + \dfrac{\pi}{2}\right)$

18. $y = -\cos \pi\left(x - \dfrac{1}{3}\right)$

19. $y = 3 \cos \dfrac{\pi}{2}\left(x - \dfrac{1}{2}\right)$

20. $y = \dfrac{1}{2} \sin\left(\dfrac{x}{2} + \pi\right)$

21. $y = 2 - \sin\left(3x - \dfrac{\pi}{5}\right)$

22. $y = -1 + \dfrac{1}{2} \cos(2x - 3\pi)$

Graph each function over a two-period interval. See Examples 1 and 2.

23. $y = \cos\left(x - \dfrac{\pi}{2}\right)$

24. $y = \sin\left(x - \dfrac{\pi}{4}\right)$

25. $y = \sin\left(x + \dfrac{\pi}{4}\right)$

26. $y = \cos\left(x - \dfrac{\pi}{3}\right)$

27. $y = 2 \cos\left(x - \dfrac{\pi}{3}\right)$

28. $y = 3 \sin\left(x - \dfrac{3\pi}{2}\right)$

Graph each function over a one-period interval. See Example 3.

29. $y = \dfrac{3}{2} \sin 2\left(x + \dfrac{\pi}{4}\right)$

30. $y = -\dfrac{1}{2} \cos 4\left(x + \dfrac{\pi}{2}\right)$

31. $y = -4 \sin(2x - \pi)$

32. $y = 3 \cos(4x + \pi)$

33. $y = \dfrac{1}{2} \cos\left(\dfrac{1}{2}x - \dfrac{\pi}{4}\right)$

34. $y = -\dfrac{1}{4} \sin\left(\dfrac{3}{4}x + \dfrac{\pi}{8}\right)$

Graph each function over a two-period interval. See Example 4.

35. $y = -3 + 2 \sin x$

36. $y = 2 - 3 \cos x$

37. $y = -1 - 2 \cos 5x$

38. $y = 1 - \dfrac{2}{3} \sin \dfrac{3}{4} x$

39. $y = 1 - 2 \cos \dfrac{1}{2} x$

40. $y = -3 + 3 \sin \dfrac{1}{2} x$

41. $y = -2 + \dfrac{1}{2} \sin 3x$

42. $y = 1 + \dfrac{2}{3} \cos \dfrac{1}{2} x$

Graph each function over a one-period interval. See Example 5.

43. $y = -3 + 2 \sin\left(x + \dfrac{\pi}{2}\right)$

44. $y = 4 - 3 \cos(x - \pi)$

45. $y = \dfrac{1}{2} + \sin 2\left(x + \dfrac{\pi}{4}\right)$

46. $y = -\dfrac{5}{2} + \cos 3\left(x - \dfrac{\pi}{6}\right)$

(Modeling)   Solve each problem. See Example 6.

47. *Average Monthly Temperature*   The average monthly temperature (in °F) in Vancouver, Canada, is shown in the table.

(a) Plot the average monthly temperature over a two-year period letting $x = 1$ correspond to the month of January during the first year. Do the data seem to indicate a translated sine graph?

(b) The highest average monthly temperature is 64°F in July, and the lowest average monthly temperature is 36°F in January. Their average is 50°F. Graph the data together with the line $y = 50$. What does this line represent with regard to temperature in Vancouver?

Month	°F	Month	°F
Jan	36	July	64
Feb	39	Aug	63
Mar	43	Sept	57
Apr	48	Oct	50
May	55	Nov	43
June	59	Dec	39

Source: Miller, A. and J. Thompson, *Elements of Meteorology, Fourth Edition,* Charles E. Merrill Publishing Co., 1983.

(c) Approximate the amplitude, period, and phase shift of the translated sine wave.

(d) Determine a function of the form $f(x) = a \sin b(x - d) + c$, where a, b, c, and d are constants, that models the data.

(e) Graph f together with the data on the same coordinate axes. How well does f model the given data?

(f) Use the sine regression capability of a graphing calculator to find the equation of a sine curve that fits these data.

48. *Average Monthly Temperature*   The average monthly temperature (in °F) in Phoenix, Arizona, is shown in the table.

(a) Predict the average yearly temperature and compare it to the actual value of 70°F.

(b) Plot the average monthly temperature over a two-year period by letting $x = 1$ correspond to January of the first year.

(c) Determine a function of the form $f(x) = a \cos b(x - d) + c$, where a, b, c, and d are constants, that models the data.

(d) Graph f together with the data on the same coordinate axes. How well does f model the data?

(e) Use the sine regression capability of a graphing calculator to find the equation of a sine curve that fits these data.

Month	°F	Month	°F
Jan	51	July	90
Feb	55	Aug	90
Mar	63	Sept	84
Apr	67	Oct	71
May	77	Nov	59
June	86	Dec	52

Source: Miller, A. and J. Thompson, *Elements of Meteorology, Fourth Edition,* Charles E. Merrill Publishing Co., 1983.

4.3 Graphs of the Other Circular Functions

Graphs of the Cosecant and Secant Functions ▪ Graphs of the Tangent and Cotangent Functions ▪ Addition of Ordinates

Graphs of the Cosecant and Secant Functions Since cosecant values are reciprocals of the corresponding sine values, the period of the function $y = \csc x$ is 2π, the same as for $y = \sin x$. When $\sin x = 1$, the value of $\csc x$ is also 1, and when $0 < \sin x < 1$, then $\csc x > 1$. Also, if $-1 < \sin x < 0$, then $\csc x < -1$. (Verify these statements.) As $|x|$ approaches 0, $|\sin x|$ approaches 0, and $|\csc x|$ gets larger and larger. The graph of $y = \csc x$ approaches the vertical line $x = 0$ but never touches it, so the line $x = 0$ is a *vertical asymptote*. In fact, the lines $x = n\pi$, where n is any integer, are all vertical asymptotes.

Using this information and plotting a few points shows that the graph takes the shape of the solid curve shown in Figure 22. To show how the two graphs are related, the graph of $y = \sin x$ is shown as a dashed curve.

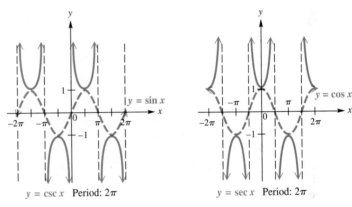

$y = \csc x$ Period: 2π
Figure 22

$y = \sec x$ Period: 2π
Figure 23

A similar analysis for the secant leads to the solid curve shown in Figure 23. The dashed curve, $y = \cos x$, is shown so that the relationship between these two reciprocal functions can be seen.

Typically, calculators do not have keys for the cosecant and secant functions. To graph $y = \csc x$ with a graphing calculator, use the fact that

$$\csc x = \frac{1}{\sin x}.$$

The graphs of $Y_1 = \sin X$ and $Y_2 = \csc X$ are shown in Figure 24. The calculator is in split screen and connected modes. Similarly, the secant function is graphed by using the identity

$$\sec x = \frac{1}{\cos x},$$

as shown in Figure 25.

Using dot mode for graphing will eliminate the vertical lines that appear in Figures 24 and 25. While they suggest asymptotes and are sometimes called *pseudo-asymptotes,* they are not actually parts of the graphs. See Figure 26 on the next page, for example. ▪

$Y_1 = \sin X$ $Y_2 = \csc X$

Trig window; connected mode

Figure 24

$Y_1 = \cos X$ $Y_2 = \sec X$

Trig window; connected mode

Figure 25

COSECANT FUNCTION $f(x) = \csc x$

Domain: $\{x \mid x \neq n\pi, \text{ where } n \text{ is an integer}\}$ Range: $(-\infty, -1] \cup [1, \infty)$

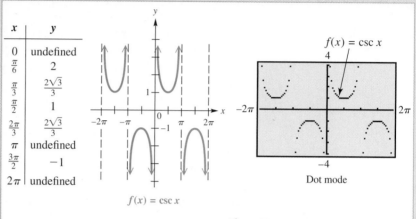

x	y
0	undefined
$\frac{\pi}{6}$	2
$\frac{\pi}{3}$	$\frac{2\sqrt{3}}{3}$
$\frac{\pi}{2}$	1
$\frac{2\pi}{3}$	$\frac{2\sqrt{3}}{3}$
π	undefined
$\frac{3\pi}{2}$	-1
2π	undefined

$f(x) = \csc x$

Dot mode

Figure 26

- The graph is discontinuous at values of x of the form $x = n\pi$ and has vertical asymptotes at these values.
- There are no x-intercepts.
- Its period is 2π.
- Its graph has no amplitude, since there are no maximum or minimum values.
- The graph is symmetric with respect to the origin, so the function is an odd function. For all x in the domain, $\csc(-x) = -\csc x$.

SECANT FUNCTION $f(x) = \sec x$

Domain: $\left\{x \mid x \neq (2n+1)\frac{\pi}{2}, \text{ where } n \text{ is an integer}\right\}$ Range: $(-\infty, -1] \cup [1, \infty)$

x	y
$-\frac{\pi}{2}$	undefined
$-\frac{\pi}{4}$	$\sqrt{2}$
0	1
$\frac{\pi}{4}$	$\sqrt{2}$
$\frac{\pi}{2}$	undefined
$\frac{3\pi}{4}$	$-\sqrt{2}$
π	-1
$\frac{3\pi}{2}$	undefined

$f(x) = \sec x$

Dot mode

Figure 27

(continued)

- The graph is discontinuous at values of x of the form $x = (2n + 1)\frac{\pi}{2}$ and has vertical asymptotes at these values.
- There are no x-intercepts.
- Its period is 2π.
- Its graph has no amplitude, since there are no maximum or minimum values.
- The graph is symmetric with respect to the y-axis, so the function is an even function. For all x in the domain, $\sec(-x) = \sec x$.

In the previous sections, we gave guidelines for sketching graphs of sine and cosine functions. We now present similar guidelines for graphing cosecant and secant functions.

Guidelines for Sketching Graphs of Cosecant and Secant Functions

To graph $y = a \csc bx$ or $y = a \sec bx$, with $b > 0$, follow these steps.

Step 1 Graph the corresponding reciprocal function as a guide, using a dashed curve.

To Graph	Use as a Guide
$y = a \csc bx$	$y = a \sin bx$
$y = a \sec bx$	$y = a \cos bx$

Step 2 Sketch the vertical asymptotes. They will have equations of the form $x = k$, where k is an x-intercept of the graph of the guide function.

Step 3 Sketch the graph of the desired function by drawing the typical U-shaped branches between the adjacent asymptotes. The branches will be above the graph of the guide function when the guide function values are positive and below the graph of the guide function when the guide function values are negative. The graph will resemble those in Figures 26 and 27 in the function boxes on the previous page.

Like graphs of the sine and cosine functions, graphs of the secant and cosecant functions may be translated vertically and horizontally. The period of both basic functions is 2π.

EXAMPLE 1 Graphing $y = a \sec bx$

Graph $y = 2 \sec \dfrac{1}{2}x$.

Solution

Step 1 This function involves the secant, so the corresponding reciprocal function will involve the cosine. The guide function to graph is

$$y = 2 \cos \frac{1}{2}x.$$

Using the guidelines of Section 4.1, we find that this guide function has amplitude 2 and one period of the graph lies along the interval that satisfies the inequality

$$0 \le \frac{1}{2}x \le 2\pi, \quad \text{or} \quad [0, 4\pi]. \quad \text{(Appendix A)}$$

Dividing this interval into four equal parts gives the key points

$$(0, 2), \qquad (\pi, 0), \qquad (2\pi, -2), \qquad (3\pi, 0), \qquad (4\pi, 2),$$

which are joined with a dashed red curve to indicate that this graph is only a guide. An additional period is graphed as seen in Figure 28(a).

(a)   (b)

Figure 28

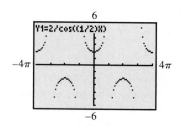

Dot mode

This is a calculator graph of the function in Example 1.

Step 2 Sketch the vertical asymptotes. These occur at *x*-values for which the guide function equals 0, such as

$$x = -3\pi, \qquad x = -\pi, \qquad x = \pi, \qquad x = 3\pi.$$

See Figure 28(a).

Step 3 Sketch the graph of $y = 2 \sec \frac{1}{2}x$ by drawing the typical U-shaped branches, approaching the asymptotes. See the solid blue graph in Figure 28(b).

Now try Exercise 7.

EXAMPLE 2 Graphing $y = a \csc(x - d)$

Graph $y = \dfrac{3}{2} \csc\left(x - \dfrac{\pi}{2}\right)$.

Solution

Step 1 Use the guidelines of Section 4.2 to graph the corresponding recipro-cal function defined by

$$y = \frac{3}{2} \sin\left(x - \frac{\pi}{2}\right),$$

shown as a red dashed curve in Figure 29.

Step 2 Sketch the vertical asymptotes through the x-intercepts of the graph of $y = \frac{3}{2} \sin\left(x - \frac{\pi}{2}\right)$. These have the form $x = (2n + 1)\frac{\pi}{2}$, where n is an integer. See the black dashed lines in Figure 29.

Step 3 Sketch the graph of $y = \frac{3}{2} \csc\left(x - \frac{\pi}{2}\right)$ by drawing the typical U-shaped branches between adjacent asymptotes. See the solid blue graph in Figure 29.

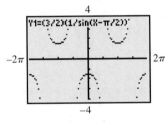

Dot mode

This is a calculator graph of the function in Example 2.

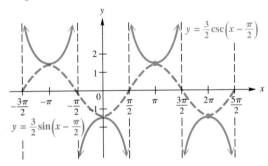

Figure 29

Now try Exercise 9.

Graphs of the Tangent and Cotangent Functions

Unlike the four functions whose graphs we studied previously, the tangent function has period π. Because $\tan x = \frac{\sin x}{\cos x}$, tangent values are 0 when sine values are 0, and un-defined when cosine values are 0. As x-values go from $-\frac{\pi}{2}$ to $\frac{\pi}{2}$, tangent values go from $-\infty$ to ∞ and increase throughout the interval. Those same values are repeated as x goes from $\frac{\pi}{2}$ to $\frac{3\pi}{2}$, $\frac{3\pi}{2}$ to $\frac{5\pi}{2}$, and so on. The graph of $y = \tan x$ from $-\pi$ to $\frac{3\pi}{2}$ is shown in Figure 30.

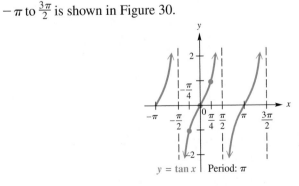

$y = \tan x$ Period: π

Figure 30

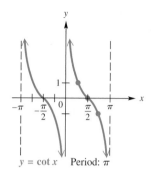

$y = \cot x$ Period: π

Figure 31

The cotangent function also has period π. Cotangent values are 0 when cosine values are 0, and undefined when sine values are 0. As x-values go from 0 to π, cotangent values go from ∞ to $-\infty$ and decrease throughout the interval. Those same values are repeated as x goes from π to 2π, 2π to 3π, and so on. The graph of $y = \cot x$ from $-\pi$ to π is shown in Figure 31.

TANGENT FUNCTION $f(x) = \tan x$

Domain: $\left\{x \mid x \neq (2n + 1)\frac{\pi}{2}, \text{ where } n \text{ is an integer}\right\}$ Range: $(-\infty, \infty)$

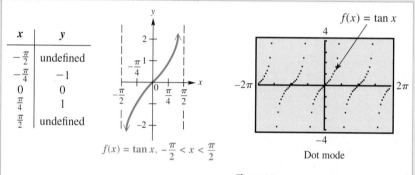

x	y
$-\frac{\pi}{2}$	undefined
$-\frac{\pi}{4}$	-1
0	0
$\frac{\pi}{4}$	1
$\frac{\pi}{2}$	undefined

$f(x) = \tan x, \ -\frac{\pi}{2} < x < \frac{\pi}{2}$

Dot mode

Figure 32

- The graph is discontinuous at values of x of the form $x = (2n + 1)\frac{\pi}{2}$ and has vertical asymptotes at these values.
- Its x-intercepts are of the form $x = n\pi$.
- Its period is π.
- Its graph has no amplitude, since there are no minimum or maximum values.
- The graph is symmetric with respect to the origin, so the function is an odd function. For all x in the domain, $\tan(-x) = -\tan x$.

COTANGENT FUNCTION $f(x) = \cot x$

Domain: $\{x \mid x \neq n\pi, \text{ where } n \text{ is an integer}\}$ Range: $(-\infty, \infty)$

x	y
0	undefined
$\frac{\pi}{4}$	1
$\frac{\pi}{2}$	0
$\frac{3\pi}{4}$	-1
π	undefined

$f(x) = \cot x, \ 0 < x < \pi$

Dot mode

Figure 33

(continued)

- The graph is discontinuous at values of x of the form $x = n\pi$ and has vertical asymptotes at these values.
- Its x-intercepts are of the form $x = (2n + 1)\frac{\pi}{2}$.
- Its period is π.
- Its graph has no amplitude, since there are no minimum or maximum values.
- The graph is symmetric with respect to the origin, so the function is an odd function. For all x in the domain, $\cot(-x) = -\cot x$.

The tangent function can be graphed directly with a graphing calculator, using the tangent key. To graph the cotangent function, however, we must use one of the identities $\cot x = \frac{1}{\tan x}$ or $\cot x = \frac{\cos x}{\sin x}$ since graphing calculators generally do not have cotangent keys. ∎

Guidelines for Sketching Graphs of Tangent and Cotangent Functions

To graph $y = a \tan bx$ or $y = a \cot bx$, with $b > 0$, follow these steps.

Step 1 Determine the period, $\frac{\pi}{b}$. To locate two adjacent vertical asymptotes, solve the following equations for x:

For $y = a \tan bx$: $\quad bx = -\dfrac{\pi}{2}$ and $bx = \dfrac{\pi}{2}$.

For $y = a \cot bx$: $\quad bx = 0$ and $bx = \pi$.

Step 2 Sketch the two vertical asymptotes found in Step 1.

Step 3 Divide the interval formed by the vertical asymptotes into four equal parts.

Step 4 Evaluate the function for the first-quarter point, midpoint, and third-quarter point, using the x-values found in Step 3.

Step 5 Join the points with a smooth curve, approaching the vertical asymptotes. Indicate additional asymptotes and periods of the graph as necessary.

EXAMPLE 3 Graphing $y = \tan bx$

Graph $y = \tan 2x$.

Solution

Step 1 The period of this function is $\frac{\pi}{2}$. To locate two adjacent vertical asymptotes, solve $2x = -\frac{\pi}{2}$ and $2x = \frac{\pi}{2}$ (since this is a tangent function). The two asymptotes have equations $x = -\frac{\pi}{4}$ and $x = \frac{\pi}{4}$.

Step 2 Sketch the two vertical asymptotes $x = \pm\frac{\pi}{4}$, as shown in Figure 34 on the next page.

Step 3 Divide the interval $\left(-\frac{\pi}{4}, \frac{\pi}{4}\right)$ into four equal parts. This gives the following key x-values.

first-quarter value: $-\dfrac{\pi}{8}$, middle value: 0, third-quarter value: $\dfrac{\pi}{8}$

Step 4 Evaluate the function for the x-values found in Step 3.

x	$-\frac{\pi}{8}$	0	$\frac{\pi}{8}$
$2x$	$-\frac{\pi}{4}$	0	$\frac{\pi}{4}$
$\tan 2x$	-1	0	1

Step 5 Join these points with a smooth curve, approaching the vertical asymptotes. See Figure 34. Another period has been graphed, one half period to the left and one half period to the right.

Now try Exercise 21.

$y = \tan 2x$ Period: $\frac{\pi}{2}$

Figure 34

EXAMPLE 4   Graphing $y = a \tan bx$

Graph $y = -3 \tan \dfrac{1}{2}x$.

Solution   The period is $\dfrac{\pi}{\frac{1}{2}} = \pi \div \frac{1}{2} = \pi \cdot \frac{2}{1} = 2\pi$. Adjacent asymptotes are at $x = -\pi$ and $x = \pi$. Dividing the interval $(-\pi, \pi)$ into four equal parts gives key x-values of $-\frac{\pi}{2}$, 0, and $\frac{\pi}{2}$. Evaluating the function at these x-values gives the following key points.

$$\left(-\frac{\pi}{2}, 3\right), \qquad (0, 0), \qquad \left(\frac{\pi}{2}, -3\right)$$

By plotting these points and joining them with a smooth curve, we obtain the graph shown in Figure 35. Because the coefficient -3 is negative, the graph is reflected across the x-axis compared to the graph of $y = 3 \tan \frac{1}{2}x$.

Now try Exercise 29.

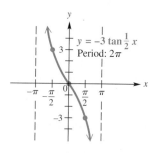

$y = -3 \tan \frac{1}{2} x$
Period: 2π

Figure 35

N O T E   The function defined by $y = -3 \tan \frac{1}{2}x$ in Example 4, graphed in Figure 35, has a graph that compares to the graph of $y = \tan x$ as follows.

1. The period is larger because $b = \frac{1}{2}$, and $\frac{1}{2} < 1$.

2. The graph is "stretched" because $a = -3$, and $|-3| > 1$.

3. Each branch of the graph goes down from left to right (that is, the function decreases) between each pair of adjacent asymptotes because $a = -3 < 0$. When $a < 0$, the graph is reflected across the x-axis compared to the graph of $y = |a| \tan bx$.

EXAMPLE 5 Graphing $y = a \cot bx$

Graph $y = \dfrac{1}{2} \cot 2x$.

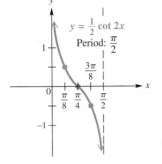

Figure 36

Solution Because this function involves the cotangent, we can locate two adjacent asymptotes by solving the equations $2x = 0$ and $2x = \pi$. The lines $x = 0$ (the y-axis) and $x = \frac{\pi}{2}$ are two such asymptotes. Divide the interval $\left(0, \frac{\pi}{2}\right)$ into four equal parts, getting key x-values of $\frac{\pi}{8}, \frac{\pi}{4}$, and $\frac{3\pi}{8}$. Evaluating the function at these x-values gives the following key points.

$$\left(\frac{\pi}{8}, \frac{1}{2}\right), \qquad \left(\frac{\pi}{4}, 0\right), \qquad \left(\frac{3\pi}{8}, -\frac{1}{2}\right)$$

Joining these points with a smooth curve approaching the asymptotes gives the graph shown in Figure 36.

Now try Exercise 31.

Like the other circular functions, the graphs of the tangent and cotangent functions may be translated horizontally and vertically.

EXAMPLE 6 Graphing a Tangent Function with a Vertical Translation

Graph $y = 2 + \tan x$.

Analytic Solution

Every value of y for this function will be 2 units more than the corresponding value of y in $y = \tan x$, causing the graph of $y = 2 + \tan x$ to be translated 2 units up compared with the graph of $y = \tan x$. See Figure 37.

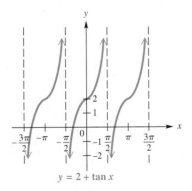

$y = 2 + \tan x$

Figure 37

Graphing Calculator Solution

To see the vertical translation, observe the coordinates displayed at the bottoms of the screens in Figures 38 and 39. For $X = \frac{\pi}{4} \approx .78539816$,

$$Y_1 = \tan X = 1,$$

while for the same X-value,

$$Y_2 = 2 + \tan X = 2 + 1 = 3.$$

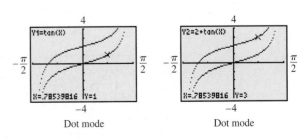

Dot mode   Dot mode

Figure 38   **Figure 39**

Now try Exercise 37.

EXAMPLE 7 Graphing a Cotangent Function with Vertical and Horizontal Translations

Graph $y = -2 - \cot\left(x - \dfrac{\pi}{4}\right)$.

Solution Here $b = 1$, so the period is π. The graph will be translated down 2 units (because $c = -2$), reflected across the x-axis (because of the negative sign in front of the cotangent), and will have a phase shift (horizontal translation) $\frac{\pi}{4}$ unit to the right $\left(\text{because of the argument } \left(x - \frac{\pi}{4}\right)\right)$. To locate adjacent asymptotes, since this function involves the cotangent, we solve the following equations:

$$x - \frac{\pi}{4} = 0, \quad \text{so } x = \frac{\pi}{4} \quad \text{and} \quad x - \frac{\pi}{4} = \pi, \quad \text{so } x = \frac{5\pi}{4}.$$

Dividing the interval $\left(\frac{\pi}{4}, \frac{5\pi}{4}\right)$ into four equal parts and evaluating the function at the three key x-values within the interval gives these points.

$$\left(\frac{\pi}{2}, -3\right), \qquad \left(\frac{3\pi}{4}, -2\right), \qquad (\pi, -1)$$

Join these points with a smooth curve. This period of the graph, along with the one in the domain interval $\left(-\frac{3\pi}{4}, \frac{\pi}{4}\right)$, is shown in Figure 40.

Figure 40

$y = -2 - \cot\left(x - \frac{\pi}{4}\right)$

Now try Exercise 45.

Addition of Ordinates New functions can be formed by adding or subtracting other functions. A function formed by combining two other functions, such as

$$y = \cos x + \sin x,$$

has historically been graphed using a method known as *addition of ordinates*. (The x-value of a point is sometimes called its *abscissa,* while its y-value is called its *ordinate.*) To apply this method to this function, we graph the functions $y = \cos x$ and $y = \sin x$. Then, for selected values of x, we add $\cos x$ and $\sin x$, and plot the points $(x, \cos x + \sin x)$. Joining the resulting points with a sinusoidal curve gives the graph of the desired function. While this method illustrates some valuable concepts involving the arithmetic of functions, it is time-consuming.

With graphing calculators, this technique is easily illustrated. Let $Y_1 = \cos X$, $Y_2 = \sin X$, and $Y_3 = Y_1 + Y_2$. Figure 41 shows the result when Y_1 and Y_2 are graphed in thin graph style, and $Y_3 = \cos X + \sin X$ is graphed in thick graph style. Notice that for $X = \frac{\pi}{6} \approx .52359878$, $Y_1 + Y_2 = Y_3$.

Figure 41

Now try Exercise 61.

4.3 Exercises

Concept Check   *In Exercises 1–6, match each function with its graph from choices A–F.*

1. $y = -\csc x$

2. $y = -\sec x$

3. $y = -\tan x$

4. $y = -\cot x$

5. $y = \tan\left(x - \dfrac{\pi}{4}\right)$

6. $y = \cot\left(x - \dfrac{\pi}{4}\right)$

A.

B.

C.

D.

E.

F.

Graph each function over a one-period interval. See Examples 1 and 2.

7. $y = 3\sec\dfrac{1}{4}x$

8. $y = -2\sec\dfrac{1}{2}x$

9. $y = -\dfrac{1}{2}\csc\left(x + \dfrac{\pi}{2}\right)$

10. $y = \dfrac{1}{2}\csc\left(x - \dfrac{\pi}{2}\right)$

11. $y = \csc\left(x - \dfrac{\pi}{4}\right)$

12. $y = \sec\left(x + \dfrac{3\pi}{4}\right)$

13. $y = \sec\left(x + \dfrac{\pi}{4}\right)$

14. $y = \csc\left(x + \dfrac{\pi}{3}\right)$

15. $y = \sec\left(\dfrac{1}{2}x + \dfrac{\pi}{3}\right)$

16. $y = \csc\left(\dfrac{1}{2}x - \dfrac{\pi}{4}\right)$

17. $y = 2 + 3\sec(2x - \pi)$

18. $y = 1 - 2\csc\left(x + \dfrac{\pi}{2}\right)$

19. $y = 1 - \dfrac{1}{2}\csc\left(x - \dfrac{3\pi}{4}\right)$

20. $y = 2 + \dfrac{1}{4}\sec\left(\dfrac{1}{2}x - \pi\right)$

Graph each function over a one-period interval. See Examples 3–5.

21. $y = \tan 4x$

22. $y = \tan\dfrac{1}{2}x$

23. $y = 2\tan x$

24. $y = 2\cot x$

25. $y = 2\tan\dfrac{1}{4}x$

26. $y = \dfrac{1}{2}\cot x$

27. $y = \cot 3x$

28. $y = -\cot\dfrac{1}{2}x$

29. $y = -2\tan\dfrac{1}{4}x$

30. $y = 3\tan\dfrac{1}{2}x$

31. $y = \dfrac{1}{2}\cot 4x$

32. $y = -\dfrac{1}{2}\cot 2x$

Graph each function over a two-period interval. See Examples 6 and 7.

33. $y = \tan(2x - \pi)$

34. $y = \tan\left(\dfrac{x}{2} + \pi\right)$

35. $y = \cot\left(3x + \dfrac{\pi}{4}\right)$

36. $y = \cot\left(2x - \dfrac{3\pi}{2}\right)$ **37.** $y = 1 + \tan x$ **38.** $y = 1 - \tan x$

39. $y = 1 - \cot x$ **40.** $y = -2 - \cot x$

41. $y = -1 + 2\tan x$ **42.** $y = 3 + \dfrac{1}{2}\tan x$

43. $y = -1 + \dfrac{1}{2}\cot(2x - 3\pi)$ **44.** $y = -2 + 3\tan(4x + \pi)$

45. $y = 1 - 2\cot 2\left(x + \dfrac{\pi}{2}\right)$ **46.** $y = \dfrac{2}{3}\tan\left(\dfrac{3}{4}x - \pi\right) - 2$

Concept Check *In Exercises 47–52, tell whether each statement is* true *or* false. *If* false, *tell why.*

47. The smallest positive number k for which $x = k$ is an asymptote for the tangent function is $\frac{\pi}{2}$.

48. The smallest positive number k for which $x = k$ is an asymptote for the cotangent function is $\frac{\pi}{2}$.

49. The tangent and secant functions are undefined for the same values.

50. The secant and cosecant functions are undefined for the same values.

51. The graph of $y = \tan x$ in Figure 32 suggests that $\tan(-x) = \tan x$ for all x in the domain of $\tan x$.

52. The graph of $y = \sec x$ in Figure 27 suggests that $\sec(-x) = \sec x$ for all x in the domain of $\sec x$.

53. *Concept Check* If c is any number, then how many solutions does the equation $c = \tan x$ have in the interval $(-2\pi, 2\pi]$?

54. *Concept Check* If c is any number such that $-1 < c < 1$, then how many solutions does the equation $c = \sec x$ have over the entire domain of the secant function?

55. Consider the function defined by $f(x) = -4\tan(2x + \pi)$. What is the domain of f? What is its range?

56. Consider the function defined by $g(x) = -2\csc(4x + \pi)$. What is the domain of g? What is its range?

(Modeling) *Solve each problem.*

57. *Distance of a Rotating Beacon* A rotating beacon is located at point A next to a long wall. (See the figure.) The beacon is 4 m from the wall. The distance d is given by

$$d = 4\tan 2\pi t,$$

where t is time measured in seconds since the beacon started rotating. (When $t = 0$, the beacon is aimed at point R. When the beacon is aimed to the right of R, the value of d is positive; d is negative if the beacon is aimed to the left of R.) Find d for each time.

(a) $t = 0$
(b) $t = .4$
(c) $t = .8$
(d) $t = 1.2$
(e) Why is .25 a meaningless value for t?

58. *Distance of a Rotating Beacon* In the figure for Exercise 57, the distance a is given by

$$a = 4|\sec 2\pi t|.$$

Find a for each time.

(a) $t = 0$ (b) $t = .86$ (c) $t = 1.24$

59. Simultaneously graph $y = \tan x$ and $y = x$ in the window $[-1, 1]$ by $[-1, 1]$ with a graphing calculator. Write a sentence or two describing the relationship of $\tan x$ and x for small x-values.

60. Between each pair of successive asymptotes, a portion of the graph of $y = \sec x$ or $y = \csc x$ resembles a parabola. Can each of these portions actually be a parabola? Explain.

Use a graphing calculator to graph Y_1, Y_2, and $Y_1 + Y_2$ on the same screen. Evaluate each of the three functions at $X = \frac{\pi}{6}$, and verify that $Y_1\left(\frac{\pi}{6}\right) + Y_2\left(\frac{\pi}{6}\right) = (Y_1 + Y_2)\left(\frac{\pi}{6}\right)$. See the discussion on addition of ordinates.

61. $Y_1 = \sin X$, $Y_2 = \sin 2X$ **62.** $Y_1 = \cos X$, $Y_2 = \sec X$

Relating Concepts

For individual or collaborative investigation
(Exercises 63–68)

Consider the function defined by $y = -2 - \cot\left(x - \frac{\pi}{4}\right)$ from Example 7. **Work these exercises in order.**

63. What is the smallest positive number for which $y = \cot x$ is undefined?

64. Let k represent the number you found in Exercise 63. Set $x - \frac{\pi}{4}$ equal to k, and solve to find a positive number for which $\cot\left(x - \frac{\pi}{4}\right)$ is undefined.

65. Based on your answer in Exercise 64 and the fact that the cotangent function has period π, give the general form of the equations of the asymptotes of the graph of $y = -2 - \cot\left(x - \frac{\pi}{4}\right)$. Let n represent any integer.

66. Use the capabilities of your calculator to find the smallest positive x-intercept of the graph of this function.

67. Use the fact that the period of this function is π to find the next positive x-intercept.

68. Give the solution set of the equation $-2 - \cot\left(x - \frac{\pi}{4}\right) = 0$ over all real numbers. Let n represent any integer.

Summary Exercises on Graphing Circular Functions

These summary exercises provide practice with the various graphing techniques presented in this chapter. Graph each function over a one-period interval.

1. $y = 2 \sin \pi x$

2. $y = 4 \cos 1.5x$

3. $y = -2 + .5 \cos \dfrac{\pi}{4} x$

4. $y = 3 \sec \dfrac{\pi x}{2}$

5. $y = -4 \csc .5x$

6. $y = 3 \tan\left(\dfrac{\pi x}{2} + \pi\right)$

Graph each function over a two-period interval.

7. $y = -5 \sin \dfrac{x}{3}$

8. $y = 10 \cos\left(\dfrac{x}{4} + \dfrac{\pi}{2}\right)$

9. $y = 3 - 4 \sin(2.5x + \pi)$

10. $y = 2 - \sec[\pi(x - 3)]$

4.4 Harmonic Motion

Simple Harmonic Motion ▪ **Damped Oscillatory Motion**

Simple Harmonic Motion   In part A of Figure 42, a spring with a weight attached to its free end is in equilibrium (or rest) position. If the weight is pulled down a units and released (part B of the figure), the spring's elasticity causes the weight to rise a units ($a > 0$) above the equilibrium position, as seen in part C, and then oscillate about the equilibrium position. If friction is neglected, this oscillatory motion is described mathematically by a sinusoid. Other applications of this type of motion include sound, electric current, and electromagnetic waves.

Figure 42

To develop a general equation for such motion, consider Figure 43. Suppose the point $P(x, y)$ moves around the circle counterclockwise at a uniform angular speed ω. Assume that at time $t = 0$, P is at $(a, 0)$. The angle swept out by ray OP at time t is given by $\theta = \omega t$. The coordinates of point P at time t are

$$x = a \cos \theta = a \cos \omega t \qquad \text{and} \qquad y = a \sin \theta = a \sin \omega t.$$

As P moves around the circle from the point $(a, 0)$, the point $Q(0, y)$ oscillates back and forth along the y-axis between the points $(0, a)$ and $(0, -a)$. Similarly, the point $R(x, 0)$ oscillates back and forth between $(a, 0)$ and $(-a, 0)$. This oscillatory motion is called **simple harmonic motion.**

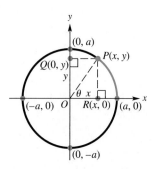

Figure 43

The amplitude of the motion is $|a|$, and the period is $\frac{2\pi}{\omega}$. The moving points P and Q or P and R complete one oscillation or cycle per period. The number of cycles per unit of time, called the **frequency,** is the reciprocal of the period, $\frac{\omega}{2\pi}$, where $\omega > 0$.

Simple Harmonic Motion

The position of a point oscillating about an equilibrium position at time t is modeled by either

$$s(t) = a \cos \omega t \qquad \text{or} \qquad s(t) = a \sin \omega t,$$

where a and ω are constants, with $\omega > 0$. The amplitude of the motion is $|a|$, the period is $\frac{2\pi}{\omega}$, and the frequency is $\frac{\omega}{2\pi}$.

EXAMPLE 1   Modeling the Motion of a Spring

Suppose that an object is attached to a coiled spring such as the one in Figure 42. It is pulled down a distance of 5 in. from its equilibrium position, and then released. The time for one complete oscillation is 4 sec.

(a) Give an equation that models the position of the object at time t.

(b) Determine the position at $t = 1.5$ sec.

(c) Find the frequency.

Solution

(a) When the object is released at $t = 0$, the distance of the object from the equilibrium position is 5 in. below equilibrium. If $s(t)$ is to model the motion, then $s(0)$ must equal -5. We use

$$s(t) = a \cos \omega t,$$

with $a = -5$. We choose the cosine function because $\cos \omega(0) = \cos 0 = 1$, and $-5 \cdot 1 = -5$. (Had we chosen the sine function, a phase shift would have been required.) The period is 4, so

$$\frac{2\pi}{\omega} = 4, \qquad \text{or} \qquad \omega = \frac{\pi}{2}. \quad \text{Solve for } \omega. \text{ (Appendix A)}$$

Thus, the motion is modeled by

$$s(t) = -5 \cos \frac{\pi}{2} t.$$

(b) After 1.5 sec, the position is

$$s(1.5) = -5 \cos \left[\frac{\pi}{2}(1.5) \right] \approx 3.54 \text{ in.} \quad \text{(Appendix C)}$$

Since $3.54 > 0$, the object is above the equilibrium position.

(c) The frequency is the reciprocal of the period, or $\frac{1}{4}$.

Now try Exercise 9.

EXAMPLE 2 Analyzing Harmonic Motion

Suppose that an object oscillates according to the model

$$s(t) = 8 \sin 3t,$$

where t is in seconds and $s(t)$ is in feet. Analyze the motion.

Solution The motion is harmonic because the model is of the form $s(t) = a \sin \omega t$. Because $a = 8$, the object oscillates 8 ft in either direction from its starting point. The period $\frac{2\pi}{3} \approx 2.1$ is the time, in seconds, it takes for one complete oscillation. The frequency is the reciprocal of the period, so the object completes $\frac{3}{2\pi} \approx .48$ oscillation per sec.

Now try Exercise 15.

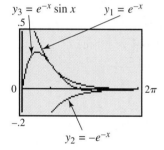

Figure 44

Damped Oscillatory Motion In the example of the stretched spring, we disregard the effect of friction. Friction causes the amplitude of the motion to diminish gradually until the weight comes to rest. In this situation, we say that the motion has been *damped* by the force of friction. Most oscillatory motions are damped, and the decrease in amplitude follows the pattern of exponential decay. A typical example of **damped oscillatory motion** is provided by the function defined by

$$s(t) = e^{-t} \sin t.$$

Figure 44 shows how the graph of $y_3 = e^{-x} \sin x$ is bounded above by the graph of $y_1 = e^{-x}$ and below by the graph of $y_2 = -e^{-x}$. The damped motion curve dips below the x-axis at $x = \pi$ but stays above the graph of y_2. Figure 45 shows a traditional graph of $s(t) = e^{-t} \sin t$, along with the graph of $y = \sin t$.

Figure 45

Shock absorbers are put on an automobile in order to damp oscillatory motion. Instead of oscillating up and down for a long while after hitting a bump or pothole, the oscillations of the car are quickly damped out for a smoother ride.

Now try Exercise 21.

4.4 Exercises

(Modeling) Springs   *Suppose that a weight on a spring has initial position s(0) and period P.*

(a) *Find a function s given by $s(t) = a \cos \omega t$ that models the displacement of the weight.*

(b) *Evaluate s(1). Is the weight moving upward, downward, or neither when $t = 1$? Support your results graphically or numerically.*

1. $s(0) = 2$ in.; $P = .5$ sec   **2.** $s(0) = 5$ in.; $P = 1.5$ sec

3. $s(0) = -3$ in.; $P = .8$ sec   **4.** $s(0) = -4$ in.; $P = 1.2$ sec

(Modeling) Music   *A note on the piano has given frequency F. Suppose the maximum displacement at the center of the piano wire is given by s(0). Find constants a and ω so that the equation $s(t) = a \cos \omega t$ models this displacement. Graph s in the viewing window $[0, .05]$ by $[-.3, .3]$.*

5. $F = 27.5$; $s(0) = .21$ **6.** $F = 110$; $s(0) = .11$

7. $F = 55$; $s(0) = .14$ **8.** $F = 220$; $s(0) = .06$

(Modeling)   *Solve each problem. See Examples 1 and 2.*

9. *Spring*   An object is attached to a coiled spring, as in Figure 42. It is pulled down a distance of 4 units from its equilibrium position, and then released. The time for one complete oscillation is 3 sec.

   (a) Give an equation that models the position of the object at time t.
   (b) Determine the position at $t = 1.25$ sec.
   (c) Find the frequency.

10. *Spring*   Repeat Exercise 9, but assume that the object is pulled down 6 units and the time for one complete oscillation is 4 sec.

11. *Particle Movement*   Write the equation and then determine the amplitude, period, and frequency of the simple harmonic motion of a particle moving uniformly around a circle of radius 2 units, with angular speed

   (a) 2 radians per sec   **(b)** 4 radians per sec.

12. *Pendulum*   What are the period P and frequency T of oscillation of a pendulum of length $\frac{1}{2}$ ft? $\left(Hint: P = 2\pi\sqrt{\frac{L}{32}}, \text{ where } L \text{ is the length of the pendulum in feet and } P \text{ is in seconds.}\right)$

13. *Pendulum*   In Exercise 12, how long should the pendulum be to have period 1 sec?

14. *Spring*   The formula for the up and down motion of a weight on a spring is given by

$$s(t) = a \sin\sqrt{\frac{k}{m}}\,t.$$

If the spring constant k is 4, what mass m must be used to produce a period of 1 sec?

15. *Spring*   The height attained by a weight attached to a spring set in motion is

$$s(t) = -4 \cos 8\pi t$$

inches after t seconds.

   (a) Find the maximum height that the weight rises above the equilibrium position of $y = 0$.
   (b) When does the weight first reach its maximum height, if $t \geq 0$?
   (c) What are the frequency and period?

16. *Spring* (See Exercise 14.) A spring with spring constant $k = 2$ and a 1-unit mass m attached to it is stretched and then allowed to come to rest.

(a) If the spring is stretched $\frac{1}{2}$ ft and released, what are the amplitude, period, and frequency of the resulting oscillatory motion?

(b) What is the equation of the motion?

17. *Spring* The position of a weight attached to a spring is

$$s(t) = -5 \cos 4\pi t$$

inches after t seconds.

(a) What is the maximum height that the weight rises above the equilibrium position?

(b) What are the frequency and period?

(c) When does the weight first reach its maximum height?

(d) Calculate and interpret $s(1.3)$.

18. *Spring* The position of a weight attached to a spring is

$$s(t) = -4 \cos 10t$$

inches after t seconds.

(a) What is the maximum height that the weight rises above the equilibrium position?

(b) What are the frequency and period?

(c) When does the weight first reach its maximum height?

(d) Calculate and interpret $s(1.466)$.

19. *Spring* A weight attached to a spring is pulled down 3 in. below the equilibrium position.

(a) Assuming that the frequency is $\frac{6}{\pi}$ cycles per sec, determine a model that gives the position of the weight at time t seconds.

(b) What is the period?

20. *Spring* A weight attached to a spring is pulled down 2 in. below the equilibrium position.

(a) Assuming that the period is $\frac{1}{3}$ sec, determine a model that gives the position of the weight at time t seconds.

(b) What is the frequency?

Use a graphing calculator to graph $y_1 = e^{-t} \sin t$, $y_2 = e^{-t}$, and $y_3 = -e^{-t}$ in the viewing window $[0, \pi]$ by $[-.5, .5]$.

21. Find the t-intercepts of the graph of y_1. Explain the relationship of these intercepts with the x-intercepts of the graph of $y = \sin x$.

22. Find any points of intersection of y_1 and y_2 or y_1 and y_3. How are these points related to the graph of $y = \sin x$?

Chapter 4 Summary

KEY TERMS

4.1 periodic function
period
sine wave (sinusoid)
amplitude

4.2 phase shift
argument
4.3 vertical asymptote
addition of ordinates

4.4 simple harmonic
motion
frequency

damped oscillatory
motion

QUICK REVIEW

CONCEPTS

EXAMPLES

4.1 Graphs of the Sine and Cosine Functions

4.2 Translations of the Graphs of the Sine and Cosine Functions

Sine and Cosine Functions

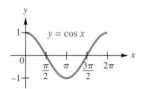

Domain: $(-\infty, \infty)$
Range: $[-1, 1]$
Amplitude: 1
Period: 2π

Domain: $(-\infty, \infty)$
Range: $[-1, 1]$
Amplitude: 1
Period: 2π

The graph of

$$y = c + a \sin b(x - d) \quad \text{or} \quad y = c + a \cos b(x - d),$$

$b > 0$, has

1. amplitude $|a|$, 2. period $\frac{2\pi}{b}$,
3. vertical translation c units up if $c > 0$ or $|c|$ units down
   if $c < 0$, and
4. phase shift d units to the right if $d > 0$ or $|d|$ units to
   the left if $d < 0$.

See pages 138 and 149 for a summary of graphing
techniques.

Graph $y = \sin 3x$.

period: $\frac{2\pi}{3}$ amplitude: 1
domain: $(-\infty, \infty)$ range: $[-1, 1]$

Graph $y = -2 \cos x$.

period: 2π amplitude: 2
domain: $(-\infty, \infty)$ range: $[-2, 2]$

CONCEPTS	EXAMPLES

4.3 Graphs of the Other Circular Functions

Cosecant and Secant Functions

Domain: $\{x \mid x \neq n\pi,$
  where n is an integer$\}$
Range: $(-\infty, -1] \cup [1, \infty)$
Period: 2π

Domain: $\{x \mid x \neq (2n + 1)\frac{\pi}{2},$
  where n is an integer$\}$
Range: $(-\infty, -1] \cup [1, \infty)$
Period: 2π

See page 157 for a summary of graphing techniques.

Graph one period of $y = \sec\left(x + \frac{\pi}{4}\right)$.

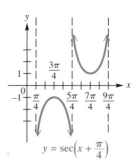

$$y = \sec\left(x + \frac{\pi}{4}\right)$$

period: 2π

phase shift: $-\frac{\pi}{4}$

domain: $\{x \mid x \neq \frac{\pi}{4} + n\pi,$
  where n is an integer$\}$

range: $(-\infty, -1] \cup [1, \infty)$

Tangent and Cotangent Functions

Domain: $\{x \mid x \neq (2n + 1)\frac{\pi}{2},$
  where n is an integer $\}$
Range: $(-\infty, \infty)$
Period: π

Domain: $\{x \mid x \neq n\pi,$
  where n is an integer$\}$
Range: $(-\infty, \infty)$
Period: π

See page 161 for a summary of graphing techniques.

Graph one period of $y = 2\tan x$.

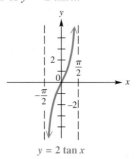

$$y = 2\tan x$$

period: π

domain: $\{x \mid x \neq (2n + 1)\frac{\pi}{2},$
  where n is an integer$\}$

range: $(-\infty, \infty)$

4.4 Harmonic Motion

Simple Harmonic Motion

The position of a point oscillating about an equilibrium position at time t is modeled by either

$$s(t) = a\cos\omega t \quad \text{or} \quad s(t) = a\sin\omega t,$$

where a and ω are constants, with $\omega > 0$. The amplitude of the motion is $|a|$, the period is $\frac{2\pi}{\omega}$, and the frequency is $\frac{\omega}{2\pi}$.

A spring oscillates according to

$$s(t) = -5\cos 6t,$$

where t is in seconds and $s(t)$ is in inches.

$$\text{amplitude} = |-5| = 5$$

$$\text{period} = \frac{2\pi}{6} = \frac{\pi}{3}$$

$$\text{frequency} = \frac{3}{\pi}$$

Chapter 4 Review Exercises

1. *Concept Check*   Which one of the following is true about the graph of $y = 4 \sin 2x$?
   A. It has amplitude 2 and period $\frac{\pi}{2}$.   B. It has amplitude 4 and period π.
   C. Its range is $[0, 4]$.   D. Its range is $[-4, 0]$.

2. *Concept Check*   Which one of the following is false about the graph of $y = -3 \cos \frac{1}{2}x$?
   A. Its range is $[-3, 3]$.
   B. Its domain is $(-\infty, \infty)$.
   C. Its amplitude is 3, and its period is 4π.
   D. Its amplitude is 3, and its period is π.

3. *Concept Check*   Which of the basic circular functions can have y-value $\frac{1}{2}$?

4. *Concept Check*   Which of the basic circular functions can have y-value 2?

For each function, give the amplitude, period, vertical translation, and phase shift, as applicable.

5. $y = 2 \sin x$

6. $y = \tan 3x$

7. $y = -\dfrac{1}{2} \cos 3x$

8. $y = 2 \sin 5x$

9. $y = 1 + 2 \sin \dfrac{1}{4}x$

10. $y = 3 - \dfrac{1}{4} \cos \dfrac{2}{3}x$

11. $y = 3 \cos\left(x + \dfrac{\pi}{2} \right)$

12. $y = -\sin\left(x - \dfrac{3\pi}{4} \right)$

13. $y = \dfrac{1}{2} \csc\left(2x - \dfrac{\pi}{4} \right)$

14. $y = 2 \sec(\pi x - 2\pi)$

15. $y = \dfrac{1}{3} \tan\left(3x - \dfrac{\pi}{3} \right)$

16. $y = \cot\left(\dfrac{x}{2} + \dfrac{3\pi}{4} \right)$

Concept Check   *Identify the circular function that satisfies each description.*

17. period is π, x-intercepts are of the form $n\pi$, where n is an integer

18. period is 2π, graph passes through the origin

19. period is 2π, graph passes through the point $\left(\frac{\pi}{2}, 0 \right)$

20. period is 2π, domain is $\{x \mid x \neq n\pi,$ where n is an integer$\}$

21. period is π, function is decreasing on the interval $(0, \pi)$

22. period is 2π, has vertical asymptotes of the form $x = (2n + 1)\frac{\pi}{2}$, where n is an integer

23. Suppose that f is a sine function with period 10 and $f(5) = 2$. Explain why $f(25) = 2$.

24. Suppose that f is a sine function with period π and $f\left(\frac{6\pi}{5}\right) = 1$. Explain why $f\left(-\frac{4\pi}{5}\right) = 1$.

Graph each function over a one-period interval.

25. $y = 3 \sin x$

26. $y = \dfrac{1}{2} \sec x$

27. $y = -\tan x$

28. $y = -2 \cos x$

29. $y = 2 + \cot x$

30. $y = -1 + \csc x$

31. $y = \sin 2x$

32. $y = \tan 3x$

33. $y = 3 \cos 2x$

34. $y = \dfrac{1}{2}\cot 3x$ **35.** $y = \cos\left(x - \dfrac{\pi}{4}\right)$ **36.** $y = \tan\left(x - \dfrac{\pi}{2}\right)$

37. $y = \sec\left(2x + \dfrac{\pi}{3}\right)$ **38.** $y = \sin\left(3x + \dfrac{\pi}{2}\right)$

39. $y = 1 + 2\cos 3x$ **40.** $y = -1 - 3\sin 2x$

41. $y = 2\sin \pi x$ **42.** $y = -\dfrac{1}{2}\cos(\pi x - \pi)$

43. Explain why a function of the form $f(x) = 2\sin(bx + c)$ has range $[-2, 2]$.

44. Explain why a function of the form $f(x) = 2\csc(bx + c)$ has range $(-\infty, -2] \cup [2, \infty)$.

Solve each problem.

45. *Viewing Angle to an Object* Let a person whose eyes are h_1 feet from the ground stand d feet from an object h_2 feet tall, where $h_2 > h_1$. Let θ be the angle of elevation to the top of the object. See the figure.

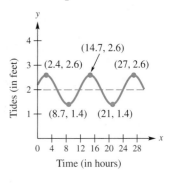

(a) Show that $d = (h_2 - h_1)\cot \theta$.
(b) Let $h_2 = 55$ and $h_1 = 5$. Graph d for the interval $0 < \theta \le \frac{\pi}{2}$.

46. *(Modeling) Tides* The figure shows a function f that models the tides in feet at Clearwater Beach, Florida, x hours after midnight starting on August 26, 1998. (*Source:* Pentcheff, D., *WWW Tide and Current Predictor.*)

(a) Find the time between high tides.
(b) What is the difference in water levels between high tide and low tide?
(c) The tides can be modeled by

$$f(x) = .6\cos[.511(x - 2.4)] + 2.$$

Estimate the tides when $x = 10$.

47. *(Modeling) Maximum Temperatures* The maximum afternoon temperature in a given city might be modeled by

$$t = 60 - 30\cos\frac{x\pi}{6},$$

where t represents the maximum afternoon temperature in month x, with $x = 0$ representing January, $x = 1$ representing February, and so on. Find the maximum afternoon temperature for each month.

(a) January   (b) April   (c) May
(d) June   (e) August   (f) October

48. *(Modeling) Average Monthly Temperature*
The average monthly temperature (in °F) in Chicago, Illinois, is shown in the table.

(a) Plot the average monthly temperature over a two-year period. Let $x = 1$ correspond to January of the first year.

(b) Determine a model function of the form $f(x) = a \sin b(x - d) + c$, where a, b, c, and d are constants.

(c) Explain the significance of each constant.

(d) Graph f together with the data on the same coordinate axes. How well does f model the data?

(e) Use the sine regression capability of a graphing calculator to find the equation of a sine curve that fits these data.

Month	°F	Month	°F
Jan	25	July	74
Feb	28	Aug	75
Mar	36	Sept	66
Apr	48	Oct	55
May	61	Nov	39
June	72	Dec	28

Source: Miller, A., J. Thompson, and R. Peterson, *Elements of Meteorology, Fourth Edition,* Charles E. Merrill Publishing Co., 1983.

49. *(Modeling) Pollution Trends*   The amount of pollution in the air is lower after heavy spring rains and higher after periods of little rain. In addition to this seasonal fluctuation, the long-term trend is upward. An idealized graph of this situation is shown in the figure. Circular functions can be used to model the fluctuating part of the pollution levels. Powers of the number e (e is the base of the natural logarithm; $e \approx 2.718282$) can be used to model long-term growth. (See Chapter 9.) The pollution level in a certain area might be given by

$$y = 7(1 - \cos 2\pi x)(x + 10) + 100e^{.2x},$$

where x is the time in years, with $x = 0$ representing January 1 of the base year. July 1 of the same year would be represented by $x = .5$, October 1 of the following year would be represented by $x = 1.75$, and so on. Find the pollution levels on each date.

(a) January 1, base year   (b) July 1, base year

(c) January 1, following year   (d) July 1, following year

50. *(Modeling) Lynx and Hare Populations*   The figure shows the populations of lynx and hares in Canada for the years 1847–1903. The hares are food for the lynx. An increase in hare population causes an increase in lynx population some time later. The increasing lynx population then causes a decline in hare population. The two graphs have the same period.

Canadian Lynx and Hare Populations

(a) Estimate the length of one period.

(b) Estimate maximum and minimum hare populations.

An object in simple harmonic motion has position function s inches from an initial point, where t is the time in seconds. Find the amplitude, period, and frequency.

51. $s(t) = 4 \sin \pi t$

52. $s(t) = 3 \cos 2t$

53. In Exercise 51, what does the frequency represent? Find the position of the object from the initial point at 1.5 sec, 2 sec, and 3.25 sec.

54. In Exercise 52, what does the period represent? What does the amplitude represent?

Chapter 4 Test

1. Consider the function defined by $y = 3 - 6 \sin\left(2x + \frac{\pi}{2}\right)$.

    (a) What is its period?
    (b) What is the amplitude of its graph?
    (c) What is its range?
    (d) What is the y-intercept of its graph?
    (e) What is its phase shift?

Graph each function over a two-period interval. Identify asymptotes when applicable.

2. $y = -\cos 2x$

3. $y = -\csc 2x$

4. $y = \tan\left(x - \frac{\pi}{2}\right)$

5. $y = -1 + 2 \sin(x + \pi)$

6. $y = -2 - \cot\left(x - \frac{\pi}{2}\right)$

7. $y = 3 \csc \pi x$

8. *(Modeling) Average Monthly Temperature* The average monthly temperature (in °F) in Austin, Texas, can be modeled using the circular function defined by

$$f(x) = 17.5 \sin\left[\frac{\pi}{6}(x - 4)\right] + 67.5,$$

where x is the month and $x = 1$ corresponds to January. (*Source:* Miller, A., J. Thompson, and R. Peterson, *Elements of Meteorology, Fourth Edition,* Charles E. Merrill Publishing Co., 1983.)

    (a) Graph f in the window $[1, 25]$ by $[45, 90]$.
    (b) Determine the amplitude, period, phase shift, and vertical translation of f.
    (c) What is the average monthly temperature for the month of December?
    (d) Determine the maximum and minimum average monthly temperatures and the months when they occur.
    (e) What would be an approximation for the average *yearly* temperature in Austin? How is this related to the vertical translation of the sine function in the formula for f?

9. *(Modeling) Spring* The height of a weight attached to a spring is

$$s(t) = -4 \cos 8\pi t$$

inches after t seconds.

    (a) Find the maximum height that the weight rises above the equilibrium position of $s(t) = 0$.
    (b) When does the weight first reach its maximum height, if $t \geq 0$?
    (c) What are the frequency and period?

10. Explain why the domains of the tangent and secant functions are the same, and then give a similar explanation for the cotangent and cosecant functions.

Chapter 4 Quantitative Reasoning

Does the fact that average monthly temperatures are periodic affect your utility bills?

In an article entitled "I Found Sinusoids in My Gas Bill" (*Mathematics Teacher*, January 2000), Cathy G. Schloemer presents the following graph that accompanied her gas bill.

Your Energy Usage

Notice that two sinusoids are suggested here: one for the behavior of the average monthly temperature and another for gas use in MCF (thousands of cubic feet).

1. If January 1997 is represented by $x = 1$, the data of estimated ordered pairs (month, temperature) are given in the list shown on the two graphing calculator screens below.

L1	L2	L3	1
1	28	------	
2	29		
3	31		
4	35		
5	51		
6	60		
7	73		
L1(1)=1			

L1	L2	L3	1
7	73		
8	70		
9	71		
10	64		
11	53		
12	37		
L1(13) =		------	

Use the sine regression feature of a graphing calculator to find a sine function that fits these data points. Then make a scatter diagram, and graph the function.

2. If January 1997 is again represented by $x = 1$, the data of estimated ordered pairs (month, gas use in MCF) are given in the list shown on the two graphing calculator screens below.

L1	L2	L3	1
1	24.2	------	
2	20		
3	18.8		
4	17.5		
5	9.2		
6	4.2		
7	4.8		
L1(1)=1			

L1	L2	L3	1
7	4.8		
8	1.8		
9	4.8		
10	4		
11	5		
12	17.5		
L1(13) =		------	

Use the sine regression feature of a graphing calculator to find a sine function that fits these data points. Then make a scatter diagram, and graph the function.

3. Answer the question posed at the top of the page, in the form of a short paragraph.

5

Trigonometric Identities

In 1831 Michael Faraday discovered that when a wire passes by a magnet, a small electric current is produced in the wire. Now we generate massive amounts of electricity by simultaneously rotating thousands of wires near large electromagnets. Because electric current alternates its direction on electrical wires, it is modeled accurately by either the sine or the cosine function.

We give many examples of applications of the trigonometric functions to electricity and other phenomena in the examples and exercises in this chapter, including a model of the wattage consumption of a toaster in Section 5.5, Example 6.

5.1 | Fundamental Identities

Negative-Angle Identities ▪ **Fundamental Identities** ▪ **Using the Fundamental Identities**

$$\sin(-\theta) = -\frac{y}{r} = -\sin\theta$$

Figure 1

Recall that an *identity* is an equation that is satisfied by *every* value in the domain of its variable.

Negative-Angle Identities As suggested by the circle shown in Figure 1, an angle θ having the point (x, y) on its terminal side has a corresponding angle $-\theta$ with the point $(x, -y)$ on its terminal side. From the definition of sine,

$$\sin(-\theta) = \frac{-y}{r} \quad \text{and} \quad \sin\theta = \frac{y}{r}, \quad \text{(Section 1.3)}$$

so $\sin(-\theta)$ and $\sin\theta$ are negatives of each other, or

$$\sin(-\theta) = -\sin\theta.$$

Figure 1 shows an angle θ in quadrant II, but the same result holds for θ in any quadrant. Also, by definition,

$$\cos(-\theta) = \frac{x}{r} \quad \text{and} \quad \cos\theta = \frac{x}{r}, \quad \text{(Section 1.3)}$$

so

$$\cos(-\theta) = \cos\theta.$$

We use the identities for $\sin(-\theta)$ and $\cos(-\theta)$ to find $\tan(-\theta)$ in terms of $\tan\theta$:

$$\tan(-\theta) = \frac{\sin(-\theta)}{\cos(-\theta)} = \frac{-\sin\theta}{\cos\theta} = -\frac{\sin\theta}{\cos\theta}$$

$$\tan(-\theta) = -\tan\theta.$$

Similar reasoning gives the following identities.

$$\csc(-\theta) = -\csc\theta, \qquad \sec(-\theta) = \sec\theta, \qquad \cot(-\theta) = -\cot\theta$$

These identities are known as the **negative-angle** or **negative-number identities.**

Fundamental Identities In Chapter 1 we used the definitions of the trigonometric functions to derive the reciprocal, quotient, and Pythagorean identities. Together with the negative-angle identities, these are called the **fundamental identities.**

Fundamental Identities

Reciprocal Identities

$$\cot\theta = \frac{1}{\tan\theta} \qquad \sec\theta = \frac{1}{\cos\theta} \qquad \csc\theta = \frac{1}{\sin\theta}$$

Quotient Identities

$$\tan\theta = \frac{\sin\theta}{\cos\theta} \qquad \cot\theta = \frac{\cos\theta}{\sin\theta}$$

(continued)

Pythagorean Identities

$$\sin^2 \theta + \cos^2 \theta = 1 \qquad \tan^2 \theta + 1 = \sec^2 \theta \qquad 1 + \cot^2 \theta = \csc^2 \theta$$

Negative-Angle Identities

$$\sin(-\theta) = -\sin \theta \qquad \cos(-\theta) = \cos \theta \qquad \tan(-\theta) = -\tan \theta$$
$$\csc(-\theta) = -\csc \theta \qquad \sec(-\theta) = \sec \theta \qquad \cot(-\theta) = -\cot \theta$$

N O T E The most commonly recognized forms of the fundamental identities are given above. Throughout this chapter you must also recognize alternative forms of these identities. For example, two other forms of $\sin^2 \theta + \cos^2 \theta = 1$ are

$$\sin^2 \theta = 1 - \cos^2 \theta \qquad \text{and} \qquad \cos^2 \theta = 1 - \sin^2 \theta.$$

Using the Fundamental Identities One way we use these identities is to find the values of other trigonometric functions from the value of a given trigonometric function. Although we could find such values using a right triangle, this is a good way to practice using the fundamental identities.

EXAMPLE 1 Finding Trigonometric Function Values Given One Value and the Quadrant

If $\tan \theta = -\frac{5}{3}$ and θ is in quadrant II, find each function value.

(a) $\sec \theta$ **(b)** $\sin \theta$ **(c)** $\cot(-\theta)$

Solution

(a) Look for an identity that relates tangent and secant.

$$\tan^2 \theta + 1 = \sec^2 \theta \qquad \text{Pythagorean identity}$$

$$\left(-\frac{5}{3}\right)^2 + 1 = \sec^2 \theta \qquad \tan \theta = -\frac{5}{3}$$

$$\frac{25}{9} + 1 = \sec^2 \theta$$

$$\frac{34}{9} = \sec^2 \theta \qquad \text{Combine terms.}$$

$$-\sqrt{\frac{34}{9}} = \sec \theta \qquad \text{Take the negative square root. (Appendix A)}$$

$$-\frac{\sqrt{34}}{3} = \sec \theta \qquad \text{Simplify the radical; } -\sqrt{\frac{34}{9}} = -\frac{\sqrt{34}}{\sqrt{9}} = -\frac{\sqrt{34}}{3}.$$

We chose the negative square root since $\sec \theta$ is negative in quadrant II.

(b)
$$\tan\theta = \frac{\sin\theta}{\cos\theta} \qquad \text{Quotient identity}$$

$$\cos\theta\tan\theta = \sin\theta \qquad \text{Multiply by } \cos\theta.$$

$$\left(\frac{1}{\sec\theta}\right)\tan\theta = \sin\theta \qquad \text{Reciprocal identity}$$

$$\left(-\frac{3\sqrt{34}}{34}\right)\left(-\frac{5}{3}\right) = \sin\theta \qquad \frac{1}{\sec\theta} = \frac{1}{-\frac{\sqrt{34}}{3}} = -\frac{3}{\sqrt{34}} = -\frac{3}{\sqrt{34}}\cdot\frac{\sqrt{34}}{\sqrt{34}} = -\frac{3\sqrt{34}}{34};$$
$$\tan\theta = -\frac{5}{3}$$

$$\sin\theta = \frac{5\sqrt{34}}{34} \qquad \text{Multiply; rewrite.}$$

(c)
$$\cot(-\theta) = \frac{1}{\tan(-\theta)} \qquad \text{Reciprocal identity}$$

$$\cot(-\theta) = \frac{1}{-\tan\theta} \qquad \text{Negative-angle identity}$$

$$\cot(-\theta) = \frac{1}{-\left(-\frac{5}{3}\right)} = \frac{3}{5} \qquad \tan\theta = -\frac{5}{3}; \text{ simplify the complex fraction.}$$

Now try Exercises 5, 7, and 9.

C A U T I O N   To avoid a common error, when taking the square root, be sure to choose the sign based on the quadrant of θ and the function being evaluated.

Any trigonometric function of a number or angle can be expressed in terms of any other function.

EXAMPLE 2 Expressing One Function in Terms of Another

Express $\cos x$ in terms of $\tan x$.

Solution Since $\sec x$ is related to both $\cos x$ and $\tan x$ by identities, start with $1 + \tan^2 x = \sec^2 x$.

$$\frac{1}{1 + \tan^2 x} = \frac{1}{\sec^2 x} \qquad \text{Take reciprocals.}$$

$$\frac{1}{1 + \tan^2 x} = \cos^2 x \qquad \text{Reciprocal identity}$$

$$\pm\sqrt{\frac{1}{1 + \tan^2 x}} = \cos x \qquad \text{Take square roots.}$$

$$\cos x = \frac{\pm 1}{\sqrt{1 + \tan^2 x}} \qquad \text{Quotient rule for radicals: } \sqrt[n]{\frac{a}{b}} = \frac{\sqrt[n]{a}}{\sqrt[n]{b}}; \text{ rewrite.}$$

$$\cos x = \frac{\pm\sqrt{1 + \tan^2 x}}{1 + \tan^2 x} \qquad \text{Rationalize the denominator.}$$

Choose the $+$ sign or the $-$ sign, depending on the quadrant of x.

Now try Exercise 43.

We can use a graphing calculator to decide whether two functions are identical. See Figure 2, which supports the identity $\sin^2 x + \cos^2 x = 1$. With an identity, you should see no difference in the two graphs. ■

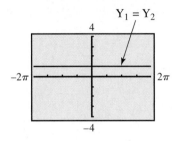

Figure 2

All other trigonometric functions can easily be expressed in terms of $\sin \theta$ and/or $\cos \theta$. We often make such substitutions in an expression to simplify it.

EXAMPLE 3 Rewriting an Expression in Terms of Sine and Cosine

Write $\tan \theta + \cot \theta$ in terms of $\sin \theta$ and $\cos \theta$, and then simplify the expression.

Solution

$$\tan \theta + \cot \theta = \frac{\sin \theta}{\cos \theta} + \frac{\cos \theta}{\sin \theta} \qquad \text{Quotient identities}$$

$$= \frac{\sin^2 \theta}{\cos \theta \sin \theta} + \frac{\cos^2 \theta}{\cos \theta \sin \theta} \qquad \begin{array}{l}\text{Write each fraction with the least}\\ \text{common denominator (LCD).}\end{array}$$

$$= \frac{\sin^2 \theta + \cos^2 \theta}{\cos \theta \sin \theta} \qquad \text{Add fractions; } \frac{a}{c} + \frac{b}{c} = \frac{a+b}{c}.$$

$$\tan \theta + \cot \theta = \frac{1}{\cos \theta \sin \theta} \qquad \text{Pythagorean identity}$$

Now try Exercise 55.

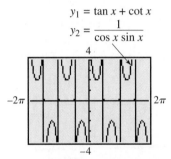

$y_1 = \tan x + \cot x$

$y_2 = \dfrac{1}{\cos x \sin x}$

The graph supports the result in Example 3. The graphs of y_1 and y_2 appear to be identical.

CAUTION When working with trigonometric expressions and identities, be sure to write the argument of the function. For example, we would *not* write $\sin^2 + \cos^2 = 1$; an argument such as θ is necessary in this identity.

5.1 Exercises

Concept Check   Fill in the blanks.

1. If $\tan x = 2.6$, then $\tan(-x) = $ _____.
2. If $\cos x = -.65$, then $\cos(-x) = $ _____.
3. If $\tan x = 1.6$, then $\cot x = $ _____.
4. If $\cos x = .8$ and $\sin x = .6$, then $\tan(-x) = $ _____.

Find $\sin s$. *See Example 1.*

5. $\cos s = \dfrac{3}{4}$, s in quadrant I

6. $\cot s = -\dfrac{1}{3}$, s in quadrant IV

7. $\cos(-s) = \dfrac{\sqrt{5}}{5}$, $\tan s < 0$

8. $\tan s = -\dfrac{\sqrt{7}}{2}$, $\sec s > 0$

9. $\sec s = \dfrac{11}{4}$, $\tan s < 0$

10. $\csc s = -\dfrac{8}{5}$

11. Why is it unnecessary to give the quadrant of s in Exercise 10?

Relating Concepts

For individual or collaborative investigation
(Exercises 12–17)

A function is called an even function if $f(-x) = f(x)$ for all x in the domain of f. A function is called an odd function if $f(-x) = -f(x)$ for all x in the domain of f. **Work Exercises 12–17 in order,** *to see the connection between the negative-angle identities and even and odd functions.*

12. Complete the statement: $\sin(-x) = $ _____.

13. Is the function defined by $f(x) = \sin x$ even or odd?

14. Complete the statement: $\cos(-x) = $ _____.

15. Is the function defined by $f(x) = \cos x$ even or odd?

16. Complete the statement: $\tan(-x) = $ _____.

17. Is the function defined by $f(x) = \tan x$ even or odd?

Concept Check   *For each graph, determine whether $f(-x) = f(x)$ or $f(-x) = -f(x)$ is true.*

18.

19.

20.

Find the remaining five trigonometric functions of θ. See Example 1.

21. $\sin \theta = \dfrac{2}{3}$, θ in quadrant II

22. $\cos \theta = \dfrac{1}{5}$, θ in quadrant I

23. $\tan \theta = -\dfrac{1}{4}$, θ in quadrant IV

24. $\csc \theta = -\dfrac{5}{2}$, θ in quadrant III

25. $\cot \theta = \dfrac{4}{3}$, $\sin \theta > 0$

26. $\sin \theta = -\dfrac{4}{5}$, $\cos \theta < 0$

27. $\sec \theta = \dfrac{4}{3}$, $\sin \theta < 0$

28. $\cos \theta = -\dfrac{1}{4}$, $\sin \theta > 0$

Concept Check *For each expression in Column I, choose the expression from Column II that completes an identity.*

I		II
29. $\dfrac{\cos x}{\sin x} = $ _____		**A.** $\sin^2 x + \cos^2 x$
30. $\tan x = $ _____		**B.** $\cot x$
31. $\cos(-x) = $ _____		**C.** $\sec^2 x$
32. $\tan^2 x + 1 = $ _____		**D.** $\dfrac{\sin x}{\cos x}$
33. $1 = $ _____		**E.** $\cos x$

Concept Check *For each expression in Column I, choose the expression from Column II that completes an identity. You may have to rewrite one or both expressions.*

I		II
34. $-\tan x \cos x = $ _____		**A.** $\dfrac{\sin^2 x}{\cos^2 x}$
35. $\sec^2 x - 1 = $ _____		**B.** $\dfrac{1}{\sec^2 x}$
36. $\dfrac{\sec x}{\csc x} = $ _____		**C.** $\sin(-x)$
37. $1 + \sin^2 x = $ _____		**D.** $\csc^2 x - \cot^2 x + \sin^2 x$
38. $\cos^2 x = $ _____		**E.** $\tan x$

39. A student writes "$1 + \cot^2 = \csc^2$." Comment on this student's work.

40. Another student makes the following claim: "Since $\sin^2 \theta + \cos^2 \theta = 1$, I should be able to also say that $\sin \theta + \cos \theta = 1$ if I take the square root of both sides." Comment on this student's statement.

41. *Concept Check* Suppose that $\cos \theta = \frac{x}{x+1}$. Find $\sin \theta$.

42. *Concept Check* Find $\tan \alpha$ if $\sec \alpha = \frac{p+4}{p}$.

Write the first trigonometric function in terms of the second trigonometric function. See Example 2.

43. $\sin x$; $\cos x$	**44.** $\cot x$; $\sin x$	**45.** $\tan x$; $\sec x$
46. $\cot x$; $\csc x$	**47.** $\csc x$; $\cos x$	**48.** $\sec x$; $\sin x$

Write each expression in terms of sine and cosine, and simplify so that no quotients appear in the final expression. See Example 3.

49. $\cot \theta \sin \theta$	**50.** $\sec \theta \cot \theta \sin \theta$	**51.** $\cos \theta \csc \theta$
52. $\cot^2 \theta (1 + \tan^2 \theta)$	**53.** $\sin^2 \theta (\csc^2 \theta - 1)$	**54.** $(\sec \theta - 1)(\sec \theta + 1)$
55. $(1 - \cos \theta)(1 + \sec \theta)$	**56.** $\dfrac{\cos \theta + \sin \theta}{\sin \theta}$	
57. $\dfrac{\cos^2 \theta - \sin^2 \theta}{\sin \theta \cos \theta}$	**58.** $\dfrac{1 - \sin^2 \theta}{1 + \cot^2 \theta}$	
59. $\sec \theta - \cos \theta$	**60.** $(\sec \theta + \csc \theta)(\cos \theta - \sin \theta)$	
61. $\sin \theta (\csc \theta - \sin \theta)$	**62.** $\dfrac{1 + \tan^2 \theta}{1 + \cot^2 \theta}$	

63. $\sin^2 \theta + \tan^2 \theta + \cos^2 \theta$

64. $\dfrac{\tan(-\theta)}{\sec \theta}$

65. *Concept Check* Let $\cos x = \frac{1}{5}$. Find all possible values of $\frac{\sec x - \tan x}{\sin x}$.

66. *Concept Check* Let $\csc x = -3$. Find all possible values of $\frac{\sin x + \cos x}{\sec x}$.

Relating Concepts

For individual or collaborative investigation
(Exercises 67–71)

In Chapter 4 we graphed functions defined by

$$y = c + a \cdot f[b(x - d)]$$

*with the assumption that $b > 0$. To see what happens when $b < 0$, **work Exercises 67–71 in order.***

67. Use a negative-angle identity to write $y = \sin(-2x)$ as a function of $2x$.

68. How does your answer to Exercise 67 relate to $y = \sin(2x)$?

69. Use a negative-angle identity to write $y = \cos(-4x)$ as a function of $4x$.

70. How does your answer to Exercise 69 relate to $y = \cos(4x)$?

71. Use your results from Exercises 67–70 to rewrite the following with a positive value of b.

(a) $\sin(-4x)$ **(b)** $\cos(-2x)$ **(c)** $-5\sin(-3x)$

Use a graphing calculator to decide whether each equation is an identity. (Hint: In Exercise 76, graph the function of x for a few different values of y (in radians).)

72. $\cos 2x = 1 - 2\sin^2 x$

73. $2 \sin s = \sin 2s$

74. $\sin x = \sqrt{1 - \cos^2 x}$

75. $\cos 2x = \cos^2 x - \sin^2 x$

76. $\cos(x - y) = \cos x - \cos y$

5.2 Verifying Trigonometric Identities

Verifying Identities by Working with One Side ▪ **Verifying Identities by Working with Both Sides**

Recall that an identity is an equation that is satisfied for all meaningful replacements of the variable. One of the skills required for more advanced work in mathematics, especially in calculus, is the ability to use identities to write expressions in alternative forms. We develop this skill by using the fundamental identities to verify that a trigonometric equation is an identity (for those values of the variable for which it is defined). Here are some hints to help you get started.

Looking Ahead to Calculus

Trigonometric identities are used in calculus to simplify trigonometric expressions, determine derivatives of trigonometric functions, and change the form of some integrals.

Hints for Verifying Identities

1. Learn the fundamental identities given in the last section. Whenever you see either side of a fundamental identity, the other side should come to mind. Also, be aware of equivalent forms of the fundamental identities. For example, $\sin^2 \theta = 1 - \cos^2 \theta$ is an alternative form of the identity $\sin^2 \theta + \cos^2 \theta = 1$.

2. Try to rewrite the more complicated side of the equation so that it is identical to the simpler side.

3. It is sometimes helpful to express all trigonometric functions in the equation in terms of sine and cosine and then simplify the result.

4. Usually, any factoring or indicated algebraic operations should be performed. For example, the expression $\sin^2 x + 2 \sin x + 1$ can be factored as $(\sin x + 1)^2$. The sum or difference of two trigonometric expressions, such as $\frac{1}{\sin \theta} + \frac{1}{\cos \theta}$, can be added or subtracted in the same way as any other rational expression.

$$\frac{1}{\sin \theta} + \frac{1}{\cos \theta} = \frac{\cos \theta}{\sin \theta \cos \theta} + \frac{\sin \theta}{\sin \theta \cos \theta}$$

$$= \frac{\cos \theta + \sin \theta}{\sin \theta \cos \theta}$$

5. As you select substitutions, keep in mind the side you are not changing, because it represents your goal. For example, to verify the identity

$$\tan^2 x + 1 = \frac{1}{\cos^2 x},$$

try to think of an identity that relates $\tan x$ to $\cos x$. In this case, since $\sec x = \frac{1}{\cos x}$ and $\sec^2 x = \tan^2 x + 1$, the secant function is the best link between the two sides.

6. If an expression contains $1 + \sin x$, multiplying both numerator and denominator by $1 - \sin x$ would give $1 - \sin^2 x$, which could be replaced with $\cos^2 x$. Similar results for $1 - \sin x$, $1 + \cos x$, and $1 - \cos x$ may be useful.

CAUTION *Verifying identities is not the same as solving equations.* Techniques used in solving equations, such as adding the same terms to both sides, or multiplying both sides by the same term, should not be used when working with identities since you are starting with a statement (to be verified) that may not be true.

Verifying Identities by Working with One Side   To avoid the temptation to use algebraic properties of equations to verify identities, *work with only one side and rewrite it to match the other side,* as shown in Examples 1–4.

EXAMPLE 1 Verifying an Identity (Working with One Side)

Verify that the following equation is an identity.

$$\cot s + 1 = \csc s(\cos s + \sin s)$$

Solution We use the fundamental identities from Section 5.1 to rewrite one side of the equation so that it is identical to the other side. Since the right side is more complicated, we work with it, using the third hint to change all functions to sine or cosine.

Steps   **Reasons**

For $s = x$,
$\cot x + 1 = \csc x (\cos x + \sin x)$

The graphs coincide, supporting the conclusion in Example 1.

Right side of
given equation

$$\overbrace{\csc s(\cos s + \sin s)} = \frac{1}{\sin s}(\cos s + \sin s) \qquad \csc s = \frac{1}{\sin s}$$

$$= \frac{\cos s}{\sin s} + \frac{\sin s}{\sin s} \qquad \begin{array}{l}\text{Distributive property;}\\ a(b + c) = ab + ac\end{array}$$

$$= \underbrace{\cot s + 1} \qquad \frac{\cos s}{\sin s} = \cot s; \frac{\sin s}{\sin s} = 1$$

Left side of
given equation

The given equation is an identity since the right side equals the left side.

Now try Exercise 33.

EXAMPLE 2 Verifying an Identity (Working with One Side)

Verify that the following equation is an identity.

$$\tan^2 x(1 + \cot^2 x) = \frac{1}{1 - \sin^2 x}$$

Solution We work with the more complicated left side, as suggested in the second hint. Again, we use the fundamental identities from Section 5.1.

$$\tan^2 x(1 + \cot^2 x) = \tan^2 x + \tan^2 x \cot^2 x \qquad \text{Distributive property}$$

$$= \tan^2 x + \tan^2 x \cdot \frac{1}{\tan^2 x} \qquad \cot^2 x = \frac{1}{\tan^2 x}$$

$$= \tan^2 x + 1 \qquad \tan^2 x \cdot \frac{1}{\tan^2 x} = 1$$

$$= \sec^2 x \qquad \tan^2 x + 1 = \sec^2 x$$

$$= \frac{1}{\cos^2 x} \qquad \sec^2 x = \frac{1}{\cos^2 x}$$

$$= \frac{1}{1 - \sin^2 x} \qquad \cos^2 x = 1 - \sin^2 x$$

$\tan^2 x (1 + \cot^2 x) = \frac{1}{1 - \sin^2 x}$

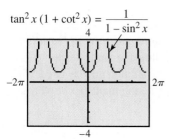

The screen supports the conclusion in Example 2.

Since the left side is identical to the right side, the given equation is an identity.

Now try Exercise 37.

EXAMPLE 3 Verifying an Identity (Working with One Side)

Verify that the following equation is an identity.

$$\frac{\tan t - \cot t}{\sin t \cos t} = \sec^2 t - \csc^2 t$$

Solution We transform the more complicated left side to match the right side.

$$\frac{\tan t - \cot t}{\sin t \cos t} = \frac{\tan t}{\sin t \cos t} - \frac{\cot t}{\sin t \cos t} \qquad \frac{a-b}{c} = \frac{a}{c} - \frac{b}{c}$$

$$= \tan t \cdot \frac{1}{\sin t \cos t} - \cot t \cdot \frac{1}{\sin t \cos t} \qquad \frac{a}{b} = a \cdot \frac{1}{b}$$

$$= \frac{\sin t}{\cos t} \cdot \frac{1}{\sin t \cos t} - \frac{\cos t}{\sin t} \cdot \frac{1}{\sin t \cos t} \qquad \tan t = \frac{\sin t}{\cos t}; \cot t = \frac{\cos t}{\sin t}$$

$$= \frac{1}{\cos^2 t} - \frac{1}{\sin^2 t}$$

$$= \sec^2 t - \csc^2 t \qquad \frac{1}{\cos^2 t} = \sec^2 t; \frac{1}{\sin^2 t} = \csc^2 t$$

The third hint about writing all trigonometric functions in terms of sine and cosine was used in the third line of the solution.

Now try Exercise 41.

EXAMPLE 4 Verifying an Identity (Working with One Side)

Verify that the following equation is an identity.

$$\frac{\cos x}{1 - \sin x} = \frac{1 + \sin x}{\cos x}$$

Solution We work on the right side, using the last hint in the list given earlier to multiply numerator and denominator on the right by $1 - \sin x$.

$$\frac{1 + \sin x}{\cos x} = \frac{(1 + \sin x)(1 - \sin x)}{\cos x (1 - \sin x)} \qquad \text{Multiply by 1 in the form } \frac{1 - \sin x}{1 - \sin x}.$$

$$= \frac{1 - \sin^2 x}{\cos x (1 - \sin x)} \qquad (x + y)(x - y) = x^2 - y^2$$

$$= \frac{\cos^2 x}{\cos x (1 - \sin x)} \qquad 1 - \sin^2 x = \cos^2 x$$

$$= \frac{\cos x}{1 - \sin x} \qquad \text{Lowest terms}$$

Now try Exercise 47.

Verifying Identities by Working with Both Sides If both sides of an identity appear to be equally complex, the identity can be verified by working independently on the left side and on the right side, until each side is changed into some common third result. *Each step, on each side, must be reversible.*

left $\qquad$ = $\qquad$ right

common third
expression

With all steps reversible, the procedure is as shown in the margin. The left side leads to a common third expression, which leads back to the right side. This procedure is just a shortcut for the procedure used in Examples 1–4: one side is changed into the other side, but by going through an intermediate step.

EXAMPLE 5 Verifying an Identity (Working with Both Sides)

Verify that the following equation is an identity.

$$\frac{\sec \alpha + \tan \alpha}{\sec \alpha - \tan \alpha} = \frac{1 + 2 \sin \alpha + \sin^2 \alpha}{\cos^2 \alpha}$$

Solution Both sides appear equally complex, so we verify the identity by changing each side into a common third expression. We work first on the left, multiplying numerator and denominator by $\cos \alpha$.

$$\frac{\sec \alpha + \tan \alpha}{\sec \alpha - \tan \alpha} = \frac{(\sec \alpha + \tan \alpha) \cos \alpha}{(\sec \alpha - \tan \alpha) \cos \alpha} \qquad \text{Multiply by 1.}$$

$$= \frac{\sec \alpha \cos \alpha + \tan \alpha \cos \alpha}{\sec \alpha \cos \alpha - \tan \alpha \cos \alpha} \qquad \text{Distributive property}$$

$$= \frac{1 + \tan \alpha \cos \alpha}{1 - \tan \alpha \cos \alpha} \qquad \sec \alpha \cos \alpha = 1$$

$$= \frac{1 + \dfrac{\sin \alpha}{\cos \alpha} \cdot \cos \alpha}{1 - \dfrac{\sin \alpha}{\cos \alpha} \cdot \cos \alpha} \qquad \tan \alpha = \frac{\sin \alpha}{\cos \alpha}$$

$$= \frac{1 + \sin \alpha}{1 - \sin \alpha}$$

On the right side of the original equation, begin by factoring.

$$\frac{1 + 2 \sin \alpha + \sin^2 \alpha}{\cos^2 \alpha} = \frac{(1 + \sin \alpha)^2}{\cos^2 \alpha} \qquad x^2 + 2xy + y^2 = (x + y)^2$$

$$= \frac{(1 + \sin \alpha)^2}{1 - \sin^2 \alpha} \qquad \cos^2 \alpha = 1 - \sin^2 \alpha$$

$$= \frac{(1 + \sin \alpha)^2}{(1 + \sin \alpha)(1 - \sin \alpha)} \qquad \begin{array}{l} \text{Factor the denominator;} \\ x^2 - y^2 = (x + y)(x - y). \end{array}$$

$$= \frac{1 + \sin \alpha}{1 - \sin \alpha} \qquad \text{Lowest terms}$$

We have shown that

$$\underbrace{\frac{\sec \alpha + \tan \alpha}{\sec \alpha - \tan \alpha}}_{\substack{\text{Left side of} \\ \text{given equation}}} = \underbrace{\frac{1 + \sin \alpha}{1 - \sin \alpha}}_{\substack{\text{Common third} \\ \text{expression}}} = \underbrace{\frac{1 + 2 \sin \alpha + \sin^2 \alpha}{\cos^2 \alpha}}_{\substack{\text{Right side of} \\ \text{given equation}}},$$

verifying that the given equation is an identity.

Now try Exercise 51.

CAUTION   Use the method of Example 5 *only* if the steps are reversible.

There are usually several ways to verify a given identity. For instance, another way to begin verifying the identity in Example 5 is to work on the left as follows.

$$\frac{\sec\alpha + \tan\alpha}{\sec\alpha - \tan\alpha} = \frac{\dfrac{1}{\cos\alpha} + \dfrac{\sin\alpha}{\cos\alpha}}{\dfrac{1}{\cos\alpha} - \dfrac{\sin\alpha}{\cos\alpha}}$$ Fundamental identities **(Section 5.1)**

$$= \frac{\dfrac{1 + \sin\alpha}{\cos\alpha}}{\dfrac{1 - \sin\alpha}{\cos\alpha}}$$ Add and subtract fractions.

$$= \frac{1 + \sin\alpha}{1 - \sin\alpha}$$ Simplify the complex fraction.

Compare this with the result shown in Example 5 for the right side to see that the two sides indeed agree.

EXAMPLE 6   Applying a Pythagorean Identity to Radios

Tuners in radios select a radio station by adjusting the frequency. A tuner may contain an inductor L and a capacitor C, as illustrated in Figure 3. The energy stored in the inductor at time t is given by

$$L(t) = k\sin^2(2\pi Ft)$$

and the energy stored in the capacitor is given by

$$C(t) = k\cos^2(2\pi Ft),$$

where F is the frequency of the radio station and k is a constant. The total energy E in the circuit is given by

$$E(t) = L(t) + C(t).$$

Show that E is a constant function. (*Source:* Weidner, R. and R. Sells, *Elementary Classical Physics,* Vol. 2, Allyn & Bacon, 1973.)

An Inductor and a Capacitor

Figure 3

Solution

$$\begin{aligned}
E(t) &= L(t) + C(t) &&\text{Given equation} \\
&= k\sin^2(2\pi Ft) + k\cos^2(2\pi Ft) &&\text{Substitute.} \\
&= k[\sin^2(2\pi Ft) + \cos^2(2\pi Ft)] &&\text{Factor out } k. \\
&= k(1) &&\sin^2\theta + \cos^2\theta = 1 \text{ (Here } \theta = 2\pi Ft.) \\
&= k
\end{aligned}$$

Since k is a constant, $E(t)$ is a constant function.

Now try Exercise 85.

5.2 Exercises

Perform each indicated operation and simplify the result.

1. $\cot \theta + \dfrac{1}{\cot \theta}$

2. $\dfrac{\sec x}{\csc x} + \dfrac{\csc x}{\sec x}$

3. $\tan s(\cot s + \csc s)$

4. $\cos \beta(\sec \beta + \csc \beta)$

5. $\dfrac{1}{\csc^2 \theta} + \dfrac{1}{\sec^2 \theta}$

6. $\dfrac{1}{\sin \alpha - 1} - \dfrac{1}{\sin \alpha + 1}$

7. $\dfrac{\cos x}{\sec x} + \dfrac{\sin x}{\csc x}$

8. $\dfrac{\cos \theta}{\sin \theta} + \dfrac{\sin \theta}{1 + \cos \theta}$

9. $(1 + \sin t)^2 + \cos^2 t$

10. $(1 + \tan s)^2 - 2 \tan s$

11. $\dfrac{1}{1 + \cos x} - \dfrac{1}{1 - \cos x}$

12. $(\sin \alpha - \cos \alpha)^2$

Factor each trigonometric expression.

13. $\sin^2 \theta - 1$

14. $\sec^2 \theta - 1$

15. $(\sin x + 1)^2 - (\sin x - 1)^2$

16. $(\tan x + \cot x)^2 - (\tan x - \cot x)^2$

17. $2 \sin^2 x + 3 \sin x + 1$

18. $4 \tan^2 \beta + \tan \beta - 3$

19. $\cos^4 x + 2 \cos^2 x + 1$

20. $\cot^4 x + 3 \cot^2 x + 2$

21. $\sin^3 x - \cos^3 x$

22. $\sin^3 \alpha + \cos^3 \alpha$

Each expression simplifies to a constant, a single function, or a power of a function. Use fundamental identities to simplify each expression.

23. $\tan \theta \cos \theta$

24. $\cot \alpha \sin \alpha$

25. $\sec r \cos r$

26. $\cot t \tan t$

27. $\dfrac{\sin \beta \tan \beta}{\cos \beta}$

28. $\dfrac{\csc \theta \sec \theta}{\cot \theta}$

29. $\sec^2 x - 1$

30. $\csc^2 t - 1$

31. $\dfrac{\sin^2 x}{\cos^2 x} + \sin x \csc x$

32. $\dfrac{1}{\tan^2 \alpha} + \cot \alpha \tan \alpha$

In Exercises 33–68, verify that each trigonometric equation is an identity. See Examples 1–5.

33. $\dfrac{\cot \theta}{\csc \theta} = \cos \theta$

34. $\dfrac{\tan \alpha}{\sec \alpha} = \sin \alpha$

35. $\dfrac{1 - \sin^2 \beta}{\cos \beta} = \cos \beta$

36. $\dfrac{\tan^2 \alpha + 1}{\sec \alpha} = \sec \alpha$

37. $\cos^2 \theta(\tan^2 \theta + 1) = 1$

38. $\sin^2 \beta(1 + \cot^2 \beta) = 1$

39. $\cot s + \tan s = \sec s \csc s$

40. $\sin^2 \alpha + \tan^2 \alpha + \cos^2 \alpha = \sec^2 \alpha$

41. $\dfrac{\cos \alpha}{\sec \alpha} + \dfrac{\sin \alpha}{\csc \alpha} = \sec^2 \alpha - \tan^2 \alpha$

42. $\dfrac{\sin^2 \theta}{\cos \theta} = \sec \theta - \cos \theta$

43. $\sin^4 \theta - \cos^4 \theta = 2 \sin^2 \theta - 1$

44. $\dfrac{\cos \theta}{\sin \theta \cot \theta} = 1$

45. $(1 - \cos^2 \alpha)(1 + \cos^2 \alpha) = 2 \sin^2 \alpha - \sin^4 \alpha$

46. $\tan^2 \alpha \sin^2 \alpha = \tan^2 \alpha + \cos^2 \alpha - 1$

47. $\dfrac{\cos \theta + 1}{\tan^2 \theta} = \dfrac{\cos \theta}{\sec \theta - 1}$

48. $\dfrac{(\sec \theta - \tan \theta)^2 + 1}{\sec \theta \csc \theta - \tan \theta \csc \theta} = 2 \tan \theta$

49. $\dfrac{1}{1 - \sin \theta} + \dfrac{1}{1 + \sin \theta} = 2 \sec^2 \theta$

50. $\dfrac{1}{\sec \alpha - \tan \alpha} = \sec \alpha + \tan \alpha$

51. $\dfrac{\tan s}{1 + \cos s} + \dfrac{\sin s}{1 - \cos s} = \cot s + \sec s \csc s$

52. $\dfrac{1 - \cos x}{1 + \cos x} = (\cot x - \csc x)^2$

53. $\dfrac{\cot \alpha + 1}{\cot \alpha - 1} = \dfrac{1 + \tan \alpha}{1 - \tan \alpha}$

54. $\dfrac{1}{\tan \alpha - \sec \alpha} + \dfrac{1}{\tan \alpha + \sec \alpha} = -2 \tan \alpha$

55. $\sin^2 \alpha \sec^2 \alpha + \sin^2 \alpha \csc^2 \alpha = \sec^2 \alpha$

56. $\dfrac{\csc \theta + \cot \theta}{\tan \theta + \sin \theta} = \cot \theta \csc \theta$

57. $\sec^4 x - \sec^2 x = \tan^4 x + \tan^2 x$

58. $\dfrac{1 - \sin \theta}{1 + \sin \theta} = \sec^2 \theta - 2 \sec \theta \tan \theta + \tan^2 \theta$

59. $\sin \theta + \cos \theta = \dfrac{\sin \theta}{1 - \dfrac{\cos \theta}{\sin \theta}} + \dfrac{\cos \theta}{1 - \dfrac{\sin \theta}{\cos \theta}}$

60. $\dfrac{\sin \theta}{1 - \cos \theta} - \dfrac{\sin \theta \cos \theta}{1 + \cos \theta} = \csc \theta (1 + \cos^2 \theta)$

61. $\dfrac{\sec^4 s - \tan^4 s}{\sec^2 s + \tan^2 s} = \sec^2 s - \tan^2 s$

62. $\dfrac{\cot^2 t - 1}{1 + \cot^2 t} = 1 - 2 \sin^2 t$

63. $\dfrac{\tan^2 t - 1}{\sec^2 t} = \dfrac{\tan t - \cot t}{\tan t + \cot t}$

64. $(1 + \sin x + \cos x)^2 = 2(1 + \sin x)(1 + \cos x)$

65. $\dfrac{1 + \cos x}{1 - \cos x} - \dfrac{1 - \cos x}{1 + \cos x} = 4 \cot x \csc x$

66. $(\sec \alpha - \tan \alpha)^2 = \dfrac{1 - \sin \alpha}{1 + \sin \alpha}$

67. $(\sec \alpha + \csc \alpha)(\cos \alpha - \sin \alpha) = \cot \alpha - \tan \alpha$

68. $\dfrac{\sin^4 \alpha - \cos^4 \alpha}{\sin^2 \alpha - \cos^2 \alpha} = 1$

69. A student claims that the equation $\cos \theta + \sin \theta = 1$ is an identity, since by letting $\theta = 90°$ (or $\frac{\pi}{2}$ radians) we get $0 + 1 = 1$, a true statement. Comment on this student's reasoning.

70. An equation that is an identity has an infinite number of solutions. If an equation has an infinite number of solutions, is it necessarily an identity? Explain.

Graph each expression and conjecture an identity. Then verify your conjecture algebraically.

71. $(\sec \theta + \tan \theta)(1 - \sin \theta)$

72. $(\csc \theta + \cot \theta)(\sec \theta - 1)$

73. $\dfrac{\cos \theta + 1}{\sin \theta + \tan \theta}$

74. $\tan \theta \sin \theta + \cos \theta$

Graph the expressions on each side of the equals sign to determine whether the equation might be an identity. (Note: Use a domain whose length is at least 2π.) If the equation looks like an identity, prove it algebraically. See Example 1.

75. $\dfrac{2 + 5 \cos s}{\sin s} = 2 \csc s + 5 \cot s$

76. $1 + \cot^2 s = \dfrac{\sec^2 s}{\sec^2 s - 1}$

77. $\dfrac{\tan s - \cot s}{\tan s + \cot s} = 2 \sin^2 s$

78. $\dfrac{1}{1 + \sin s} + \dfrac{1}{1 - \sin s} = \sec^2 s$

By substituting a number for s or t, show that the equation is not an identity.

79. $\sin(\csc s) = 1$

80. $\sqrt{\cos^2 s} = \cos s$

81. $\csc t = \sqrt{1 + \cot^2 t}$

82. $\cos t = \sqrt{1 - \sin^2 t}$

83. When is $\sin x = \sqrt{1 - \cos^2 x}$ a true statement?

(Modeling) Work each problem.

84. *Intensity of a Lamp* According to Lambert's law, the intensity of light from a single source on a flat surface at point P is given by

$$I = k \cos^2 \theta,$$

where k is a constant. (*Source:* Winter, C., *Solar Power Plants,* Springer-Verlag, 1991.)

(a) Write I in terms of the sine function.

(b) Why does the maximum value of I occur when $\theta = 0$?

85. *Oscillating Spring* The distance or displacement y of a weight attached to an oscillating spring from its natural position is modeled by

$$y = 4 \cos(2\pi t),$$

where t is time in seconds. Potential energy is the energy of position and is given by

$$P = ky^2,$$

where k is a constant. The weight has the greatest potential energy when the spring is stretched the most. (*Source:* Weidner, R. and R. Sells, *Elementary Classical Physics,* Vol. 1, Allyn & Bacon, 1973.)

(a) Write an expression for P that involves the cosine function.

(b) Use a fundamental identity to write P in terms of $\sin(2\pi t)$.

86. *Radio Tuners* Refer to Example 6. Let the energy stored in the inductor be given by

$$L(t) = 3 \cos^2(6,000,000t)$$

and the energy in the capacitor be given by

$$C(t) = 3 \sin^2(6,000,000t),$$

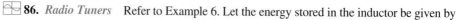

where t is time in seconds. The total energy E in the circuit is given by $E(t) = L(t) + C(t)$.

(a) Graph L, C, and E in the window $[0, 10^{-6}]$ by $[-1, 4]$, with Xscl $= 10^{-7}$ and Yscl $= 1$. Interpret the graph.

(b) Make a table of values for L, C, and E starting at $t = 0$, incrementing by 10^{-7}. Interpret your results.

(c) Use a fundamental identity to derive a simplified expression for $E(t)$.

5.3 Sum and Difference Identities for Cosine

Difference Identity for Cosine ▪ **Sum Identity for Cosine** ▪ **Cofunction Identities** ▪ **Applying the Sum and Difference Identities**

Difference Identity for Cosine   Several examples presented earlier should have convinced you by now that $\cos(A - B)$ *does not equal* $\cos A - \cos B$. For example, if $A = \frac{\pi}{2}$ and $B = 0$, then

$$\cos(A - B) = \cos\left(\frac{\pi}{2} - 0\right) = \cos\frac{\pi}{2} = 0,$$

while

$$\cos A - \cos B = \cos\frac{\pi}{2} - \cos 0 = 0 - 1 = -1.$$

We can now derive a formula for $\cos(A - B)$. We start by locating angles A and B in standard position on a unit circle, with $B < A$. Let S and Q be the points where the terminal sides of angles A and B, respectively, intersect the circle. Locate point R on the unit circle so that angle POR equals the difference $A - B$. See Figure 4.

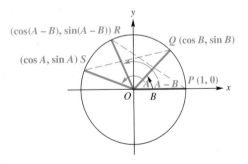

Figure 4

Point Q is on the unit circle, so by the work with circular functions in Chapter 3, the x-coordinate of Q is the cosine of angle B, while the y-coordinate of Q is the sine of angle B.

$$Q \text{ has coordinates } (\cos B, \sin B).$$

In the same way,

$$S \text{ has coordinates } (\cos A, \sin A),$$

and

$$R \text{ has coordinates } (\cos(A - B), \sin(A - B)).$$

Angle SOQ also equals $A - B$. Since the central angles SOQ and POR are equal, chords PR and SQ are equal. By the distance formula, since $PR = SQ$,

$$\sqrt{[\cos(A - B) - 1]^2 + [\sin(A - B) - 0]^2}$$
$$= \sqrt{(\cos A - \cos B)^2 + (\sin A - \sin B)^2}. \quad \text{(Appendix B)}$$

Squaring both sides and clearing parentheses gives

$$\cos^2(A - B) - 2\cos(A - B) + 1 + \sin^2(A - B)$$
$$= \cos^2 A - 2\cos A \cos B + \cos^2 B + \sin^2 A - 2\sin A \sin B + \sin^2 B.$$

Since $\sin^2 x + \cos^2 x = 1$ for any value of x, we can rewrite the equation as

$$2 - 2\cos(A - B) = 2 - 2\cos A \cos B - 2\sin A \sin B$$
$$\cos(A - B) = \cos A \cos B + \sin A \sin B. \quad \text{Subtract 2; divide by } -2.$$

Although Figure 4 shows angles A and B in the second and first quadrants, respectively, this result is the same for any values of these angles.

Sum Identity for Cosine To find a similar expression for $\cos(A + B)$, rewrite $A + B$ as $A - (-B)$ and use the identity for $\cos(A - B)$.

$$\cos(A + B) = \cos[A - (-B)] \qquad .$$
$$= \cos A \cos(-B) + \sin A \sin(-B) \quad \text{Cosine difference identity}$$
$$= \cos A \cos B + \sin A(-\sin B) \quad \text{Negative-angle identities (Section 5.1)}$$
$$\cos(A + B) = \cos A \cos B - \sin A \sin B$$

Cosine of a Sum or Difference

$$\mathbf{\cos(A + B) = \cos A \cos B - \sin A \sin B}$$
$$\mathbf{\cos(A - B) = \cos A \cos B + \sin A \sin B}$$

These identities are important in calculus and useful in certain applications. Although a calculator can be used to find an approximation for $\cos 15°$, for example, the method shown below can be applied to get an exact value, as well as to practice using the sum and difference identities.

EXAMPLE 1 Finding Exact Cosine Function Values

Find the *exact* value of each expression.

(a) $\cos 15°$ **(b)** $\cos \dfrac{5\pi}{12}$ **(c)** $\cos 87° \cos 93° - \sin 87° \sin 93°$

Solution

(a) To find $\cos 15°$, we write $15°$ as the sum or difference of two angles with known function values. Since we know the exact trigonometric function values of $45°$ and $30°$, we write $15°$ as $45° - 30°$. (We could also use $60° - 45°$.) Then we use the cosine difference identity.

$$\cos 15° = \cos(45° - 30°)$$

$$= \cos 45° \cos 30° + \sin 45° \sin 30° \quad \text{Cosine difference identity}$$

$$= \frac{\sqrt{2}}{2} \cdot \frac{\sqrt{3}}{2} + \frac{\sqrt{2}}{2} \cdot \frac{1}{2} \quad \text{(Section 2.1)}$$

$$= \frac{\sqrt{6} + \sqrt{2}}{4}$$

(b) $\cos \dfrac{5\pi}{12} = \cos\left(\dfrac{\pi}{6} + \dfrac{\pi}{4}\right)$ \qquad $\frac{\pi}{6} = \frac{2\pi}{12}; \frac{\pi}{4} = \frac{3\pi}{12}$

$$= \cos \frac{\pi}{6} \cos \frac{\pi}{4} - \sin \frac{\pi}{6} \sin \frac{\pi}{4} \quad \text{Cosine sum identity}$$

$$= \frac{\sqrt{3}}{2} \cdot \frac{\sqrt{2}}{2} - \frac{1}{2} \cdot \frac{\sqrt{2}}{2} \quad \text{(Section 3.1)}$$

$$= \frac{\sqrt{6} - \sqrt{2}}{4}$$

(c) $\cos 87° \cos 93° - \sin 87° \sin 93° = \cos(87° + 93°)$

$$\text{Cosine sum identity}$$

$$= \cos 180°$$

$$= -1 \quad \text{(Section 1.3)}$$

```
cos(5π/12)
        .2588190451
(√(6)-√(2))/4
        .2588190451
```

The screen supports the solution in Example 1(b) by showing that

$$\cos \frac{5\pi}{12} = \frac{\sqrt{6} - \sqrt{2}}{4}.$$

Now try Exercises 7, 9, and 11.

Cofunction Identities We can use the identity for the cosine of the difference of two angles and the fundamental identities to derive the *cofunction identities*, presented in Section 2.1 for values of θ in the interval $[0°, 90°]$.

Cofunction Identities

$$\cos(90° - \theta) = \sin \theta \qquad \cot(90° - \theta) = \tan \theta$$

$$\sin(90° - \theta) = \cos \theta \qquad \sec(90° - \theta) = \csc \theta$$

$$\tan(90° - \theta) = \cot \theta \qquad \csc(90° - \theta) = \sec \theta$$

Similar identities can be obtained for a real number domain by replacing $90°$ with $\frac{\pi}{2}$.

Substituting $90°$ for A and θ for B in the identity for $\cos(A - B)$ gives

$$\cos(90° - \theta) = \cos 90° \cos \theta + \sin 90° \sin \theta$$

$$= 0 \cdot \cos \theta + 1 \cdot \sin \theta$$

$$= \sin \theta.$$

This result is true for *any* value of θ since the identity for $\cos(A - B)$ is true for any values of A and B.

EXAMPLE 2 Using Cofunction Identities to Find θ

Find an angle θ that satisfies each of the following.

(a) $\cot \theta = \tan 25°$ **(b)** $\sin \theta = \cos(-30°)$ **(c)** $\csc \dfrac{3\pi}{4} = \sec \theta$

Solution

(a) Since tangent and cotangent are cofunctions, $\tan(90° - \theta) = \cot \theta$.

$$\cot \theta = \tan 25°$$
$$\tan(90° - \theta) = \tan 25° \quad \text{Cofunction identity}$$
$$90° - \theta = 25° \quad \text{Set angle measures equal.}$$
$$\theta = 65°$$

(b)
$$\sin \theta = \cos(-30°)$$
$$\cos(90° - \theta) = \cos(-30°) \quad \text{Cofunction identity}$$
$$90° - \theta = -30°$$
$$\theta = 120°$$

(c)
$$\csc \frac{3\pi}{4} = \sec \theta$$
$$\sec\left(\frac{\pi}{2} - \frac{3\pi}{4}\right) = \sec \theta \quad \text{Cofunction identity}$$
$$\sec\left(-\frac{\pi}{4}\right) = \sec \theta \quad \text{Combine terms.}$$
$$-\frac{\pi}{4} = \theta$$

Now try Exercises 33 and 37.

N O T E Because trigonometric (circular) functions are periodic, the solutions in Example 2 are not unique. We give only one of infinitely many possibilities.

Applying the Sum and Difference Identities   If one of the angles A or B in the identities for $\cos(A + B)$ and $\cos(A - B)$ is a quadrantal angle, then the identity allows us to write the expression in terms of a single function of A or B.

EXAMPLE 3 Reducing $\cos(A - B)$ to a Function of a Single Variable

Write $\cos(180° - \theta)$ as a trigonometric function of θ.

Solution $\cos(180° - \theta) = \cos 180° \cos \theta + \sin 180° \sin \theta$

                             Cosine difference identity

$$= (-1) \cos \theta + (0) \sin \theta \quad \text{(Section 1.3)}$$
$$= -\cos \theta$$

Now try Exercise 39.

EXAMPLE 4 Finding cos(s + t) Given Information about s and t

Suppose that $\sin s = \frac{3}{5}$, $\cos t = -\frac{12}{13}$, and both s and t are in quadrant II. Find $\cos(s + t)$.

Solution By the cosine sum identity, $\cos(s + t) = \cos s \cos t - \sin s \sin t$. The values of $\sin s$ and $\cos t$ are given, so we can find $\cos(s + t)$ if we know the values of $\cos s$ and $\sin t$. To find $\cos s$ and $\sin t$, we sketch two angles in the second quadrant, one with $\sin s = \frac{3}{5}$ and the other with $\cos t = -\frac{12}{13}$. See Figure 5.

In Figure 5(a), since $\sin s = \frac{3}{5} = \frac{y}{r}$, we let $y = 3$ and $r = 5$. Substituting in the Pythagorean theorem, we get $x^2 + 3^2 = 5^2$ and solve to find $x = -4$. Thus, $\cos s = -\frac{4}{5}$. In Figure 5(b), $\cos t = -\frac{12}{13} = \frac{x}{r}$, so we let $x = -12$ and $r = 13$. Then $(-12)^2 + y^2 = 13^2$; we solve to get $y = 5$. Thus, $\sin t = \frac{5}{13}$. Now we can find $\cos(s + t)$.

$$
\begin{aligned}
\cos(s + t) &= \cos s \cos t - \sin s \sin t \quad &\text{Cosine sum identity}\\
&= -\frac{4}{5}\left(-\frac{12}{13}\right) - \frac{3}{5} \cdot \frac{5}{13} \quad &\text{Substitute.}\\
&= \frac{48}{65} - \frac{15}{65} = \frac{33}{65}
\end{aligned}
$$

(a)

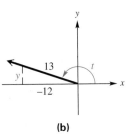

(b)

Figure 5

Now try Exercise 47.

NOTE In Example 4, the values of $\cos s$ and $\sin t$ could also be found by using the Pythagorean identities. The problem could then be solved using the identity for $\cos(s + t)$ as shown in the example.

EXAMPLE 5 Applying the Cosine Difference Identity to Voltage

Common household electric current is called *alternating current* because the current alternates direction within the wires. The voltage V in a typical 115-volt outlet can be expressed by the function

$$V(t) = 163 \sin \omega t,$$

where ω is the angular speed (in radians per second) of the rotating generator at the electrical plant and t is time measured in seconds. (*Source:* Bell, D., *Fundamentals of Electric Circuits*, Fourth Edition, Prentice-Hall, 1988.)

(a) It is essential for electric generators to rotate at precisely 60 cycles per sec so household appliances and computers will function properly. Determine ω for these electric generators.

(b) Graph V in the window $[0, .05]$ by $[-200, 200]$.

(c) Determine a value of ϕ so that the graph of $V(t) = 163 \cos(\omega t - \phi)$ is the same as the graph of $V(t) = 163 \sin \omega t$.

Solution

(a) Each cycle is 2π radians at 60 cycles per sec, so the angular speed is $\omega = 60(2\pi) = 120\pi$ radians per sec.

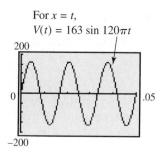

For $x = t$,
$V(t) = 163 \sin 120\pi t$

Figure 6

(b) $V(t) = 163 \sin \omega t = 163 \sin 120\pi t$. Because the amplitude of the function is 163 (from Section 4.1), $[-200, 200]$ is an appropriate interval for the range, as shown in Figure 6.

(c) Using the negative-angle identity for cosine and a cofunction identity,

$$\cos\left(x - \frac{\pi}{2}\right) = \cos\left[-\left(\frac{\pi}{2} - x\right)\right] = \cos\left(\frac{\pi}{2} - x\right) = \sin x.$$

Therefore, if $\phi = \frac{\pi}{2}$, then

$$V(t) = 163\cos(\omega t - \phi) = 163\cos\left(\omega t - \frac{\pi}{2}\right) = 163 \sin \omega t.$$

Now try Exercise 71.

5.3 Exercises

Concept Check *Match each expression in Column I with the correct expression in Column II to form an identity.*

I	II
1. $\cos(x + y) =$ _____	**A.** $\cos x \cos y + \sin x \sin y$
2. $\cos(x - y) =$ _____	**B.** $\cos x$
3. $\cos(90° - x) =$ _____	**C.** $\cos x + \sin x$
4. $\sin(90° - x) =$ _____	**D.** $\cos x - \sin x$
	E. $\sin x$
	F. $\cos x \cos y - \sin x \sin y$

Use identities to find each exact value. (Do not use a calculator.) See Example 1.

5. $\cos 75°$

6. $\cos(-15°)$

7. $\cos 105°$
(*Hint:* $105° = 60° + 45°$)

8. $\cos(-105°)$
(*Hint:* $-105° = -60° + (-45°)$)

9. $\cos \dfrac{7\pi}{12}$

10. $\cos\left(-\dfrac{\pi}{12}\right)$

11. $\cos 40° \cos 50° - \sin 40° \sin 50°$

12. $\cos \dfrac{7\pi}{9} \cos \dfrac{2\pi}{9} - \sin \dfrac{7\pi}{9} \sin \dfrac{2\pi}{9}$

Use a graphing calculator to support your answer for each of the following. See Example 1.

13. Exercise 11

14. Exercise 12

Write each function value in terms of the cofunction of a complementary angle. See Example 2.

15. $\tan 87°$

16. $\sin 15°$

17. $\cos \dfrac{\pi}{12}$

18. $\sin \dfrac{2\pi}{5}$

19. $\csc(-14° \, 24')$

20. $\sin 142° \, 14'$

21. $\sin \dfrac{5\pi}{8}$

22. $\cot \dfrac{9\pi}{10}$

23. $\sec 146° \, 42'$

24. $\tan 174° \, 3'$

25. $\cot 176.9814°$

26. $\sin 98.0142°$

Use identities to fill in each blank with the appropriate trigonometric function name. See Example 2.

27. $\cot \dfrac{\pi}{3} =$ _____ $\dfrac{\pi}{6}$

28. $\sin \dfrac{2\pi}{3} =$ _____ $\left(-\dfrac{\pi}{6}\right)$

29. _____ $33° = \sin 57°$

30. _____ $72° = \cot 18°$

31. $\cos 70° = \dfrac{1}{\underline{\hspace{0.6cm}} 20°}$

32. $\tan 24° = \dfrac{1}{\underline{\hspace{0.6cm}} 66°}$

Find an angle θ that makes each statement true. See Example 2.

33. $\tan \theta = \cot(45° + 2\theta)$

34. $\sin \theta = \cos(2\theta - 10°)$

35. $\sec \theta = \csc\left(\dfrac{\theta}{2} + 20°\right)$

36. $\cos \theta = \sin(3\theta + 10°)$

37. $\sin(3\theta - 15°) = \cos(\theta + 25°)$

38. $\cot(\theta - 10°) = \tan(2\theta + 20°)$

Use identities to write each expression as a function of θ. See Example 3.

39. $\cos(0° - \theta)$ **40.** $\cos(90° - \theta)$ **41.** $\cos(180° - \theta)$ **42.** $\cos(270° - \theta)$

43. $\cos(0° + \theta)$ **44.** $\cos(90° + \theta)$ **45.** $\cos(180° + \theta)$ **46.** $\cos(270° + \theta)$

Find $\cos(s + t)$ and $\cos(s - t)$. See Example 4.

47. $\cos s = -\dfrac{1}{5}$ and $\sin t = \dfrac{3}{5}$, s and t in quadrant II

48. $\sin s = \dfrac{2}{3}$ and $\sin t = -\dfrac{1}{3}$, s in quadrant II and t in quadrant IV

49. $\sin s = \dfrac{3}{5}$ and $\sin t = -\dfrac{12}{13}$, s in quadrant I and t in quadrant III

50. $\cos s = -\dfrac{8}{17}$ and $\cos t = -\dfrac{3}{5}$, s and t in quadrant III

51. $\sin s = \dfrac{\sqrt{5}}{7}$ and $\sin t = \dfrac{\sqrt{6}}{8}$, s and t in quadrant I

52. $\cos s = \dfrac{\sqrt{2}}{4}$ and $\sin t = -\dfrac{\sqrt{5}}{6}$, s and t in quadrant IV

Concept Check   *Tell whether each statement is* true *or* false.

53. $\cos 42° = \cos(30° + 12°)$

54. $\cos(-24°) = \cos 16° - \cos 40°$

55. $\cos 74° = \cos 60° \cos 14° + \sin 60° \sin 14°$

56. $\cos 140° = \cos 60° \cos 80° - \sin 60° \sin 80°$

57. $\cos \dfrac{\pi}{3} = \cos \dfrac{\pi}{12} \cos \dfrac{\pi}{4} - \sin \dfrac{\pi}{12} \sin \dfrac{\pi}{4}$

58. $\cos \dfrac{2\pi}{3} = \cos \dfrac{11\pi}{12} \cos \dfrac{\pi}{4} + \sin \dfrac{11\pi}{12} \sin \dfrac{\pi}{4}$

59. $\cos 70° \cos 20° - \sin 70° \sin 20° = 0$

60. $\cos 85° \cos 40° + \sin 85° \sin 40° = \dfrac{\sqrt{2}}{2}$

61. $\tan\left(\theta - \dfrac{\pi}{2}\right) = \cot \theta$

62. $\sin\left(\theta - \dfrac{\pi}{2}\right) = \cos \theta$

Verify that each equation is an identity.

63. $\cos\left(\dfrac{\pi}{2} + x\right) = -\sin x$

64. $\sec(\pi - x) = -\sec x$

65. $\cos 2x = \cos^2 x - \sin^2 x$ (*Hint:* $\cos 2x = \cos(x + x)$.)

66. $1 + \cos 2x - \cos^2 x = \cos^2 x$ (*Hint:* Use the result from Exercise 65.)

Relating Concepts

For individual or collaborative investigation
(Exercises 67–70)

The identities for $\cos(A + B)$ *and* $\cos(A - B)$ *can be used to find exact values of expressions like* $\cos 195°$ *and* $\cos 255°$, *where the angle is not in the first quadrant.* **Work Exercises 67–70 in order,** *to see how this is done.*

67. By writing 195° as 180° + 15°, use the identity for $\cos(A + B)$ to express $\cos 195°$ as $-\cos 15°$.

68. Use the identity for $\cos(A - B)$ to find $-\cos 15°$.

69. By the results of Exercises 67 and 68, $\cos 195° = $ _____.

70. Find each exact value using the method shown in Exercises 67–69.

   (a) $\cos 255°$ **(b)** $\cos \dfrac{11\pi}{12}$

(Modeling) *Solve each problem.*

71. *Electric Current* Refer to Example 5.

   (a) How many times does the current oscillate in .05 sec?

   (b) What are the maximum and minimum voltages in this outlet? Is the voltage always equal to 115 volts?

72. *Sound Waves* Sound is a result of waves applying pressure to a person's eardrum. For a pure sound wave radiating outward in a spherical shape, the trigonometric function defined by

$$P = \frac{a}{r} \cos\left(\frac{2\pi r}{\lambda} - ct\right)$$

can be used to model the sound pressure at a radius of r feet from the source, where t is time in seconds, λ is length of the sound wave in feet, c is speed of sound in feet per second, and a is maximum sound pressure at the source measured in pounds per square foot. (*Source:* Beranek, L., *Noise and Vibration Control,* Institute of Noise Control Engineering, Washington, D.C., 1988.) Let $\lambda = 4.9$ ft and $c = 1026$ ft per sec.

   (a) Let $a = .4$ lb per ft^2. Graph the sound pressure at distance $r = 10$ ft from its source in the window $[0, .05]$ by $[-.05, .05]$. Describe P at this distance.

   (b) Now let $a = 3$ and $t = 10$. Graph the sound pressure in the window $[0, 20]$ by $[-2, 2]$. What happens to pressure P as radius r increases?

   (c) Suppose a person stands at a radius r so that $r = n\lambda$, where n is a positive integer. Use the difference identity for cosine to simplify P in this situation.

5.4 | Sum and Difference Identities for Sine and Tangent

Sum and Difference Identities for Sine ▪ **Sum and Difference Identities for Tangent** ▪ **Applying the Sum and Difference Identities**

Sum and Difference Identities for Sine We can use the cosine sum and difference identities to derive similar identities for sine and tangent. Since $\sin \theta = \cos(90° - \theta)$, we replace θ with $A + B$ to get

$$\sin(A + B) = \cos[90° - (A + B)] \qquad \text{Cofunction identity (Section 5.3)}$$
$$= \cos[(90° - A) - B]$$
$$= \cos(90° - A) \cos B + \sin(90° - A) \sin B$$
$$\qquad\qquad \text{Cosine difference identity (Section 5.3)}$$
$$\sin(A + B) = \sin A \cos B + \cos A \sin B. \quad \text{Cofunction identities}$$

Now we write $\sin(A - B)$ as $\sin[A + (-B)]$ and use the identity for $\sin(A + B)$.

$$\sin(A - B) = \sin[A + (-B)]$$
$$= \sin A \cos(-B) + \cos A \sin(-B) \quad \text{Sine sum identity}$$
$$\sin(A - B) = \sin A \cos B - \cos A \sin B \qquad \text{Negative-angle identities}$$
$$\qquad\qquad\qquad\qquad\qquad\qquad\qquad\qquad \text{(Section 5.1)}$$

Sine of a Sum or Difference

$$\sin(A + B) = \sin A \cos B + \cos A \sin B$$

$$\sin(A - B) = \sin A \cos B - \cos A \sin B$$

Sum and Difference Identities for Tangent To derive the identity for $\tan(A + B)$, we start with

$$\tan(A + B) = \frac{\sin(A + B)}{\cos(A + B)} \qquad \text{Fundamental identity (Section 5.1)}$$

$$= \frac{\sin A \cos B + \cos A \sin B}{\cos A \cos B - \sin A \sin B}. \qquad \text{Sum identities}$$

We express this result in terms of the tangent function by multiplying both numerator and denominator by $\frac{1}{\cos A \cos B}$.

$$\tan(A + B) = \frac{\dfrac{\sin A \cos B + \cos A \sin B}{1}}{\dfrac{\cos A \cos B - \sin A \sin B}{1}} \cdot \frac{\dfrac{1}{\cos A \cos B}}{\dfrac{1}{\cos A \cos B}} \qquad \begin{array}{l}\text{Simplify the}\\\text{complex fraction.}\end{array}$$

$$= \frac{\dfrac{\sin A}{\cos A} + \dfrac{\sin B}{\cos B}}{1 - \dfrac{\sin A}{\cos A} \cdot \dfrac{\sin B}{\cos B}} \qquad \begin{array}{l}\text{Multiply numerators;}\\\text{multiply denominators;}\\\text{simplify.}\end{array}$$

Since $\frac{\sin \theta}{\cos \theta} = \tan \theta$, we have

$$\tan(A + B) = \frac{\tan A + \tan B}{1 - \tan A \tan B}.$$

Replacing B with $-B$ and using the fact that $\tan(-B) = -\tan B$ give the identity for the tangent of the difference of two angles.

Tangent of a Sum or Difference

$$\tan(A + B) = \frac{\tan A + \tan B}{1 - \tan A \tan B} \qquad \tan(A - B) = \frac{\tan A - \tan B}{1 + \tan A \tan B}$$

Applying the Sum and Difference Identities

EXAMPLE 1 Finding Exact Sine and Tangent Function Values

Find the *exact* value of each expression.

(a) $\sin 75°$ **(b)** $\tan \dfrac{7\pi}{12}$ **(c)** $\sin 40° \cos 160° - \cos 40° \sin 160°$

Solution

(a) $\sin 75° = \sin(45° + 30°)$

$\qquad\qquad = \sin 45° \cos 30° + \cos 45° \sin 30°$ Sine sum identity

$\qquad\qquad = \dfrac{\sqrt{2}}{2} \cdot \dfrac{\sqrt{3}}{2} + \dfrac{\sqrt{2}}{2} \cdot \dfrac{1}{2}$ (Section 2.1)

$\qquad\qquad = \dfrac{\sqrt{6} + \sqrt{2}}{4}$

(b) $\tan \dfrac{7\pi}{12} = \tan\left(\dfrac{\pi}{3} + \dfrac{\pi}{4}\right)$

$\qquad\qquad = \dfrac{\tan \frac{\pi}{3} + \tan \frac{\pi}{4}}{1 - \tan \frac{\pi}{3} \tan \frac{\pi}{4}}$ Tangent sum identity

$\qquad\qquad = \dfrac{\sqrt{3} + 1}{1 - \sqrt{3} \cdot 1}$ (Section 3.1)

$\qquad\qquad = \dfrac{\sqrt{3} + 1}{1 - \sqrt{3}} \cdot \dfrac{1 + \sqrt{3}}{1 + \sqrt{3}}$ Rationalize the denominator.

$\qquad\qquad = \dfrac{\sqrt{3} + 3 + 1 + \sqrt{3}}{1 - 3}$ Multiply.

$\qquad\qquad = \dfrac{4 + 2\sqrt{3}}{-2}$ Combine terms.

$\qquad\qquad = \dfrac{2\left(2 + \sqrt{3}\right)}{-2}$ Factor out 2 in the numerator.

$\qquad\qquad = -2 - \sqrt{3}$ Lowest terms

(c) $\sin 40° \cos 160° - \cos 40° \sin 160° = \sin(40° - 160°)$

Sine difference identity

$$= \sin(-120°)$$

$$= -\sin 120° \qquad \text{Negative-angle identity}$$

$$= -\frac{\sqrt{3}}{2} \qquad \text{(Section 2.2)}$$

Now try Exercises 9, 11, and 15.

EXAMPLE 2 Writing Functions as Expressions Involving Functions of θ

Write each function as an expression involving functions of θ.

(a) $\sin(30° + \theta)$ **(b)** $\tan(45° - \theta)$ **(c)** $\sin(180° + \theta)$

Solution

(a) Using the identity for $\sin(A + B)$,

$$\sin(30° + \theta) = \sin 30° \cos \theta + \cos 30° \sin \theta$$

$$= \frac{1}{2} \cos \theta + \frac{\sqrt{3}}{2} \sin \theta.$$

(b) $\tan(45° - \theta) = \dfrac{\tan 45° - \tan \theta}{1 + \tan 45° \tan \theta} = \dfrac{1 - \tan \theta}{1 + \tan \theta}$

(c) $\sin(180° + \theta) = \sin 180° \cos \theta + \cos 180° \sin \theta$

$$= 0 \cdot \cos \theta + (-1) \sin \theta \qquad \text{(Section 1.3)}$$

$$= -\sin \theta$$

Now try Exercises 29 and 33.

EXAMPLE 3 Finding Function Values and the Quadrant of $A + B$

Suppose that A and B are angles in standard position, with $\sin A = \frac{4}{5}$, $\frac{\pi}{2} < A < \pi$, and $\cos B = -\frac{5}{13}$, $\pi < B < \frac{3\pi}{2}$. Find each of the following.

(a) $\sin(A + B)$ **(b)** $\tan(A + B)$ **(c)** the quadrant of $A + B$

Solution

(a) The identity for $\sin(A + B)$ requires $\sin A$, $\cos A$, $\sin B$, and $\cos B$. We are given values of $\sin A$ and $\cos B$. We must find values of $\cos A$ and $\sin B$.

$$\sin^2 A + \cos^2 A = 1 \qquad \text{Fundamental identity (Section 5.1)}$$

$$\frac{16}{25} + \cos^2 A = 1 \qquad \sin A = \tfrac{4}{5}$$

$$\cos^2 A = \frac{9}{25} \qquad \text{Subtract } \tfrac{16}{25}.$$

$$\cos A = -\frac{3}{5} \qquad \begin{array}{l}\text{Take square roots (\textbf{Appendix A}); since} \\ A \text{ is in quadrant II, } \cos A < 0.\end{array}$$

In the same way, $\sin B = -\frac{12}{13}$. Now use the formula for $\sin(A + B)$.

$$\sin(A + B) = \frac{4}{5}\left(-\frac{5}{13}\right) + \left(-\frac{3}{5}\right)\left(-\frac{12}{13}\right)$$

$$= -\frac{20}{65} + \frac{36}{65} = \frac{16}{65}$$

(b) To find $\tan(A + B)$, first use the values of sine and cosine from part (a) to get $\tan A = -\frac{4}{3}$ and $\tan B = \frac{12}{5}$.

$$\tan(A + B) = \frac{-\frac{4}{3} + \frac{12}{5}}{1 - \left(-\frac{4}{3}\right)\left(\frac{12}{5}\right)} = \frac{\frac{16}{15}}{1 + \frac{48}{15}} = \frac{\frac{16}{15}}{\frac{63}{15}} = \frac{16}{63}$$

(c) From parts (a) and (b), $\sin(A + B) = \frac{16}{65}$ and $\tan(A + B) = \frac{16}{63}$, both positive. Therefore, $A + B$ must be in quadrant I, since it is the only quadrant in which both sine and tangent are positive.

Now try Exercise 41.

EXAMPLE 4 Verifying an Identity Using Sum and Difference Identities

Verify that the equation is an identity.

$$\sin\left(\frac{\pi}{6} + s\right) + \cos\left(\frac{\pi}{3} + s\right) = \cos s$$

Solution Work on the left side, using the sum identities for $\sin(A + B)$ and $\cos(A + B)$.

$$\sin\left(\frac{\pi}{6} + s\right) + \cos\left(\frac{\pi}{3} + s\right)$$

$$= \left(\sin\frac{\pi}{6}\cos s + \cos\frac{\pi}{6}\sin s\right) \qquad \text{Sine sum identity}$$

$$+ \left(\cos\frac{\pi}{3}\cos s - \sin\frac{\pi}{3}\sin s\right) \qquad \text{Cosine sum identity (Section 5.3)}$$

$$= \left(\frac{1}{2}\cos s + \frac{\sqrt{3}}{2}\sin s\right) \qquad \sin\frac{\pi}{6} = \frac{1}{2};\ \cos\frac{\pi}{6} = \frac{\sqrt{3}}{2}$$

$$+ \left(\frac{1}{2}\cos s - \frac{\sqrt{3}}{2}\sin s\right) \qquad \cos\frac{\pi}{3} = \frac{1}{2};\ \sin\frac{\pi}{3} = \frac{\sqrt{3}}{2}$$

$$= \frac{1}{2}\cos s + \frac{1}{2}\cos s$$

$$= \cos s$$

Now try Exercise 59.

5.4 Exercises

1. Compare the formulas for $\sin(A - B)$ and $\sin(A + B)$. How do they differ? How are they alike?

2. Compare the formulas for $\tan(A - B)$ and $\tan(A + B)$. How do they differ? How are they alike?

Concept Check Match each expression in Column I with its value in Column II. See Example 1.

I	II
3. $\sin 15°$	**A.** $\dfrac{\sqrt{6} + \sqrt{2}}{4}$
4. $\sin 105°$	**B.** $\dfrac{-\sqrt{6} - \sqrt{2}}{4}$
5. $\tan 15°$	**C.** $\dfrac{\sqrt{6} - \sqrt{2}}{4}$
6. $\tan 105°$	**D.** $2 + \sqrt{3}$
7. $\sin(-105°)$	**E.** $2 - \sqrt{3}$
8. $\tan(-105°)$	**F.** $-2 - \sqrt{3}$

Use identities to find each exact value. See Example 1.

9. $\sin \dfrac{5\pi}{12}$

10. $\tan \dfrac{5\pi}{12}$

11. $\tan \dfrac{\pi}{12}$

12. $\sin \dfrac{\pi}{12}$

13. $\sin\left(-\dfrac{7\pi}{12}\right)$

14. $\tan\left(-\dfrac{7\pi}{12}\right)$

15. $\sin 76° \cos 31° - \cos 76° \sin 31°$

16. $\sin 40° \cos 50° + \cos 40° \sin 50°$

17. $\dfrac{\tan 80° + \tan 55°}{1 - \tan 80° \tan 55°}$

18. $\dfrac{\tan 80° - \tan(-55°)}{1 + \tan 80° \tan(-55°)}$

19. $\dfrac{\tan 100° + \tan 80°}{1 - \tan 100° \tan 80°}$

20. $\sin 100° \cos 10° - \cos 100° \sin 10°$

21. $\sin \dfrac{\pi}{5} \cos \dfrac{3\pi}{10} + \cos \dfrac{\pi}{5} \sin \dfrac{3\pi}{10}$

22. $\dfrac{\tan \frac{5\pi}{12} + \tan \frac{\pi}{4}}{1 - \tan \frac{5\pi}{12} \tan \frac{\pi}{4}}$

Use identities to write each expression as a single function of x or θ. See Example 2.

23. $\cos(30° + \theta)$

24. $\cos(45° - \theta)$

25. $\cos(60° + \theta)$

26. $\cos(\theta - 30°)$

27. $\cos\left(\dfrac{3\pi}{4} - x\right)$

28. $\sin(45° + \theta)$

29. $\tan(\theta + 30°)$

30. $\tan\left(\dfrac{\pi}{4} + x\right)$

31. $\sin\left(\dfrac{\pi}{4} + x\right)$

32. $\sin(180° - \theta)$

33. $\sin(270° - \theta)$

34. $\tan(180° + \theta)$

35. $\tan(360° - \theta)$

36. $\sin(\pi + \theta)$

37. $\tan(\pi - \theta)$

38. Why is it not possible to follow Example 2 to find a formula for $\tan(270° - \theta)$?

39. What happens when you try to evaluate $\dfrac{\tan 65.902° + \tan 24.098°}{1 - \tan 65.902° \tan 24.098°}$?

40. Show that if A, B, and C are the angles of a triangle, then $\sin(A + B + C) = 0$.

*Use the given information to find (**a**) $\sin(s + t)$, (**b**) $\tan(s + t)$, and (**c**) the quadrant of $s + t$. See Example 3.*

41. $\cos s = \dfrac{3}{5}$ and $\sin t = \dfrac{5}{13}$, s and t in quadrant I

42. $\cos s = -\dfrac{1}{5}$ and $\sin t = \dfrac{3}{5}$, s and t in quadrant II

43. $\sin s = \dfrac{2}{3}$ and $\sin t = -\dfrac{1}{3}$, s in quadrant II and t in quadrant IV

44. $\sin s = \dfrac{3}{5}$ and $\sin t = -\dfrac{12}{13}$, s in quadrant I and t in quadrant III

45. $\cos s = -\dfrac{8}{17}$ and $\cos t = -\dfrac{3}{5}$, s and t in quadrant III

46. $\cos s = -\dfrac{15}{17}$ and $\sin t = \dfrac{4}{5}$, s in quadrant II and t in quadrant I

Find each exact value. Use the technique developed in Section 5.3, Exercises 67–70.

47. $\sin 165°$ **48.** $\tan 165°$ **49.** $\sin 255°$

50. $\tan 285°$ **51.** $\tan \dfrac{11\pi}{12}$ **52.** $\sin\left(-\dfrac{13\pi}{12}\right)$

Graph each expression and use the graph to conjecture an identity. Then verify your conjecture algebraically.

53. $\sin\left(\dfrac{\pi}{2} + x\right)$ **54.** $\sin\left(\dfrac{3\pi}{2} + x\right)$ **55.** $\tan\left(\dfrac{\pi}{2} + x\right)$ **56.** $\dfrac{1 + \tan x}{1 - \tan x}$

Verify that each equation is an identity. See Example 4.

57. $\sin 2x = 2 \sin x \cos x$ (*Hint:* $\sin 2x = \sin(x + x)$)

58. $\sin(x + y) + \sin(x - y) = 2 \sin x \cos y$

59. $\sin(210° + x) - \cos(120° + x) = 0$

60. $\tan(x - y) - \tan(y - x) = \dfrac{2(\tan x - \tan y)}{1 + \tan x \tan y}$

61. $\dfrac{\cos(\alpha - \beta)}{\cos \alpha \sin \beta} = \tan \alpha + \cot \beta$ **62.** $\dfrac{\sin(s + t)}{\cos s \cos t} = \tan s + \tan t$

63. $\dfrac{\sin(x - y)}{\sin(x + y)} = \dfrac{\tan x - \tan y}{\tan x + \tan y}$ **64.** $\dfrac{\sin(x + y)}{\cos(x - y)} = \dfrac{\cot x + \cot y}{1 + \cot x \cot y}$

65. $\dfrac{\sin(s - t)}{\sin t} + \dfrac{\cos(s - t)}{\cos t} = \dfrac{\sin s}{\sin t \cos t}$ **66.** $\dfrac{\tan(\alpha + \beta) - \tan \beta}{1 + \tan(\alpha + \beta) \tan \beta} = \tan \alpha$

Relating Concepts

For individual or collaborative investigation
(Exercises 67–72)

Refer to the figure on the left below. By the definition of $\tan \theta$, $m = \tan \theta$, *where m is the slope and* θ *is the angle of inclination of the line. The following exercises, which depend on properties of triangles, refer to triangle ABC in the figure on the right below.* **Work Exercises 67–72 in order.** *Assume that all angles are measured in degrees.*

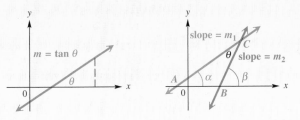

67. In terms of β, what is the measure of angle ABC?

68. Use the fact that the sum of the angles in a triangle is $180°$ to express θ in terms of α and β.

69. Apply the formula for $\tan(A - B)$ to obtain an expression for $\tan \theta$ in terms of $\tan \alpha$ and $\tan \beta$.

70. Replace $\tan \alpha$ with m_1 and $\tan \beta$ with m_2 to obtain $\tan \theta = \dfrac{m_2 - m_1}{1 + m_1 m_2}$.

 In Exercises 71 and 72, use the result from Exercise 70 to find the angle between each pair of lines. Use a calculator and round to the nearest tenth of a degree.

71. $x + y = 9$, $\quad 2x + y = -1$ 72. $5x - 2y + 4 = 0$, $\quad 3x + 5y = 6$

(Modeling)   *Solve each problem.*

73. *Back Stress*   If a person bends at the waist with a straight back making an angle of θ degrees with the horizontal, then the force F exerted on the back muscles can be modeled by the equation

$$F = \frac{.6W \sin(\theta + 90°)}{\sin 12°},$$

where W is the weight of the person. (*Source:* Metcalf, H., *Topics in Classical Biophysics,* Prentice-Hall, 1980.)

(a) Calculate F when $W = 170$ lb and $\theta = 30°$.

(b) Use an identity to show that F is approximately equal to $2.9W \cos \theta$.

(c) For what value of θ is F maximum?

74. *Back Stress*   Refer to Exercise 73.

(a) Suppose a 200-lb person bends at the waist so that $\theta = 45°$. Estimate the force exerted on the person's back muscles.

(b) Approximate graphically the value of θ that results in the back muscles exerting a force of 400 lb.

75. *Voltage*   A coil of wire rotating in a magnetic field induces a voltage

$$e = 20 \sin\left(\frac{\pi t}{4} - \frac{\pi}{2}\right).$$

Use an identity from this section to express this in terms of $\cos \frac{\pi t}{4}$.

76. *Voltage of a Circuit*   When the two voltages

$$V_1 = 30 \sin 120\pi t \qquad \text{and} \qquad V_2 = 40 \cos 120\pi t$$

are applied to the same circuit, the resulting voltage V will be equal to their sum. (*Source:* Bell, D., *Fundamentals of Electric Circuits,* Second Edition, Reston Publishing Company, 1981.)

(a) Graph the sum in the window $[0, .05]$ by $[-60, 60]$.

(b) Use the graph to estimate values for a and ϕ so that $V = a \sin(120\pi t + \phi)$.

(c) Use identities to verify that your expression for V is valid.

5.5 Double-Angle Identities

Double-Angle Identities ▪ Product-to-Sum and Sum-to-Product Identities

Double-Angle Identities   When $A = B$ in the identities for the sum of two angles, these identities are called the **double-angle identities.** For example, to derive an expression for $\cos 2A$, we let $B = A$ in the identity $\cos(A + B) = \cos A \cos B - \sin A \sin B$.

$$\cos 2A = \cos(A + A)$$
$$= \cos A \cos A - \sin A \sin A \qquad \text{Cosine sum identity (Section 5.3)}$$
$$\cos 2A = \cos^2 A - \sin^2 A$$

Two other useful forms of this identity can be obtained by substituting either $\cos^2 A = 1 - \sin^2 A$ or $\sin^2 A = 1 - \cos^2 A$. Replace $\cos^2 A$ with the expression $1 - \sin^2 A$ to get

$$\cos 2A = \cos^2 A - \sin^2 A$$
$$= (1 - \sin^2 A) - \sin^2 A \qquad \text{Fundamental identity (Section 5.1)}$$
$$\cos 2A = 1 - 2\sin^2 A,$$

or replace $\sin^2 A$ with $1 - \cos^2 A$ to get

$$\cos 2A = \cos^2 A - \sin^2 A$$
$$= \cos^2 A - (1 - \cos^2 A) \qquad \text{Fundamental identity}$$
$$= \cos^2 A - 1 + \cos^2 A$$
$$\cos 2A = 2\cos^2 A - 1.$$

We find $\sin 2A$ with the identity $\sin(A + B) = \sin A \cos B + \cos A \sin B$, letting $B = A$.

$$\sin 2A = \sin(A + A)$$
$$= \sin A \cos A + \cos A \sin A \quad \text{Sine sum identity (Section 5.4)}$$
$$\sin 2A = 2 \sin A \cos A$$

Using the identity for $\tan(A + B)$, we find $\tan 2A$.

$$\tan 2A = \tan(A + A)$$
$$= \frac{\tan A + \tan A}{1 - \tan A \tan A} \quad \text{Tangent sum identity (Section 5.4)}$$
$$\tan 2A = \frac{2 \tan A}{1 - \tan^2 A}$$

Looking Ahead to Calculus

The identities

$$\cos 2A = 1 - 2 \sin^2 A$$

and $\quad \cos 2A = 2 \cos^2 A - 1$

can be rewritten as

$$\sin^2 A = \frac{1}{2}(1 - \cos 2A)$$

and $\quad \cos^2 A = \frac{1}{2}(1 + \cos 2A)$.

These identities are used to integrate the functions $f(A) = \sin^2 A$ and $g(A) = \cos^2 A$.

Double-Angle Identities

$$\cos 2A = \cos^2 A - \sin^2 A \qquad \cos 2A = 1 - 2 \sin^2 A$$
$$\cos 2A = 2 \cos^2 A - 1 \qquad \sin 2A = 2 \sin A \cos A$$
$$\tan 2A = \frac{2 \tan A}{1 - \tan^2 A}$$

EXAMPLE 1 Finding Function Values of 2θ Given Information about θ

Given $\cos \theta = \frac{3}{5}$ and $\sin \theta < 0$, find $\sin 2\theta$, $\cos 2\theta$, and $\tan 2\theta$.

Solution To find $\sin 2\theta$, we must first find the value of $\sin \theta$.

$$\sin^2 \theta + \left(\frac{3}{5}\right)^2 = 1 \qquad \sin^2 \theta + \cos^2 \theta = 1;\ \cos \theta = \frac{3}{5}$$

$$\sin^2 \theta = \frac{16}{25} \qquad \text{Simplify.}$$

$$\sin \theta = -\frac{4}{5} \qquad \begin{array}{l}\text{Take square roots (Appendix A); choose the}\\ \text{negative square root since } \sin \theta < 0.\end{array}$$

Using the double-angle identity for sine,

$$\sin 2\theta = 2 \sin \theta \cos \theta$$
$$= 2\left(-\frac{4}{5}\right)\left(\frac{3}{5}\right) = -\frac{24}{25}. \qquad \sin \theta = -\frac{4}{5};\ \cos \theta = \frac{3}{5}$$

Now we find $\cos 2\theta$, using the first of the double-angle identities for cosine. (Any of the three forms may be used.)

$$\cos 2\theta = \cos^2 \theta - \sin^2 \theta$$
$$= \frac{9}{25} - \frac{16}{25} = -\frac{7}{25}$$

The value of tan 2θ can be found in either of two ways. We can use the double-angle identity and the fact that $\tan \theta = \frac{\sin \theta}{\cos \theta} = \frac{-\frac{4}{5}}{\frac{3}{5}} = -\frac{4}{5} \div \frac{3}{5} = -\frac{4}{5} \cdot \frac{5}{3} = -\frac{4}{3}.$

$$\tan 2\theta = \frac{2 \tan \theta}{1 - \tan^2 \theta} = \frac{2\left(-\frac{4}{3}\right)}{1 - \frac{16}{9}} = \frac{-\frac{8}{3}}{-\frac{7}{9}} = \frac{24}{7}$$

Simplify the complex fraction.

Alternatively, we can find tan 2θ by finding the quotient of sin 2θ and cos 2θ.

$$\tan 2\theta = \frac{\sin 2\theta}{\cos 2\theta} = \frac{-\frac{24}{25}}{-\frac{7}{25}} = \frac{24}{7}$$

Now try Exercise 15.

EXAMPLE 2 Finding Function Values of θ Given Information about 2θ

Find the values of the six trigonometric functions of θ if $\cos 2\theta = \frac{4}{5}$ and $90° < \theta < 180°$.

Solution Use one of the double-angle identities for cosine to get a trigonometric function value for θ.

$$\cos 2\theta = 1 - 2 \sin^2 \theta$$

$$\frac{4}{5} = 1 - 2 \sin^2 \theta \qquad \cos 2\theta = \frac{4}{5}$$

$$-\frac{1}{5} = -2 \sin^2 \theta \qquad \text{Subtract 1.}$$

$$\frac{1}{10} = \sin^2 \theta \qquad \text{Multiply by } -\frac{1}{2}.$$

$$\sin \theta = \sqrt{\frac{1}{10}} \qquad \begin{array}{l}\text{Take square roots; choose the positive square} \\ \text{root since } \theta \text{ terminates in quadrant II.}\end{array}$$

$$\sin \theta = \frac{\sqrt{1}}{\sqrt{10}} \cdot \frac{\sqrt{10}}{\sqrt{10}} = \frac{\sqrt{10}}{10} \qquad \text{Quotient rule; rationalize the denominator.}$$

Now find values of cos θ and tan θ using the fundamental identities or by sketching and labeling a right triangle in quadrant II. From the triangle in Figure 7,

$$\cos \theta = \frac{-3}{\sqrt{10}} = -\frac{3\sqrt{10}}{10}, \qquad \text{and} \qquad \tan \theta = \frac{1}{-3} = -\frac{1}{3}. \quad \text{(Section 1.3)}$$

Figure 7

Find the other three functions using reciprocals.

$$\csc \theta = \frac{1}{\sin \theta} = \sqrt{10}, \qquad \sec \theta = \frac{1}{\cos \theta} = -\frac{\sqrt{10}}{3}, \qquad \cot \theta = \frac{1}{\tan \theta} = -3$$

Now try Exercise 9.

EXAMPLE 3 Verifying a Double-Angle Identity

Verify that the following equation is an identity.

$$\cot x \sin 2x = 1 + \cos 2x$$

Solution We start by working on the left side, using the hint from Section 5.1 about writing all functions in terms of sine and cosine.

$$\cot x \sin 2x = \frac{\cos x}{\sin x} \cdot \sin 2x \qquad \text{Quotient identity}$$

$$= \frac{\cos x}{\sin x}(2 \sin x \cos x) \quad \text{Double-angle identity}$$

$$= 2 \cos^2 x$$

$$= 1 + \cos 2x \qquad \cos 2x = 2\cos^2 x - 1, \text{ so}$$
$$\qquad\qquad\qquad\qquad\quad 2\cos^2 x = 1 + \cos 2x$$

The final step illustrates the importance of being able to recognize alternative forms of identities.

Now try Exercise 37.

EXAMPLE 4 Simplifying Expressions Using Double-Angle Identities

Simplify each expression.

(a) $\cos^2 7x - \sin^2 7x$ **(b)** $\sin 15° \cos 15°$

Solution

(a) This expression suggests one of the double-angle identities for cosine: $\cos 2A = \cos^2 A - \sin^2 A$. Substituting $7x$ for A gives

$$\cos^2 7x - \sin^2 7x = \cos 2(7x) = \cos 14x.$$

(b) If this expression were $2 \sin 15° \cos 15°$, we could apply the identity for $\sin 2A$ directly since $\sin 2A = 2 \sin A \cos A$. We can still apply the identity with $A = 15°$ by writing the multiplicative identity element 1 as $\frac{1}{2}(2)$.

$$\sin 15° \cos 15° = \frac{1}{2}(2) \sin 15° \cos 15° \qquad \text{Multiply by 1 in the form } \tfrac{1}{2}(2).$$

$$= \frac{1}{2}(2 \sin 15° \cos 15°) \qquad \text{Associative property}$$

$$= \frac{1}{2}\sin(2 \cdot 15°) \qquad 2 \sin A \cos A = \sin 2A, \text{ with } A = 15°$$

$$= \frac{1}{2}\sin 30°$$

$$= \frac{1}{2} \cdot \frac{1}{2} = \frac{1}{4} \qquad \sin 30° = \tfrac{1}{2} \text{ (Section 2.1)}$$

Now try Exercises 17 and 19.

Identities involving larger multiples of the variable can be derived by repeated use of the double-angle identities and other identities.

EXAMPLE 5 Deriving a Multiple-Angle Identity

Write $\sin 3x$ in terms of $\sin x$.

Solution

$$\sin 3x = \sin(2x + x)$$

$$= \sin 2x \cos x + \cos 2x \sin x \qquad \text{Sine sum identity (Section 5.4)}$$

$$= (2 \sin x \cos x) \cos x + (\cos^2 x - \sin^2 x) \sin x \qquad \text{Double-angle identities}$$

$$= 2 \sin x \cos^2 x + \cos^2 x \sin x - \sin^3 x \qquad \text{Multiply.}$$

$$= 2 \sin x(1 - \sin^2 x) + (1 - \sin^2 x) \sin x - \sin^3 x \qquad \cos^2 x = 1 - \sin^2 x$$

$$= 2 \sin x - 2 \sin^3 x + \sin x - \sin^3 x - \sin^3 x \qquad \text{Distributive property}$$

$$= 3 \sin x - 4 \sin^3 x \qquad \text{Combine terms.}$$

Now try Exercise 29.

The next example applies a multiple-angle identity to answer a question about electric current.

EXAMPLE 6 Determining Wattage Consumption

If a toaster is plugged into a common household outlet, the wattage consumed is not constant. Instead, it varies at a high frequency according to the model

$$W = \frac{V^2}{R},$$

where V is the voltage and R is a constant that measures the resistance of the toaster in ohms. (*Source:* Bell, D., *Fundamentals of Electric Circuits, Fourth Edition*, Prentice-Hall, 1998.) Graph the wattage W consumed by a typical toaster with $R = 15$ and $V = 163 \sin 120\pi t$ in the window $[0, .05]$ by $[-500, 2000]$. How many oscillations are there?

Solution Substituting the given values into the wattage equation gives

$$W = \frac{V^2}{R} = \frac{(163 \sin 120\pi t)^2}{15}.$$

To determine the range of W, we note that $\sin 120\pi t$ has maximum value 1, so the expression for W has maximum value $\frac{163^2}{15} \approx 1771$. The minimum value is 0. The graph in Figure 8 shows that there are six oscillations.

Now try Exercise 65.

For $x = t$,
$$W(t) = \frac{(163 \sin 120\pi t)^2}{15}$$

Figure 8

Product-to-Sum and Sum-to-Product Identities
Because they make it possible to rewrite a product as a sum, the identities for $\cos(A + B)$ and $\cos(A - B)$ are used to derive a group of identities useful in calculus.

Adding the identities for $\cos(A + B)$ and $\cos(A - B)$ gives

$$\cos(A + B) = \cos A \cos B - \sin A \sin B$$
$$\cos(A - B) = \cos A \cos B + \sin A \sin B$$
$$\overline{\cos(A + B) + \cos(A - B) = 2 \cos A \cos B}$$

or

$$\cos A \cos B = \frac{1}{2}[\cos(A + B) + \cos(A - B)].$$

Similarly, subtracting $\cos(A + B)$ from $\cos(A - B)$ gives

$$\sin A \sin B = \frac{1}{2}[\cos(A - B) - \cos(A + B)].$$

Using the identities for $\sin(A + B)$ and $\sin(A - B)$ in the same way, we get two more identities. Those and the previous ones are now summarized.

Looking Ahead to Calculus

The product-to-sum identities are used in calculus to find integrals of functions that are products of trigonometric functions. One classic calculus text includes the following example:

Evaluate $\int \cos 5x \cos 3x\, dx$.

The first solution line reads: "We may write

$$\cos 5x \cos 3x = \frac{1}{2}[\cos 8x + \cos 2x].\text{"}$$

Product-to-Sum Identities

$$\cos A \cos B = \frac{1}{2}[\cos(A + B) + \cos(A - B)]$$

$$\sin A \sin B = \frac{1}{2}[\cos(A - B) - \cos(A + B)]$$

$$\sin A \cos B = \frac{1}{2}[\sin(A + B) + \sin(A - B)]$$

$$\cos A \sin B = \frac{1}{2}[\sin(A + B) - \sin(A - B)]$$

EXAMPLE 7   Using a Product-to-Sum Identity

Write $\cos 2\theta \sin \theta$ as the sum or difference of two functions.

Solution   Use the identity for $\cos A \sin B$, with $2\theta = A$ and $\theta = B$.

$$\cos 2\theta \sin \theta = \frac{1}{2}[\sin(2\theta + \theta) - \sin(2\theta - \theta)]$$

$$= \frac{1}{2} \sin 3\theta - \frac{1}{2} \sin \theta$$

Now try Exercise 55.

Another group of identities allows us to write sums as products.

Sum-to-Product Identities

$$\sin A + \sin B = 2 \sin\left(\frac{A + B}{2}\right) \cos\left(\frac{A - B}{2}\right)$$

$$\sin A - \sin B = 2 \cos\left(\frac{A + B}{2}\right) \sin\left(\frac{A - B}{2}\right)$$

(continued)

Sum-to-Product Identities *(continued)*

$$\cos A + \cos B = 2 \cos\left(\frac{A + B}{2}\right) \cos\left(\frac{A - B}{2}\right)$$

$$\cos A - \cos B = -2 \sin\left(\frac{A + B}{2}\right) \sin\left(\frac{A - B}{2}\right)$$

EXAMPLE 8   Using a Sum-to-Product Identity

Write $\sin 2\theta - \sin 4\theta$ as a product of two functions.

Solution   Use the identity for $\sin A - \sin B$, with $2\theta = A$ and $4\theta = B$.

$$\sin 2\theta - \sin 4\theta = 2 \cos\left(\frac{2\theta + 4\theta}{2}\right) \sin\left(\frac{2\theta - 4\theta}{2}\right)$$

$$= 2 \cos \frac{6\theta}{2} \sin\left(\frac{-2\theta}{2}\right)$$

$$= 2 \cos 3\theta \sin(-\theta)$$

$$= -2 \cos 3\theta \sin \theta \quad \sin(-\theta) = -\sin \theta \text{ (Section 5.1)}$$

Now try Exercise 59.

5.5 Exercises

Concept Check   *Match each expression in Column I with its value in Column II.*

I	II
1. $2 \cos^2 15° - 1$	**A.** $\dfrac{1}{2}$
2. $\dfrac{2 \tan 15°}{1 - \tan^2 15°}$	**B.** $\dfrac{\sqrt{2}}{2}$
3. $2 \sin 22.5° \cos 22.5°$	**C.** $\dfrac{\sqrt{3}}{2}$
4. $\cos^2 \dfrac{\pi}{6} - \sin^2 \dfrac{\pi}{6}$	**D.** $-\sqrt{3}$
5. $2 \sin \dfrac{\pi}{3} \cos \dfrac{\pi}{3}$	**E.** $\dfrac{\sqrt{3}}{3}$
6. $\dfrac{2 \tan \frac{\pi}{3}}{1 - \tan^2 \frac{\pi}{3}}$	

Use identities to find values of the sine and cosine functions for each angle measure. See Examples 1 and 2.

7. θ, given $\cos 2\theta = \dfrac{3}{5}$ and θ terminates in quadrant I

8. θ, given $\cos 2\theta = \dfrac{3}{4}$ and θ terminates in quadrant III

9. θ, given $\cos 2\theta = -\dfrac{5}{12}$ and $\dfrac{\pi}{2} < \theta < \pi$

10. x, given $\cos 2x = \dfrac{2}{3}$ and $\dfrac{\pi}{2} < x < \pi$

11. 2θ, given $\sin \theta = \dfrac{2}{5}$ and $\cos \theta < 0$

12. 2θ, given $\cos \theta = -\dfrac{12}{13}$ and $\sin \theta > 0$

13. $2x$, given $\tan x = 2$ and $\cos x > 0$

14. $2x$, given $\tan x = \dfrac{5}{3}$ and $\sin x < 0$

15. 2θ, given $\sin \theta = -\dfrac{\sqrt{5}}{7}$ and $\cos \theta > 0$

16. 2θ, given $\cos \theta = \dfrac{\sqrt{3}}{5}$ and $\sin \theta > 0$

Use an identity to write each expression as a single trigonometric function value or as a single number. See Example 4.

17. $\cos^2 15° - \sin^2 15°$

18. $\dfrac{2 \tan 15°}{1 - \tan^2 15°}$

19. $1 - 2 \sin^2 15°$

20. $1 - 2 \sin^2 22\dfrac{1}{2}°$

21. $2 \cos^2 67\dfrac{1}{2}° - 1$

22. $\cos^2 \dfrac{\pi}{8} - \dfrac{1}{2}$

23. $\dfrac{\tan 51°}{1 - \tan^2 51°}$

24. $\dfrac{\tan 34°}{2(1 - \tan^2 34°)}$

25. $\dfrac{1}{4} - \dfrac{1}{2} \sin^2 47.1°$

26. $\dfrac{1}{8} \sin 29.5° \cos 29.5°$

27. $\sin^2 \dfrac{2\pi}{5} - \cos^2 \dfrac{2\pi}{5}$

28. $\cos^2 2x - \sin^2 2x$

Express each function as a trigonometric function of x. See Example 5.

29. $\cos 3x$

30. $\sin 4x$

31. $\tan 3x$

32. $\cos 4x$

⊞ *Graph each expression and use the graph to conjecture an identity. Then verify your conjecture algebraically.*

33. $\cos^4 x - \sin^4 x$

34. $\dfrac{4 \tan x \cos^2 x - 2 \tan x}{1 - \tan^2 x}$

35. $\dfrac{2 \tan x}{2 - \sec^2 x}$

36. $\dfrac{\cot^2 x - 1}{2 \cot x}$

Verify that each equation is an identity. See Example 3.

37. $(\sin x + \cos x)^2 = \sin 2x + 1$

38. $\sec 2x = \dfrac{\sec^2 x + \sec^4 x}{2 + \sec^2 x - \sec^4 x}$

39. $\tan 8k - \tan 8k \tan^2 4k = 2 \tan 4k$

40. $\sin 2x = \dfrac{2 \tan x}{1 + \tan^2 x}$

41. $\cos 2y = \dfrac{2 - \sec^2 y}{\sec^2 y}$

42. $-\tan 2\theta = \dfrac{2 \tan \theta}{\sec^2 \theta - 2}$

43. $\sin 4x = 4 \sin x \cos x \cos 2x$

44. $\dfrac{1 + \cos 2x}{\sin 2x} = \cot x$

45. $\tan(\theta - 45°) + \tan(\theta + 45°) = 2 \tan 2\theta$

46. $\cot 4\theta = \dfrac{1 - \tan^2 2\theta}{2 \tan 2\theta}$

47. $\dfrac{2 \cos 2\theta}{\sin 2\theta} = \cot \theta - \tan \theta$

48. $\sin 4x = 4 \sin x \cos x - 8 \sin^3 x \cos x$

49. $\sin 2A \cos 2A = \sin 2A - 4 \sin^3 A \cos A$

50. $\cos 2x = \dfrac{1 - \tan^2 x}{1 + \tan^2 x}$

51. $\tan s + \cot s = 2 \csc 2s$

52. $\dfrac{\cot A - \tan A}{\cot A + \tan A} = \cos 2A$

53. $1 + \tan x \tan 2x = \sec 2x$

54. $\cot \theta \tan(\theta + \pi) - \sin(\pi - \theta) \cos\left(\dfrac{\pi}{2} - \theta\right) = \cos^2 \theta$

Write each expression as a sum or difference of trigonometric functions. See Example 7.

55. $2 \sin 58° \cos 102°$

56. $2 \cos 85° \sin 140°$

57. $5 \cos 3x \cos 2x$

58. $\sin 4x \sin 5x$

Write each expression as a product of trigonometric functions. See Example 8.

59. $\cos 4x - \cos 2x$

60. $\cos 5x + \cos 8x$

61. $\sin 25° + \sin(-48°)$

62. $\sin 102° - \sin 95°$

63. $\cos 4x + \cos 8x$

64. $\sin 9x - \sin 3x$

(Modeling) Solve each problem.

65. *Wattage Consumption* Refer to Example 6. Use an identity to determine values of a, c, and ω so that $W = a \cos(\omega t) + c$. Check your answer by graphing both expressions for W on the same coordinate axes.

66. *Amperage, Wattage, and Voltage* Amperage is a measure of the amount of electricity that is moving through a circuit, whereas voltage is a measure of the force pushing the electricity. The wattage W consumed by an electrical device can be determined by calculating the product of the amperage I and voltage V. (*Source:* Wilcox, G. and C. Hesselberth, *Electricity for Engineering Technology,* Allyn & Bacon, 1970.)

(a) A household circuit has voltage

$$V = 163 \sin 120\pi t$$

when an incandescent lightbulb is turned on with amperage

$$I = 1.23 \sin 120\pi t.$$

Graph the wattage $W = VI$ consumed by the lightbulb in the window $[0, .05]$ by $[-50, 300]$.

(b) Determine the maximum and minimum wattages used by the lightbulb.

(c) Use identities to determine values for a, c, and ω so that $W = a \cos(\omega t) + c$.

(d) Check your answer by graphing both expressions for W on the same coordinate axes.

(e) Use the graph to estimate the average wattage used by the light. For how many watts do you think this incandescent lightbulb is rated?

5.6 | Half-Angle Identities

Half-Angle Identities ▪ **Applying the Half-Angle Identities**

Half-Angle Identities   From the alternative forms of the identity for $\cos 2A$, we derive three additional identities for $\sin \frac{A}{2}$, $\cos \frac{A}{2}$, and $\tan \frac{A}{2}$. These are known as **half-angle identities.**

To derive the identity for $\sin \frac{A}{2}$, start with the following double-angle identity for cosine and solve for $\sin x$.

$$\cos 2x = 1 - 2 \sin^2 x$$

$$2 \sin^2 x = 1 - \cos 2x \qquad \text{Add } 2 \sin^2 x; \text{ subtract } \cos 2x.$$

$$\sin x = \pm \sqrt{\frac{1 - \cos 2x}{2}} \qquad \text{Divide by 2; take square roots. (Appendix A)}$$

$$\sin \frac{A}{2} = \pm \sqrt{\frac{1 - \cos A}{2}} \qquad \text{Let } 2x = A, \text{ so } x = \tfrac{A}{2}; \text{ substitute.}$$

The $\pm$ sign in this identity indicates that the appropriate sign is chosen depending on the quadrant of $\frac{A}{2}$. For example, if $\frac{A}{2}$ is a quadrant III angle, we choose the negative sign since the sine function is negative in quadrant III.

We derive the identity for $\cos \frac{A}{2}$ using the double-angle identity $\cos 2x = 2 \cos^2 x - 1$.

$$1 + \cos 2x = 2 \cos^2 x \qquad \text{Add 1.}$$

$$\cos^2 x = \frac{1 + \cos 2x}{2} \qquad \text{Rewrite; divide by 2.}$$

$$\cos x = \pm \sqrt{\frac{1 + \cos 2x}{2}} \qquad \text{Take square roots.}$$

$$\cos \frac{A}{2} = \pm \sqrt{\frac{1 + \cos A}{2}} \qquad \text{Replace } x \text{ with } \tfrac{A}{2}.$$

An identity for $\tan \frac{A}{2}$ comes from the identities for $\sin \frac{A}{2}$ and $\cos \frac{A}{2}$.

$$\tan \frac{A}{2} = \frac{\sin \frac{A}{2}}{\cos \frac{A}{2}} = \frac{\pm \sqrt{\dfrac{1 - \cos A}{2}}}{\pm \sqrt{\dfrac{1 + \cos A}{2}}} = \pm \sqrt{\frac{1 - \cos A}{1 + \cos A}}$$

We derive an alternative identity for $\tan \frac{A}{2}$ using double-angle identities.

$$\tan \frac{A}{2} = \frac{\sin \frac{A}{2}}{\cos \frac{A}{2}} = \frac{2 \sin \frac{A}{2} \cos \frac{A}{2}}{2 \cos^2 \frac{A}{2}} \qquad \text{Multiply by } 2 \cos \tfrac{A}{2} \text{ in numerator and denominator.}$$

$$= \frac{\sin 2\left(\frac{A}{2}\right)}{1 + \cos 2\left(\frac{A}{2}\right)} \qquad \text{Double-angle identities (Section 5.5)}$$

$$\tan \frac{A}{2} = \frac{\sin A}{1 + \cos A}$$

From the identity $\tan \dfrac{A}{2} = \dfrac{\sin A}{1 + \cos A}$, we can also derive $\tan \dfrac{A}{2} = \dfrac{1 - \cos A}{\sin A}$.

Half-Angle Identities

$$\cos \frac{A}{2} = \pm\sqrt{\frac{1 + \cos A}{2}} \qquad \sin \frac{A}{2} = \pm\sqrt{\frac{1 - \cos A}{2}}$$

$$\tan \frac{A}{2} = \pm\sqrt{\frac{1 - \cos A}{1 + \cos A}} \qquad \tan \frac{A}{2} = \frac{\sin A}{1 + \cos A} \qquad \tan \frac{A}{2} = \frac{1 - \cos A}{\sin A}$$

N O T E The last two identities for $\tan \frac{A}{2}$ do not require a sign choice. When using the other half-angle identities, select the plus or minus sign according to the quadrant in which $\frac{A}{2}$ terminates. For example, if an angle $A = 324°$, then $\frac{A}{2} = 162°$, which lies in quadrant II. In quadrant II, $\cos \frac{A}{2}$ and $\tan \frac{A}{2}$ are negative, while $\sin \frac{A}{2}$ is positive.

Applying the Half-Angle Identities

EXAMPLE 1 Using a Half-Angle Identity to Find an Exact Value

Find the exact value of $\cos 15°$ using the half-angle identity for cosine.

Solution

$$\cos 15° = \cos \frac{1}{2}(30°) = \sqrt{\frac{1 + \cos 30°}{2}}$$

Choose the positive square root.

$$= \sqrt{\frac{1 + \frac{\sqrt{3}}{2}}{2}} = \sqrt{\frac{\left(1 + \frac{\sqrt{3}}{2}\right) \cdot 2}{2 \cdot 2}} = \frac{\sqrt{2 + \sqrt{3}}}{2}$$

Simplify the radicals.

Now try Exercise 11.

EXAMPLE 2 Using a Half-Angle Identity to Find an Exact Value

Find the exact value of $\tan 22.5°$ using the identity $\tan \frac{A}{2} = \frac{\sin A}{1 + \cos A}$.

Solution Since $22.5° = \frac{1}{2}(45°)$, replace A with $45°$.

$$\tan 22.5° = \tan \frac{45°}{2} = \frac{\sin 45°}{1 + \cos 45°} = \frac{\frac{\sqrt{2}}{2}}{1 + \frac{\sqrt{2}}{2}} = \frac{\frac{\sqrt{2}}{2}}{1 + \frac{\sqrt{2}}{2}} \cdot \frac{2}{2}$$

$$= \frac{\sqrt{2}}{2 + \sqrt{2}} = \frac{\sqrt{2}}{2 + \sqrt{2}} \cdot \frac{2 - \sqrt{2}}{2 - \sqrt{2}} = \frac{2\sqrt{2} - 2}{2}$$ Rationalize the denominator.

$$= \frac{2(\sqrt{2} - 1)}{2} = \sqrt{2} - 1$$ Factor out 2.

Now try Exercise 13.

EXAMPLE 3 Finding Function Values of $\frac{s}{2}$ Given Information about s

Given $\cos s = \frac{2}{3}$, with $\frac{3\pi}{2} < s < 2\pi$, find $\cos \frac{s}{2}$, $\sin \frac{s}{2}$, and $\tan \frac{s}{2}$.

Solution Since

$$\frac{3\pi}{2} < s < 2\pi$$

and

$$\frac{3\pi}{4} < \frac{s}{2} < \pi, \quad \text{Divide by 2. (Appendix A)}$$

Figure 9

$\frac{s}{2}$ terminates in quadrant II. See Figure 9. In quadrant II, the values of $\cos \frac{s}{2}$ and $\tan \frac{s}{2}$ are negative and the value of $\sin \frac{s}{2}$ is positive. Now use the appropriate half-angle identities and simplify the radicals.

$$\sin \frac{s}{2} = \sqrt{\frac{1 - \frac{2}{3}}{2}} = \sqrt{\frac{1}{6}} = \frac{\sqrt{1}}{\sqrt{6}} \cdot \frac{\sqrt{6}}{\sqrt{6}} = \frac{\sqrt{6}}{6}$$

$$\cos \frac{s}{2} = -\sqrt{\frac{1 + \frac{2}{3}}{2}} = -\sqrt{\frac{5}{6}} = -\frac{\sqrt{5}}{\sqrt{6}} \cdot \frac{\sqrt{6}}{\sqrt{6}} = -\frac{\sqrt{30}}{6}$$

$$\tan \frac{s}{2} = \frac{\sin \frac{s}{2}}{\cos \frac{s}{2}} = \frac{\frac{\sqrt{6}}{6}}{-\frac{\sqrt{30}}{6}} = \frac{\sqrt{6}}{-\sqrt{30}} = -\frac{\sqrt{6}}{\sqrt{30}} \cdot \frac{\sqrt{30}}{\sqrt{30}} = -\frac{\sqrt{180}}{30} = -\frac{6\sqrt{5}}{6 \cdot 5} = -\frac{\sqrt{5}}{5}$$

Notice that it is not necessary to use a half-angle identity for $\tan \frac{s}{2}$ once we find $\sin \frac{s}{2}$ and $\cos \frac{s}{2}$. However, using this identity would provide an excellent check.

Now try Exercise 19.

EXAMPLE 4 Simplifying Expressions Using the Half-Angle Identities

Simplify each expression.

(a) $\pm \sqrt{\dfrac{1 + \cos 12x}{2}}$

(b) $\dfrac{1 - \cos 5\alpha}{\sin 5\alpha}$

Solution

(a) This matches part of the identity for $\cos \frac{A}{2}$.

$$\cos \frac{A}{2} = \pm \sqrt{\frac{1 + \cos A}{2}}$$

Replace A with $12x$ to get

$$\pm \sqrt{\frac{1 + \cos 12x}{2}} = \cos \frac{12x}{2} = \cos 6x.$$

(b) Use the third identity for $\tan \frac{A}{2}$ given earlier with $A = 5\alpha$ to get

$$\frac{1 - \cos 5\alpha}{\sin 5\alpha} = \tan \frac{5\alpha}{2}.$$

Now try Exercises 37 and 39.

EXAMPLE 5 Verifying an Identity

Verify that the following equation is an identity.

$$\left(\sin \frac{x}{2} + \cos \frac{x}{2} \right)^2 = 1 + \sin x$$

Solution We work on the more complicated left side.

$$\left(\sin \frac{x}{2} + \cos \frac{x}{2} \right)^2$$

$$= \sin^2 \frac{x}{2} + 2 \sin \frac{x}{2} \cos \frac{x}{2} + \cos^2 \frac{x}{2} \quad (a + b)^2 = a^2 + 2ab + b^2$$

$$= 1 + 2 \sin \frac{x}{2} \cos \frac{x}{2} \qquad\qquad \sin^2 \frac{x}{2} + \cos^2 \frac{x}{2} = 1$$

$$= 1 + \sin 2\left(\frac{x}{2} \right) \qquad\qquad 2 \sin \frac{x}{2} \cos \frac{x}{2} = \sin 2\left(\frac{x}{2} \right)$$

$$= 1 + \sin x$$

Now try Exercise 51.

5.6 Exercises

Concept Check *Determine whether the positive or negative square root should be selected.*

1. $\sin 195° = \pm \sqrt{\dfrac{1 - \cos 390°}{2}}$

2. $\cos 58° = \pm \sqrt{\dfrac{1 + \cos 116°}{2}}$

3. $\tan 225° = \pm \sqrt{\dfrac{1 - \cos 450°}{1 + \cos 450°}}$

4. $\sin(-10°) = \pm \sqrt{\dfrac{1 - \cos(-20°)}{2}}$

Match each expression in Column I with its value in Column II. See Examples 1 and 2.

I		II	
5. $\sin 15°$	**6.** $\tan 15°$	**A.** $2 - \sqrt{3}$	**B.** $\dfrac{\sqrt{2 - \sqrt{2}}}{2}$
7. $\cos \dfrac{\pi}{8}$	**8.** $\tan\left(-\dfrac{\pi}{8} \right)$	**C.** $\dfrac{\sqrt{2 - \sqrt{3}}}{2}$	**D.** $\dfrac{\sqrt{2 + \sqrt{2}}}{2}$
9. $\tan 67.5°$	**10.** $\cos 67.5°$	**E.** $1 - \sqrt{2}$	**F.** $1 + \sqrt{2}$

Use a half-angle identity to find each exact value. See Examples 1 and 2.

11. $\sin 67.5°$ **12.** $\sin 195°$ **13.** $\cos 195°$

14. $\tan 195°$ **15.** $\cos 165°$ **16.** $\sin 165°$

17. Explain how you could use an identity of this section to find the exact value of $\sin 7.5°$. (*Hint:* $7.5 = \frac{1}{2}(\frac{1}{2})(30)$.)

18. The identity $\tan \frac{A}{2} = \pm\sqrt{\frac{1 - \cos A}{1 + \cos A}}$ can be used to find $\tan 22.5° = \sqrt{3 - 2\sqrt{2}}$, and the identity $\tan \frac{A}{2} = \frac{\sin A}{1 + \cos A}$ can be used to find $\tan 22.5° = \sqrt{2} - 1$. Show that these answers are the same, without using a calculator. (*Hint:* If $a > 0$ and $b > 0$ and $a^2 = b^2$, then $a = b$.)

Find each of the following. See Example 3.

19. $\cos \frac{x}{2}$, given $\cos x = \frac{1}{4}$, with $0 < x < \frac{\pi}{2}$

20. $\sin \frac{x}{2}$, given $\cos x = -\frac{5}{8}$, with $\frac{\pi}{2} < x < \pi$

21. $\tan \frac{\theta}{2}$, given $\sin \theta = \frac{3}{5}$, with $90° < \theta < 180°$

22. $\cos \frac{\theta}{2}$, given $\sin \theta = -\frac{1}{5}$, with $180° < \theta < 270°$

23. $\sin \frac{x}{2}$, given $\tan x = 2$, with $0 < x < \frac{\pi}{2}$

24. $\cos \frac{x}{2}$, given $\cot x = -3$, with $\frac{\pi}{2} < x < \pi$

25. $\tan \frac{\theta}{2}$, given $\tan \theta = \frac{\sqrt{7}}{3}$, with $180° < \theta < 270°$

26. $\cot \frac{\theta}{2}$, given $\tan \theta = -\frac{\sqrt{5}}{2}$, with $90° < \theta < 180°$

27. $\sin \theta$, given $\cos 2\theta = \frac{3}{5}$ and θ terminates in quadrant I

28. $\cos \theta$, given $\cos 2\theta = \frac{1}{2}$ and θ terminates in quadrant II

29. $\cos x$, given $\cos 2x = -\frac{5}{12}$, with $\frac{\pi}{2} < x < \pi$

30. $\sin x$, given $\cos 2x = \frac{2}{3}$, with $\pi < x < \frac{3\pi}{2}$

31. *Concept Check*   If $\cos x \approx .9682$ and $\sin x \approx .25$, then $\tan \frac{x}{2} \approx$ _____ .

32. *Concept Check*   If $\cos x \approx -.75$ and $\sin x \approx .6614$, then $\tan \frac{x}{2} \approx$ _____ .

Use an identity to write each expression as a single trigonometric function. See Example 4.

33. $\sqrt{\dfrac{1 - \cos 40°}{2}}$

34. $\sqrt{\dfrac{1 + \cos 76°}{2}}$

35. $\sqrt{\dfrac{1 - \cos 147°}{1 + \cos 147°}}$

36. $\sqrt{\dfrac{1 + \cos 165°}{1 - \cos 165°}}$

37. $\dfrac{1 - \cos 59.74°}{\sin 59.74°}$

38. $\dfrac{\sin 158.2°}{1 + \cos 158.2°}$

39. $\pm\sqrt{\dfrac{1 + \cos 18x}{2}}$

40. $\pm\sqrt{\dfrac{1 + \cos 20\alpha}{2}}$

41. $\pm\sqrt{\dfrac{1 - \cos 8\theta}{1 + \cos 8\theta}}$

42. $\pm\sqrt{\dfrac{1 - \cos 5A}{1 + \cos 5A}}$

43. $\pm\sqrt{\dfrac{1 + \cos \frac{x}{4}}{2}}$

44. $\pm\sqrt{\dfrac{1 - \cos \frac{3\theta}{5}}{2}}$

Graph each expression and use the graph to conjecture an identity. Then verify your conjecture algebraically.

45. $\dfrac{\sin x}{1 + \cos x}$

46. $\dfrac{1 - \cos x}{\sin x}$

47. $\dfrac{\tan \frac{x}{2} + \cot \frac{x}{2}}{\cot \frac{x}{2} - \tan \frac{x}{2}}$

48. $1 - 8 \sin^2 \dfrac{x}{2} \cos^2 \dfrac{x}{2}$

Verify that each equation is an identity. See Example 5.

49. $\sec^2 \dfrac{x}{2} = \dfrac{2}{1 + \cos x}$

50. $\cot^2 \dfrac{x}{2} = \dfrac{(1 + \cos x)^2}{\sin^2 x}$

51. $\sin^2 \dfrac{x}{2} = \dfrac{\tan x - \sin x}{2 \tan x}$

52. $\dfrac{\sin 2x}{2 \sin x} = \cos^2 \dfrac{x}{2} - \sin^2 \dfrac{x}{2}$

53. $\dfrac{2}{1 + \cos x} - \tan^2 \dfrac{x}{2} = 1$

54. $\tan \dfrac{\theta}{2} = \csc \theta - \cot \theta$

55. $1 - \tan^2 \dfrac{\theta}{2} = \dfrac{2 \cos \theta}{1 + \cos \theta}$

56. $\cos x = \dfrac{1 - \tan^2 \frac{x}{2}}{1 + \tan^2 \frac{x}{2}}$

57. Use the identity $\tan \frac{A}{2} = \frac{\sin A}{1 + \cos A}$ to derive the equivalent identity $\tan \frac{A}{2} = \frac{1 - \cos A}{\sin A}$ by multiplying both the numerator and denominator by $1 - \cos A$.

(Modeling) Mach Number An airplane flying faster than sound sends out sound waves that form a cone, as shown in the figure. The cone intersects the ground to form a hyperbola. As this hyperbola passes over a particular point on the ground, a sonic boom is heard at that point. If θ is the angle at the vertex of the cone, then

$$\sin \dfrac{\theta}{2} = \dfrac{1}{m},$$

where m is the Mach number for the speed of the plane. (We assume m > 1.) The Mach number is the ratio of the speed of the plane and the speed of sound. Thus, a speed of Mach 1.4 means that the plane is flying at 1.4 times the speed of sound. In Exercises 58–61, one of the values θ or m is given. Find the other value.

58. $m = \dfrac{3}{2}$ **59.** $m = \dfrac{5}{4}$ **60.** $\theta = 30°$ **61.** $\theta = 60°$

62. *(Modeling) Railroad Curves* In the United States, circular railroad curves are designated by the *degree of curvature*, the central angle subtended by a chord of 100 ft. See the figure. (*Source:* Hay, W. W., *Railroad Engineering*, John Wiley & Sons, 1982.)

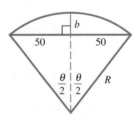

(a) Use the figure to write an expression for $\cos \frac{\theta}{2}$.

(b) Use the result of part (a) and the third half-angle identity for tangent to write an expression for $\tan \frac{\theta}{4}$.

(c) If $b = 12$, what is the measure of angle θ to the nearest degree?

Relating Concepts

For individual or collaborative investigation
(Exercises 63–70)

These exercises use results from plane geometry, instead of the half-angle formulas, to obtain exact values of the trigonometric functions of 15°. **Work Exercises 63–70 in order.**

   Start with a right triangle ACB having a 60° angle at A and a 30° angle at B. Let the hypotenuse of this triangle have length 2. Extend side BC and draw a semicircle with diameter along BC extended, center at B, and radius AB. Draw segment AE. (See the figure.) Since any angle inscribed in a semicircle is a right angle, triangle EAD is a right triangle.

63. Why does $AB = BD$? Conclude that triangle ABD is isosceles.

64. Why does angle ABD have measure 150°?

65. Why do angles DAB and ADB both have measures of 15°?

66. What is the length DC?

67. Use the Pythagorean theorem to show that the length AD is $\sqrt{6} + \sqrt{2}$. (*Note:* $\left(\sqrt{6} + \sqrt{2}\right)^2 = 8 + 4\sqrt{3}$.)

68. Use angle ADB of triangle EAD to find cos 15°.

69. Show that AE has length $\sqrt{6} - \sqrt{2}$ and find sin 15°.

70. Use triangle ACD to find tan 15°.

Summary Exercises on Verifying Trigonometric Identities

These summary exercises provide practice with the various types of trigonometric identities presented in this chapter. Verify that each equation is an identity.

1. $\tan \theta + \cot \theta = \sec \theta \csc \theta$

2. $\csc \theta \cos^2 \theta + \sin \theta = \csc \theta$

3. $\tan \dfrac{x}{2} = \csc x - \cot x$

4. $\sec(\pi - x) = -\sec x$

5. $\dfrac{\sin t}{1 + \cos t} = \dfrac{1 - \cos t}{\sin t}$

6. $\dfrac{1 - \sin t}{\cos t} = \dfrac{1}{\sec t + \tan t}$

7. $\sin 2\theta = \dfrac{2 \tan \theta}{1 + \tan^2 \theta}$

8. $\dfrac{2}{1 + \cos x} - \tan^2 \dfrac{x}{2} = 1$

9. $\cot\theta - \tan\theta = \dfrac{2\cos^2\theta - 1}{\sin\theta\cos\theta}$

10. $\dfrac{1}{\sec t - 1} + \dfrac{1}{\sec t + 1} = 2\cot t\csc t$

11. $\dfrac{\sin(x+y)}{\cos(x-y)} = \dfrac{\cot x + \cot y}{1 + \cot x\cot y}$

12. $1 - \tan^2\dfrac{\theta}{2} = \dfrac{2\cos\theta}{1 + \cos\theta}$

13. $\dfrac{\sin\theta + \tan\theta}{1 + \cos\theta} = \tan\theta$

14. $\csc^4 x - \cot^4 x = \dfrac{1 + \cos^2 x}{1 - \cos^2 x}$

15. $\cos x = \dfrac{1 - \tan^2\frac{x}{2}}{1 + \tan^2\frac{x}{2}}$

16. $\cos 2x = \dfrac{2 - \sec^2 x}{\sec^2 x}$

17. $\dfrac{\tan^2 t + 1}{\tan t\csc^2 t} = \tan t$

18. $\dfrac{\sin s}{1 + \cos s} + \dfrac{1 + \cos s}{\sin s} = 2\csc s$

19. $\tan 4\theta = \dfrac{2\tan 2\theta}{2 - \sec^2 2\theta}$

20. $\tan\left(\dfrac{x}{2} + \dfrac{\pi}{4}\right) = \sec x + \tan x$

21. $\dfrac{\cot s - \tan s}{\cos s + \sin s} = \dfrac{\cos s - \sin s}{\sin s\cos s}$

22. $\dfrac{\tan\theta - \cot\theta}{\tan\theta + \cot\theta} = 1 - 2\cos^2\theta$

23. $\dfrac{\tan(x+y) - \tan y}{1 + \tan(x+y)\tan y} = \tan x$

24. $2\cos^2\dfrac{x}{2}\tan x = \tan x + \sin x$

25. $\dfrac{\cos^4 x - \sin^4 x}{\cos^2 x} = 1 - \tan^2 x$

26. $\dfrac{\csc t + 1}{\csc t - 1} = (\sec t + \tan t)^2$

27. $\dfrac{2(\sin x - \sin^3 x)}{\cos x} = \sin 2x$

28. $\dfrac{1}{2}\cot\dfrac{x}{2} - \dfrac{1}{2}\tan\dfrac{x}{2} = \cot x$

Chapter 5 Summary

QUICK REVIEW

CONCEPTS	EXAMPLES

5.1 Fundamental Identities

Reciprocal Identities

$$\cot \theta = \frac{1}{\tan \theta} \qquad \sec \theta = \frac{1}{\cos \theta} \qquad \csc \theta = \frac{1}{\sin \theta}$$

Quotient Identities

$$\tan \theta = \frac{\sin \theta}{\cos \theta} \qquad \cot \theta = \frac{\cos \theta}{\sin \theta}$$

Pythagorean Identities

$$\sin^2 \theta + \cos^2 \theta = 1 \qquad \tan^2 \theta + 1 = \sec^2 \theta$$
$$1 + \cot^2 \theta = \csc^2 \theta$$

Negative-Angle Identities

$$\sin(-\theta) = -\sin \theta \quad \cos(-\theta) = \cos \theta \quad \tan(-\theta) = -\tan \theta$$
$$\csc(-\theta) = -\csc \theta \quad \sec(-\theta) = \sec \theta \quad \cot(-\theta) = -\cot \theta$$

If θ is in quadrant IV and $\sin \theta = -\frac{3}{5}$, find $\csc \theta$, $\cos \theta$, and $\sin(-\theta)$.

$$\csc \theta = \frac{1}{\sin \theta} = \frac{1}{-\frac{3}{5}} = -\frac{5}{3}$$

$$\sin^2 \theta + \cos^2 \theta = 1$$

$$\left(-\frac{3}{5}\right)^2 + \cos^2 \theta = 1$$

$$\cos^2 \theta = 1 - \frac{9}{25} = \frac{16}{25}$$

$$\cos \theta = +\sqrt{\frac{16}{25}} = \frac{4}{5} \qquad \begin{array}{l}\cos \theta \text{ is positive} \\ \text{in quadrant IV.}\end{array}$$

$$\sin(-\theta) = -\sin \theta = \frac{3}{5}$$

5.2 Verifying Trigonometric Identities

See the box titled Hints for Verifying Identities on page 189.

5.3 Sum and Difference Identities for Cosine
5.4 Sum and Difference Identities for Sine and Tangent

Cofunction Identities

$$\cos(90° - \theta) = \sin \theta \qquad \cot(90° - \theta) = \tan \theta$$
$$\sin(90° - \theta) = \cos \theta \qquad \sec(90° - \theta) = \csc \theta$$
$$\tan(90° - \theta) = \cot \theta \qquad \csc(90° - \theta) = \sec \theta$$

Find a value of θ such that $\tan \theta = \cot 78°$.

$$\tan \theta = \cot 78°$$
$$\cot(90° - \theta) = \cot 78°$$
$$90° - \theta = 78°$$
$$\theta = 12°$$

Sum and Difference Identities

$$\cos(A - B) = \cos A \cos B + \sin A \sin B$$
$$\cos(A + B) = \cos A \cos B - \sin A \sin B$$
$$\sin(A + B) = \sin A \cos B + \cos A \sin B$$
$$\sin(A - B) = \sin A \cos B - \cos A \sin B$$
$$\tan(A + B) = \frac{\tan A + \tan B}{1 - \tan A \tan B}$$
$$\tan(A - B) = \frac{\tan A - \tan B}{1 + \tan A \tan B}$$

Find the exact value of $\cos(-15°)$.

$$\cos(-15°) = \cos(30° - 45°)$$
$$= \cos 30° \cos 45° + \sin 30° \sin 45°$$
$$= \frac{\sqrt{3}}{2} \cdot \frac{\sqrt{2}}{2} + \frac{1}{2} \cdot \frac{\sqrt{2}}{2}$$
$$= \frac{\sqrt{6}}{4} + \frac{\sqrt{2}}{4} = \frac{\sqrt{6} + \sqrt{2}}{4}$$

Write $\tan\left(\frac{\pi}{4} + \theta\right)$ in terms of $\tan \theta$.

$$\tan\left(\frac{\pi}{4} + \theta\right) = \frac{\tan \frac{\pi}{4} + \tan \theta}{1 - \tan \frac{\pi}{4} \tan \theta}$$

$$= \frac{1 + \tan \theta}{1 - \tan \theta} \qquad \tan \frac{\pi}{4} = 1$$

CONCEPTS	EXAMPLES

5.5 Double-Angle Identities

Double-Angle Identities

$$\cos 2A = \cos^2 A - \sin^2 A \qquad \cos 2A = 1 - 2\sin^2 A$$

$$\cos 2A = 2\cos^2 A - 1 \qquad \sin 2A = 2\sin A \cos A$$

$$\tan 2A = \frac{2\tan A}{1 - \tan^2 A}$$

Given $\cos\theta = -\frac{5}{13}$ and $\sin\theta > 0$, find $\sin 2\theta$.

Sketch a triangle in quadrant II and use it to find $\sin\theta$:
$\sin\theta = \frac{12}{13}$.

$$\sin 2\theta = 2\sin\theta\cos\theta$$

$$= 2\left(\frac{12}{13}\right)\left(-\frac{5}{13}\right) = -\frac{120}{169}$$

Product-to-Sum Identities

$$\cos A \cos B = \frac{1}{2}[\cos(A + B) + \cos(A - B)]$$

$$\sin A \sin B = \frac{1}{2}[\cos(A - B) - \cos(A + B)]$$

$$\sin A \cos B = \frac{1}{2}[\sin(A + B) + \sin(A - B)]$$

$$\cos A \sin B = \frac{1}{2}[\sin(A + B) - \sin(A - B)]$$

Write $\sin(-\theta)\sin 2\theta$ as the difference of two functions.

$$\sin(-\theta)\sin 2\theta = \frac{1}{2}[\cos(-\theta - 2\theta) - \cos(-\theta + 2\theta)]$$

$$= \frac{1}{2}[\cos(-3\theta) - \cos\theta]$$

$$= \frac{1}{2}\cos(-3\theta) - \frac{1}{2}\cos\theta$$

$$= \frac{1}{2}\cos 3\theta - \frac{1}{2}\cos\theta$$

Sum-to-Product Identities

$$\sin A + \sin B = 2\sin\left(\frac{A + B}{2}\right)\cos\left(\frac{A - B}{2}\right)$$

$$\sin A - \sin B = 2\cos\left(\frac{A + B}{2}\right)\sin\left(\frac{A - B}{2}\right)$$

$$\cos A + \cos B = 2\cos\left(\frac{A + B}{2}\right)\cos\left(\frac{A - B}{2}\right)$$

$$\cos A - \cos B = -2\sin\left(\frac{A + B}{2}\right)\sin\left(\frac{A - B}{2}\right)$$

Write $\cos\theta + \cos 3\theta$ as a product of two functions.

$$\cos\theta + \cos 3\theta = 2\cos\left(\frac{\theta + 3\theta}{2}\right)\cos\left(\frac{\theta - 3\theta}{2}\right)$$

$$= 2\cos\left(\frac{4\theta}{2}\right)\cos\left(\frac{-2\theta}{2}\right)$$

$$= 2\cos 2\theta\cos(-\theta)$$

$$= 2\cos 2\theta\cos\theta$$

5.6 Half-Angle Identities

Half-Angle Identities

$$\cos\frac{A}{2} = \pm\sqrt{\frac{1 + \cos A}{2}} \qquad \sin\frac{A}{2} = \pm\sqrt{\frac{1 - \cos A}{2}}$$

$$\tan\frac{A}{2} = \pm\sqrt{\frac{1 - \cos A}{1 + \cos A}} \qquad \tan\frac{A}{2} = \frac{\sin A}{1 + \cos A}$$

$$\tan\frac{A}{2} = \frac{1 - \cos A}{\sin A}$$

$\left(\text{In the first three identities, the sign is chosen based on the quadrant of }\frac{A}{2}.\right)$

Find the exact value of $\tan 67.5°$.

We choose the last form with $A = 135°$.

$$\tan 67.5° = \tan\frac{135°}{2} = \frac{1 - \cos 135°}{\sin 135°} = \frac{1 - \left(-\frac{\sqrt{2}}{2}\right)}{\frac{\sqrt{2}}{2}}$$

$$= \frac{1 + \frac{\sqrt{2}}{2}}{\frac{\sqrt{2}}{2}} \cdot \frac{2}{2} = \frac{2 + \sqrt{2}}{\sqrt{2}} \quad \text{or} \quad \sqrt{2} + 1$$

Rationalize the denominator; simplify.

Chapter 5 Review Exercises

Concept Check *For each expression in Column I, choose the expression from Column II that completes an identity.*

I

1. $\sec x =$ _____ **2.** $\csc x =$ _____

3. $\tan x =$ _____ **4.** $\cot x =$ _____

5. $\tan^2 x =$ _____ **6.** $\sec^2 x =$ _____

II

A. $\dfrac{1}{\sin x}$ **B.** $\dfrac{1}{\cos x}$

C. $\dfrac{\sin x}{\cos x}$ **D.** $\dfrac{1}{\cot^2 x}$

E. $\dfrac{1}{\cos^2 x}$ **F.** $\dfrac{\cos x}{\sin x}$

Use identities to write each expression in terms of $\sin \theta$ *and* $\cos \theta$, *and simplify.*

7. $\sec^2 \theta - \tan^2 \theta$ **8.** $\dfrac{\cot \theta}{\sec \theta}$ **9.** $\tan^2 \theta (1 + \cot^2 \theta)$

10. $\csc \theta + \cot \theta$ **11.** $\tan \theta - \sec \theta \csc \theta$ **12.** $\csc^2 \theta + \sec^2 \theta$

13. Use the trigonometric identities to find $\sin x$, $\tan x$, and $\cot(-x)$, given $\cos x = \frac{3}{5}$ and x in quadrant IV.

14. Given $\tan x = -\frac{5}{4}$, where $\frac{\pi}{2} < x < \pi$, use the trigonometric identities to find $\cot x$, $\csc x$, and $\sec x$.

15. Find the exact values of the six trigonometric functions of $165°$.

16. Find the exact values of $\sin x$, $\cos x$, and $\tan x$, for $x = \frac{\pi}{12}$, using

  (a) difference identities **(b)** half-angle identities.

Concept Check *For each expression in Column I, use an identity to choose an expression from Column II with the same value.*

I

17. $\cos 210°$ **18.** $\sin 35°$

19. $\tan(-35°)$ **20.** $-\sin 35°$

21. $\cos 35°$ **22.** $\cos 75°$

23. $\sin 75°$ **24.** $\sin 300°$

25. $\cos 300°$ **26.** $\cos(-55°)$

II

A. $\sin(-35°)$ **B.** $\cos 55°$

C. $\sqrt{\dfrac{1 + \cos 150°}{2}}$ **D.** $2 \sin 150° \cos 150°$

E. $\cos 150° \cos 60° - \sin 150° \sin 60°$

F. $\cot(-35°)$

G. $\cos^2 150° - \sin^2 150°$

H. $\sin 15° \cos 60° + \cos 15° \sin 60°$

I. $\cos(-35°)$ **J.** $\cot 125°$

For each of the following, find $\sin(x + y)$, $\cos(x - y)$, $\tan(x + y)$, *and the quadrant of* $x + y$.

27. $\sin x = -\dfrac{1}{4}$, $\cos y = -\dfrac{4}{5}$, x and y in quadrant III

28. $\sin x = \dfrac{1}{10}$, $\cos y = \dfrac{4}{5}$, x in quadrant I, y in quadrant IV

29. $\sin x = -\dfrac{1}{2}$, $\cos y = -\dfrac{2}{5}$, x and y in quadrant III

30. $\sin y = -\dfrac{2}{3}$, $\cos x = -\dfrac{1}{5}$, x in quadrant II, y in quadrant III

31. $\sin x = \dfrac{1}{10}$, $\cos y = \dfrac{4}{5}$, x in quadrant I, y in quadrant IV

32. $\cos x = \dfrac{2}{9}$, $\sin y = -\dfrac{1}{2}$, x in quadrant IV, y in quadrant III

Find sine and cosine of each of the following.

33. θ, given $\cos 2\theta = -\dfrac{3}{4}$, $90° < 2\theta < 180°$

34. B, given $\cos 2B = \dfrac{1}{8}$, B in quadrant IV

35. $2x$, given $\tan x = 3$, $\sin x < 0$ **36.** $2y$, given $\sec y = -\dfrac{5}{3}$, $\sin y > 0$

Find each of the following.

37. $\cos \dfrac{\theta}{2}$, given $\cos \theta = -\dfrac{1}{2}$, with $90° < \theta < 180°$

38. $\sin \dfrac{A}{2}$, given $\cos A = -\dfrac{3}{4}$, with $90° < A < 180°$

39. $\tan x$, given $\tan 2x = 2$, with $\pi < x < \dfrac{3\pi}{2}$

40. $\sin y$, given $\cos 2y = -\dfrac{1}{3}$, with $\dfrac{\pi}{2} < y < \pi$

41. $\tan \dfrac{x}{2}$, given $\sin x = .8$, with $0 < x < \dfrac{\pi}{2}$

42. $\sin 2x$, given $\sin x = .6$, with $\dfrac{\pi}{2} < x < \pi$

Graph each expression and use the graph to conjecture an identity. Then verify your conjecture algebraically.

43. $-\dfrac{\sin 2x + \sin x}{\cos 2x - \cos x}$ **44.** $\dfrac{1 - \cos 2x}{\sin 2x}$ **45.** $\dfrac{\sin x}{1 - \cos x}$

46. $\dfrac{\cos x \sin 2x}{1 + \cos 2x}$ **47.** $\dfrac{2(\sin x - \sin^3 x)}{\cos x}$ **48.** $\csc x - \cot x$

Verify that each equation is an identity.

49. $\sin^2 x - \sin^2 y = \cos^2 y - \cos^2 x$ **50.** $2\cos^3 x - \cos x = \dfrac{\cos^2 x - \sin^2 x}{\sec x}$

51. $\dfrac{\sin^2 x}{2 - 2\cos x} = \cos^2 \dfrac{x}{2}$ **52.** $\dfrac{\sin 2x}{\sin x} = \dfrac{2}{\sec x}$

53. $2\cos A - \sec A = \cos A - \dfrac{\tan A}{\csc A}$ **54.** $\dfrac{2\tan B}{\sin 2B} = \sec^2 B$

55. $1 + \tan^2 \alpha = 2\tan \alpha \csc 2\alpha$ **56.** $\dfrac{2\cot x}{\tan 2x} = \csc^2 x - 2$

57. $\tan \theta \sin 2\theta = 2 - 2\cos^2 \theta$ **58.** $\csc A \sin 2A - \sec A = \cos 2A \sec A$

59. $2\tan x \csc 2x - \tan^2 x = 1$ **60.** $2\cos^2 \theta - 1 = \dfrac{1 - \tan^2 \theta}{1 + \tan^2 \theta}$

61. $\tan\theta\cos^2\theta = \dfrac{2\tan\theta\cos^2\theta - \tan\theta}{1 - \tan^2\theta}$

62. $\sec^2\alpha - 1 = \dfrac{\sec 2\alpha - 1}{\sec 2\alpha + 1}$

63. $2\cos^3 x - \cos x = \dfrac{\cos^2 x - \sin^2 x}{\sec x}$

64. $\sin^3\theta = \sin\theta - \cos^2\theta\sin\theta$

65. $\tan 4\theta = \dfrac{2\tan 2\theta}{2 - \sec^2 2\theta}$

66. $2\cos^2\dfrac{x}{2}\tan x = \tan x + \sin x$

67. $\tan\left(\dfrac{x}{2} + \dfrac{\pi}{4}\right) = \sec x + \tan x$

68. $\dfrac{1}{2}\cot\dfrac{x}{2} - \dfrac{1}{2}\tan\dfrac{x}{2} = \cot x$

69. $-\cot\dfrac{x}{2} = \dfrac{\sin 2x + \sin x}{\cos 2x - \cos x}$

70. $\dfrac{\sin 3t + \sin 2t}{\sin 3t - \sin 2t} = \dfrac{\tan\frac{5t}{2}}{\tan\frac{t}{2}}$

(Modeling) Solve each problem.

71. *Distance Traveled by a Stone* The distance D of an object thrown (or propelled) from height h (feet) at angle θ with initial velocity v is modeled by the formula

$$D = \dfrac{v^2\sin\theta\cos\theta + v\cos\theta\sqrt{(v\sin\theta)^2 + 64h}}{32}.$$

See the figure. (*Source:* Kreighbaum, E. and K. Barthels, *Biomechanics,* Allyn & Bacon, 1996.) Also see the Chapter 2 Quantitative Reasoning.

(a) Find D when $h = 0$; that is, when the object is propelled from the ground.

(b) Suppose a car driving over loose gravel kicks up a small stone at a velocity of 36 ft per sec (about 25 mph) and an angle $\theta = 30°$. How far will the stone travel?

72. *Amperage, Wattage, and Voltage* Suppose that for an electric heater, voltage is given by $V = a\sin 2\pi\omega t$ and amperage by $I = b\sin 2\pi\omega t$, where t is time in seconds.

(a) Find the period of the graph for the voltage.

(b) Show that the graph of the wattage $W = VI$ will have half the period of the voltage. Interpret this result.

Chapter 5 Test

1. Given $\tan x = -\frac{5}{6}$, $\frac{3\pi}{2} < x < 2\pi$, use trigonometric identities to find $\sin x$ and $\cos x$.

2. Express $\tan^2 x - \sec^2 x$ in terms of $\sin x$ and $\cos x$, and simplify.

3. Find $\sin(x + y)$, $\cos(x - y)$, and $\tan(x + y)$, if $\sin x = -\frac{1}{3}$, $\cos y = -\frac{2}{5}$, x is in quadrant III, and y is in quadrant II.

4. Use a half-angle identity to find $\sin(-22.5°)$.

Graph each expression and use the graph to conjecture an identity. Then verify your conjecture algebraically.

5. $\sec x - \sin x\tan x$

6. $\cot\dfrac{x}{2} - \cot x$

Verify that each equation is an identity.

7. $\sec^2 B = \dfrac{1}{1 - \sin^2 B}$

8. $\cos 2A = \dfrac{\cot A - \tan A}{\csc A\sec A}$

9. $\dfrac{\tan x - \cot x}{\tan x + \cot x} = 2 \sin^2 x - 1$ **10.** $\tan^2 x - \sin^2 x = (\tan x \sin x)^2$

11. Use an identity to write each expression as a trigonometric function of θ alone.

    (a) $\cos(270° - \theta)$ **(b)** $\sin(\pi + \theta)$

12. *(Modeling) Voltage*   The voltage in common household current is expressed as $V = 163 \sin \omega t$, where ω is the angular speed (in radians per second) of the generator at the electrical plant and t is time (in seconds).

    (a) Use an identity to express V in terms of cosine.

    (b) If $\omega = 120\pi$, what is the maximum voltage? Give the smallest positive value of t when the maximum voltage occurs.

Chapter 5 Quantitative Reasoning

How does NASA calculate rotation and roll of a spacecraft?

With the spectacular success of the Mars explorers *Spirit* and *Opportunity*, interest in space travel has been revived. The drawing on the left below shows the three quantities that determine the motion of a spacecraft. A conventional three-dimensional spacecraft coordinate system is shown on the right below.

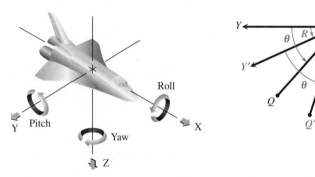

Angle $YOQ = \theta$ and $OQ = r$. The coordinates of Q are (x, y, z), where

$$y = r \cos \theta \qquad \text{and} \qquad z = r \sin \theta.$$

When the spacecraft performs a rotation, it is necessary to find the coordinates in the spacecraft system after the rotation takes place. For example, suppose the spacecraft undergoes roll through angle R. The coordinates (x, y, z) of point Q become (x', y', z'), the coordinates of the corresponding point Q'. In the new reference system, $OQ' = r$ and, since the roll is around the x-axis and angle $Y'OQ' = YOQ = \theta$,

$$x' = x, \qquad y' = r \cos(\theta + R), \qquad \text{and} \qquad z' = r \sin(\theta + R).$$

Use the cosine and sine sum identities to write expressions for y' and z' in terms of y, z, and R. (*Source:* Kastner, B., *Space Mathematics,* NASA.)

6

Inverse Circular Functions and Trigonometric Equations

The association between music and mathematics has a long history. In 500 B.C., Pythagoras' discovery of mathematical relationships between lengths of strings and musical intervals was the beginning of the science of musical sound. In the Middle Ages, music was studied with arithmetic, geometry, and astronomy as part of the liberal arts curriculum. Today, computers are used to create complex musical sounds. (*Source:* Pierce, J., *The Science of Musical Sound,* Scientific American Books, 1992.)

In Section 6.2, Example 6, and Section 6.3, Example 5, various aspects of music are analyzed using trigonometric equations and graphs.

235

6.1 | Inverse Circular Functions

Inverse Functions ▪ **Inverse Sine Function** ▪ **Inverse Cosine Function** ▪ **Inverse Tangent Function** ▪ **Remaining Inverse Circular Functions** ▪ **Inverse Function Values**

$f(x) = x^3$ is a one-to-one function. It satisfies the conditions of the horizontal line test.

$g(x) = x^2$ is not one-to-one. It does not satisfy the conditions of the horizontal line test.

Figure 1

Inverse Functions Recall that for a function f, every element x in the domain corresponds to one and only one element y, or $f(x)$, in the range. (Appendix C) This means that if point (a, b) lies on the graph of f, then there is no other point on the graph that has a as first coordinate. However, there may be other points having b as second coordinate, since the definition of function allows range elements to be used more than once.

If a function is defined so that each range element is used only once, then it is called a **one-to-one function.** For example, the function $f(x) = x^3$ is a one-to-one function because every real number has exactly one real cube root. On the other hand, $g(x) = x^2$ is not one-to-one because, for example, $f(2) = 4$ and $f(-2) = 4$. There are two domain elements, 2 and -2, that correspond to the range element 4.

The *horizontal line test* helps determine graphically whether a function is one-to-one.

Horizontal Line Test

Any horizontal line will intersect the graph of a one-to-one function in at most one point.

This test is applied to the graphs of $f(x) = x^3$ and $g(x) = x^2$ in Figure 1.

By interchanging the components of the ordered pairs of a one-to-one function f, we obtain a new set of ordered pairs that satisfies the definition of function. This new function, called the *inverse function,* or *inverse,* of f, is symbolized f^{-1}.

Inverse Function

The **inverse function** of the one-to-one function f is defined as

$$f^{-1} = \{(y, x) \,|\, (x, y) \text{ belongs to } f\}.$$

C A U T I O N   The -1 in $f^{-1}(x)$ is not an exponent. That is,

$$f^{-1}(x) \neq \frac{1}{f(x)}.$$

The following statements summarize our discussion of inverse functions. Figure 2 illustrates some of the concepts.

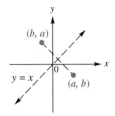

(b, a) is the reflection of (a, b) across the line $y = x$.

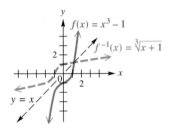

The graph of f^{-1} is the reflection of the graph of f across the line $y = x$.

Figure 2

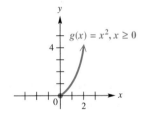

This is a one-to-one function.

Figure 3

Looking Ahead to Calculus

The inverse circular functions are used in calculus to solve certain types of related-rates problems and to integrate certain rational functions.

Summary of Inverse Functions

1. In a one-to-one function, each x-value corresponds to only one y-value and each y-value corresponds to only one x-value.

2. If a function f is one-to-one, then f has an inverse function f^{-1}.

3. The domain of f is the range of f^{-1}, and the range of f is the domain of f^{-1}.

4. The graphs of f and f^{-1} are reflections of each other across the line $y = x$.

5. To find $f^{-1}(x)$ from $f(x)$, follow these steps.

Step 1 Replace $f(x)$ with y and interchange x and y.

Step 2 Solve for y.

Step 3 Replace y with $f^{-1}(x)$.

We often restrict the domain of a function that is not one-to-one to make it one-to-one, without changing the range. For example, we saw in Figure 1 that $g(x) = x^2$, with its natural domain $(-\infty, \infty)$, is not one-to-one. However, if we restrict its domain to the set of nonnegative numbers $[0, \infty)$, we obtain a new function f that is one-to-one and has the same range as g, $[0, \infty)$. See Figure 3.

NOTE We could have chosen to restrict the domain of $g(x) = x^2$, $x \geq 0$ to $(-\infty, 0]$ to obtain a different one-to-one function. For the trigonometric functions, such choices are made based on general agreement by mathematicians.

In the remainder of this section, we develop the inverse circular (trigonometric) functions.

Inverse Sine Function From Figure 4 and the horizontal line test, we see that $y = \sin x$ does not define a one-to-one function. If we restrict the domain to the interval $\left[-\frac{\pi}{2}, \frac{\pi}{2} \right]$, which is the part of the graph in Figure 4 shown in color, this restricted function is one-to-one and has an inverse function. The range of $y = \sin x$ is $[-1, 1]$, so the domain of the inverse function will be $[-1, 1]$, and its range will be $\left[-\frac{\pi}{2}, \frac{\pi}{2} \right]$.

Restricted domain
$\left[-\frac{\pi}{2}, \frac{\pi}{2} \right]$

Figure 4

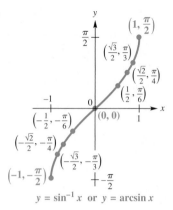

$y = \sin^{-1} x$ or $y = \arcsin x$

Figure 5

Reflecting the graph of $y = \sin x$ on the restricted domain across the line $y = x$ gives the graph of the inverse function, shown in Figure 5. Some key points are labeled on the graph. The equation of the inverse of $y = \sin x$ is found by interchanging x and y to get $x = \sin y$. This equation is solved for y by writing $y = \sin^{-1} x$ (read "inverse sine of x"). As Figure 5 shows, the domain of $y = \sin^{-1} x$ is $[-1, 1]$, while the restricted domain of $y = \sin x$, $\left[-\frac{\pi}{2}, \frac{\pi}{2}\right]$, is the range of $y = \sin^{-1} x$. An alternative notation for $\sin^{-1} x$ is arcsin x.

Inverse Sine Function

$$y = \sin^{-1} x \text{ or } y = \arcsin x \text{ means that } x = \sin y, \text{ for } -\tfrac{\pi}{2} \le y \le \tfrac{\pi}{2}.$$

We can think of $y = \sin^{-1} x$ or $y = \arcsin x$ as "y is the number in the interval $\left[-\frac{\pi}{2}, \frac{\pi}{2}\right]$ whose sine is x." Thus, we can write $y = \sin^{-1} x$ as $\sin y = x$ to evaluate it. We must pay close attention to the domain and range intervals.

EXAMPLE 1 Finding Inverse Sine Values

Find y in each equation.

(a) $y = \arcsin \dfrac{1}{2}$ **(b)** $y = \sin^{-1}(-1)$ **(c)** $y = \sin^{-1}(-2)$

Algebraic Solution

(a) The graph of the function defined by $y = \arcsin x$ (Figure 5) includes the point $\left(\frac{1}{2}, \frac{\pi}{6}\right)$. Thus,

$$\arcsin \frac{1}{2} = \frac{\pi}{6}.$$

Alternatively, we can think of $y = \arcsin \frac{1}{2}$ as "y is the number in $\left[-\frac{\pi}{2}, \frac{\pi}{2}\right]$ whose sine is $\frac{1}{2}$." Then we can write the given equation as $\sin y = \frac{1}{2}$. Since $\sin \frac{\pi}{6} = \frac{1}{2}$ and $\frac{\pi}{6}$ is in the range of the arcsine function, $y = \frac{\pi}{6}$.

(b) Writing the equation $y = \sin^{-1}(-1)$ in the form $\sin y = -1$ shows that $y = -\frac{\pi}{2}$. This can be verified by noticing that the point $\left(-1, -\frac{\pi}{2}\right)$ is on the graph of $y = \sin^{-1} x$.

(c) Because -2 is not in the domain of the inverse sine function, $\sin^{-1}(-2)$ does not exist.

Graphing Calculator Solution

To find these values with a graphing calculator, we graph $Y_1 = \sin^{-1} X$ and locate the points with X-values $\frac{1}{2}$ and -1. Figure 6(a) shows that when $X = \frac{1}{2}$, $Y = \frac{\pi}{6} \approx .52359878$. Similarly, Figure 6(b) shows that when $X = -1$, $Y = -\frac{\pi}{2} \approx -1.570796$.

(a) (b)

Figure 6

Since $\sin^{-1}(-2)$ does not exist, a calculator will give an error message for this input.

Now try Exercises 13 and 23.

CAUTION In Example 1(b), it is tempting to give the value of $\sin^{-1}(-1)$ as $\frac{3\pi}{2}$, since $\sin \frac{3\pi}{2} = -1$. Notice, however, that $\frac{3\pi}{2}$ is not in the range of the inverse sine function. Be certain that the number given for an inverse function value is in the range of the particular inverse function being considered.

Our observations about the inverse sine function from Figure 5 lead to the following generalizations.

INVERSE SINE FUNCTION
$$y = \sin^{-1}x \quad \text{or} \quad y = \arcsin x$$

Domain: $[-1,1]$ Range: $\left[-\frac{\pi}{2},\frac{\pi}{2}\right]$

Figure 7

- The inverse sine function is increasing and continuous on its domain $[-1,1]$.
- Its x-intercept is 0, and its y-intercept is 0.
- Its graph is symmetric with respect to the origin; it is an odd function.

Inverse Cosine Function   The function $y = \cos^{-1}x$ (or $y = \arccos x$) is defined by restricting the domain of the function $y = \cos x$ to the interval $[0, \pi]$ as in Figure 8, and then interchanging the roles of x and y. The graph of $y = \cos^{-1}x$ is shown in Figure 9. Again, some key points are shown on the graph.

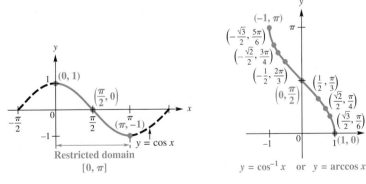

Figure 8

Figure 9

Inverse Cosine Function

$y = \cos^{-1}x$ or $y = \arccos x$ means that $x = \cos y$, for $0 \le y \le \pi$.

These screens support the results of Example 2, since $\frac{3\pi}{4} \approx 2.3561945$.

EXAMPLE 2 Finding Inverse Cosine Values

Find y in each equation.

(a) $y = \arccos 1$

(b) $y = \cos^{-1}\left(-\dfrac{\sqrt{2}}{2}\right)$

Solution

(a) Since the point $(1, 0)$ lies on the graph of $y = \arccos x$ in Figure 9 on the previous page, the value of y is 0. Alternatively, we can think of $y = \arccos 1$ as "y is the number in $[0, \pi]$ whose cosine is 1," or $\cos y = 1$. Then $y = 0$, since $\cos 0 = 1$ and 0 is in the range of the arccosine function.

(b) We must find the value of y that satisfies $\cos y = -\dfrac{\sqrt{2}}{2}$, where y is in the interval $[0, \pi]$, the range of the function $y = \cos^{-1} x$. The only value for y that satisfies these conditions is $\frac{3\pi}{4}$. Again, this can be verified from the graph in Figure 9.

Now try Exercises 15 and 21.

Our observations about the inverse cosine function lead to the following generalizations.

INVERSE COSINE FUNCTION
$y = \cos^{-1} x$ or $y = \arccos x$

Domain: $[-1, 1]$ Range: $[0, \pi]$

x	y
-1	π
$-\dfrac{\sqrt{2}}{2}$	$\dfrac{3\pi}{4}$
0	$\dfrac{\pi}{2}$
$\dfrac{\sqrt{2}}{2}$	$\dfrac{\pi}{4}$
1	0

Figure 10

- The inverse cosine function is decreasing and continuous on its domain $[-1, 1]$.
- Its x-intercept is 1, and its y-intercept is $\frac{\pi}{2}$.
- Its graph is not symmetric with respect to the y-axis or the origin.

Inverse Tangent Function Restricting the domain of the function $y = \tan x$ to the open interval $\left(-\frac{\pi}{2}, \frac{\pi}{2}\right)$ yields a one-to-one function. By interchanging the roles of x and y, we obtain the inverse tangent function given by $y = \tan^{-1} x$ or $y = \arctan x$. Figure 11 shows the graph of the restricted tangent function. Figure 12 gives the graph of $y = \tan^{-1} x$.

Figure 11

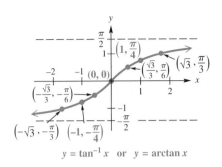

$y = \tan^{-1} x$ or $y = \arctan x$

Figure 12

Inverse Tangent Function

$y = \tan^{-1}x$ or $y = \arctan x$ means that $x = \tan y$, for $-\frac{\pi}{2} < y < \frac{\pi}{2}$.

INVERSE TANGENT FUNCTION
$$y = \tan^{-1} x \quad \text{or} \quad y = \arctan x$$

Domain: $(-\infty, \infty)$ Range: $\left(-\frac{\pi}{2}, \frac{\pi}{2}\right)$

x	y
-1	$-\frac{\pi}{4}$
$-\frac{\sqrt{3}}{3}$	$-\frac{\pi}{6}$
0	0
$\frac{\sqrt{3}}{3}$	$\frac{\pi}{6}$
1	$\frac{\pi}{4}$

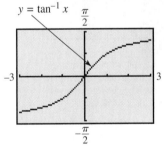

Figure 13

- The inverse tangent function is increasing and continuous on its domain $(-\infty, \infty)$.
- Its x-intercept is 0, and its y-intercept is 0.
- Its graph is symmetric with respect to the origin; it is an odd function.
- The lines $y = \frac{\pi}{2}$ and $y = -\frac{\pi}{2}$ are horizontal asymptotes.

Remaining Inverse Circular Functions The remaining three inverse trigonometric functions are defined similarly; their graphs are shown in Figure 14 on the next page. All six inverse trigonometric functions with their domains and ranges are given in the table that follows.

		Range	
Inverse Function	**Domain**	**Interval**	**Quadrants of the Unit Circle**
$y = \sin^{-1} x$	$[-1, 1]$	$\left[-\frac{\pi}{2}, \frac{\pi}{2}\right]$	I and IV
$y = \cos^{-1} x$	$[-1, 1]$	$[0, \pi]$	I and II
$y = \tan^{-1} x$	$(-\infty, \infty)$	$\left(-\frac{\pi}{2}, \frac{\pi}{2}\right)$	I and IV
$y = \cot^{-1} x$	$(-\infty, \infty)$	$(0, \pi)$	I and II
$y = \sec^{-1} x$	$(-\infty, -1] \cup [1, \infty)$	$[0, \pi], y \neq \frac{\pi}{2}$*	I and II
$y = \csc^{-1} x$	$(-\infty, -1] \cup [1, \infty)$	$\left[-\frac{\pi}{2}, \frac{\pi}{2}\right], y \neq 0$*	I and IV

Inverse Function Values The inverse circular functions are formally defined with real number ranges. However, there are times when it may be convenient to find degree-measured angles equivalent to these real number values. It is also often convenient to think in terms of the unit circle and choose the inverse function values based on the quadrants given in the preceding table.

The first three screens show the graphs of the three remaining inverse circular functions. The last screen shows how they are defined.

Figure 14

EXAMPLE 3 Finding Inverse Function Values (Degree-Measured Angles)

Find the *degree measure* of θ in the following.

(a) $\theta = \arctan 1$　　　　　　　**(b)** $\theta = \sec^{-1} 2$

Solution

(a) Here θ must be in $(-90°, 90°)$, but since $1 > 0$, θ must be in quadrant I. The alternative statement, $\tan \theta = 1$, leads to $\theta = 45°$.

(b) Write the equation as $\sec \theta = 2$. For $\sec^{-1} x$, θ is in quadrant I or II. Because 2 is positive, θ is in quadrant I and $\theta = 60°$, since $\sec 60° = 2$. Note that $60°$ (the degree equivalent of $\frac{\pi}{3}$) is in the range of the inverse secant function.

Now try Exercises 33 and 39.

The inverse trigonometric function keys on a calculator give results in the proper quadrant for the inverse sine, inverse cosine, and inverse tangent functions, according to the definitions of these functions. For example, on a calculator, in degrees, $\sin^{-1} .5 = 30°$, $\sin^{-1}(-.5) = -30°$, $\tan^{-1}(-1) = -45°$, and $\cos^{-1}(-.5) = 120°$.

Finding $\cot^{-1} x$, $\sec^{-1} x$, and $\csc^{-1} x$ with a calculator is not as straightforward, because these functions must be expressed in terms of $\tan^{-1} x$, $\cos^{-1} x$, and $\sin^{-1} x$, respectively. If $y = \sec^{-1} x$, for example, then $\sec y = x$, which must be written as a cosine function as follows:

$$\text{If } \sec y = x, \text{ then } \frac{1}{\cos y} = x \quad \text{or} \quad \cos y = \frac{1}{x}, \quad \text{and} \quad y = \cos^{-1} \frac{1}{x}.$$

---

*The inverse secant and inverse cosecant functions are sometimes defined with different ranges. We use intervals that match their reciprocal functions (except for one missing point).

In summary, to find $\sec^{-1} x$, we find $\cos^{-1} \frac{1}{x}$. Similar statements apply to $\csc^{-1} x$ and $\cot^{-1} x$. There is one additional consideration with $\cot^{-1} x$. Since we take the inverse tangent of the reciprocal to find inverse cotangent, the calculator gives values of inverse cotangent with the same range as inverse tangent, $\left(-\frac{\pi}{2}, \frac{\pi}{2}\right)$, which is not the correct range for inverse cotangent. For inverse cotangent, the proper range must be considered and the results adjusted accordingly.

EXAMPLE 4   Finding Inverse Function Values with a Calculator

(a) Find y in radians if $y = \csc^{-1}(-3)$.

(b) Find θ in degrees if $\theta = \text{arccot}(-.3541)$.

Solution

Figure 15

(a) With the calculator in radian mode, enter $\csc^{-1}(-3)$ as $\sin^{-1}\left(-\frac{1}{3}\right)$ to get $y \approx -.3398369095$. See Figure 15.

(b) Set the calculator to degree mode. A calculator gives the inverse tangent value of a negative number as a quadrant IV angle. The restriction on the range of arccotangent implies that θ must be in quadrant II, so enter

$$\text{arccot}(-.3541) \quad \text{as} \quad \tan^{-1}\left(\frac{1}{-.3541}\right) + 180°.$$

As shown in Figure 15, $\theta \approx 109.4990544°$.

Now try Exercises **43** and **49**.

EXAMPLE 5   Finding Function Values Using Definitions of the Trigonometric Functions

Evaluate each expression without using a calculator.

(a) $\sin\left(\tan^{-1} \dfrac{3}{2}\right)$ **(b)** $\tan\left(\cos^{-1}\left(-\dfrac{5}{13}\right)\right)$

Solution

$\theta = \tan^{-1} \frac{3}{2}$

Figure 16

(a) Let $\theta = \tan^{-1} \frac{3}{2}$, so $\tan\theta = \frac{3}{2}$. The inverse tangent function yields values only in quadrants I and IV, and since $\frac{3}{2}$ is positive, θ is in quadrant I. Sketch θ in quadrant I, and label a triangle, as shown in Figure 16. By the Pythagorean theorem, the hypotenuse is $\sqrt{13}$. The value of sine is the quotient of the side opposite and the hypotenuse, so

$$\sin\left(\tan^{-1}\frac{3}{2}\right) = \sin\theta = \frac{3}{\sqrt{13}} = \frac{3}{\sqrt{13}} \cdot \frac{\sqrt{13}}{\sqrt{13}} = \frac{3\sqrt{13}}{13}. \quad \text{(Section 2.1)}$$

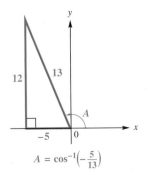

$A = \cos^{-1}\left(-\frac{5}{13}\right)$

Figure 17

(b) Let $A = \cos^{-1}\left(-\frac{5}{13}\right)$. Then, $\cos A = -\frac{5}{13}$. Since $\cos^{-1} x$ for a negative value of x is in quadrant II, sketch A in quadrant II, as shown in Figure 17.

$$\tan\left(\cos^{-1}\left(-\frac{5}{13}\right)\right) = \tan A = -\frac{12}{5}$$

Now try Exercises **63** and **65**.

EXAMPLE 6　Finding Function Values Using Identities

Evaluate each expression without using a calculator.

(a) $\cos\left(\arctan\sqrt{3} + \arcsin\dfrac{1}{3}\right)$　　**(b)** $\tan\left(2\arcsin\dfrac{2}{5}\right)$

Solution

(a) Let $A = \arctan\sqrt{3}$ and $B = \arcsin\frac{1}{3}$, so $\tan A = \sqrt{3}$ and $\sin B = \frac{1}{3}$. Sketch both A and B in quadrant I, as shown in Figure 18. Now, use the cosine sum identity.

$$\cos(A + B) = \cos A \cos B - \sin A \sin B \quad \text{(Section 5.3)}$$

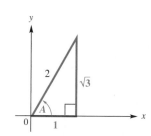

$$\cos\left(\arctan\sqrt{3} + \arcsin\frac{1}{3}\right) = \cos\left(\arctan\sqrt{3}\right)\cos\left(\arcsin\frac{1}{3}\right)$$

$$- \sin\left(\arctan\sqrt{3}\right)\sin\left(\arcsin\frac{1}{3}\right) \quad (1)$$

From Figure 18,

$$\cos\left(\arctan\sqrt{3}\right) = \cos A = \frac{1}{2}, \quad \cos\left(\arcsin\frac{1}{3}\right) = \cos B = \frac{2\sqrt{2}}{3},$$

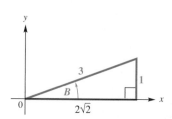

$$\sin\left(\arctan\sqrt{3}\right) = \sin A = \frac{\sqrt{3}}{2}, \quad \sin\left(\arcsin\frac{1}{3}\right) = \sin B = \frac{1}{3}.$$

Substitute these values into equation (1) to get

$$\cos\left(\arctan\sqrt{3} + \arcsin\frac{1}{3}\right) = \frac{1}{2}\cdot\frac{2\sqrt{2}}{3} - \frac{\sqrt{3}}{2}\cdot\frac{1}{3} = \frac{2\sqrt{2} - \sqrt{3}}{6}.$$

Figure 18

(b) Let $\arcsin\frac{2}{5} = B$. Then, from the double-angle tangent identity,

$$\tan\left(2\arcsin\frac{2}{5}\right) = \tan 2B = \frac{2\tan B}{1 - \tan^2 B}. \quad \text{(Section 5.5)}$$

Since $\arcsin\frac{2}{5} = B$, $\sin B = \frac{2}{5}$. Sketch a triangle in quadrant I, find the length of the third side, and then find $\tan B$. From the triangle in Figure 19, $\tan B = \frac{2}{\sqrt{21}}$, and

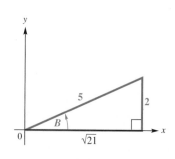

$$\tan\left(2\arcsin\frac{2}{5}\right) = \frac{2\left(\frac{2}{\sqrt{21}}\right)}{1 - \left(\frac{2}{\sqrt{21}}\right)^2} = \frac{\frac{4}{\sqrt{21}}}{1 - \frac{4}{21}} = \frac{\frac{4}{\sqrt{21}}\cdot\frac{\sqrt{21}}{\sqrt{21}}}{\frac{17}{21}} = \frac{\frac{4\sqrt{21}}{21}}{\frac{17}{21}} = \frac{4\sqrt{21}}{17}.$$

Figure 19

Simplify the complex fraction.

Now try Exercises 69 and 75.

While the work shown in Examples 5 and 6 does not rely on a calculator, we can support our algebraic work with one. By entering $\cos\left(\arctan\sqrt{3} + \arcsin\frac{1}{3}\right)$ from Example 6(a) into a calculator, we get the approximation .1827293862, the same approximation as when we enter $\frac{2\sqrt{2} - \sqrt{3}}{6}$ (the exact value we obtained algebraically). Similarly, we obtain the same approximation when we evaluate $\tan\left(2\arcsin\frac{2}{5}\right)$ and $\frac{4\sqrt{21}}{17}$, supporting our answer in Example 6(b).

EXAMPLE 7 Writing Function Values in Terms of u

Write each trigonometric expression as an algebraic expression in u.

(a) $\sin(\tan^{-1} u)$

(b) $\cos(2 \sin^{-1} u)$

Solution

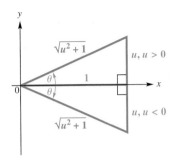

Figure 20

(a) Let $\theta = \tan^{-1} u$, so $\tan \theta = u$. Here, u may be positive or negative. Since $-\frac{\pi}{2} < \tan^{-1} u < \frac{\pi}{2}$, sketch θ in quadrants I and IV and label two triangles, as shown in Figure 20. Since sine is given by the quotient of the side opposite and the hypotenuse,

$$\sin(\tan^{-1} u) = \sin \theta = \frac{u}{\sqrt{u^2 + 1}} = \frac{u}{\sqrt{u^2 + 1}} \cdot \frac{\sqrt{u^2 + 1}}{\sqrt{u^2 + 1}} = \frac{u\sqrt{u^2 + 1}}{u^2 + 1}.$$

Rationalize the denominator.

The result is positive when u is positive and negative when u is negative.

(b) Let $\theta = \sin^{-1} u$, so $\sin \theta = u$. To find $\cos 2\theta$, use the double-angle identity $\cos 2\theta = 1 - 2 \sin^2 \theta$.

$$\cos(2 \sin^{-1} u) = \cos 2\theta = 1 - 2 \sin^2 \theta = 1 - 2u^2 \quad \text{(Section 5.5)}$$

Now try Exercises 83 and 85.

EXAMPLE 8 Finding the Optimal Angle of Elevation of a Shot Put

The optimal angle of elevation θ a shot-putter should aim for to throw the greatest distance depends on the velocity v of the throw and the initial height h of the shot. See Figure 21. One model for θ that achieves this greatest distance is

$$\theta = \arcsin\left(\sqrt{\frac{v^2}{2v^2 + 64h}} \right).$$

(*Source:* Townend, M. S., *Mathematics in Sport*, Chichester, Ellis Horwood Limited, 1984.)

Figure 21

Suppose a shot-putter can consistently throw the steel ball with $h = 6.6$ ft and $v = 42$ ft per sec. At what angle should he throw the ball to maximize distance?

Solution To find this angle, substitute and use a calculator in degree mode.

$$\theta = \arcsin\left(\sqrt{\frac{42^2}{2(42^2) + 64(6.6)}} \right) \approx 41.9° \quad h = 6.6, v = 42$$

Now try Exercise 93.

6.1 Exercises

Concept Check *Complete each statement.*

1. For a function to have an inverse, it must be _____.
2. The domain of $y = \arcsin x$ equals the _____ of $y = \sin x$.
3. The range of $y = \cos^{-1} x$ equals the _____ of $y = \cos x$.
4. The point $\left(\frac{\pi}{4}, 1\right)$ lies on the graph of $y = \tan x$. Therefore, the point _____ lies on the graph of _____.
5. If a function f has an inverse and $f(\pi) = -1$, then $f^{-1}(-1) =$ _____.
6. How can the graph of f^{-1} be sketched if the graph of f is known?

Concept Check *In Exercises 7–10, write short answers.*

7. Consider the inverse sine function, defined by $y = \sin^{-1} x$ or $y = \arcsin x$.
   (a) What is its domain? (b) What is its range?
   (c) Is this function increasing or decreasing?
   (d) Why is $\arcsin(-2)$ not defined?

8. Consider the inverse cosine function, defined by $y = \cos^{-1} x$ or $y = \arccos x$.
   (a) What is its domain? (b) What is its range?
   (c) Is this function increasing or decreasing?
   (d) $\text{Arccos}\left(-\frac{1}{2}\right) = \frac{2\pi}{3}$. Why is $\arccos\left(-\frac{1}{2}\right)$ not equal to $-\frac{4\pi}{3}$?

9. Consider the inverse tangent function, defined by $y = \tan^{-1} x$ or $y = \arctan x$.
   (a) What is its domain? (b) What is its range?
   (c) Is this function increasing or decreasing?
   (d) Is there any real number x for which $\arctan x$ is not defined? If so, what is it (or what are they)?

10. Give the domain and range of the three other inverse trigonometric functions, as defined in this section.
    (a) inverse cosecant function (b) inverse secant function
    (c) inverse cotangent function

11. *Concept Check* Is $\sec^{-1} a$ calculated as $\cos^{-1} \frac{1}{a}$ or as $\frac{1}{\cos^{-1} a}$?

12. *Concept Check* For positive values of a, $\cot^{-1} a$ is calculated as $\tan^{-1} \frac{1}{a}$. How is $\cot^{-1} a$ calculated for negative values of a?

Find the exact value of each real number y. Do not use a calculator. See Examples 1 and 2.

13. $y = \sin^{-1} 0$

14. $y = \tan^{-1} 1$

15. $y = \cos^{-1}(-1)$

16. $y = \arctan(-1)$

17. $y = \sin^{-1}(-1)$

18. $y = \cos^{-1} \frac{1}{2}$

19. $y = \arctan 0$

20. $y = \arcsin\left(-\frac{\sqrt{3}}{2}\right)$

21. $y = \arccos 0$

22. $y = \tan^{-1}(-1)$

23. $y = \sin^{-1} \frac{\sqrt{2}}{2}$

24. $y = \cos^{-1}\left(-\frac{1}{2}\right)$

25. $y = \arccos\left(-\frac{\sqrt{3}}{2}\right)$

26. $y = \arcsin\left(-\frac{\sqrt{2}}{2}\right)$

27. $y = \cot^{-1}(-1)$

28. $y = \sec^{-1}\left(-\sqrt{2}\right)$

29. $y = \csc^{-1}(-2)$

30. $y = \text{arccot}\left(-\sqrt{3}\right)$

31. $y = \text{arcsec} \frac{2\sqrt{3}}{3}$

32. $y = \csc^{-1} \sqrt{2}$

Give the degree measure of θ. Do not use a calculator. See Example 3.

33. $\theta = \arctan(-1)$

34. $\theta = \arccos\left(-\dfrac{1}{2}\right)$

35. $\theta = \arcsin\left(-\dfrac{\sqrt{3}}{2}\right)$

36. $\theta = \arcsin\left(-\dfrac{\sqrt{2}}{2}\right)$

37. $\theta = \cot^{-1}\left(-\dfrac{\sqrt{3}}{3}\right)$

38. $\theta = \csc^{-1}(-2)$

39. $\theta = \sec^{-1}(-2)$

40. $\theta = \csc^{-1}(-1)$

Use a calculator to give each value in decimal degrees. See Example 4.

41. $\theta = \sin^{-1}(-.13349122)$

42. $\theta = \cos^{-1}(-.13348816)$

43. $\theta = \arccos(-.39876459)$

44. $\theta = \arcsin .77900016$

45. $\theta = \csc^{-1} 1.9422833$

46. $\theta = \cot^{-1} 1.7670492$

Use a calculator to give each real number value. (Be sure the calculator is in radian mode.) See Example 4.

47. $y = \arctan 1.1111111$

48. $y = \arcsin .81926439$

49. $y = \cot^{-1}(-.92170128)$

50. $y = \sec^{-1}(-1.2871684)$

51. $y = \arcsin .92837781$

52. $y = \arccos .44624593$

Graph each inverse function as defined in the text.

53. $y = \cot^{-1} x$

54. $y = \csc^{-1} x$

55. $y = \sec^{-1} x$

56. $y = \text{arccsc } 2x$

57. $y = \text{arcsec } \dfrac{1}{2} x$

58. Explain why attempting to find $\sin^{-1} 1.003$ on your calculator will result in an error message.

59. Explain why you are able to find $\tan^{-1} 1.003$ on your calculator. Why is this situation different from the one described in Exercise 58?

Relating Concepts

For individual or collaborative investigation
(Exercises 60–62)*

60. Consider the function defined by $f(x) = 3x - 2$ and its inverse $f^{-1}(x) = \dfrac{x+2}{3}$. Simplify $f[f^{-1}(x)]$ and $f^{-1}[f(x)]$. What do you notice in each case? What would the graph look like in each case?

61. Use a graphing calculator to graph $y = \tan(\tan^{-1} x)$ in the standard viewing window, using radian mode. How does this compare to the graph you described in Exercise 60?

62. Use a graphing calculator to graph $y = \tan^{-1}(\tan x)$ in the standard viewing window, using radian and dot modes. Why does this graph not agree with the graph you found in Exercise 61?

*The authors wish to thank Carol Walker of Hinds Community College for making a suggestion on which these exercises are based.

Give the exact value of each expression without using a calculator. See Examples 5 and 6.

63. $\tan\left(\arccos\dfrac{3}{4}\right)$

64. $\sin\left(\arccos\dfrac{1}{4}\right)$

65. $\cos(\tan^{-1}(-2))$

66. $\sec\left(\sin^{-1}\left(-\dfrac{1}{5}\right)\right)$

67. $\sin\left(2\tan^{-1}\dfrac{12}{5}\right)$

68. $\cos\left(2\sin^{-1}\dfrac{1}{4}\right)$

69. $\cos\left(2\arctan\dfrac{4}{3}\right)$

70. $\tan\left(2\cos^{-1}\dfrac{1}{4}\right)$

71. $\sin\left(2\cos^{-1}\dfrac{1}{5}\right)$

72. $\cos(2\tan^{-1}(-2))$

73. $\sec(\sec^{-1}2)$

74. $\csc\left(\csc^{-1}\sqrt{2}\right)$

75. $\cos\left(\tan^{-1}\dfrac{5}{12}-\tan^{-1}\dfrac{3}{4}\right)$

76. $\cos\left(\sin^{-1}\dfrac{3}{5}+\cos^{-1}\dfrac{5}{13}\right)$

77. $\sin\left(\sin^{-1}\dfrac{1}{2}+\tan^{-1}(-3)\right)$

78. $\tan\left(\cos^{-1}\dfrac{\sqrt{3}}{2}-\sin^{-1}\left(-\dfrac{3}{5}\right)\right)$

Use a calculator to find each value. Give answers as real numbers.

79. $\cos(\tan^{-1}.5)$

80. $\sin(\cos^{-1}.25)$

81. $\tan(\arcsin .12251014)$

82. $\cot(\arccos .58236841)$

Write each expression as an algebraic (nontrigonometric) expression in u, u > 0. See Example 7.

83. $\sin(\arccos u)$

84. $\tan(\arccos u)$

85. $\cos(\arcsin u)$

86. $\cot(\arcsin u)$

87. $\sin\left(\sec^{-1}\dfrac{u}{2}\right)$

88. $\cos\left(\tan^{-1}\dfrac{3}{u}\right)$

89. $\tan\left(\sin^{-1}\dfrac{u}{\sqrt{u^2+2}}\right)$

90. $\sec\left(\cos^{-1}\dfrac{u}{\sqrt{u^2+5}}\right)$

91. $\sec\left(\operatorname{arccot}\dfrac{\sqrt{4-u^2}}{u}\right)$

92. $\csc\left(\arctan\dfrac{\sqrt{9-u^2}}{u}\right)$

(Modeling) Solve each problem.

93. *Angle of Elevation of a Shot Put* Refer to Example 8.

    (a) What is the optimal angle when $h = 0$?

    (b) Fix h at 6 ft and regard θ as a function of v. As v gets larger and larger, the graph approaches an asymptote. Find the equation of that asymptote.

94. *Landscaping Formula* A shrub is planted in a 100-ft-wide space between buildings measuring 75 ft and 150 ft tall. The location of the shrub determines how much sun it receives each day. Show that if θ is the angle in the figure and x is the distance of the shrub from the taller building, then the value of θ (in radians) is given by

$$\theta = \pi - \arctan\left(\frac{75}{100-x}\right) - \arctan\left(\frac{150}{x}\right).$$

95. *Observation of a Painting* A painting 1 m high and 3 m from the floor will cut off an angle θ to an observer, where

$$\theta = \tan^{-1}\left(\frac{x}{x^2 + 2}\right).$$

Assume that the observer is x meters from the wall where the painting is displayed and that the eyes of the observer are 2 m above the ground. (See the figure.) Find the value of θ for the following values of x. Round to the nearest degree.

(a) 1 (b) 2 (c) 3

(d) Derive the formula given above. (*Hint:* Use the identity for $\tan(\theta + \alpha)$. Use right triangles.)

(e) Graph the function for θ with a graphing calculator, and determine the distance that maximizes the angle.

(f) The idea in part (e) was first investigated in 1471 by the astronomer Regiomontanus. (*Source:* Maor, E., *Trigonometric Delights,* Princeton University Press, 1998.) If the bottom of the picture is a meters above eye level and the top of the picture is b meters above eye level, then the optimum value of x is $\sqrt{ab}$ meters. Use this result to find the exact answer to part (e).

96. *Communications Satellite Coverage* The figure shows a stationary communications satellite positioned 20,000 mi above the equator. What percent of the equator can be seen from the satellite? The diameter of Earth is 7927 mi at the equator.

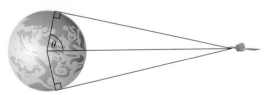

6.2 Trigonometric Equations I

Solving by Linear Methods ▪ **Solving by Factoring** ▪ **Solving by Quadratic Methods** ▪ **Solving by Using Trigonometric Identities**

Looking Ahead to Calculus

There are many instances in calculus where it is necessary to solve trigonometric equations. Examples include solving related-rates problems and optimization problems.

In Chapter 5, we studied trigonometric equations that were identities. We now consider trigonometric equations that are *conditional;* that is, equations that are satisfied by some values but not others.

Solving by Linear Methods Conditional equations with trigonometric (or circular) functions can usually be solved using algebraic methods and trigonometric identities.

(a)

(b)

Figure 22

EXAMPLE 1 Solving a Trigonometric Equation by Linear Methods

Solve $2 \sin \theta - 1 = 0$ over the interval $[0°, 360°)$.

Solution Because $\sin \theta$ is the first power of a trigonometric function, we use the same method as we would to solve the linear equation $2x + 1 = 0$.

$$2 \sin \theta + 1 = 0$$

$$2 \sin \theta = -1 \qquad \text{Subtract 1. (Appendix A)}$$

$$\sin \theta = -\frac{1}{2} \qquad \text{Divide by 2.}$$

To find values of θ that satisfy $\sin \theta = -\frac{1}{2}$, we observe that θ must be in either quadrant III or IV since the sine function is negative only in these two quadrants. Furthermore, the reference angle must be $30°$ since $\sin 30° = \frac{1}{2}$. The graphs in Figure 22 show the two possible values of θ, $210°$ and $330°$. The solution set is $\{210°, 330°\}$.

Alternatively, we could determine the solutions by referring to Figure 12 in Section 3.3 on page 109.

Now try Exercise 11.

Solving by Factoring

EXAMPLE 2 Solving a Trigonometric Equation by Factoring

Solve $\sin x \tan x = \sin x$ over the interval $[0°, 360°)$.

Solution
$$\sin x \tan x = \sin x$$
$$\sin x \tan x - \sin x = 0 \qquad \qquad \text{Subtract } \sin x.$$
$$\sin x(\tan x - 1) = 0 \qquad \qquad \text{Factor out } \sin x.$$
$$\sin x = 0 \qquad \text{or} \qquad \tan x - 1 = 0 \quad \text{Zero-factor property (Appendix A)}$$
$$\tan x = 1$$
$$x = 0° \quad \text{or} \quad x = 180° \qquad x = 45° \quad \text{or} \quad x = 225°$$

The solution set is $\{0°, 45°, 180°, 225°\}$.

Now try Exercise 31.

C A U T I O N There are four solutions in Example 2. Trying to solve the equation by dividing each side by $\sin x$ would lead to just $\tan x = 1$, which would give $x = 45°$ or $x = 225°$. The other two solutions would not appear. The missing solutions are the ones that make the divisor, $\sin x$, equal 0. For this reason, we avoid dividing by a variable expression.

Solving by Quadratic Methods
An equation in the form $au^2 + bu + c = 0$, where u is an algebraic expression, is solved by quadratic methods. (Appendix A) The expression u may be a trigonometric function, as in the equation $\tan^2 x + \tan x - 2 = 0$ which we solve in the next example.

EXAMPLE 3   Solving a Trigonometric Equation by Factoring

Solve $\tan^2 x + \tan x - 2 = 0$ over the interval $[0, 2\pi)$.

Solution   This equation is quadratic in form and can be solved by factoring.

$$\tan^2 x + \tan x - 2 = 0$$

$$(\tan x - 1)(\tan x + 2) = 0 \qquad \text{Factor.}$$

$$\tan x - 1 = 0 \quad \text{or} \quad \tan x + 2 = 0 \qquad \text{Zero-factor property}$$

$$\tan x = 1 \quad \text{or} \quad \tan x = -2$$

The solutions for $\tan x = 1$ over the interval $[0, 2\pi)$ are $x = \frac{\pi}{4}$ and $x = \frac{5\pi}{4}$.

To solve $\tan x = -2$ over that interval, we use a scientific calculator set in *radian* mode. We find that $\tan^{-1}(-2) \approx -1.1071487$. This is a quadrant IV number, based on the range of the inverse tangent function. (Refer to Figure 12 in Section 3.3 on page 109.) However, since we want solutions over the interval $[0, 2\pi)$, we must first add π to -1.1071487, and then add 2π.

$$x \approx -1.1071487 + \pi \approx 2.0344439$$

$$x \approx -1.1071487 + 2\pi \approx 5.1760366$$

The solutions over the required interval form the solution set

$$\left\{ \underbrace{\frac{\pi}{4}, \quad \frac{5\pi}{4},}_{\substack{\text{Exact} \\ \text{values}}} \quad \underbrace{2.0, \quad 5.2}_{\substack{\text{Approximate values} \\ \text{to the nearest tenth}}} \right\}.$$

Now try Exercise 21.

EXAMPLE 4   Solving a Trigonometric Equation Using the Quadratic Formula

Find all solutions of $\cot x(\cot x + 3) = 1$.

Solution   We multiply the factors on the left and subtract 1 to get the equation in standard quadratic form.

$$\cot^2 x + 3 \cot x - 1 = 0 \quad \text{(Appendix A)}$$

Since this equation cannot be solved by factoring, we use the quadratic formula, with $a = 1$, $b = 3$, $c = -1$, and $\cot x$ as the variable.

$$\cot x = \frac{-3 \pm \sqrt{9 + 4}}{2} = \frac{-3 \pm \sqrt{13}}{2} \qquad \begin{array}{l} \text{Quadratic formula with} \\ a = 1, b = 3, c = -1 \\ \text{(Appendix A)} \end{array}$$

$$\cot x \approx -3.302775638 \quad \text{or} \quad \cot x \approx .3027756377 \qquad \text{Use a calculator.}$$

We cannot find inverse cotangent values directly on a calculator, so we use the fact that $\cot x = \frac{1}{\tan x}$, and take reciprocals to get

$$\tan x \approx -.3027756377 \quad \text{or} \quad \tan x \approx 3.302775638$$

$$x \approx -.2940013018 \quad \text{or} \quad x \approx 1.276795025.$$

To find *all* solutions, we add integer multiples of the period of the tangent function, which is π, to each solution found above. Thus, all solutions of the equation are written as

$$-.2940013018 + n\pi \quad \text{and} \quad 1.276795025 + n\pi, \qquad \text{where } n \text{ is any integer.}*$$

Now try Exercise 43.

Solving by Using Trigonometric Identities   Recall that squaring both sides of an equation, such as $\sqrt{x + 4} = x + 2$, will yield all solutions but may also give extraneous values. (In this equation, 0 is a solution, while -3 is extraneous. Verify this.) The same situation may occur when trigonometric equations are solved in this manner.

EXAMPLE 5   Solving a Trigonometric Equation by Squaring

Solve $\tan x + \sqrt{3} = \sec x$ over the interval $[0, 2\pi)$.

Solution   Since the tangent and secant functions are related by the identity $1 + \tan^2 x = \sec^2 x$, square both sides and express $\sec^2 x$ in terms of $\tan^2 x$.

$$\tan x + \sqrt{3} = \sec x$$

$$\tan^2 x + 2\sqrt{3} \tan x + 3 = \sec^2 x \qquad (x + y)^2 = x^2 + 2xy + y^2$$

$$\tan^2 x + 2\sqrt{3} \tan x + 3 = 1 + \tan^2 x \qquad \text{Pythagorean identity (Section 5.1)}$$

$$2\sqrt{3} \tan x = -2 \qquad \text{Subtract } 3 + \tan^2 x.$$

$$\tan x = -\frac{1}{\sqrt{3}} = -\frac{\sqrt{3}}{3} \qquad \begin{array}{l}\text{Divide by } 2\sqrt{3}; \text{ rationalize the}\\\text{denominator.}\end{array}$$

The possible solutions are $\frac{5\pi}{6}$ and $\frac{11\pi}{6}$. Now check them. Try $\frac{5\pi}{6}$ first.

$$\text{Left side:} \quad \tan x + \sqrt{3} = \tan \frac{5\pi}{6} + \sqrt{3} = -\frac{\sqrt{3}}{3} + \sqrt{3} = \frac{2\sqrt{3}}{3}$$

$$\text{Right side:} \quad \sec x = \sec \frac{5\pi}{6} = -\frac{2\sqrt{3}}{3} \quad \longleftarrow \text{Not equal}$$

The check shows that $\frac{5\pi}{6}$ is not a solution. Now check $\frac{11\pi}{6}$.

$$\text{Left side:} \quad \tan \frac{11\pi}{6} + \sqrt{3} = -\frac{\sqrt{3}}{3} + \sqrt{3} = \frac{2\sqrt{3}}{3}$$

$$\text{Right side:} \quad \sec \frac{11\pi}{6} = \frac{2\sqrt{3}}{3} \quad \longleftarrow \text{Equal}$$

This solution satisfies the equation, so $\left\{\frac{11\pi}{6}\right\}$ is the solution set.

Now try Exercise 41.

$y = \tan x + \sqrt{3} - \sec x$

Dot mode; radian mode

The graph shows that on the interval $[0, 2\pi)$, the only x-intercept of the graph of $y = \tan x + \sqrt{3} - \sec x$ is 5.7595865, which is an approximation for $\frac{11\pi}{6}$, the solution found in Example 5.

*We usually give solutions of equations as solution sets, except when we ask for all solutions of a trigonometric equation.

The methods for solving trigonometric equations illustrated in the examples can be summarized as follows.

Solving a Trigonometric Equation

1. Decide whether the equation is linear or quadratic in form, so you can determine the solution method.

2. If only one trigonometric function is present, first solve the equation for that function.

3. If more than one trigonometric function is present, rearrange the equation so that one side equals 0. Then try to factor and set each factor equal to 0 to solve.

4. If the equation is quadratic in form, but not factorable, use the quadratic formula. Check that solutions are in the desired interval.

5. Try using identities to change the form of the equation. It may be helpful to square both sides of the equation first. If this is done, check for extraneous solutions.

EXAMPLE 6 Describing a Musical Tone from a Graph

A basic component of music is a pure tone. The graph in Figure 23 models the sinusoidal pressure $y = P$ in pounds per square foot from a pure tone at time $x = t$ in seconds.

(a) The frequency of a pure tone is often measured in hertz. One hertz is equal to one cycle per second and is abbreviated Hz. What is the frequency f in hertz of the pure tone shown in the graph?

(b) The time for the tone to produce one complete cycle is called the *period*. Approximate the period T in seconds of the pure tone.

(c) An equation for the graph is $y = .004 \sin 300\pi x$. Use a calculator to estimate all solutions to the equation that make $y = .004$ over the interval $[0, .02]$.

Solution

(a) From the graph in Figure 23, we see that there are 6 cycles in .04 sec. This is equivalent to $\frac{6}{.04} = 150$ cycles per sec. The pure tone has a frequency of $f = 150$ Hz.

(b) Six periods cover a time of .04 sec. One period would be equal to $T = \frac{.04}{6} = \frac{1}{150}$ or $.00\overline{6}$ sec.

(c) If we reproduce the graph in Figure 23 on a calculator as Y_1 and also graph a second function as $Y_2 = .004$, we can determine that the approximate values of x at the points of intersection of the graphs over the interval $[0, .02]$ are

$$.0017, \quad .0083, \quad \text{and} \quad .015.$$

The first value is shown in Figure 24.

$y = .004 \sin 300\pi x$

Figure 23

$Y_2 = .004$

$Y_1 = .004 \sin 300\pi X$

Figure 24

Now try Exercise 53.

6.2 Exercises

Concept Check *Refer to the summary box on solving a trigonometric equation. Decide on the appropriate technique to begin the solution of each equation. Do not solve the equation.*

1. $2 \cot x + 1 = -1$
2. $\sin x + 2 = 3$
3. $5 \sec^2 x = 6 \sec x$
4. $2 \cos^2 x - \cos x = 1$
5. $9 \sin^2 x - 5 \sin x = 1$
6. $\tan^2 x - 4 \tan x + 2 = 0$
7. $\tan x - \cot x = 0$
8. $\cos^2 x = \sin^2 x + 1$

9. Suppose in solving an equation over the interval $[0°, 360°)$, you reach the step $\sin \theta = -\frac{1}{2}$. Why is $\sin^{-1}\left(-\frac{1}{2}\right) = -30°$ not a correct answer?

10. Lindsay solved the equation $\sin x = 1 - \cos x$ by squaring both sides to get $\sin^2 x = 1 - 2 \cos x + \cos^2 x$. Several steps later, using correct algebra, she found that the solution set for solutions over the interval $[0, 2\pi)$ was $\left\{0, \frac{\pi}{2}, \frac{3\pi}{2}\right\}$. Explain why this is not the correct solution set.

Solve each equation for exact solutions over the interval $[0, 2\pi)$. See Examples 1–3.

11. $2 \cot x + 1 = -1$
12. $\sin x + 2 = 3$
13. $2 \sin x + 3 = 4$
14. $2 \sec x + 1 = \sec x + 3$
15. $\tan^2 x + 3 = 0$
16. $\sec^2 x + 2 = -1$
17. $(\cot x - 1)\left(\sqrt{3} \cot x + 1\right) = 0$
18. $(\csc x + 2)\left(\csc x - \sqrt{2}\right) = 0$
19. $\cos^2 x + 2 \cos x + 1 = 0$
20. $2 \cos^2 x - \sqrt{3} \cos x = 0$
21. $-2 \sin^2 x = 3 \sin x + 1$
22. $2 \cos^2 x - \cos x = 1$

Solve each equation for solutions over the interval $[0°, 360°)$. Give solutions to the nearest tenth as appropriate. See Examples 2–5.

23. $\left(\cot \theta - \sqrt{3}\right)\left(2 \sin \theta + \sqrt{3}\right) = 0$
24. $(\tan \theta - 1)(\cos \theta - 1) = 0$
25. $2 \sin \theta - 1 = \csc \theta$
26. $\tan \theta + 1 = \sqrt{3} + \sqrt{3} \cot \theta$
27. $\tan \theta - \cot \theta = 0$
28. $\cos^2 \theta = \sin^2 \theta + 1$
29. $\csc^2 \theta - 2 \cot \theta = 0$
30. $\sin^2 \theta \cos \theta = \cos \theta$
31. $2 \tan^2 \theta \sin \theta - \tan^2 \theta = 0$
32. $\sin^2 \theta \cos^2 \theta = 0$
33. $\sec^2 \theta \tan \theta = 2 \tan \theta$
34. $\cos^2 \theta - \sin^2 \theta = 0$
35. $9 \sin^2 \theta - 6 \sin \theta = 1$
36. $4 \cos^2 \theta + 4 \cos \theta = 1$
37. $\tan^2 \theta + 4 \tan \theta + 2 = 0$
38. $3 \cot^2 \theta - 3 \cot \theta - 1 = 0$
39. $\sin^2 \theta - 2 \sin \theta + 3 = 0$
40. $2 \cos^2 \theta + 2 \cos \theta - 1 = 0$
41. $\cot \theta + 2 \csc \theta = 3$
42. $2 \sin \theta = 1 - 2 \cos \theta$

Determine all solutions of each equation in radians (for x) or degrees (for θ) to the nearest tenth as appropriate. See Example 4.

43. $3 \sin^2 x - \sin x - 1 = 0$
44. $2 \cos^2 x + \cos x = 1$
45. $4 \cos^2 x - 1 = 0$
46. $2 \cos^2 x + 5 \cos x + 2 = 0$
47. $5 \sec^2 \theta = 6 \sec \theta$
48. $3 \sin^2 \theta - \sin \theta = 2$
49. $\dfrac{2 \tan \theta}{3 - \tan^2 \theta} = 1$
50. $\sec^2 \theta = 2 \tan \theta + 4$

The following equations cannot be solved by algebraic methods. Use a graphing calculator to find all solutions over the interval $[0, 2\pi)$. *Express solutions to four decimal places.*

51. $x^2 + \sin x - x^3 - \cos x = 0$

52. $x^3 - \cos^2 x = \dfrac{1}{2}x - 1$

(Modeling)   *Solve each problem.*

53. *Pressure on the Eardrum*   See Example 6. No musical instrument can generate a true pure tone. A pure tone has a unique, constant frequency and amplitude that sounds rather dull and uninteresting. The pressures caused by pure tones on the eardrum are sinusoidal. The change in pressure P in pounds per square foot on a person's eardrum from a pure tone at time t in seconds can be modeled using the equation

$$P = A \sin(2\pi f t + \phi),$$

where f is the frequency in cycles per second, and ϕ is the phase angle. When P is positive, there is an increase in pressure and the eardrum is pushed inward; when P is negative, there is a decrease in pressure and the eardrum is pushed outward. (*Source:* Roederer, J., *Introduction to the Physics and Psychophysics of Music,* Second Edition, Springer-Verlag, 1975.) A graph of the tone middle C is shown in the figure.

(a) Determine algebraically the values of t for which $P = 0$ over $[0, .005]$.

(b) From the graph and your answer in part (a), determine the interval for which $P \leq 0$ over $[0, .005]$.

(c) Would an eardrum hearing this tone be vibrating outward or inward when $P < 0$?

For $x = t$,
$$P(t) = .004 \sin\left[2\pi(261.63)t + \frac{\pi}{7}\right]$$

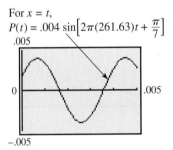

54. *Accident Reconstruction*   The model

$$.342D \cos \theta + h \cos^2 \theta = \frac{16D^2}{V_0^2}$$

is used to reconstruct accidents in which a vehicle vaults into the air after hitting an obstruction. V_0 is velocity in feet per second of the vehicle when it hits, D is distance (in feet) from the obstruction to the landing point, and h is the difference in height (in feet) between landing point and takeoff point. Angle θ is the takeoff angle, the angle between the horizontal and the path of the vehicle. Find θ to the nearest degree if $V_0 = 60$, $D = 80$, and $h = 2$.

55. *Electromotive Force*   In an electric circuit, let

$$V = \cos 2\pi t$$

model the electromotive force in volts at t seconds. Find the smallest positive value of t where $0 \leq t \leq \frac{1}{2}$ for each value of V.

(a) $V = 0$ (b) $V = .5$ (c) $V = .25$

56. *Voltage Induced by a Coil of Wire*   A coil of wire rotating in a magnetic field induces a voltage modeled by

$$e = 20 \sin\left(\frac{\pi t}{4} - \frac{\pi}{2}\right),$$

where t is time in seconds. Find the smallest positive time to produce each voltage.

(a) 0   (b) $10\sqrt{3}$

57. *Movement of a Particle*   A particle moves along a straight line. The distance of the particle from the origin at time t is modeled by

$$s(t) = \sin t + 2 \cos t.$$

Find a value of t that satisfies each equation.

(a) $s(t) = \dfrac{2 + \sqrt{3}}{2}$

(b) $s(t) = \dfrac{3\sqrt{2}}{2}$

58. Explain what is wrong with the following solution for all x over the interval $[0, 2\pi)$ of the equation $\sin^2 x - \sin x = 0$.

$$\sin^2 x - \sin x = 0$$

$$\sin x - 1 = 0 \qquad \text{Divide by } \sin x.$$

$$\sin x = 1 \qquad \text{Add 1.}$$

$$x = \frac{\pi}{2}$$

The solution set is $\left\{\frac{\pi}{2}\right\}$.

6.3 | Trigonometric Equations II

Equations with Half-Angles ▪ **Equations with Multiple Angles**

In this section, we discuss trigonometric equations that involve functions of half-angles and multiple angles. Solving these equations often requires adjusting solution intervals to fit given domains.

Equations with Half-Angles

EXAMPLE 1   Solving an Equation Using a Half-Angle Identity

Solve $2 \sin \dfrac{x}{2} = 1$

(a) over the interval $[0, 2\pi)$, and   **(b)** give all solutions.

Solution

(a) Write the interval $[0, 2\pi)$ as the inequality

$$0 \le x < 2\pi.$$

The corresponding interval for $\frac{x}{2}$ is

$$0 \le \frac{x}{2} < \pi. \qquad \text{Divide by 2. (Appendix A)}$$

To find all values of $\frac{x}{2}$ over the interval $[0, \pi)$ that satisfy the given equation, first solve for $\sin \frac{x}{2}$.

$$2 \sin \frac{x}{2} = 1$$

$$\sin \frac{x}{2} = \frac{1}{2} \qquad \text{Divide by 2.}$$

$y = 2 \sin \frac{x}{2} - 1$

The x-intercepts are the solutions found in Example 1.

Using Xscl $= \frac{\pi}{3}$ makes it possible to support the exact solutions by counting the tick marks from 0 on the graph.

The two numbers over the interval $[0, \pi)$ with sine value $\frac{1}{2}$ are $\frac{\pi}{6}$ and $\frac{5\pi}{6}$, so

$$\frac{x}{2} = \frac{\pi}{6} \quad \text{or} \quad \frac{x}{2} = \frac{5\pi}{6}$$

$$x = \frac{\pi}{3} \quad \text{or} \quad x = \frac{5\pi}{3}. \quad \text{Multiply by 2.}$$

The solution set over the given interval is $\left\{ \frac{\pi}{3}, \frac{5\pi}{3} \right\}$.

(b) Since this is a sine function with period 4π, all solutions are given by the expressions

$$\frac{\pi}{3} + 4n\pi \quad \text{and} \quad \frac{5\pi}{3} + 4n\pi, \quad \text{where } n \text{ is any integer.}$$

Now try Exercise 15.

Equations with Multiple Angles

EXAMPLE 2 Solving an Equation with a Double Angle

Solve $\cos 2x = \cos x$ over the interval $[0, 2\pi)$.

Solution First change $\cos 2x$ to a trigonometric function of x. Use the identity $\cos 2x = 2 \cos^2 x - 1$ so the equation involves only $\cos x$. Then factor.

$$\cos 2x = \cos x$$

$$2 \cos^2 x - 1 = \cos x \qquad \text{Substitute; double-angle identity (Section 5.5)}$$

$$2 \cos^2 x - \cos x - 1 = 0 \qquad \text{Subtract } \cos x.$$

$$(2 \cos x + 1)(\cos x - 1) = 0 \qquad \text{Factor.}$$

$$2 \cos x + 1 = 0 \quad \text{or} \quad \cos x - 1 = 0 \qquad \text{Zero-factor property (Appendix A)}$$

$$\cos x = -\frac{1}{2} \quad \text{or} \quad \cos x = 1$$

Over the required interval,

$$x = \frac{2\pi}{3} \quad \text{or} \quad x = \frac{4\pi}{3} \quad \text{or} \quad x = 0.$$

The solution set is $\left\{ 0, \frac{2\pi}{3}, \frac{4\pi}{3} \right\}$.

Now try Exercise 17.

C A U T I O N   In the solution of Example 2, $\cos 2x$ cannot be changed to $\cos x$ by dividing by 2 since 2 is not a factor of $\cos 2x$, that is, $\frac{\cos 2x}{2} \neq \cos x$. The only way to change $\cos 2x$ to a trigonometric function of x is by using one of the identities for $\cos 2x$.

EXAMPLE 3 Solving an Equation Using a Multiple-Angle Identity

Solve $4 \sin \theta \cos \theta = \sqrt{3}$ over the interval $[0°, 360°)$.

Solution The identity $2 \sin \theta \cos \theta = \sin 2\theta$ is useful here.

$$4 \sin \theta \cos \theta = \sqrt{3}$$

$$2(2 \sin \theta \cos \theta) = \sqrt{3} \qquad 4 = 2 \cdot 2$$

$$2 \sin 2\theta = \sqrt{3} \qquad 2 \sin \theta \cos \theta = \sin 2\theta \text{ (Section 5.5)}$$

$$\sin 2\theta = \frac{\sqrt{3}}{2} \qquad \text{Divide by 2.}$$

From the given interval $0° \le \theta < 360°$, the interval for 2θ is $0° \le 2\theta < 720°$. List all solutions over this interval.

$$2\theta = 60°, 120°, 420°, 480°$$

or $\qquad\qquad\qquad \theta = 30°, 60°, 210°, 240° \qquad \text{Divide by 2.}$

The final two solutions for 2θ were found by adding $360°$ to $60°$ and $120°$, respectively, giving the solution set $\{30°, 60°, 210°, 240°\}$.

Now try Exercise 37.

EXAMPLE 4 Solving an Equation with a Multiple Angle

Solve $\tan 3x + \sec 3x = 2$ over the interval $[0, 2\pi)$.

Solution Since the tangent and secant functions are related by the identity $1 + \tan^2 \theta = \sec^2 \theta$, one way to begin is to express everything in terms of secant.

$$\tan 3x + \sec 3x = 2$$

$$\tan 3x = 2 - \sec 3x \qquad \text{Subtract } \sec 3x.$$

$$\tan^2 3x = 4 - 4 \sec 3x + \sec^2 3x \qquad \begin{array}{l}\text{Square both sides;}\\ (x - y)^2 = x^2 - 2xy + y^2.\end{array}$$

$$\sec^2 3x - 1 = 4 - 4 \sec 3x + \sec^2 3x \qquad \begin{array}{l}\text{Replace } \tan^2 3x \text{ with } \sec^2 3x - 1. \\ \text{(Section 5.1)}\end{array}$$

$$4 \sec 3x = 5 \qquad \text{Simplify.}$$

$$\sec 3x = \frac{5}{4} \qquad \text{Divide by 4.}$$

$$\frac{1}{\cos 3x} = \frac{5}{4} \qquad \sec \theta = \frac{1}{\cos \theta} \text{ (Section 5.1)}$$

$$\cos 3x = \frac{4}{5} \qquad \text{Use reciprocals.}$$

$Y_2 = 2$ $Y_1 = \tan 3X + \sec 3X$

Intersection
X=2.3088955 Y=2

Connected mode; radian mode

The screen shows that one solution is approximately 2.3089. An advantage of using a graphing calculator is that extraneous values do not appear.

Multiply each term of the inequality $0 \le x < 2\pi$ by 3 to find the interval for $3x$: $[0, 6\pi)$. Using a calculator and the fact that cosine is positive in quadrants I and IV,

$$3x \approx .6435, 5.6397, 6.9267, 11.9229, 13.2099, 18.2061$$

$$x \approx .2145, 1.8799, 2.3089, 3.9743, 4.4033, 6.0687. \qquad \text{Divide by 3.}$$

Since both sides of the equation were squared, each proposed solution must be checked. Verify by substitution in the given equation that the solution set is {.2145, 2.3089, 4.4033}.

Now try Exercise 33.

A piano string can vibrate at more than one frequency when it is struck. It produces a complex wave that can mathematically be modeled by a sum of several pure tones. If a piano key with a frequency of f_1 is played, then the corresponding string will not only vibrate at f_1 but it will also vibrate at the higher frequencies of $2f_1$, $3f_1$, $4f_1$, ..., nf_1. f_1 is called the *fundamental frequency* of the string, and higher frequencies are called the *upper harmonics*. The human ear will hear the sum of these frequencies as one complex tone. (*Source:* Roederer, J., *Introduction to the Physics and Psychophysics of Music,* Second Edition, Springer-Verlag, 1975.)

EXAMPLE 5 Analyzing Pressures of Upper Harmonics

Suppose that the A key above middle C is played. Its fundamental frequency is $f_1 = 440$ Hz, and its associated pressure is expressed as

$$P_1 = .002 \sin 880\pi t.$$

The string will also vibrate at

$$f_2 = 880, \quad f_3 = 1320, \quad f_4 = 1760, \quad f_5 = 2200, \ldots \text{Hz}.$$

The corresponding pressures of these upper harmonics are

$$P_2 = \frac{.002}{2} \sin 1760\pi t, \qquad P_3 = \frac{.002}{3} \sin 2640\pi t,$$

$$P_4 = \frac{.002}{4} \sin 3520\pi t, \qquad \text{and} \qquad P_5 = \frac{.002}{5} \sin 4400\pi t.$$

The graph of

$$P = P_1 + P_2 + P_3 + P_4 + P_5,$$

shown in Figure 25, is "saw-toothed."

(a) What is the maximum value of P?

(b) At what values of x does this maximum occur over the interval $[0, .01]$?

Solution

(a) A graphing calculator shows that the maximum value of P is approximately .00317. See Figure 26.

(b) The maximum occurs at $x \approx .000188$, .00246, .00474, .00701, and .00928. Figure 26 shows how the second value is found; the others are found similarly.

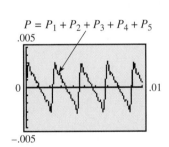

$P = P_1 + P_2 + P_3 + P_4 + P_5$

Figure 25

$P = P_1 + P_2 + P_3 + P_4 + P_5$

Maximum
X=.00246233 Y=.0031657

Figure 26

Now try Exercise 43.

6.3 Exercises

Concept Check   *Answer each question.*

1. Suppose you are solving a trigonometric equation for solutions over the interval $[0, 2\pi)$, and your work leads to $2x = \frac{2\pi}{3}, 2\pi, \frac{8\pi}{3}$. What are the corresponding values of x?

2. Suppose you are solving a trigonometric equation for solutions over the interval $[0, 2\pi)$, and your work leads to $\frac{1}{2}x = \frac{\pi}{16}, \frac{5\pi}{12}, \frac{5\pi}{8}$. What are the corresponding values of x?

3. Suppose you are solving a trigonometric equation for solutions over the interval $[0°, 360°)$, and your work leads to $3\theta = 180°, 630°, 720°, 930°$. What are the corresponding values of θ?

4. Suppose you are solving a trigonometric equation for solutions over the interval $[0°, 360°)$, and your work leads to $\frac{1}{3}\theta = 45°, 60°, 75°, 90°$. What are the corresponding values of θ?

5. Explain what is wrong with the following solution.
   Solve $\tan 2\theta = 2$ over the interval $[0, 2\pi)$.

$$\tan 2\theta = 2$$

$$\frac{\tan 2\theta}{2} = \frac{2}{2}$$

$$\tan \theta = 1$$

$$\theta = \frac{\pi}{4} \quad \text{or} \quad \theta = \frac{5\pi}{4}$$

The solution set is $\left\{\frac{\pi}{4}, \frac{5\pi}{4}\right\}$.

6. The equation $\cot \frac{x}{2} - \csc \frac{x}{2} - 1 = 0$ has no solution over the interval $[0, 2\pi)$. Using this information, what can be said about the graph of

$$y = \cot \frac{x}{2} - \csc \frac{x}{2} - 1$$

over this interval? Confirm your answer by actually graphing the function over the interval.

Solve each equation for exact solutions over the interval $[0, 2\pi)$. See Examples 1–4.

7. $\cos 2x = \frac{\sqrt{3}}{2}$

8. $\cos 2x = -\frac{1}{2}$

9. $\sin 3x = -1$

10. $\sin 3x = 0$

11. $3 \tan 3x = \sqrt{3}$

12. $\cot 3x = \sqrt{3}$

13. $\sqrt{2} \cos 2x = -1$

14. $2\sqrt{3} \sin 2x = \sqrt{3}$

15. $\sin \frac{x}{2} = \sqrt{2} - \sin \frac{x}{2}$

16. $\tan 4x = 0$

17. $\sin x = \sin 2x$

18. $\cos 2x - \cos x = 0$

19. $8 \sec^2 \frac{x}{2} = 4$

20. $\sin^2 \frac{x}{2} - 2 = 0$

21. $\sin \frac{x}{2} = \cos \frac{x}{2}$

22. $\sec \frac{x}{2} = \cos \frac{x}{2}$

23. $\cos 2x + \cos x = 0$

24. $\sin x \cos x = \frac{1}{4}$

Solve each equation in Exercises 25–32 for exact solutions over the interval $[0°, 360°)$. In Exercises 33–40, give all exact solutions. See Examples 1–4.

25. $\sqrt{2} \sin 3\theta - 1 = 0$

26. $-2 \cos 2\theta = \sqrt{3}$

27. $\cos \frac{\theta}{2} = 1$

28. $\sin \dfrac{\theta}{2} = 1$ **29.** $2\sqrt{3} \sin \dfrac{\theta}{2} = 3$ **30.** $2\sqrt{3} \cos \dfrac{\theta}{2} = -3$

31. $2 \sin \theta = 2 \cos 2\theta$ **32.** $\cos \theta - 1 = \cos 2\theta$ **33.** $1 - \sin \theta = \cos 2\theta$

34. $\sin 2\theta = 2 \cos^2 \theta$ **35.** $\csc^2 \dfrac{\theta}{2} = 2 \sec \theta$ **36.** $\cos \theta = \sin^2 \dfrac{\theta}{2}$

37. $2 - \sin 2\theta = 4 \sin 2\theta$ **38.** $4 \cos 2\theta = 8 \sin \theta \cos \theta$

39. $2 \cos^2 2\theta = 1 - \cos 2\theta$ **40.** $\sin \theta - \sin 2\theta = 0$

The following equations cannot be solved by algebraic methods. Use a graphing calculator to find all solutions over the interval $[0, 2\pi)$. Express solutions to four decimal places.

41. $2 \sin 2x - x^3 + 1 = 0$ **42.** $3 \cos \dfrac{x}{2} + \sqrt{x} - 2 = -\dfrac{1}{2}x + 2$

(Modeling)   *Solve each problem. See Example 5.*

43. *Pressure of a Plucked String*   If a string with a fundamental frequency of 110 Hz is plucked in the middle, it will vibrate at the odd harmonics of 110, 330, 550, . . . Hz but not at the even harmonics of 220, 440, 660, . . . Hz. The resulting pressure P caused by the string can be modeled by the equation

$$P = .003 \sin 220\pi t + \frac{.003}{3} \sin 660\pi t + \frac{.003}{5} \sin 1100\pi t + \frac{.003}{7} \sin 1540\pi t.$$

(*Source:* Benade, A., *Fundamentals of Musical Acoustics,* Dover Publications, 1990; Roederer, J., *Introduction to the Physics and Psychophysics of Music,* Second Edition, Springer-Verlag, 1975.)

(a) Graph P in the window $[0, .03]$ by $[-.005, .005]$.

(b) Use the graph to describe the shape of the sound wave that is produced.

(c) See Section 6.2, Exercise 53. At lower frequencies, the inner ear will hear a tone only when the eardrum is moving outward. Determine the times over the interval $[0, .03]$ when this will occur.

44. *Hearing Beats in Music*   Musicians sometimes tune instruments by playing the same tone on two different instruments and listening for a phenomenon known as *beats.* Beats occur when two tones vary in frequency by only a few hertz. When the two instruments are in tune, the beats disappear. The ear hears beats because the pressure slowly rises and falls as a result of this slight variation in the frequency. This phenomenon can be seen using a graphing calculator. (*Source:* Pierce, J., *The Science of Musical Sound,* Scientific American Books, 1992.)

(a) Consider two tones with frequencies of 220 and 223 Hz and pressures $P_1 = .005 \sin 440\pi t$ and $P_2 = .005 \sin 446\pi t$, respectively. Graph the pressure $P = P_1 + P_2$ felt by an eardrum over the 1-sec interval $[.15, 1.15]$. How many beats are there in 1 sec?

(b) Repeat part (a) with frequencies of 220 and 216 Hz.

(c) Determine a simple way to find the number of beats per second if the frequency of each tone is given.

45. *Hearing Difference Tones*   Small speakers like those found in older radios and telephones often cannot vibrate slower than 200 Hz—yet 35 keys on a piano have frequencies below 200 Hz. When a musical instrument creates a tone of 110 Hz, it also creates tones at 220, 330, 440, 550, 660, . . . Hz. A small speaker cannot reproduce the 110-Hz vibration but it can reproduce the higher frequencies, which are called the upper harmonics. The low tones can still be heard because the speaker produces *difference tones* of the upper harmonics. The difference between consecutive frequencies is 110 Hz, and this difference tone will be heard by a listener. We can model this phenomenon using a graphing calculator. (*Source:* Benade, A., *Fundamentals of Musical Acoustics,* Dover Publications, 1990.)

(a) In the window $[0, .03]$ by $[-1, 1]$, graph the upper harmonics represented by the pressure

$$P = \frac{1}{2} \sin[2\pi(220)t] + \frac{1}{3} \sin[2\pi(330)t] + \frac{1}{4} \sin[2\pi(440)t].$$

(b) Estimate all t-coordinates where P is maximum.

(c) What does a person hear in addition to the frequencies of 220, 330, and 440 Hz?

(d) Graph the pressure produced by a speaker that can vibrate at 110 Hz and above.

46. *Daylight Hours in New Orleans* The seasonal variation in length of daylight can be modeled by a sine function. For example, the daily number of hours of daylight in New Orleans is given by

$$h = \frac{35}{3} + \frac{7}{3} \sin \frac{2\pi x}{365},$$

where x is the number of days after March 21 (disregarding leap year). (*Source:* Bushaw, D., et al., *A Sourcebook of Applications of School Mathematics.* Copyright © 1980 by The Mathematical Association of America.)

(a) On what date will there be about 14 hr of daylight?

(b) What date has the least number of hours of daylight?

(c) When will there be about 10 hr of daylight?

(Modeling) Alternating Electric Current The study of alternating electric current requires the solutions of equations of the form

$$i = I_{max} \sin 2\pi ft,$$

for time t in seconds, where i is instantaneous current in amperes, I_{max} is maximum current in amperes, and f is the number of cycles per second. (Source: Hannon, R. H., Basic Technical Mathematics with Calculus, W. B. Saunders Company, 1978.) Find the smallest positive value of t, given the following data.

47. $i = 40, I_{max} = 100, f = 60$ **48.** $i = 50, I_{max} = 100, f = 120$

49. $i = I_{max}, f = 60$ **50.** $i = \frac{1}{2} I_{max}, f = 60$

6.4 Equations Involving Inverse Trigonometric Functions

Solving for *x* in Terms of *y* Using Inverse Functions ▪ Solving Inverse Trigonometric Equations

Until now, the equations in this chapter have involved trigonometric functions of angles or real numbers. Now we examine equations involving *inverse* trigonometric functions.

Solving for *x* in Terms of *y* Using Inverse Functions

EXAMPLE 1   Solving an Equation for a Variable Using Inverse Notation

Solve $y = 3 \cos 2x$ for x.

Solution   We want $\cos 2x$ alone on one side of the equation so we can solve for $2x$, and then for x.

$$y = 3 \cos 2x$$

$$\frac{y}{3} = \cos 2x \qquad \text{Divide by 3. (Appendix A)}$$

$$2x = \arccos \frac{y}{3} \qquad \text{Definition of arccosine (Section 6.1)}$$

$$x = \frac{1}{2} \arccos \frac{y}{3} \qquad \text{Multiply by } \tfrac{1}{2}.$$

> Now try Exercise 7.

Solving Inverse Trigonometric Equations

EXAMPLE 2   Solving an Equation Involving an Inverse Trigonometric Function

Solve $2 \arcsin x = \pi$.

Solution   First solve for $\arcsin x$, and then for x.

$$2 \arcsin x = \pi$$

$$\arcsin x = \frac{\pi}{2} \qquad \text{Divide by 2.}$$

$$x = \sin \frac{\pi}{2} \qquad \text{Definition of arcsine (Section 6.1)}$$

$$x = 1 \qquad \text{(Section 3.3)}$$

Verify that the solution satisfies the given equation. The solution set is $\{1\}$.

> Now try Exercise 23.

EXAMPLE 3   Solving an Equation Involving Inverse Trigonometric Functions

Solve $\cos^{-1} x = \sin^{-1} \frac{1}{2}$.

Solution   Let $\sin^{-1} \frac{1}{2} = u$. Then $\sin u = \frac{1}{2}$ and for u in quadrant I, we have

$$\cos^{-1} x = u$$

$$\cos u = x. \qquad \text{Alternative form (Section 6.1)}$$

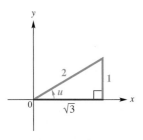

Figure 27

Sketch a triangle and label it using the facts that u is in quadrant I and $\sin u = \frac{1}{2}$. See Figure 27. Since $x = \cos u$, $x = \frac{\sqrt{3}}{2}$, and the solution set is $\left\{ \frac{\sqrt{3}}{2} \right\}$. *Check.*

> Now try Exercise 29.

EXAMPLE 4   Solving an Inverse Trigonometric Equation Using an Identity

Solve $\arcsin x - \arccos x = \dfrac{\pi}{6}$.

Solution   Isolate one inverse function on one side of the equation.

$$\arcsin x - \arccos x = \frac{\pi}{6}$$

$$\arcsin x = \arccos x + \frac{\pi}{6} \qquad \text{Add } \arccos x. \quad (1)$$

$$\sin\left(\arccos x + \frac{\pi}{6}\right) = x \qquad \text{Definition of arcsine}$$

Let $u = \arccos x$, so $0 \le u \le \pi$ by definition.

$$\sin\left(u + \frac{\pi}{6}\right) = x \qquad \text{Substitute.} \quad (2)$$

$$\sin\left(u + \frac{\pi}{6}\right) = \sin u \cos \frac{\pi}{6} + \cos u \sin \frac{\pi}{6}$$

Sine sum identity (**Section 5.4**)

Substitute this result into equation (2) to get

$$\sin u \cos \frac{\pi}{6} + \cos u \sin \frac{\pi}{6} = x. \quad (3)$$

From equation (1) and by the definition of the arcsine function,

$$-\frac{\pi}{2} \le \arccos x + \frac{\pi}{6} \le \frac{\pi}{2}$$

$$-\frac{2\pi}{3} \le \arccos x \le \frac{\pi}{3}. \qquad \text{Subtract } \tfrac{\pi}{6}. \text{ (Appendix A)}$$

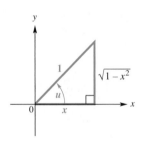

Figure 28

Since $0 \le \arccos x \le \pi$, we must have $0 \le \arccos x \le \frac{\pi}{3}$. Thus, $x > 0$, and we can sketch the triangle in Figure 28. From this triangle we find that $\sin u = \sqrt{1 - x^2}$. Now substitute into equation (3) using $\sin u = \sqrt{1 - x^2}$, $\sin \frac{\pi}{6} = \frac{1}{2}$, $\cos \frac{\pi}{6} = \frac{\sqrt{3}}{2}$, and $\cos u = x$.

$$\sin u \cos \frac{\pi}{6} + \cos u \sin \frac{\pi}{6} = x \qquad (3)$$

$$\left(\sqrt{1 - x^2}\right)\frac{\sqrt{3}}{2} + x \cdot \frac{1}{2} = x$$

$$\left(\sqrt{1 - x^2}\right)\sqrt{3} + x = 2x \qquad \text{Multiply by 2.}$$

$$\left(\sqrt{3}\right)\sqrt{1 - x^2} = x \qquad \text{Subtract } x.$$

$$3(1 - x^2) = x^2 \qquad \text{Square both sides.}$$

$$3 - 3x^2 = x^2 \qquad \text{Distributive property}$$

$$3 = 4x^2 \qquad \text{Add } 3x^2.$$

$$x = \sqrt{\frac{3}{4}}$$ Solve for x (**Appendix A**); choose the positive square root because $x > 0$.

$$x = \frac{\sqrt{3}}{2}$$ Quotient rule; $\sqrt{\frac{a}{b}} = \frac{\sqrt{a}}{\sqrt{b}}$

To *check*, replace x with $\frac{\sqrt{3}}{2}$ in the original equation:

$$\arcsin \frac{\sqrt{3}}{2} - \arccos \frac{\sqrt{3}}{2} = \frac{\pi}{3} - \frac{\pi}{6} = \frac{\pi}{6},$$

as required. The solution set is $\left\{\frac{\sqrt{3}}{2}\right\}$.

Now try Exercise 31.

6.4 Exercises

Concept Check   Answer each question.

1. Which one of the following equations has solution 0?

    A. $\arctan 1 = x$ **B.** $\arccos 0 = x$ **C.** $\arcsin 0 = x$

2. Which one of the following equations has solution $\frac{\pi}{4}$?

    A. $\arcsin \frac{\sqrt{2}}{2} = x$ **B.** $\arccos\left(-\frac{\sqrt{2}}{2}\right) = x$ **C.** $\arctan \frac{\sqrt{3}}{3} = x$

3. Which one of the following equations has solution $\frac{3\pi}{4}$?

    A. $\arctan 1 = x$ **B.** $\arcsin \frac{\sqrt{2}}{2} = x$ **C.** $\arccos\left(-\frac{\sqrt{2}}{2}\right) = x$

4. Which one of the following equations has solution $-\frac{\pi}{6}$?

    A. $\arctan \frac{\sqrt{3}}{3} = x$ **B.** $\arccos\left(-\frac{1}{2}\right) = x$ **C.** $\arcsin\left(-\frac{1}{2}\right) = x$

Solve each equation for x. See Example 1.

5. $y = 5 \cos x$ 6. $4y = \sin x$ 7. $2y = \cot 3x$

8. $6y = \frac{1}{2} \sec x$ 9. $y = 3 \tan 2x$ 10. $y = 3 \sin \frac{x}{2}$

11. $y = 6 \cos \frac{x}{4}$ 12. $y = -\sin \frac{x}{3}$ 13. $y = -2 \cos 5x$

14. $y = 3 \cot 5x$ **15.** $y = \cos(x + 3)$ 16. $y = \tan(2x - 1)$

17. $y = \sin x - 2$ 18. $y = \cot x + 1$ 19. $y = 2 \sin x - 4$

20. $y = 4 + 3 \cos x$

21. Refer to Exercise 17. A student attempting to solve this equation wrote as the first step $y = \sin(x - 2)$, inserting parentheses as shown. Explain why this is incorrect.

22. Explain why the equation $\sin^{-1} x = \cos^{-1} 2$ cannot have a solution. (No work is required.)

Solve each equation for exact solutions. See Examples 2 and 3.

23. $\dfrac{4}{3} \cos^{-1} \dfrac{y}{4} = \pi$

24. $4\pi + 4 \tan^{-1} y = \pi$

25. $2 \arccos\left(\dfrac{y - \pi}{3}\right) = 2\pi$

26. $\arccos\left(y - \dfrac{\pi}{3}\right) = \dfrac{\pi}{6}$

27. $\arcsin x = \arctan \dfrac{3}{4}$

28. $\arctan x = \arccos \dfrac{5}{13}$

29. $\cos^{-1} x = \sin^{-1} \dfrac{3}{5}$

30. $\cot^{-1} x = \tan^{-1} \dfrac{4}{3}$

Solve each equation for exact solutions. See Example 4.

31. $\sin^{-1} x - \tan^{-1} 1 = -\dfrac{\pi}{4}$

32. $\sin^{-1} x + \tan^{-1} \sqrt{3} = \dfrac{2\pi}{3}$

33. $\arccos x + 2 \arcsin \dfrac{\sqrt{3}}{2} = \pi$

34. $\arccos x + 2 \arcsin \dfrac{\sqrt{3}}{2} = \dfrac{\pi}{3}$

35. $\arcsin 2x + \arccos x = \dfrac{\pi}{6}$

36. $\arcsin 2x + \arcsin x = \dfrac{\pi}{2}$

37. $\cos^{-1} x + \tan^{-1} x = \dfrac{\pi}{2}$

38. $\sin^{-1} x + \tan^{-1} x = 0$

39. Provide graphical support for the solution in Example 4 by showing that the graph of $y = \arcsin x - \arccos x - \dfrac{\pi}{6}$ has x-intercept $\dfrac{\sqrt{3}}{2} \approx .8660254$.

40. Provide graphical support for the solution in Example 4 by showing that the x-coordinate of the point of intersection of the graphs of $Y_1 = \arcsin X - \arccos X$ and $Y_2 = \dfrac{\pi}{6}$ is $\dfrac{\sqrt{3}}{2} \approx .8660254$.

The following equations cannot be solved by algebraic methods. Use a graphing calculator to find all solutions over the interval $[0, 6]$. Express solutions to four decimal places.

41. $(\arctan x)^3 - x + 2 = 0$

42. $\pi \sin^{-1}(.2x) - 3 = -\sqrt{x}$

(Modeling) Solve each problem.

43. *Tone Heard by a Listener* When two sources located at different positions produce the same pure tone, the human ear will often hear one sound that is equal to the sum of the individual tones. Since the sources are at different locations, they will have different phase angles ϕ. If two speakers located at different positions produce pure tones $P_1 = A_1 \sin(2\pi ft + \phi_1)$ and $P_2 = A_2 \sin(2\pi ft + \phi_2)$, where $-\dfrac{\pi}{4} \leq \phi_1$, $\phi_2 \leq \dfrac{\pi}{4}$, then the resulting tone heard by a listener can be written as $P = A \sin(2\pi ft + \phi)$, where

$$A = \sqrt{(A_1 \cos \phi_1 + A_2 \cos \phi_2)^2 + (A_1 \sin \phi_1 + A_2 \sin \phi_2)^2}$$

and $\quad \phi = \arctan\left(\dfrac{A_1 \sin \phi_1 + A_2 \sin \phi_2}{A_1 \cos \phi_1 + A_2 \cos \phi_2}\right)$.

(*Source:* Fletcher, N. and T. Rossing, *The Physics of Musical Instruments,* Second Edition, Springer-Verlag, 1998.)

(a) Calculate A and ϕ if $A_1 = .0012$, $\phi_1 = .052$, $A_2 = .004$, and $\phi_2 = .61$. Also find an expression for $P = A \sin(2\pi ft + \phi)$ if $f = 220$.

(b) Graph $Y_1 = P$ and $Y_2 = P_1 + P_2$ on the same coordinate axes over the interval $[0, .01]$. Are the two graphs the same?

44. *Tone Heard by a Listener* Repeat Exercise 43, using $A_1 = .0025$, $\phi_1 = \frac{\pi}{7}$, $A_2 = .001$, $\phi_2 = \frac{\pi}{6}$, and $f = 300$.

45. *Depth of Field* When a large-view camera is used to take a picture of an object that is not parallel to the film, the lens board should be tilted so that the planes containing the subject, the lens board, and the film intersect in a line. This gives the best "depth of field." See the figure. (*Source:* Bushaw, D., et al., *A Sourcebook of Applications of School Mathematics.* Copyright © 1980 by The Mathematical Association of America.)

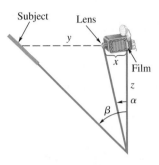

(a) Write two equations, one relating α, x, and z, and the other relating β, x, y, and z.
(b) Eliminate z from the equations in part (a) to get one equation relating α, β, x, and y.
(c) Solve the equation from part (b) for α.
(d) Solve the equation from part (b) for β.

46. *Programming Language for Inverse Functions* In Visual Basic, a widely used programming language for PCs, the only inverse trigonometric function available is arctangent. The other inverse trigonometric functions can be expressed in terms of arctangent as follows.

(a) Let $u = \arcsin x$. Solve the equation for x in terms of u.
(b) Use the result of part (a) to label the three sides of the triangle in the figure in terms of x.

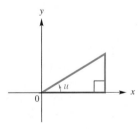

(c) Use the triangle from part (b) to write an equation for $\tan u$ in terms of x.
(d) Solve the equation from part (c) for u.

47. *Alternating Electric Current* In the study of alternating electric current, instantaneous voltage is modeled by

$$e = E_{\max} \sin 2\pi f t,$$

where f is the number of cycles per second, $E_{\max}$ is the maximum voltage, and t is time in seconds.

(a) Solve the equation for t.
(b) Find the smallest positive value of t if $E_{\max} = 12$, $e = 5$, and $f = 100$. Use a calculator.

48. *Viewing Angle of an Observer* While visiting a museum, Marsha Langlois views a painting that is 3 ft high and hangs 6 ft above the ground. See the figure. Assume her eyes are 5 ft above the ground, and let x be the distance from the spot where she is standing to the wall displaying the painting.

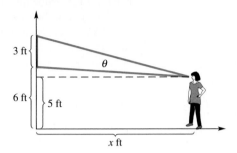

(a) Show that θ, the viewing angle subtended by the painting, is given by

$$\theta = \tan^{-1}\left(\frac{4}{x}\right) - \tan^{-1}\left(\frac{1}{x}\right).$$

(b) Find the value of x for each value of θ.

(i) $\theta = \dfrac{\pi}{6}$ (ii) $\theta = \dfrac{\pi}{8}$

(c) Find the value of θ for each value of x.

(i) $x = 4$ (ii) $x = 3$

49. *Movement of an Arm* In the Section 4.1 Exercises we found the equation

$$y = \frac{1}{3} \sin \frac{4\pi t}{3},$$

where t is time (in seconds) and y is the angle formed by a rhythmically moving arm.

(a) Solve the equation for t.

(b) At what time(s) does the arm form an angle of .3 radian?

50. The function $y = \sec^{-1} x$ is not found on graphing calculators. However, with some models it can be graphed as

$$y = \frac{\pi}{2} - ((x > 0) - (x < 0))\left(\frac{\pi}{2} - \tan^{-1}\left(\sqrt{(x^2 - 1)}\right)\right).$$

(This formula appears as Y_1 in the screen here.) Use the formula to obtain the graph of $y = \sec^{-1} x$ in the window $[-4, 4]$ by $[0, \pi]$.

Chapter 6 Summary

KEY TERMS

6.1 one-to-one function
inverse function

NEW SYMBOLS

f^{-1}	inverse of function f
$\sin^{-1}x$ (**arcsin** x)	inverse sine of x
$\cos^{-1}x$ (**arccos** x)	inverse cosine of x
$\tan^{-1}x$ (**arctan** x)	inverse tangent of x

$\cot^{-1}x$ (**arccot** x)	inverse cotangent of x
$\sec^{-1}x$ (**arcsec** x)	inverse secant of x
$\csc^{-1}x$ (**arccsc** x)	inverse cosecant of x

QUICK REVIEW

CONCEPTS

EXAMPLES

6.1 Inverse Circular Functions

Inverse Function	Domain	Range Interval	Range Quadrants of the Unit Circle
$y = \sin^{-1}x$	$[-1, 1]$	$\left[-\frac{\pi}{2}, \frac{\pi}{2}\right]$	I and IV
$y = \cos^{-1}x$	$[-1, 1]$	$[0, \pi]$	I and II
$y = \tan^{-1}x$	$(-\infty, \infty)$	$\left(-\frac{\pi}{2}, \frac{\pi}{2}\right)$	I and IV
$y = \cot^{-1}x$	$(-\infty, \infty)$	$(0, \pi)$	I and II
$y = \sec^{-1}x$	$(-\infty, -1] \cup [1, \infty)$	$[0, \pi], y \neq \frac{\pi}{2}$	I and II
$y = \csc^{-1}x$	$(-\infty, -1] \cup [1, \infty)$	$\left[-\frac{\pi}{2}, \frac{\pi}{2}\right], y \neq 0$	I and IV

Evaluate $y = \cos^{-1}0$.

Write $y = \cos^{-1}0$ as $\cos y = 0$. Then $y = \frac{\pi}{2}$, because $\cos\frac{\pi}{2} = 0$ and $\frac{\pi}{2}$ is in the range of $\cos^{-1}x$.

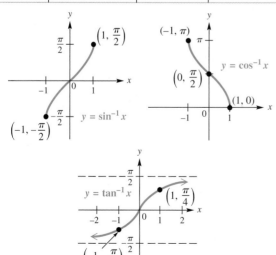

Evaluate $\sin\left(\tan^{-1}\left(-\frac{3}{4}\right)\right)$.

Let $u = \tan^{-1}\left(-\frac{3}{4}\right)$. Then $\tan u = -\frac{3}{4}$. Since $\tan^{-1}x$ is negative in quadrant IV, sketch a triangle as shown.

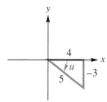

We want $\sin\left(\tan^{-1}\left(-\frac{3}{4}\right)\right) = \sin u$. From the triangle, $\sin u = -\frac{3}{5}$.

See page 242 for graphs of the other inverse circular (trigonometric) functions.

CONCEPTS	EXAMPLES

6.2 Trigonometric Equations I
6.3 Trigonometric Equations II

Solving a Trigonometric Equation

1. Decide whether the equation is linear or quadratic in form, so you can determine the solution method.
2. If only one trigonometric function is present, first solve the equation for that function.
3. If more than one trigonometric function is present, re-arrange the equation so that one side equals 0. Then try to factor and set each factor equal to 0 to solve.
4. If the equation is quadratic in form, but not factorable, use the quadratic formula. Check that solutions are in the desired interval.
5. Try using identities to change the form of the equation. It may be helpful to square both sides of the equation first. If this is done, check for extraneous solutions.

Solve $\tan \theta + \sqrt{3} = 2\sqrt{3}$ over the interval $[0°, 360°)$. Use a linear method.

$$\tan \theta + \sqrt{3} = 2\sqrt{3}$$
$$\tan \theta = \sqrt{3}$$
$$\theta = 60°$$

Another solution over $[0°, 360°)$ is

$$\theta = 60° + 180° = 240°.$$

The solution set is $\{60°, 240°\}$.

Solve $2 \cos^2 x = 1$ over the interval $[0, 2\pi)$ using a double-angle identity.

$$2 \cos^2 x = 1$$
$$2 \cos^2 x - 1 = 0$$
$$\cos 2x = 0 \quad \text{Cosine double-angle identity}$$
$$2x = \frac{\pi}{2}, \frac{3\pi}{2}, \frac{5\pi}{2}, \frac{7\pi}{2} \quad \begin{array}{l} 0 \le x < 2\pi, \text{ so} \\ 0 \le 2x < 4\pi. \end{array}$$
$$x = \frac{\pi}{4}, \frac{3\pi}{4}, \frac{5\pi}{4}, \frac{7\pi}{4}$$

The solution set is $\left\{\frac{\pi}{4}, \frac{3\pi}{4}, \frac{5\pi}{4}, \frac{7\pi}{4}\right\}$.

6.4 Equations Involving Inverse Trigonometric Functions

We solve equations of the form $y = f(x)$ where $f(x)$ is a trigonometric function using inverse trigonometric functions.

Solve $y = 2 \sin 3x$ for x.

$$y = 2 \sin 3x$$
$$\frac{y}{2} = \sin 3x \qquad \text{Divide by 2.}$$
$$3x = \arcsin \frac{y}{2} \qquad \text{Rewrite using arcsin.}$$
$$x = \frac{1}{3} \arcsin \frac{y}{2}$$

Techniques introduced in this section also show how to solve equations that involve inverse functions.

Solve $4 \tan^{-1} x = \pi$.

$$\tan^{-1} x = \frac{\pi}{4} \qquad \text{Divide by 4.}$$
$$x = \tan \frac{\pi}{4} = 1 \qquad \text{Definition of inverse tangent}$$

The solution set is $\{1\}$.

Chapter 6 Review Exercises

Concept Check   Tell whether each statement is true *or* false. *If false, tell why.*

1. The ranges of the inverse sine and inverse cosine functions are the same.

2. The ranges of the inverse tangent and inverse cotangent functions are the same.

3. It is true that $\sin \frac{11\pi}{6} = -\frac{1}{2}$, and therefore $\arcsin\left(-\frac{1}{2}\right) = \frac{11\pi}{6}$.

4. For all x, $\tan(\tan^{-1} x) = x$.

Give the exact real number value of y. Do not use a calculator.

5. $y = \sin^{-1} \dfrac{\sqrt{2}}{2}$

6. $y = \arccos\left(-\dfrac{1}{2}\right)$

7. $y = \tan^{-1}\left(-\sqrt{3}\right)$

8. $y = \arcsin(-1)$

9. $y = \cos^{-1}\left(-\dfrac{\sqrt{2}}{2}\right)$

10. $y = \arctan \dfrac{\sqrt{3}}{3}$

11. $y = \sec^{-1}(-2)$

12. $y = \text{arccsc} \dfrac{2\sqrt{3}}{3}$

13. $y = \text{arccot}(-1)$

Give the degree measure of θ. Do not use a calculator.

14. $\theta = \arccos \dfrac{1}{2}$

15. $\theta = \arcsin\left(-\dfrac{\sqrt{3}}{2}\right)$

16. $\theta = \tan^{-1} 0$

Use a calculator to give the degree measure of θ.

17. $\theta = \arctan 1.7804675$

18. $\theta = \sin^{-1}(-.66045320)$

19. $\theta = \cos^{-1} .80396577$

20. $\theta = \cot^{-1} 4.5046388$

21. $\theta = \text{arcsec}\, 3.4723155$

22. $\theta = \csc^{-1} 7.4890096$

Evaluate the following without using a calculator.

23. $\cos(\arccos(-1))$

24. $\sin\left(\arcsin\left(-\dfrac{\sqrt{3}}{2}\right)\right)$

25. $\arccos\left(\cos \dfrac{3\pi}{4}\right)$

26. $\text{arcsec}(\sec \pi)$

27. $\tan^{-1}\left(\tan \dfrac{\pi}{4}\right)$

28. $\cos^{-1}(\cos 0)$

29. $\sin\left(\arccos \dfrac{3}{4}\right)$

30. $\cos(\arctan 3)$

31. $\cos(\csc^{-1}(-2))$

32. $\sec\left(2 \sin^{-1}\left(-\dfrac{1}{3}\right)\right)$

33. $\tan\left(\arcsin \dfrac{3}{5} + \arccos \dfrac{5}{7}\right)$

Write each of the following as an algebraic (nontrigonometric) expression in u.

34. $\cos\left(\arctan \dfrac{u}{\sqrt{1 - u^2}}\right)$

35. $\tan\left(\text{arcsec} \dfrac{\sqrt{u^2 + 1}}{u}\right)$

Solve each equation for solutions over the interval $[0, 2\pi)$.

36. $\sin^2 x = 1$

37. $2 \tan x - 1 = 0$

38. $3 \sin^2 x - 5 \sin x + 2 = 0$

39. $\tan x = \cot x$

40. $\sec^4 2x = 4$

41. $\tan^2 2x - 1 = 0$

Give all solutions for each equation.

42. $\sec \dfrac{x}{2} = \cos \dfrac{x}{2}$ **43.** $\cos 2x + \cos x = 0$ **44.** $4 \sin x \cos x = \sqrt{3}$

Solve each equation for solutions over the interval $[0°, 360°)$. If necessary, express solutions to the nearest tenth of a degree.

45. $\sin^2 \theta + 3 \sin \theta + 2 = 0$ **46.** $2 \tan^2 \theta = \tan \theta + 1$

47. $\sin 2\theta = \cos 2\theta + 1$ **48.** $2 \sin 2\theta = 1$

49. $3 \cos^2 \theta + 2 \cos \theta - 1 = 0$ **50.** $5 \cot^2 \theta - \cot \theta - 2 = 0$

Solve each equation in Exercises 51–57 for x.

51. $4y = 2 \sin x$ **52.** $y = 3 \cos \dfrac{x}{2}$ **53.** $2y = \tan(3x + 2)$

54. $5y = 4 \sin x - 3$ **55.** $\dfrac{4}{3} \arctan \dfrac{x}{2} = \pi$ **56.** $\arccos x = \arcsin \dfrac{2}{7}$

57. $\arccos x + \arctan 1 = \dfrac{11\pi}{12}$

58. Solve $d = 550 + 450 \cos\left(\dfrac{\pi}{50} t\right)$ for t in terms of d.

(Modeling)   Solve each problem.

59. *Viewing Angle of an Observer*   A 10-ft-wide chalkboard is situated 5 ft from the left wall of a classroom. See the figure. A student sitting next to the wall x feet from the front of the classroom has a viewing angle of θ radians.

(a) Show that the value of θ is given by the function defined by

$$f(x) = \arctan\left(\dfrac{15}{x}\right) - \arctan\left(\dfrac{5}{x}\right).$$

(b) Graph $f(x)$ with a graphing calculator to estimate the value of x that maximizes the viewing angle.

60. *Snell's Law*   Recall Snell's law from Exercises 57 and 58 of Section 2.3:

$$\dfrac{c_1}{c_2} = \dfrac{\sin \theta_1}{\sin \theta_2},$$

where c_1 is the speed of light in one medium, c_2 is the speed of light in a second medium, and θ_1 and θ_2 are the angles shown in the figure. Suppose a light is shining up through water into the air as in the figure. As θ_1 increases, θ_2 approaches 90°, at which point no light will emerge from the water. Assume the ratio $\dfrac{c_1}{c_2}$ in this case is .752. For what value of θ_1 does $\theta_2 = 90°$? This value of θ_1 is called the *critical angle* for water.

61. *Snell's Law*   Refer to Exercise 60. What happens when θ_1 is greater than the critical angle?

62. *British Nautical Mile* The British nautical mile is defined as the length of a minute of arc of a meridian. Since Earth is flat at its poles, the nautical mile, in feet, is given by

A nautical mile is the length on any of the meridians cut by a central angle of measure 1 minute.

$$L = 6077 - 31 \cos 2\theta,$$

where θ is the latitude in degrees. See the figure. (*Source:* Bushaw, D., et al., *A Sourcebook of Applications of School Mathematics.* Copyright © 1980 by The Mathematical Association of America.)

(a) Find the latitude between $0°$ and $90°$ at which the nautical mile is 6074 ft.

(b) At what latitude between $0°$ and $180°$ is the nautical mile 6108 ft?

(c) In the United States, the nautical mile is defined everywhere as 6080.2 ft. At what latitude between $0°$ and $90°$ does this agree with the British nautical mile?

63. The function $y = \csc^{-1} x$ is not found on graphing calculators. However, with some models it can be graphed as

$$y = ((x > 0) - (x < 0))\left(\frac{\pi}{2} - \tan^{-1}\left(\sqrt{(x^2 - 1)}\right)\right).$$

(This formula appears as Y_1 in the screen here.) Use the formula to obtain the graph of $y = \csc^{-1} x$ in the window $[-4, 4]$ by $\left[-\frac{\pi}{2}, \frac{\pi}{2}\right]$.

64. (a) Use the graph of $y = \sin^{-1} x$ to approximate $\sin^{-1} .4$.

(b) Use the inverse sine key of a graphing calculator to approximate $\sin^{-1} .4$.

Chapter 6 Test

1. Graph $y = \sin^{-1} x$, and indicate the coordinates of three points on the graph. Give the domain and range.

2. Find the exact value of y for each equation.

(a) $y = \arccos\left(-\frac{1}{2}\right)$

(b) $y = \sin^{-1}\left(-\frac{\sqrt{3}}{2}\right)$

(c) $y = \tan^{-1} 0$

(d) $y = \text{arcsec}(-2)$

3. Find each exact value.

(a) $\cos\left(\arcsin \frac{2}{3}\right)$

(b) $\sin\left(2 \cos^{-1} \frac{1}{3}\right)$

4. Explain why $\sin^{-1} 3$ cannot be defined.

5. Explain why $\arcsin\left(\sin \frac{5\pi}{6}\right) \neq \frac{5\pi}{6}$.

6. Write $\tan(\arcsin u)$ as an algebraic (nontrigonometric) expression in u.

7. Solve $\sin^2 \theta = \cos^2 \theta + 1$ for solutions over the interval $[0°, 360°)$.

8. Solve $\csc^2 \theta - 2 \cot \theta = 4$ for solutions over the interval $[0°, 360°)$. Express approximate solutions to the nearest tenth of a degree.

9. Solve $\cos x = \cos 2x$ for solutions over the interval $[0, 2\pi)$.

10. Solve $2\sqrt{3} \sin \dfrac{x}{2} = 3$, giving all solutions in radians.

11. Solve each equation for x.

    (a) $y = \cos 3x$ **(b)** $\arcsin x = \arctan \dfrac{4}{3}$

12. *(Modeling) Movement of a Runner's Arm*   A runner's arm swings rhythmically according to the model

$$y = \frac{\pi}{8} \cos\left[\pi\left(t - \frac{1}{3} \right) \right],$$

where y represents the angle between the actual position of the upper arm and the downward vertical position and t represents time in seconds. At what times over the interval $[0, 3)$ is the angle y equal to 0?

Chapter 6 Quantitative Reasoning

How can we determine the amount of oil in a submerged storage tank?

The level of oil in a storage tank buried in the ground can be found in much the same way a dipstick is used to determine the oil level in an automobile crankcase. The person in the figure on the left has lowered a calibrated rod into an oil storage tank. When the rod is removed, the reading on the rod can be used with the dimensions of the storage tank to calculate the amount of oil in the tank.

    Suppose the ends of the cylindrical storage tank in the figure are circles of radius 3 ft and the cylinder is 20 ft long. Determine the volume of oil in the tank if the rod shows a depth of 2 ft. (*Hint:* The volume will be 20 times the area of the shaded segment of the circle shown in the figure on the right.)

7

Applications of Trigonometry and Vectors

Aerial photography first began in 1858 when French photographer Gaspard Tournachon took pictures of Paris from a hot-air balloon that had a makeshift darkroom. Since then, hot-air balloons have been replaced by airplanes, helicopters, and satellites, and aerial photography has been used in surveying, road design, weather forecasting, military surveillance, topographic maps, and even archaeology. The first archaeological aerial photographs were taken of Stonehenge in 1906. By searching these photographs for unusual soil and marks caused by structures lying below the ground, Stonehenge Avenue was discovered. (*Source:* Brooks, R. and D. Johannes, *Phytoarchaeology*, Dioscorides Press, 1990.)

Being able to determine the measurements of the sides and angles in a triangle is essential to solving applications involving aerial photography, as seen in the Connections box in Section 7.1.

7.1	Oblique Triangles and the Law of Sines
7.2	The Ambiguous Case of the Law of Sines
7.3	The Law of Cosines
7.4	Vectors, Operations, and the Dot Product
7.5	Applications of Vectors

7.1 Oblique Triangles and The Law of Sines

Congruency and Oblique Triangles ▪ Derivation of the Law of Sines ▪ Solving SAA and ASA Triangles (Case 1) ▪ Area of a Triangle

Until now, our work with triangles has been limited to right triangles. However, the concepts developed in Chapter 2 can be extended to *all* triangles. Every triangle has three sides and three angles. If any three of the six measures of a triangle are known (provided at least one measure is the length of a side), then the other three measures can be found. This is called *solving the triangle*.

Congruency and Oblique Triangles   The following axioms from geometry allow us to prove that two triangles are congruent (that is, their corresponding sides and angles are equal).

Congruence Axioms

Side-Angle-Side (SAS)	If two sides and the included angle of one triangle are equal, respectively, to two sides and the included angle of a second triangle, then the triangles are congruent.
Angle-Side-Angle (ASA)	If two angles and the included side of one triangle are equal, respectively, to two angles and the included side of a second triangle, then the triangles are congruent.
Side-Side-Side (SSS)	If three sides of one triangle are equal, respectively, to three sides of a second triangle, then the triangles are congruent.

Keep in mind that whenever SAS, ASA, or SSS is given, the triangle is unique. We continue to label triangles as we did earlier with right triangles: side a opposite angle A, side b opposite angle B, and side c opposite angle C.

A triangle that is not a right triangle is called an **oblique triangle.** The measures of the three sides and the three angles of a triangle can be found if at least one side and any other two measures are known. There are four possible cases.

Data Required for Solving Oblique Triangles

Case 1   One side and two angles are known (SAA or ASA).

Case 2   Two sides and one angle not included between the two sides are known (SSA). This case may lead to more than one triangle.

Case 3   Two sides and the angle included between the two sides are known (SAS).

Case 4   Three sides are known (SSS).

N O T E   If we know three angles of a triangle, we cannot find unique side lengths since AAA assures us only of similarity, not congruence. For example, there are infinitely many triangles ABC with $A = 35°$, $B = 65°$, and $C = 80°$.

Case 1, discussed in this section, and Case 2, discussed in Section 7.2, require the *law of sines*. Cases 3 and 4, discussed in Section 7.3, require the *law of cosines*.

Derivation of the Law of Sines

To derive the law of sines, we start with an oblique triangle, such as the *acute triangle* in Figure 1(a) or the *obtuse triangle* in Figure 1(b). This discussion applies to both triangles. First, construct the perpendicular from B to side AC (or its extension). Let h be the length of this perpendicular. Then c is the hypotenuse of right triangle ADB, and a is the hypotenuse of right triangle BDC.

Acute triangle *ABC*

(a)

Obtuse triangle *ABC*

(b)

Figure 1

In triangle ADB, $\qquad \sin A = \dfrac{h}{c} \quad$ or $\quad h = c \sin A.$

(Section 2.1)

In triangle BDC, $\qquad \sin C = \dfrac{h}{a} \quad$ or $\quad h = a \sin C.$

Since $h = c \sin A$ and $h = a \sin C,$

$$a \sin C = c \sin A$$

$$\frac{a}{\sin A} = \frac{c}{\sin C}. \qquad \text{Divide each side by } \sin A \sin C.$$

In a similar way, by constructing perpendicular lines from the other vertices, it can be shown that

$$\frac{a}{\sin A} = \frac{b}{\sin B} \qquad \text{and} \qquad \frac{b}{\sin B} = \frac{c}{\sin C}.$$

This discussion proves the following theorem.

Law of Sines

In any triangle ABC, with sides a, b, and c,

$$\frac{a}{\sin A} = \frac{b}{\sin B}, \qquad \frac{a}{\sin A} = \frac{c}{\sin C}, \qquad \text{and} \qquad \frac{b}{\sin B} = \frac{c}{\sin C}.$$

This can be written in compact form as

$$\frac{a}{\sin A} = \frac{b}{\sin B} = \frac{c}{\sin C}.$$

That is, according to the law of sines, the lengths of the sides in a triangle are proportional to the sines of the measures of the angles opposite them.

When solving for an angle, we use an alternative form of the law of sines,

$$\frac{\sin A}{a} = \frac{\sin B}{b} = \frac{\sin C}{c}.$$

NOTE When using the law of sines, a good strategy is to select an equation so that the unknown variable is in the numerator and all other variables are known.

Solving SAA and ASA Triangles (Case 1) If two angles and one side of a triangle are known (Case 1, SAA or ASA), then the law of sines can be used to solve the triangle.

EXAMPLE 1 Using the Law of Sines to Solve a Triangle (SAA)

Solve triangle ABC if $A = 32.0°$, $B = 81.8°$, and $a = 42.9$ cm.

Solution Start by drawing a triangle, roughly to scale, and labeling the given parts as in Figure 2. Since the values of A, B, and a are known, use the form of the law of sines that involves these variables, and then solve for b.

$$\frac{a}{\sin A} = \frac{b}{\sin B} \qquad \text{Law of sines}$$

$$\frac{42.9}{\sin 32.0°} = \frac{b}{\sin 81.8°} \qquad \text{Substitute the given values.}$$

$$b = \frac{42.9 \sin 81.8°}{\sin 32.0°} \qquad \text{Multiply by } \sin 81.8°; \text{ rewrite.}$$

$$b \approx 80.1 \text{ cm} \qquad \text{Approximate with a calculator.}$$

To find C, use the fact that the sum of the angles of any triangle is $180°$.

$$A + B + C = 180° \qquad \text{(Section 1.2)}$$
$$C = 180° - A - B$$
$$C = 180° - 32.0° - 81.8° = 66.2°$$

Use the law of sines to find c. (Why does the Pythagorean theorem not apply?)

$$\frac{a}{\sin A} = \frac{c}{\sin C}$$

$$\frac{42.9}{\sin 32.0°} = \frac{c}{\sin 66.2°}$$

$$c = \frac{42.9 \sin 66.2°}{\sin 32.0°}$$

$$c \approx 74.1 \text{ cm}$$

Now try Exercise 7.

C A U T I O N When solving oblique triangles such as the one in Example 1, be sure to label a sketch carefully to help set up the correct equation.

EXAMPLE 2 Using the Law of Sines in an Application (ASA)

Jerry Keefe wishes to measure the distance across the Big Muddy River. See Figure 3. He determines that $C = 112.90°$, $A = 31.10°$, and $b = 347.6$ ft. Find the distance a across the river.

Figure 3

Solution To use the law of sines, one side and the angle opposite it must be known. Since b is the only side whose length is given, angle B must be found before the law of sines can be used.

$$B = 180° - A - C$$
$$B = 180° - 31.10° - 112.90° = 36.00°$$

Now use the form of the law of sines involving A, B, and b to find a.

$$\frac{a}{\sin A} = \frac{b}{\sin B} \qquad \text{Law of sines}$$

$$\frac{a}{\sin 31.10°} = \frac{347.6}{\sin 36.00°} \qquad \text{Substitute.}$$

$$a = \frac{347.6 \sin 31.10°}{\sin 36.00°} \qquad \text{Multiply by } \sin 31.10°.$$

$$a \approx 305.5 \text{ ft} \qquad \text{Use a calculator.}$$

Now try Exercise 25.

The next example involves the concept of bearing, first discussed in Chapter 2.

EXAMPLE 3 Using the Law of Sines in an Application (ASA)

Two ranger stations are on an east-west line 110 mi apart. A forest fire is located on a bearing of N 42° E from the western station at A and a bearing of N 15° E from the eastern station at B. How far is the fire from the western station?

Solution Figure 4 shows the two stations at points A and B and the fire at point C.

Figure 4

Angle $BAC = 90° - 42° = 48°$, the obtuse angle at B equals $90° + 15° = 105°$, and the third angle, C, equals $180° - 105° - 48° = 27°$. We use the law of sines to find side b.

$$\frac{b}{\sin B} = \frac{c}{\sin C} \qquad \text{Law of sines}$$

$$\frac{b}{\sin 105°} = \frac{110}{\sin 27°} \qquad \text{Substitute.}$$

$$b = \frac{110 \sin 105°}{\sin 27°} \qquad \text{Multiply by } \sin 105°.$$

$$b \approx 234 \text{ mi}$$

Now try Exercise 27.

Acute triangle ABC

(a)

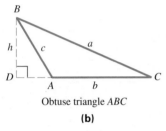

Obtuse triangle ABC

(b)

Figure 5

Area of a Triangle The method used to derive the law of sines can also be used to derive a formula to find the area of a triangle. A familiar formula for the area of a triangle is $\mathcal{A} = \frac{1}{2}bh$, where $\mathcal{A}$ represents area, b base, and h height. This formula cannot always be used easily since in practice, h is often unknown. To find another formula, refer to acute triangle ABC in Figure 5(a) or obtuse triangle ABC in Figure 5(b).

A perpendicular has been drawn from B to the base of the triangle (or the extension of the base). This perpendicular forms two right triangles. Using triangle ABD,

$$\sin A = \frac{h}{c} \qquad \text{or} \qquad h = c \sin A.$$

Substituting into the formula $\mathcal{A} = \frac{1}{2}bh$,

$$\mathcal{A} = \frac{1}{2}b(c \sin A) \qquad \text{or} \qquad \mathcal{A} = \frac{1}{2}bc \sin A.$$

Any other pair of sides and the angle between them could have been used.

Area of a Triangle (SAS)

In any triangle ABC, the area $\mathcal{A}$ is given by the following formulas:

$$\mathcal{A} = \frac{1}{2}bc \sin A, \quad \mathcal{A} = \frac{1}{2}ab \sin C, \quad \text{and} \quad \mathcal{A} = \frac{1}{2}ac \sin B.$$

That is, the area is half the product of the lengths of two sides and the sine of the angle included between them.

NOTE If the included angle measures $90°$, its sine is 1, and the formula becomes the familiar $\mathcal{A} = \frac{1}{2}bh$.

EXAMPLE 4 Finding the Area of a Triangle (SAS)

Find the area of triangle ABC in Figure 6.

Solution We are given $B = 55°$, $a = 34$ ft, and $c = 42$ ft, so

$$\mathcal{A} = \frac{1}{2}ac \sin B = \frac{1}{2}(34)(42) \sin 55° \approx 585 \text{ ft}^2.$$

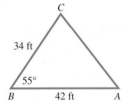

Figure 6

Now try Exercise 43.

EXAMPLE 5 Finding the Area of a Triangle (ASA)

Find the area of triangle ABC if $A = 24° \, 40'$, $b = 27.3$ cm, and $C = 52° \, 40'$.

Solution Before the area formula can be used, we must find either a or c. Since the sum of the measures of the angles of any triangle is $180°$,

$$B = 180° - 24° \, 40' - 52° \, 40' = 102° \, 40'.$$

We can use the law of sines to find a.

$$\frac{a}{\sin A} = \frac{b}{\sin B}$$

$$\frac{a}{\sin 24° 40'} = \frac{27.3}{\sin 102° 40'}$$

$$a = \frac{27.3 \sin 24° 40'}{\sin 102° 40'}$$

$$a \approx 11.7 \text{ cm}$$

Now, we find the area.

$$\mathcal{A} = \frac{1}{2} ab \sin C = \frac{1}{2}(11.7)(27.3) \sin 52° 40' \approx 127 \text{ cm}^2$$

Now try Exercise 49.

N O T E   Whenever possible, use given values in solving triangles or finding areas rather than values obtained in intermediate steps to avoid possible rounding errors.

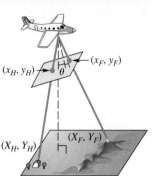

Figure 7

CONNECTIONS   As suggested in the chapter introduction, aerial photographs can be used to provide coordinates of ordered pairs to determine distances on the ground. Suppose we assign coordinates as shown in Figure 7. If an object's photographic coordinates are (x, y), then its ground coordinates (X, Y) in feet can be computed using the following formulas.

$$X = \frac{(a - h)x}{f \sec \theta - y \sin \theta}, \qquad Y = \frac{(a - h)y \cos \theta}{f \sec \theta - y \sin \theta}$$

Here, f is focal length of the camera in inches, a is altitude in feet of the airplane, and h is elevation in feet of the object. Suppose that a house has photographic coordinates $(x_H, y_H) = (.9, 3.5)$ with elevation 150 ft, while a nearby forest fire has photographic coordinates $(x_F, y_F) = (2.1, -2.4)$ and is at elevation 690 ft. If the photograph was taken at 7400 ft by a camera with focal length 6 in. and tilt angle $\theta = 4.1°$, we can use these formulas to find the distance on the ground in feet between the house and the forest fire. (*Source:* Moffitt, F. and E. Mikhail, *Photogrammetry,* Third Edition, Harper & Row, 1980.)

For Discussion or Writing

1. Use the formulas to find the ground coordinates of the house and the fire.

2. Use the distance formula given in Appendix B to find the required distance on the ground to the nearest tenth of a foot.

7.1 Exercises

1. *Concept Check* Consider the oblique triangle *ABC*. Which one of the following proportions is *not* valid?

A. $\dfrac{a}{b} = \dfrac{\sin A}{\sin B}$ **B.** $\dfrac{a}{\sin A} = \dfrac{b}{\sin B}$

C. $\dfrac{\sin A}{a} = \dfrac{b}{\sin B}$ **D.** $\dfrac{\sin A}{a} = \dfrac{\sin B}{b}$

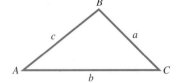

2. *Concept Check* Which two of the following situations do not provide sufficient information for solving a triangle by the law of sines?

  A. You are given two angles and the side included between them.
  B. You are given two angles and a side opposite one of them.
  C. You are given two sides and the angle included between them.
  D. You are given three sides.

Find the length of each side a. Do not use a calculator.

3.

4.

Determine the remaining sides and angles of each triangle ABC. See Example 1.

5.

6.

7.

8.

9. $A = 68.41°$, $B = 54.23°$, $a = 12.75$ ft

10. $C = 74.08°$, $B = 69.38°$, $c = 45.38$ m

11. $A = 87.2°$, $b = 75.9$ yd, $C = 74.3°$

12. $B = 38° \, 40'$, $a = 19.7$ cm, $C = 91° \, 40'$

13. $B = 20° \, 50'$, $C = 103° \, 10'$, $AC = 132$ ft

14. $A = 35.3°$, $B = 52.8°$, $AC = 675$ ft

15. $A = 39.70°$, $C = 30.35°$, $b = 39.74$ m

16. $C = 71.83°$, $B = 42.57°$, $a = 2.614$ cm

17. $B = 42.88°$, $C = 102.40°$, $b = 3974$ ft

18. $A = 18.75°$, $B = 51.53°$, $c = 2798$ yd

19. $A = 39° 54'$, $a = 268.7$ m, $B = 42° 32'$

20. $C = 79° 18'$, $c = 39.81$ mm, $A = 32° 57'$

21. Explain why the law of sines cannot be used to solve a triangle if we are given the lengths of the three sides of the triangle.

22. *Concept Check* In Example 1, we asked the question, "Why does the Pythagorean theorem not apply?" Answer this question.

23. Terry Harris, a perceptive trigonometry student, makes the statement, "If we know *any* two angles and one side of a triangle, then the triangle is uniquely determined." Is this a valid statement? Explain, referring to the congruence axioms given in this section.

24. *Concept Check* If a is twice as long as b, is A necessarily twice as large as B?

Solve each problem. See Examples 2 and 3.

25. *Distance Across a River* To find the distance AB across a river, a distance $BC = 354$ m is laid off on one side of the river. It is found that $B = 112° 10'$ and $C = 15° 20'$. Find AB. See the figure.

26. *Distance Across a Canyon* To determine the distance RS across a deep canyon, Rhonda lays off a distance $TR = 582$ yd. She then finds that $T = 32° 50'$ and $R = 102° 20'$. Find RS.

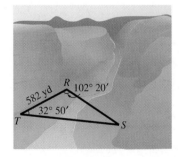

27. *Distance a Ship Travels* A ship is sailing due north. At a certain point the bearing of a lighthouse 12.5 km distant is N 38.8° E. Later on, the captain notices that the bearing of the lighthouse has become S 44.2° E. How far did the ship travel between the two observations of the lighthouse?

28. *Distance Between Radio Direction Finders* Radio direction finders are placed at points A and B, which are 3.46 mi apart on an east-west line, with A west of B. From A the bearing of a certain radio transmitter is 47.7°, and from B the bearing is 302.5°. Find the distance of the transmitter from A.

29. *Measurement of a Folding Chair* A folding chair is to have a seat 12.0 in. deep with angles as shown in the figure. How far down from the seat should the crossing legs be joined? (Find *x* in the figure.)

30. *Distance Across a River* Mark notices that the bearing of a tree on the opposite bank of a river flowing north is 115.45°. Lisa is on the same bank as Mark, but 428.3 m away. She notices that the bearing of the tree is 45.47°. The two banks are parallel. What is the distance across the river?

31. *Angle Formed by Radii of Gears* Three gears are arranged as shown in the figure. Find angle θ.

32. *Distance Between Atoms* Three atoms with atomic radii of 2.0, 3.0, and 4.5 are arranged as in the figure. Find the distance between the centers of atoms *A* and *C*.

33. *Distance Between a Ship and a Lighthouse* The bearing of a lighthouse from a ship was found to be N 37° E. After the ship sailed 2.5 mi due south, the new bearing was N 25° E. Find the distance between the ship and the lighthouse at each location.

34. *Height of a Balloon* A balloonist is directly above a straight road 1.5 mi long that joins two villages. She finds that the town closer to her is at an angle of depression of 35°, and the farther town is at an angle of depression of 31°. How high above the ground is the balloon?

Not to scale

35. *Distance to the Moon*   Since the moon is a relatively close celestial object, its distance can be measured directly by taking two different photographs at precisely the same time from two different locations. The moon will have a different angle of elevation at each location. On April 29, 1976, at 11:35 A.M., the lunar angles of elevation during a partial solar eclipse at Bochum in upper Germany and at Donaueschingen in lower Germany were measured as 52.6997° and 52.7430°, respectively. The two cities are 398 km apart. Calculate the distance to the moon from Bochum on this day, and compare it with the actual value of 406,000 km. Disregard the curvature of Earth in this calculation. (*Source:* Scholosser, W., T. Schmidt-Kaler, and E. Milone, *Challenges of Astronomy,* Springer-Verlag, 1991.)

36. *Ground Distances Measured by Aerial Photography*   The distance covered by an aerial photograph is determined by both the focal length of the camera and the tilt of the camera from the perpendicular to the ground. Although the tilt is usually small, both archaeological and Canadian photographs often use larger tilts. A camera lens with a 12-in. focal length will have an angular coverage of 60°. If an aerial photograph is taken with this camera tilted $\theta = 35°$ at an altitude of 5000 ft, calculate the ground distance d in feet that will be shown in this photograph. (*Source:* Brooks, R. and D. Johannes, *Phytoarchaeology,* Dioscorides Press, 1990; Moffitt, F. and E. Mikhail, *Photogrammetry,* Third Edition, Harper & Row, 1980.)

37. *Ground Distances Measured by Aerial Photography*   Refer to Exercise 36. A camera lens with a 6-in. focal length has an angular coverage of 86°. Suppose an aerial photograph is taken vertically with no tilt at an altitude of 3500 ft over ground with an increasing slope of 5°, as shown in the figure. Calculate the ground distance CB that would appear in the resulting photograph. (*Source:* Moffitt, F. and E. Mikhail, *Photogrammetry,* Third Edition, Harper & Row, 1980.)

38. *Ground Distances Measured by Aerial Photography*   Repeat Exercise 37 if the camera lens has an 8.25-in. focal length with an angular coverage of 72°.

Find the area of each triangle using the formula $\mathcal{A} = \frac{1}{2}bh$, and then verify that the formula $\mathcal{A} = \frac{1}{2}ab \sin C$ gives the same result.

39.

40.

41.

42.

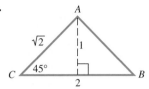

Find the area of each triangle ABC. See Examples 4 and 5.

43. $A = 42.5°$, $b = 13.6$ m, $c = 10.1$ m

44. $C = 72.2°$, $b = 43.8$ ft, $a = 35.1$ ft

45. $B = 124.5°$, $a = 30.4$ cm, $c = 28.4$ cm

46. $C = 142.7°$, $a = 21.9$ km, $b = 24.6$ km

47. $A = 56.80°$, $b = 32.67$ in., $c = 52.89$ in.

48. $A = 34.97°$, $b = 35.29$ m, $c = 28.67$ m

49. $A = 30.50°$, $b = 13.00$ cm, $C = 112.60°$

50. $A = 59.80°$, $b = 15.00$ m, $C = 53.10°$

Solve each problem.

51. *Area of a Metal Plate* A painter is going to apply a special coating to a triangular metal plate on a new building. Two sides measure 16.1 m and 15.2 m. She knows that the angle between these sides is 125°. What is the area of the surface she plans to cover with the coating?

52. *Area of a Triangular Lot* A real estate agent wants to find the area of a triangular lot. A surveyor takes measurements and finds that two sides are 52.1 m and 21.3 m, and the angle between them is 42.2°. What is the area of the triangular lot?

53. *Triangle Inscribed in a Circle* For a triangle inscribed in a circle of radius r, each law of sines ratio $\frac{a}{\sin A}$, $\frac{b}{\sin B}$, and $\frac{c}{\sin C}$ has value $2r$. The circle in the figure has diameter 1. What are the values of a, b, and c? (*Note:* This result provides an alternative way to define the sine function for angles between 0° and 180°. It was used nearly 2000 yr ago by the mathematician Ptolemy to construct one of the earliest trigonometric tables.)

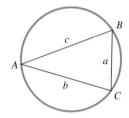

54. *Theorem of Ptolemy* The following theorem is also attributed to Ptolemy: *In a quadrilateral inscribed in a circle, the product of the diagonals is equal to the sum of the products of the opposite sides.* (*Source:* Eves, H., *An Introduction to the History of Mathematics,* Sixth Edition, Saunders College Publishing, 1990.) The circle in the figure has diameter 1. Explain why the lengths of the line segments are as shown, and then apply Ptolemy's theorem to derive the formula for the sine of the sum of two angles.

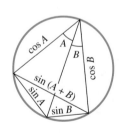

55. Several of the exercises on right triangle applications involved a figure similar to the one shown here, in which angles α and β and the length of line segment AB are known, and the length of side CD is to be determined. Use the law of sines to obtain x in terms of α, β, and d.

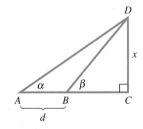

7.2 | The Ambiguous Case of the Law of Sines

Description of the Ambiguous Case ▪ **Solving SSA Triangles (Case 2)** ▪ **Analyzing Data for Possible Number of Triangles**

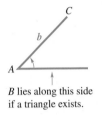

B lies along this side if a triangle exists.

Description of the Ambiguous Case   We used the law of sines to solve triangles involving Case 1, SAA or ASA, in Section 7.1. If we are given the lengths of two sides and the angle opposite one of them (Case 2, SSA), then zero, one, or two such triangles may exist. (There is no SSA axiom.)

Suppose we know the measure of acute angle *A* of triangle *ABC*, the length of side *a*, and the length of side *b*, as shown in the margin. Now we must draw the side of length *a* opposite angle *A*. The table shows possible outcomes. This situation (SSA) is called the **ambiguous case** of the law of sines.

If angle *A* is acute, there are four possible outcomes.

Number of Possible Triangles	Sketch	Conditions Necessary for Case to Hold
0		$\sin B > 1,$ $a < h < b$
1		$\sin B = 1,$ $a = h < b$
1		$0 < \sin B < 1,$ $a \geq b$
2		$0 < \sin B_2 < 1,$ $h < a < b$

If angle *A* is obtuse, there are two possible outcomes.

Number of Possible Triangles	Sketch	Conditions Necessary for Case to Hold
0		$\sin B \geq 1,$ $a \leq b$
1		$0 < \sin B < 1,$ $a > b$

We can apply the law of sines to the values of a, b, and A and use some basic properties of geometry and trigonometry to determine which situation applies. The following basic facts should be kept in mind.

1. For any angle θ of a triangle, $0 < \sin \theta \leq 1$. If $\sin \theta = 1$, then $\theta = 90°$ and the triangle is a right triangle.

2. $\sin \theta = \sin(180° - \theta)$ (Supplementary angles have the same sine value.)

3. The smallest angle is opposite the shortest side, the largest angle is opposite the longest side, and the middle-valued angle is opposite the intermediate side (assuming the triangle has sides that are all of different lengths).

Solving SSA Triangles (Case 2)

EXAMPLE 1 Solving the Ambiguous Case (No Such Triangle)

Solve triangle ABC if $B = 55° 40'$, $b = 8.94$ m, and $a = 25.1$ m.

Solution Since we are given B, b, and a, we can use the law of sines to find A.

b = 8.94

a = 25.1

55° 40'

B

Figure 8

$$\frac{\sin A}{a} = \frac{\sin B}{b} \qquad \text{Law of sines (alternative form) (Section 7.1)}$$

$$\frac{\sin A}{25.1} = \frac{\sin 55° 40'}{8.94} \qquad \text{Substitute the given values.}$$

$$\sin A = \frac{25.1 \sin 55° 40'}{8.94} \qquad \text{Multiply by 25.1.}$$

$$\sin A \approx 2.3184379 \qquad \text{Use a calculator.}$$

Since $\sin A$ cannot be greater than 1, there can be no such angle A and thus, no triangle with the given information. An attempt to sketch such a triangle leads to the situation shown in Figure 8.

Now try Exercise 17.

N O T E In the ambiguous case, we are given two sides and an angle opposite one of the sides (SSA). For example, suppose b, c, and angle C are given. This situation represents the ambiguous case because angle C is opposite side c.

EXAMPLE 2 Solving the Ambiguous Case (Two Triangles)

Solve triangle ABC if $A = 55.3°$, $a = 22.8$ ft, and $b = 24.9$ ft.

Solution To begin, use the law of sines to find angle B.

$$\frac{\sin A}{a} = \frac{\sin B}{b}$$

$$\frac{\sin 55.3°}{22.8} = \frac{\sin B}{24.9}$$

$$\sin B = \frac{24.9 \sin 55.3°}{22.8}$$

$$\sin B \approx .8978678$$

There are two angles B between $0°$ and $180°$ that satisfy this condition. Since $\sin B \approx .8978678$, to the nearest tenth one value of B is

$$B_1 = 63.9°. \quad \text{Use the inverse sine function. (Section 6.1)}$$

Supplementary angles have the same sine value, so another *possible* value of B is

$$B_2 = 180° - 63.9° = 116.1°. \quad \text{(Section 1.1)}$$

To see if $B_2 = 116.1°$ is a valid possibility, add $116.1°$ to the measure of A, $55.3°$. Since $116.1° + 55.3° = 171.4°$, and this sum is less than $180°$, it is a valid angle measure for this triangle.

Now separately solve triangles AB_1C_1 and AB_2C_2 shown in Figure 9. Begin with AB_1C_1. Find C_1 first.

$$C_1 = 180° - A - B_1 \quad \text{(Section 1.2)}$$
$$C_1 = 180° - 55.3° - 63.9°$$
$$C_1 = 60.8°$$

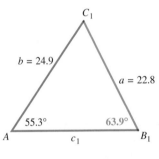

Now, use the law of sines to find c_1.

$$\frac{a}{\sin A} = \frac{c_1}{\sin C_1}$$

$$\frac{22.8}{\sin 55.3°} = \frac{c_1}{\sin 60.8°}$$

$$c_1 = \frac{22.8 \sin 60.8°}{\sin 55.3°}$$

$$c_1 \approx 24.2 \text{ ft}$$

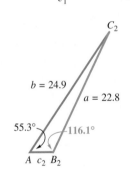

Figure 9

To solve triangle AB_2C_2, first find C_2.

$$C_2 = 180° - A - B_2$$
$$C_2 = 180° - 55.3° - 116.1°$$
$$C_2 = 8.6°$$

By the law of sines,

$$\frac{a}{\sin A} = \frac{c_2}{\sin C_2}$$

$$\frac{22.8}{\sin 55.3°} = \frac{c_2}{\sin 8.6°}$$

$$c_2 = \frac{22.8 \sin 8.6°}{\sin 55.3°}$$

$$c_2 \approx 4.15 \text{ ft.}$$

Now try Exercise 25.

The ambiguous case results in zero, one, or two triangles. The following guidelines can be used to determine how many triangles there are.

Number of Triangles Satisfying the Ambiguous Case (SSA)

Let sides a and b and angle A be given in triangle ABC. (The law of sines can be used to calculate the value of $\sin B$.)

1. If applying the law of sines results in an equation having $\sin B > 1$, then *no triangle* satisfies the given conditions.

2. If $\sin B = 1$, then *one triangle* satisfies the given conditions and $B = 90°$.

3. If $0 < \sin B < 1$, then either *one or two triangles* satisfy the given conditions.

    (a) If $\sin B = k$, then let $B_1 = \sin^{-1} k$ and use B_1 for B in the first triangle.

    (b) Let $B_2 = 180° - B_1$. If $A + B_2 < 180°$, then a second triangle exists. In this case, use B_2 for B in the second triangle.

EXAMPLE 3 Solving the Ambiguous Case (One Triangle)

Solve triangle ABC, given $A = 43.5°$, $a = 10.7$ in., and $c = 7.2$ in.

Solution To find angle C, use an alternative form of the law of sines.

$$\frac{\sin C}{c} = \frac{\sin A}{a}$$

$$\frac{\sin C}{7.2} = \frac{\sin 43.5°}{10.7}$$

$$\sin C = \frac{7.2 \sin 43.5°}{10.7} \approx .46319186$$

$$C \approx 27.6° \quad \text{Use the inverse sine function.}$$

There is another angle C that has sine value .46319186; it is $C = 180° - 27.6° = 152.4°$. However, notice in the given information that $c < a$, meaning that in the triangle, angle C must have measure *less than* angle A. Notice also that when we add this obtuse value to the given angle $A = 43.5°$, we obtain

$$152.4° + 43.5° = 195.9°,$$

which is greater than $180°$. So either of these approaches shows that there can be only one triangle. See Figure 10. Then

$$B = 180° - 27.6° - 43.5° = 108.9°,$$

and we can find side b with the law of sines.

$$\frac{b}{\sin B} = \frac{a}{\sin A}$$

$$\frac{b}{\sin 108.9°} = \frac{10.7}{\sin 43.5°}$$

$$b = \frac{10.7 \sin 108.9°}{\sin 43.5°}$$

$$b \approx 14.7 \text{ in.}$$

Figure 10

Now try Exercise 21.

Analyzing Data for Possible Number of Triangles

EXAMPLE 4   Analyzing Data Involving an Obtuse Angle

Without using the law of sines, explain why $A = 104°$, $a = 26.8$ m, and $b = 31.3$ m cannot be valid for a triangle ABC.

Solution   Since A is an obtuse angle, it is the largest angle and so the largest side of the triangle must be a. However, we are given $b > a$; thus, $B > A$, which is impossible if A is obtuse. Therefore, no such triangle ABC exists.

Now try Exercise 33.

7.2 Exercises

1. *Concept Check*   Which one of the following sets of data does not determine a unique triangle?

   A. $A = 40°, B = 60°, C = 80°$ **B.** $a = 5, b = 12, c = 13$

   C. $a = 3, b = 7, C = 50°$ **D.** $a = 2, b = 2, c = 2$

2. *Concept Check*   Which one of the following sets of data determines a unique triangle?

   A. $A = 50°, B = 50°, C = 80°$ **B.** $a = 3, b = 5, c = 20$

   C. $A = 40°, B = 20°, C = 30°$ **D.** $a = 7, b = 24, c = 25$

Concept Check   *In each figure, a line of length h is to be drawn from the given point to the positive x-axis in order to form a triangle. For what value(s) of h can you draw the following?*

(a) *two triangles*   **(b)** *exactly one triangle*   **(c)** *no triangle*

3.

4.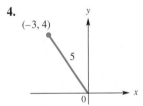

Determine the number of triangles ABC possible with the given parts. See Examples 1–4.

5. $a = 50, b = 26, A = 95°$ **6.** $b = 60, a = 82, B = 100°$

7. $a = 31, b = 26, B = 48°$ **8.** $a = 35, b = 30, A = 40°$

9. $a = 50, b = 61, A = 58°$ **10.** $B = 54°, c = 28, b = 23$

Find each angle B. Do not use a calculator.

11.

12.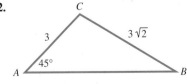

Find the unknown angles in triangle ABC for each triangle that exists. See Examples 1–3.

13. $A = 29.7°$, $b = 41.5$ ft, $a = 27.2$ ft

14. $B = 48.2°$, $a = 890$ cm, $b = 697$ cm

15. $C = 41° 20'$, $b = 25.9$ m, $c = 38.4$ m

16. $B = 48° 50'$, $a = 3850$ in., $b = 4730$ in.

17. $B = 74.3°$, $a = 859$ m, $b = 783$ m

18. $C = 82.2°$, $a = 10.9$ km, $c = 7.62$ km

19. $A = 142.13°$, $b = 5.432$ ft, $a = 7.297$ ft

20. $B = 113.72°$, $a = 189.6$ yd, $b = 243.8$ yd

Solve each triangle ABC that exists. See Examples 1–3.

21. $A = 42.5°$, $a = 15.6$ ft, $b = 8.14$ ft   **22.** $C = 52.3°$, $a = 32.5$ yd, $c = 59.8$ yd

23. $B = 72.2°$, $b = 78.3$ m, $c = 145$ m   **24.** $C = 68.5°$, $c = 258$ cm, $b = 386$ cm

25. $A = 38° 40'$, $a = 9.72$ km, $b = 11.8$ km

26. $C = 29° 50'$, $a = 8.61$ m, $c = 5.21$ m

27. $A = 96.80°$, $b = 3.589$ ft, $a = 5.818$ ft

28. $C = 88.70°$, $b = 56.87$ yd, $c = 112.4$ yd

29. $B = 39.68°$, $a = 29.81$ m, $b = 23.76$ m

30. $A = 51.20°$, $c = 7986$ cm, $a = 7208$ cm

31. Apply the law of sines to the following: $a = \sqrt{5}$, $c = 2\sqrt{5}$, $A = 30°$. What is the value of $\sin C$? What is the measure of C? Based on its angle measures, what kind of triangle is triangle ABC?

32. Explain the condition that must exist to determine that there is no triangle satisfying the given values of a, b, and B, once the value of $\sin B$ is found.

33. Without using the law of sines, explain why no triangle ABC exists satisfying $A = 103° 20'$, $a = 14.6$ ft, $b = 20.4$ ft.

34. Apply the law of sines to the data given in Example 4. Describe in your own words what happens when you try to find the measure of angle B using a calculator.

35. *Property Survey* A surveyor reported the following data about a piece of property: "The property is triangular in shape, with dimensions as shown in the figure." Use the law of sines to see whether such a piece of property could exist.

Can such a triangle exist?

36. *Property Survey* The surveyor tries again: "A second triangular piece of property has dimensions as shown." This time it turns out that the surveyor did not consider every possible case. Use the law of sines to show why.

Use the law of sines to prove that each statement is true for any triangle ABC, with corresponding sides a, b, and c.

37. $\dfrac{a + b}{b} = \dfrac{\sin A + \sin B}{\sin B}$

38. $\dfrac{a - b}{a + b} = \dfrac{\sin A - \sin B}{\sin A + \sin B}$

7.3 | **The Law of Cosines**

Derivation of the Law of Cosines ▪ **Solving SAS and SSS Triangles (Cases 3 and 4)** ▪ **Heron's Formula for the Area of a Triangle**

As mentioned in Section 7.1, if we are given two sides and the included angle (Case 3) or three sides (Case 4) of a triangle, then a unique triangle is determined. These are the SAS and SSS cases, respectively. Both cases require using the *law of cosines*. Remember the following property of triangles when applying the law of cosines to solve a triangle.

> ## **Triangle Side Length Restriction**
>
> In any triangle, the sum of the lengths of any two sides must be greater than the length of the remaining side.

For example, it would be impossible to construct a triangle with sides of lengths 3, 4, and 10. See Figure 11.

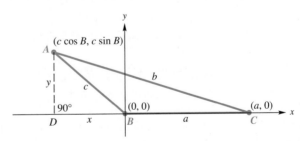

No triangle is formed.

Figure 11

Derivation of the Law of Cosines   To derive the law of cosines, let *ABC* be any oblique triangle. Choose a coordinate system so that vertex *B* is at the origin and side *BC* is along the positive *x*-axis. See Figure 12.

Figure 12

Let (x, y) be the coordinates of vertex *A* of the triangle. Verify that for angle *B*, whether obtuse or acute,

$$\sin B = \frac{y}{c} \quad \text{and} \quad \cos B = \frac{x}{c}. \quad \text{(Section 2.2)}$$

(Here *x* is negative when *B* is obtuse.) From these results

$$y = c \sin B \quad \text{and} \quad x = c \cos B,$$

so the coordinates of point *A* become $(c \cos B, c \sin B)$.

Point C in Figure 12 has coordinates $(a, 0)$, and AC has length b. Since point A has coordinates $(c \cos B, c \sin B)$, by the distance formula,

$$b = \sqrt{(c \cos B - a)^2 + (c \sin B - 0)^2} \qquad \text{(Appendix B)}$$

$$b^2 = (c \cos B - a)^2 + (c \sin B)^2 \qquad \text{Square both sides.}$$

$$= c^2 \cos^2 B - 2ac \cos B + a^2 + c^2 \sin^2 B \qquad \begin{array}{l}\text{Multiply;} \\ (x-y)^2 = x^2 - 2xy + y^2\end{array}$$

$$= a^2 + c^2(\cos^2 B + \sin^2 B) - 2ac \cos B \qquad \text{Properties of real numbers}$$

$$= a^2 + c^2(1) - 2ac \cos B \qquad \text{Fundamental identity (Section 5.1)}$$

$$b^2 = a^2 + c^2 - 2ac \cos B.$$

This result is one form of the law of cosines. In our work, we could just as easily have placed A or C at the origin. This would have given the same result, but with the variables rearranged.

Law of Cosines

In any triangle ABC, with sides a, b, and c,

$$a^2 = b^2 + c^2 - 2bc \cos A,$$
$$b^2 = a^2 + c^2 - 2ac \cos B,$$
$$c^2 = a^2 + b^2 - 2ab \cos C.$$

That is, according to the law of cosines, the square of a side of a triangle is equal to the sum of the squares of the other two sides, minus twice the product of those two sides and the cosine of the angle included between them.

N O T E   If we let $C = 90°$ in the third form of the law of cosines, then $\cos C = \cos 90° = 0$, and the formula becomes $c^2 = a^2 + b^2$, the Pythagorean theorem (Appendix B). The Pythagorean theorem is a special case of the law of cosines.

Solving SAS and SSS Triangles (Cases 3 and 4)

EXAMPLE 1   Using the Law of Cosines in an Application (SAS)

A surveyor wishes to find the distance between two inaccessible points A and B on opposite sides of a lake. While standing at point C, she finds that $AC = 259$ m, $BC = 423$ m, and angle ACB measures $132° 40'$. Find the distance AB. See Figure 13.

Solution   The law of cosines can be used here since we know the lengths of two sides of the triangle and the measure of the included angle.

$$AB^2 = 259^2 + 423^2 - 2(259)(423) \cos 132° 40'$$

$$AB^2 \approx 394{,}510.6 \quad \text{Use a calculator.}$$

$$AB \approx 628 \quad \text{Take the square root of each side. (Appendix A)}$$

The distance between the points is approximately 628 m.

259 m

C

132° 40′

423 m

B

Figure 13

Now try Exercise 39.

EXAMPLE 2 Using the Law of Cosines to Solve a Triangle (SAS)

Solve triangle ABC if $A = 42.3°$, $b = 12.9$ m, and $c = 15.4$ m. See Figure 14.

Solution We start by finding a with the law of cosines.

$$a^2 = b^2 + c^2 - 2bc \cos A$$
$$a^2 = 12.9^2 + 15.4^2 - 2(12.9)(15.4) \cos 42.3°$$
$$a^2 \approx 109.7$$
$$a \approx 10.47 \text{ m}$$

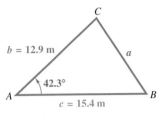

Figure 14

Of the two remaining angles B and C, B must be the smaller since it is opposite the shorter of the two sides b and c. Therefore, B cannot be obtuse.

$$\frac{\sin A}{a} = \frac{\sin B}{b} \qquad \text{Law of sines (Section 7.1)}$$

$$\frac{\sin 42.3°}{10.47} = \frac{\sin B}{12.9}$$

$$\sin B = \frac{12.9 \sin 42.3°}{10.47} \qquad \text{Multiply by 10.47; rewrite.}$$

$$B \approx 56.0° \qquad \text{Use the inverse sine function. (Section 6.1)}$$

The easiest way to find C is to subtract the measures of A and B from $180°$.

$$C = 180° - 42.3° - 56.0° = 81.7° \quad \text{(Section 1.2)}$$

Now try Exercise 19.

C A U T I O N   Had we used the law of sines to find C rather than B in Example 2, we would not have known whether C equals $81.7°$ or its supplement, $98.3°$.

EXAMPLE 3 Using the Law of Cosines to Solve a Triangle (SSS)

Solve triangle ABC if $a = 9.47$ ft, $b = 15.9$ ft, and $c = 21.1$ ft.

Solution We can use the law of cosines to solve for any angle of the triangle. We solve for C, the largest angle. We will know that C is obtuse if $\cos C < 0$.

$$c^2 = a^2 + b^2 - 2ab \cos C \qquad \text{Law of cosines}$$

$$\cos C = \frac{a^2 + b^2 - c^2}{2ab} \qquad \text{Solve for } \cos C.$$

$$\cos C = \frac{9.47^2 + 15.9^2 - 21.1^2}{2(9.47)(15.9)} \qquad \text{Substitute.}$$

$$\cos C \approx -.34109402 \qquad \text{Use a calculator.}$$

$$C \approx 109.9° \qquad \text{Use the inverse cosine function. (Section 6.1)}$$

We can use either the law of sines or the law of cosines to find $B \approx 45.1°$. (Verify this.) Since $A = 180° - B - C$, we obtain $A \approx 25.0°$.

Now try Exercise 23.

(a)

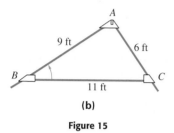

(b)

Figure 15

Trusses are frequently used to support roofs on buildings, as illustrated in Figure 15(a). The simplest type of roof truss is a triangle, as shown in Figure 15(b). One basic task when constructing a roof truss is to cut the ends of the rafters so that the roof has the correct slope. (*Source:* Riley, W., L. Sturges, and D. Morris, *Statics and Mechanics of Materials,* John Wiley and Sons, 1995.)

EXAMPLE 4 Designing a Roof Truss (SSS)

Find angle B for the truss shown in Figure 15(b).

Solution

$$b^2 = a^2 + c^2 - 2ac \cos B \qquad \text{Law of cosines}$$

$$\cos B = \frac{a^2 + c^2 - b^2}{2ac} \qquad \text{Solve for } \cos B.$$

$$\cos B = \frac{11^2 + 9^2 - 6^2}{2(11)(9)} \qquad \text{Let } a = 11, b = 6, \text{ and } c = 9.$$

$$\cos B \approx .8384 \qquad \text{Use a calculator.}$$

$$B \approx 33° \qquad \text{Use the inverse cosine function.}$$

Now try Exercise 45.

Four possible cases can occur when solving an oblique triangle, as summarized in the following table. In all four cases, it is assumed that the given information actually produces a triangle.

Oblique Triangle	Suggested Procedure for Solving
Case 1: One side and two angles are known. (SAA or ASA)	*Step 1* Find the remaining angle using the angle sum formula $(A + B + C = 180°)$. *Step 2* Find the remaining sides using the law of sines.
Case 2: Two sides and one angle (not included between the two sides) are known. (SSA)	*This is the ambiguous case; there may be no triangle, one triangle, or two triangles.* *Step 1* Find an angle using the law of sines. *Step 2* Find the remaining angle using the angle sum formula. *Step 3* Find the remaining side using the law of sines. *If two triangles exist, repeat Steps 2 and 3.*
Case 3: Two sides and the included angle are known. (SAS)	*Step 1* Find the third side using the law of cosines. *Step 2* Find the smaller of the two remaining angles using the law of sines. *Step 3* Find the remaining angle using the angle sum formula.
Case 4: Three sides are known. (SSS)	*Step 1* Find the largest angle using the law of cosines. *Step 2* Find either remaining angle using the law of sines. *Step 3* Find the remaining angle using the angle sum formula.

Heron of Alexandria

Heron's Formula for the Area of a Triangle The law of cosines can be used to derive a formula for the area of a triangle given the lengths of the three sides. This formula is known as **Heron's formula,** named after the Greek mathematician Heron of Alexandria, who lived around A.D. 75. It is found in his work *Metrica*. Heron's formula can be used for the case SSS.

Heron's Area Formula (SSS)

If a triangle has sides of lengths a, b, and c, with **semiperimeter**

$$s = \frac{1}{2}(a + b + c),$$

then the area of the triangle is

$$\mathscr{A} = \sqrt{s(s - a)(s - b)(s - c)}.$$

That is, according to Heron's formula, the area of a triangle is the square root of the product of four factors: (1) the semiperimeter, (2) the semiperimeter minus the first side, (3) the semiperimeter minus the second side, and (4) the semiperimeter minus the third side.

EXAMPLE 5 Using Heron's Formula to Find an Area (SSS)

The distance "as the crow flies" from Los Angeles to New York is 2451 mi, from New York to Montreal is 331 mi, and from Montreal to Los Angeles is 2427 mi. What is the area of the triangular region having these three cities as vertices? (Ignore the curvature of Earth.)

Solution In Figure 16 we let $a = 2451$, $b = 331$, and $c = 2427$.

Figure 16

The semiperimeter s is

$$s = \frac{1}{2}(2451 + 331 + 2427) = 2604.5.$$

Using Heron's formula, the area $\mathscr{A}$ is

$$\mathscr{A} = \sqrt{s(s - a)(s - b)(s - c)}$$
$$\mathscr{A} = \sqrt{2604.5(2604.5 - 2451)(2604.5 - 331)(2604.5 - 2427)}$$
$$\mathscr{A} \approx 401{,}700 \text{ mi}^2.$$

Now try Exercise 73.

CONNECTIONS We have introduced two new formulas for the area of a triangle in this chapter. You should now be able to find the area $\mathcal{A}$ of a triangle using one of three formulas:

(a) $\mathcal{A} = \frac{1}{2}bh$

(b) $\mathcal{A} = \frac{1}{2}ab \sin C$ $\left(\text{or } \mathcal{A} = \frac{1}{2}ac \sin B \text{ or } \mathcal{A} = \frac{1}{2}bc \sin A\right)$

(c) $\mathcal{A} = \sqrt{s(s-a)(s-b)(s-c)}$.

Another area formula can be used when the coordinates of the vertices of a triangle are given. If the vertices are the ordered pairs (x_1, y_1), (x_2, y_2), and (x_3, y_3), then

$$\mathcal{A} = \frac{1}{2}\left|(x_1y_2 - y_1x_2 + x_2y_3 - y_2x_3 + x_3y_1 - y_3x_1)\right|.$$

For Discussion or Writing

Consider triangle PQR with vertices $P(2, 5)$, $Q(-1, 3)$, and $R(4, 0)$. (*Hint:* Draw a sketch first.)

1. Find the area of the triangle using the new formula just introduced.

2. Find the area of the triangle using (c) above. Use the distance formula to find the lengths of the three sides.

3. Find the area of the triangle using (b) above. First use the law of cosines to find the measure of an angle.

7.3 Exercises

Concept Check *Assume triangle ABC has standard labeling and complete the following.*

(a) *Determine whether SAA, ASA, SSA, SAS, or SSS is given.*

(b) *Decide whether the law of sines or the law of cosines should be used to begin solving the triangle.*

1. a, b, and C **2.** A, C, and c **3.** a, b, and A

4. a, b, and c **5.** A, B, and c **6.** a, c, and A

7. a, B, and C **8.** b, c, and A

Find the length of the remaining side of each triangle. Do not use a calculator.

9.

10.

Find the value of θ in each triangle. Do not use a calculator.

11.

12.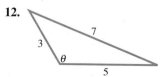

Solve each triangle. Approximate values to the nearest tenth.

13.

14.

15.

16.

17.

18.

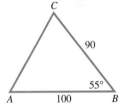

Solve each triangle. See Examples 2 and 3.

19. $A = 41.4°$, $b = 2.78$ yd, $c = 3.92$ yd

20. $C = 28.3°$, $b = 5.71$ in., $a = 4.21$ in.

21. $C = 45.6°$, $b = 8.94$ m, $a = 7.23$ m

22. $A = 67.3°$, $b = 37.9$ km, $c = 40.8$ km

23. $a = 9.3$ cm, $b = 5.7$ cm, $c = 8.2$ cm

24. $a = 28$ ft, $b = 47$ ft, $c = 58$ ft

25. $a = 42.9$ m, $b = 37.6$ m, $c = 62.7$ m

26. $a = 189$ yd, $b = 214$ yd, $c = 325$ yd

27. $AB = 1240$ ft, $AC = 876$ ft, $BC = 965$ ft

28. $AB = 298$ m, $AC = 421$ m, $BC = 324$ m

29. $A = 80° 40'$, $b = 143$ cm, $c = 89.6$ cm

30. $C = 72° 40'$, $a = 327$ ft, $b = 251$ ft

31. $B = 74.80°$, $a = 8.919$ in., $c = 6.427$ in.

32. $C = 59.70°$, $a = 3.725$ mi, $b = 4.698$ mi

33. $A = 112.8°$, $b = 6.28$ m, $c = 12.2$ m

34. $B = 168.2°$, $a = 15.1$ cm, $c = 19.2$ cm

35. $a = 3.0$ ft, $b = 5.0$ ft, $c = 6.0$ ft

36. $a = 4.0$ ft, $b = 5.0$ ft, $c = 8.0$ ft

37. Refer to Figure 11. If you attempt to find any angle of a triangle with the values $a = 3$, $b = 4$, and $c = 10$ with the law of cosines, what happens?

38. "The shortest distance between two points is a straight line." Explain how this relates to the geometric property that states that the sum of the lengths of any two sides of a triangle must be greater than the length of the remaining side.

Solve each problem. See Examples 1–4.

39. *Distance Across a Lake*   Points A and B are on opposite sides of Lake Yankee. From a third point, C, the angle between the lines of sight to A and B is 46.3°. If AC is 350 m long and BC is 286 m long, find AB.

40. *Diagonals of a Parallelogram*   The sides of a parallelogram are 4.0 cm and 6.0 cm. One angle is 58° while another is 122°. Find the lengths of the diagonals of the parallelogram.

41. *Flight Distance*   Airports A and B are 450 km apart, on an east-west line. Tom flies in a northeast direction from A to airport C. From C he flies 359 km on a bearing of 128° 40′ to B. How far is C from A?

42. *Distance Between Two Ships*   Two ships leave a harbor together, traveling on courses that have an angle of 135° 40′ between them. If they each travel 402 mi, how far apart are they?

43. *Distance Between a Ship and a Rock*   A ship is sailing east. At one point, the bearing of a submerged rock is 45° 20′. After the ship has sailed 15.2 mi, the bearing of the rock has become 308° 40′. Find the distance of the ship from the rock at the latter point.

44. *Distance Between a Ship and a Submarine*   From an airplane flying over the ocean, the angle of depression to a submarine lying under the surface is 24° 10′. At the same moment, the angle of depression from the airplane to a battleship is 17° 30′. (See the figure.) The distance from the airplane to the battleship is 5120 ft. Find the distance between the battleship and the submarine. (Assume the airplane, submarine, and battleship are in a vertical plane.)

45. *Truss Construction*   A triangular truss is shown in the figure. Find angle θ.

46. *Distance Between Points on a Crane*   A crane with a counterweight is shown in the figure. Find the horizontal distance between points A and B.

47. *Distance Between a Beam and Cables* A weight is supported by cables attached to both ends of a balance beam, as shown in the figure. What angles are formed between the beam and the cables?

48. *Length of a Tunnel* To measure the distance through a mountain for a proposed tunnel, a point C is chosen that can be reached from each end of the tunnel. (See the figure.) If $AC = 3800$ m, $BC = 2900$ m, and angle $C = 110°$, find the length of the tunnel.

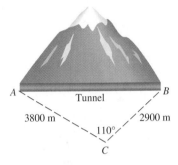

49. *Distance on a Baseball Diamond* A baseball diamond is a square, 90.0 ft on a side, with home plate and the three bases as vertices. The pitcher's rubber is 60.5 ft from home plate. Find the distance from the pitcher's rubber to each of the bases.

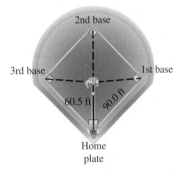

50. *Distance Between Ends of the Vietnam Memorial* The Vietnam Veterans' Memorial in Washington, D.C., is V-shaped with equal sides of length 246.75 ft. The angle between these sides measures 125° 12′. Find the distance between the ends of the two sides. (*Source:* Pamphlet obtained at Vietnam Veterans' Memorial.)

51. *Distance Between a Ship and a Point* Starting at point A, a ship sails 18.5 km on a bearing of 189°, then turns and sails 47.8 km on a bearing of 317°. Find the distance of the ship from point A.

52. *Distance Between Two Factories* Two factories blow their whistles at exactly 5:00. A man hears the two blasts at 3 sec and 6 sec after 5:00, respectively. The angle between his lines of sight to the two factories is 42.2°. If sound travels 344 m per sec, how far apart are the factories?

53. *Bearing of One Town to Another* Two towns 21 mi apart are separated by a dense forest. (See the figure.) To travel from town *A* to town *B*, a person must go 17 mi on a bearing of 325°, then turn and continue for 9 mi to reach town *B*. Find the bearing of *B* from *A*.

54. *Distance of a Plane* An airplane flies 180 mi from point *X* at a bearing of 125°, and then turns and flies at a bearing of 230° for 100 mi. How far is the plane from point *X*?

55. *Measurement Using Triangulation* Surveyors are often confronted with obstacles, such as trees, when measuring the boundary of a lot. One technique used to obtain an accurate measurement is the so-called *triangulation method.* In this technique, a triangle is constructed around the obstacle and one angle and two sides of the triangle are measured. Use this technique to find the length of the property line (the straight line between the two markers) in the figure. (*Source:* Kavanagh, B., *Surveying Principles and Applications,* Sixth Edition, Prentice-Hall, 2003.)

56. *Path of a Ship* A ship sailing due east in the North Atlantic has been warned to change course to avoid icebergs. The captain turns and sails on a bearing of 62°, then changes course again to a bearing of 115° until the ship reaches its original course. (See the figure.) How much farther did the ship have to travel to avoid the icebergs?

57. *Angle in a Parallelogram* A parallelogram has sides of length 25.9 cm and 32.5 cm. The longer diagonal has length 57.8 cm. Find the angle opposite the longer diagonal.

58. *Distance Between an Airplane and a Mountain* A person in a plane flying straight north observes a mountain at a bearing of 24.1°. At that time, the plane is 7.92 km from the mountain. A short time later, the bearing to the mountain becomes 32.7°. How far is the airplane from the mountain when the second bearing is taken?

59. *Layout for a Playhouse* The layout for a child's playhouse has the dimensions given in the figure. Find x.

60. *Distance Between Two Towns* To find the distance between two small towns, an electronic distance measuring (EDM) instrument is placed on a hill from which both towns are visible. The distance to each town from the EDM and the angle between the two lines of sight are measured. (See the figure.) Find the distance between the towns.

Find the measure of each angle θ to two decimal places.

61.

62.

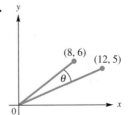

Find the exact area of each triangle using the formula $\mathcal{A} = \frac{1}{2}bh$, and then verify that Heron's formula gives the same result.

63.

64.

Find the area of each triangle ABC. See Example 5.

65. $a = 12$ m, $b = 16$ m, $c = 25$ m

66. $a = 22$ in., $b = 45$ in., $c = 31$ in.

67. $a = 154$ cm, $b = 179$ cm, $c = 183$ cm

68. $a = 25.4$ yd, $b = 38.2$ yd, $c = 19.8$ yd

69. $a = 76.3$ ft, $b = 109$ ft, $c = 98.8$ ft

70. $a = 15.89$ in., $b = 21.74$ in., $c = 10.92$ in.

Volcano Movement To help predict eruptions from the volcano Mauna Loa on the island of Hawaii, scientists keep track of the volcano's movement by using a "super triangle" with vertices on the three volcanoes shown on the map at the right. (For example, in one year, Mauna Loa moved 6 in., a result of increasing internal pressure.) Refer to the map to work Exercises 71 and 72.

71. $AB = 22.47928$ mi, $AC = 28.14276$ mi, $A = 58.56989°$; find BC

72. $AB = 22.47928$ mi, $BC = 25.24983$ mi, $A = 58.56989°$; find B

Solve each problem. See Example 5.

73. *Area of the Bermuda Triangle* Find the area of the Bermuda Triangle if the sides of the triangle have approximate lengths 850 mi, 925 mi, and 1300 mi.

74. *Required Amount of Paint* A painter needs to cover a triangular region 75 m by 68 m by 85 m. A can of paint covers 75 m² of area. How many cans (to the next higher number of cans) will be needed?

75. *Perfect Triangles* A *perfect triangle* is a triangle whose sides have whole number lengths and whose area is numerically equal to its perimeter. Show that the triangle with sides of length 9, 10, and 17 is perfect.

76. *Heron Triangles* A *Heron triangle* is a triangle having integer sides and area. Show that each of the following is a Heron triangle.

  (a) $a = 11, b = 13, c = 20$ **(b)** $a = 13, b = 14, c = 15$
  (c) $a = 7, b = 15, c = 20$

77. Consider triangle ABC shown here.

  (a) Use the law of sines to find candidates for the value of angle C.
  (b) Rework part (a) using the law of cosines.
  (c) Why is the law of cosines a better method in this case?

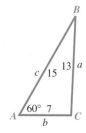

78. Show that the measure of angle A is twice the measure of angle B. (*Hint:* Use the law of cosines to find $\cos A$ and $\cos B$, and then show that $\cos A = 2 \cos^2 B - 1$.)

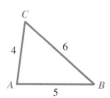

79. Let point D on side AB of triangle ABC be such that CD bisects angle C. Show that $\frac{AD}{DB} = \frac{b}{a}$.

80. In addition to the law of sines and the law of cosines, there is a **law of tangents**. In any triangle ABC,

$$\frac{\tan \frac{1}{2}(A - B)}{\tan \frac{1}{2}(A + B)} = \frac{a - b}{a + b}.$$

Verify this law for the triangle ABC with $a = 2, b = 2\sqrt{3}, A = 30°$, and $B = 60°$.

7.4 Vectors, Operations, and the Dot Product

Basic Terminology ▪ **Algebraic Interpretation of Vectors** ▪ **Operations with Vectors** ▪ **Dot Product and the Angle Between Vectors**

Figure 17

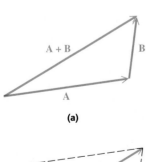

Vector **OP**   Vector **PO**

Figure 18

Basic Terminology   Many quantities involve magnitudes, such as 45 lb or 60 mph. These quantities are called **scalars** and can be represented by real numbers. Other quantities, called **vector quantities,** involve both magnitude and direction. Typical vector quantities are velocity, acceleration, and force. For example, traveling 50 mph *east* represents a vector quantity.

A vector quantity is often represented with a directed line segment (a segment that uses an arrowhead to indicate direction), called a **vector.** The length of the vector represents the **magnitude** of the vector quantity. The direction of the vector, indicated by the arrowhead, represents the direction of the quantity. For example, the vector in Figure 17 represents a force of 10 lb applied at an angle of 30° from the horizontal.

The symbol for a vector is often printed in boldface type. When writing vectors by hand, it is customary to use an arrow over the letter or letters. Thus **OP** and $\overrightarrow{\text{OP}}$ both represent the vector **OP**. Vectors may be named with either one lowercase or uppercase letter, or two uppercase letters. When two letters are used, the first indicates the **initial point** and the second indicates the **terminal point** of the vector. Knowing these points gives the direction of the vector. For example, vectors **OP** and **PO** in Figure 18 are not the same vector. They have the same magnitude, but *opposite* directions. The magnitude of vector **OP** is written |**OP**|.

Two vectors are equal if and only if they have the same direction and the same magnitude. In Figure 19, vectors **A** and **B** are equal, as are vectors **C** and **D**. As Figure 19 shows, equal vectors need not coincide, but they must be parallel. Vectors **A** and **E** are unequal because they do not have the same direction, while **A** ≠ **F** because they have different magnitudes.

Figure 19

Now try Exercise 1.

(a)

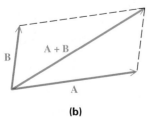

(b)

Figure 20

To find the sum of two vectors **A** and **B**, we place the initial point of vector **B** at the terminal point of vector **A**, as shown in Figure 20(a). The vector with the same initial point as **A** and the same terminal point as **B** is the sum **A** + **B**. The sum of two vectors is also a vector.

Another way to find the sum of two vectors is to use the **parallelogram rule.** Place vectors **A** and **B** so that their initial points coincide, as in Figure 20(b). Then, complete a parallelogram that has **A** and **B** as two sides. The diagonal of the parallelogram with the same initial point as **A** and **B** is the sum **A** + **B** found by the definition. Compare Figures 20(a) and (b). Parallelograms can be used to show that vector **B** + **A** is the same as vector **A** + **B**, or that **A** + **B** = **B** + **A**, so vector addition is commutative. The vector sum **A** + **B** is called the **resultant** of vectors **A** and **B**.

Vectors **v** and **−v**
are opposites.

Figure 21

For every vector **v** there is a vector **−v** that has the same magnitude as **v** but opposite direction. Vector **−v** is called the **opposite** of **v**. (See Figure 21.) The sum of **v** and **−v** has magnitude 0 and is called the **zero vector.** As with real numbers, to subtract vector **B** from vector **A**, find the vector sum **A** + (**−B**). (See Figure 22.)

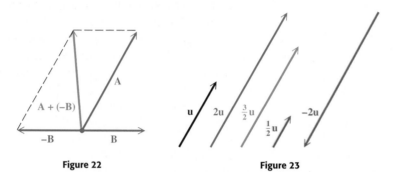

Figure 22

Figure 23

The **scalar product** of a real number (or scalar) k and a vector **u** is the vector $k \cdot$ **u**, which has magnitude $|k|$ times the magnitude of **u**. As suggested by Figure 23, the vector $k \cdot$ **u** has the same direction as **u** if $k > 0$, and opposite direction if $k < 0$.

> **Now try Exercises 5, 7, 9, and 11.**

Algebraic Interpretation of Vectors We now consider vectors in a rectangular coordinate system. A vector with its initial point at the origin is called a **position vector.** A position vector **u** with its endpoint at the point (a, b) is written $\langle a, b \rangle$, so

$$\mathbf{u} = \langle a, b \rangle.$$

This means that every vector in the real plane corresponds to an ordered pair of real numbers. Thus, geometrically a vector is a directed line segment; algebraically, it is an ordered pair. The numbers a and b are the **horizontal component** and **vertical component,** respectively, of vector **u**. Figure 24 shows the vector $\mathbf{u} = \langle a, b \rangle$. The positive angle between the x-axis and a position vector is the **direction angle** for the vector. In Figure 24, θ is the direction angle for vector **u**.

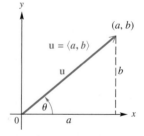

Figure 24

From Figure 24, we can see that the magnitude and direction of a vector are related to its horizontal and vertical components.

Looking Ahead to Calculus

In addition to two-dimensional vectors in a plane, calculus courses introduce three-dimensional vectors in space. The magnitude of the two-dimensional vector $\langle a, b \rangle$ is given by $\sqrt{a^2 + b^2}$. If we extend this to the three-dimensional vector $\langle a, b, c \rangle$, the expression becomes $\sqrt{a^2 + b^2 + c^2}$. Similar extensions are made for other concepts.

Magnitude and Direction Angle of a Vector $\langle a, b \rangle$

The magnitude (length) of vector $\mathbf{u} = \langle a, b \rangle$ is given by

$$|\mathbf{u}| = \sqrt{a^2 + b^2}.$$

The direction angle θ satisfies $\tan \theta = \frac{b}{a}$, where $a \neq 0$.

EXAMPLE 1 Finding Magnitude and Direction Angle

Find the magnitude and direction angle for $\mathbf{u} = \langle 3, -2 \rangle$.

Algebraic Solution

The magnitude is $|\mathbf{u}| = \sqrt{3^2 + (-2)^2} = \sqrt{13}$. To find the direction angle θ, start with $\tan \theta = \frac{b}{a} = \frac{-2}{3} = -\frac{2}{3}$. Vector $\mathbf{u}$ has a positive horizontal component and a negative vertical component, placing the position vector in quadrant IV. A calculator gives $\tan^{-1}\left(-\frac{2}{3}\right) \approx -33.7°$. Adding $360°$ yields the direction angle $\theta = 326.3°$. See Figure 25.

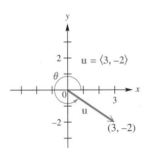

Figure 25

Graphing Calculator Solution

A calculator returns the magnitude and direction angle, given the horizontal and vertical components. An approximation for $\sqrt{13}$ is given, and the direction angle has a measure with smallest possible absolute value. We must add $360°$ to the value of θ to obtain the positive direction angle. See Figure 26.

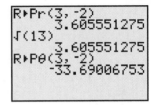

Figure 26

For more information, see your owner's manual or the graphing calculator manual that accompanies this text.

Now try Exercise 33.

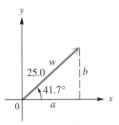

Figure 27

Horizontal and Vertical Components

The horizontal and vertical components, respectively, of a vector $\mathbf{u}$ having magnitude $|\mathbf{u}|$ and direction angle θ are given by

$$a = |\mathbf{u}| \cos \theta \quad \text{and} \quad b = |\mathbf{u}| \sin \theta.$$

That is, $\mathbf{u} = \langle a, b \rangle = \langle |\mathbf{u}| \cos \theta, |\mathbf{u}| \sin \theta \rangle$.

EXAMPLE 2 Finding Horizontal and Vertical Components

Vector $\mathbf{w}$ in Figure 27 has magnitude 25.0 and direction angle 41.7°. Find the horizontal and vertical components.

Algebraic Solution

Use the two formulas in the box, with $|\mathbf{w}| = 25.0$ and $\theta = 41.7°$.

$$a = 25.0 \cos 41.7° \qquad b = 25.0 \sin 41.7°$$
$$a \approx 18.7 \qquad\qquad b \approx 16.6$$

Therefore, $\mathbf{w} = \langle 18.7, 16.6 \rangle$. The horizontal component is 18.7, and the vertical component is 16.6 (rounded to the nearest tenth).

Graphing Calculator Solution

See Figure 28. The results support the algebraic solution.

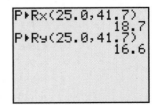

Figure 28

Now try Exercise 37.

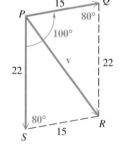

Figure 29

EXAMPLE 3 Writing Vectors in the Form $\langle a, b \rangle$

Write each vector in Figure 29 in the form $\langle a, b \rangle$.

Solution

$$\mathbf{u} = \langle 5 \cos 60°, 5 \sin 60° \rangle = \left\langle 5 \cdot \frac{1}{2}, 5 \cdot \frac{\sqrt{3}}{2} \right\rangle = \left\langle \frac{5}{2}, \frac{5\sqrt{3}}{2} \right\rangle$$

$$\mathbf{v} = \langle 2 \cos 180°, 2 \sin 180° \rangle = \langle 2(-1), 2(0) \rangle = \langle -2, 0 \rangle$$

$$\mathbf{w} = \langle 6 \cos 280°, 6 \sin 280° \rangle \approx \langle 1.0419, -5.9088 \rangle \quad \text{Use a calculator.}$$

Now try Exercise 43.

The following properties of parallelograms are helpful when studying vectors.

1. A parallelogram is a quadrilateral whose opposite sides are parallel.

2. The opposite sides and opposite angles of a parallelogram are equal, and adjacent angles of a parallelogram are supplementary.

3. The diagonals of a parallelogram bisect each other, but do not necessarily bisect the angles of the parallelogram.

EXAMPLE 4 Finding the Magnitude of a Resultant

Two forces of 15 and 22 newtons act on a point in the plane. (A *newton* is a unit of force that equals .225 lb.) If the angle between the forces is 100°, find the magnitude of the resultant force.

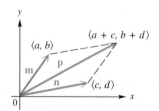

Figure 30

Solution As shown in Figure 30, a parallelogram that has the forces as adjacent sides can be formed. The angles of the parallelogram adjacent to angle P measure 80°, since adjacent angles of a parallelogram are supplementary. Opposite sides of the parallelogram are equal in length. The resultant force divides the parallelogram into two triangles. Use the law of cosines with either triangle.

$$|\mathbf{v}|^2 = 15^2 + 22^2 - 2(15)(22) \cos 80° \quad \text{Law of cosines (Section 7.3)}$$

$$\approx 225 + 484 - 115$$

$$|\mathbf{v}|^2 \approx 594$$

$$|\mathbf{v}| \approx 24 \quad \text{Take square roots. (Appendix A)}$$

To the nearest unit, the magnitude of the resultant force is 24 newtons.

Now try Exercise 49.

Operations with Vectors In Figure 31, $\mathbf{m} = \langle a, b \rangle$, $\mathbf{n} = \langle c, d \rangle$, and $\mathbf{p} = \langle a + c, b + d \rangle$. Using geometry, we can show that the endpoints of the three vectors and the origin form a parallelogram. Since a diagonal of this parallelogram gives the resultant of $\mathbf{m}$ and $\mathbf{n}$, we have $\mathbf{p} = \mathbf{m} + \mathbf{n}$ or

$$\langle a + c, b + d \rangle = \langle a, b \rangle + \langle c, d \rangle.$$

Figure 31

Similarly, we could verify the following vector operations.

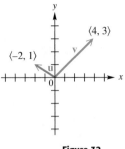

Figure 32

Vector Operations

For any real numbers a, b, c, d, and k,

$$\langle a, b \rangle + \langle c, d \rangle = \langle a + c, b + d \rangle$$
$$k \cdot \langle a, b \rangle = \langle ka, kb \rangle.$$
If $\mathbf{a} = \langle a_1, a_2 \rangle$, then $-\mathbf{a} = \langle -a_1, -a_2 \rangle$.
$$\langle a, b \rangle - \langle c, d \rangle = \langle a, b \rangle + (-\langle c, d \rangle) = \langle a - c, b - d \rangle$$

EXAMPLE 5 Performing Vector Operations

Let $\mathbf{u} = \langle -2, 1 \rangle$ and $\mathbf{v} = \langle 4, 3 \rangle$. (See Figure 32.) Find the following: **(a)** $\mathbf{u} + \mathbf{v}$, **(b)** $-2\mathbf{u}$, **(c)** $4\mathbf{u} - 3\mathbf{v}$.

Algebraic Solution

(a) $\mathbf{u} + \mathbf{v} = \langle -2, 1 \rangle + \langle 4, 3 \rangle$
$\qquad = \langle -2 + 4, 1 + 3 \rangle$
$\qquad = \langle 2, 4 \rangle$

(b) $-2\mathbf{u} = -2 \cdot \langle -2, 1 \rangle$
$\qquad = \langle -2(-2), -2(1) \rangle$
$\qquad = \langle 4, -2 \rangle$

(c) $4\mathbf{u} - 3\mathbf{v} = 4 \cdot \langle -2, 1 \rangle - 3 \cdot \langle 4, 3 \rangle$
$\qquad = \langle -8, 4 \rangle - \langle 12, 9 \rangle$
$\qquad = \langle -8 - 12, 4 - 9 \rangle$
$\qquad = \langle -20, -5 \rangle$

Graphing Calculator Solution

Vector arithmetic can be performed with a graphing calculator, as shown in Figure 33.

```
{-2,1}+{4,3}
                {2 4}
-2{-2,1}
                {4 -2}
4{-2,1}-3{4,3}
              {-20 -5}
```

Figure 33

Now try Exercises 59, 61, and 63.

A **unit vector** is a vector that has magnitude 1. Two very useful unit vectors are defined as follows and shown in Figure 34(a).

$$\mathbf{i} = \langle 1, 0 \rangle \qquad \mathbf{j} = \langle 0, 1 \rangle$$

(a)

(b)

Figure 34

With the unit vectors $\mathbf{i}$ and $\mathbf{j}$, we can express any other vector $\langle a, b \rangle$ in the form $a\mathbf{i} + b\mathbf{j}$, as shown in Figure 34(b), where $\langle 3, 4 \rangle = 3\mathbf{i} + 4\mathbf{j}$. The vector operations previously given can be restated, using $a\mathbf{i} + b\mathbf{j}$ notation.

i, j Form for Vectors

If $\mathbf{v} = \langle a, b \rangle$, then $\mathbf{v} = a\mathbf{i} + b\mathbf{j}$.

Dot Product and the Angle Between Vectors The *dot product* of two vectors is a real number, not a vector. It is also known as the *inner product*. Dot products are used to determine the angle between two vectors, derive geometric theorems, and solve physics problems.

Dot Product

The **dot product** of the two vectors $\mathbf{u} = \langle a, b \rangle$ and $\mathbf{v} = \langle c, d \rangle$ is denoted $\mathbf{u} \cdot \mathbf{v}$, read "**u** dot **v**," and given by

$$\mathbf{u} \cdot \mathbf{v} = ac + bd.$$

That is, the dot product of two vectors is the sum of the product of their first components and the product of their second components.

EXAMPLE 6 Finding Dot Products

Find each dot product.

(a) $\langle 2, 3 \rangle \cdot \langle 4, -1 \rangle$ **(b)** $\langle 6, 4 \rangle \cdot \langle -2, 3 \rangle$

Solution

(a) $\langle 2, 3 \rangle \cdot \langle 4, -1 \rangle = 2(4) + 3(-1) = 5$

(b) $\langle 6, 4 \rangle \cdot \langle -2, 3 \rangle = 6(-2) + 4(3) = 0$

Now try Exercise 71.

The following properties of dot products are easily verified.

Properties of the Dot Product

For all vectors $\mathbf{u}$, $\mathbf{v}$, and $\mathbf{w}$ and real numbers k,

(a) $\mathbf{u} \cdot \mathbf{v} = \mathbf{v} \cdot \mathbf{u}$ **(b)** $\mathbf{u} \cdot (\mathbf{v} + \mathbf{w}) = \mathbf{u} \cdot \mathbf{v} + \mathbf{u} \cdot \mathbf{w}$

(c) $(\mathbf{u} + \mathbf{v}) \cdot \mathbf{w} = \mathbf{u} \cdot \mathbf{w} + \mathbf{v} \cdot \mathbf{w}$ **(d)** $(k\mathbf{u}) \cdot \mathbf{v} = k(\mathbf{u} \cdot \mathbf{v}) = \mathbf{u} \cdot (k\mathbf{v})$

(e) $\mathbf{0} \cdot \mathbf{u} = 0$ **(f)** $\mathbf{u} \cdot \mathbf{u} = |\mathbf{u}|^2.$

For example, to prove the first part of (d), we let $\mathbf{u} = \langle a, b \rangle$ and $\mathbf{v} = \langle c, d \rangle$. Then,

$$(k\mathbf{u}) \cdot \mathbf{v} = (k\langle a, b \rangle) \cdot \langle c, d \rangle = \langle ka, kb \rangle \cdot \langle c, d \rangle$$
$$= kac + kbd = k(ac + bd)$$
$$= k(\langle a, b \rangle \cdot \langle c, d \rangle) = k(\mathbf{u} \cdot \mathbf{v}).$$

The proofs of the remaining properties are similar.

Figure 35

The dot product of two vectors can be positive, 0, or negative. A geometric interpretation of the dot product explains when each of these cases occurs. This interpretation involves the angle between the two vectors. Consider the vectors $\mathbf{u} = \langle a, b \rangle$ and $\mathbf{v} = \langle c, d \rangle$, as shown in Figure 35. The **angle θ between u and v** is defined to be the angle having the two vectors as its sides for which $0° \leq \theta \leq 180°$. The following theorem relates the dot product to the angle between the vectors. Its proof is outlined in Exercise 32 in Section 7.5.

> ## Geometric Interpretation of Dot Product
>
> If θ is the angle between the two nonzero vectors $\mathbf{u}$ and $\mathbf{v}$, where $0° \leq \theta \leq 180°$, then
>
> $$\mathbf{u} \cdot \mathbf{v} = |\mathbf{u}||\mathbf{v}| \cos \theta \quad \text{or, equivalently,} \quad \cos \theta = \frac{\mathbf{u} \cdot \mathbf{v}}{|\mathbf{u}||\mathbf{v}|}.$$

EXAMPLE 7 Finding the Angle Between Two Vectors

Find the angle θ between the two vectors $\mathbf{u} = \langle 3, 4 \rangle$ and $\mathbf{v} = \langle 2, 1 \rangle$.

Solution By the geometric interpretation,

$$\cos \theta = \frac{\mathbf{u} \cdot \mathbf{v}}{|\mathbf{u}||\mathbf{v}|} = \frac{\langle 3, 4 \rangle \cdot \langle 2, 1 \rangle}{|\langle 3, 4 \rangle||\langle 2, 1 \rangle|}$$

$$= \frac{3(2) + 4(1)}{\sqrt{9 + 16} \sqrt{4 + 1}}$$

$$= \frac{10}{5\sqrt{5}} \approx .894427191.$$

Therefore, $\quad \theta \approx \cos^{-1} .894427191 \approx 26.57°.$ (Section 6.1)

Now try Exercise 77.

For angles θ between 0° and 180°, $\cos \theta$ is positive, 0, or negative when θ is less than, equal to, or greater than 90°, respectively. Therefore, the dot product is positive, 0, or negative according to this table.

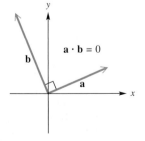

Figure 36

Dot Product	Angle Between Vectors
Positive	Acute
0	Right
Negative	Obtuse

N O T E If $\mathbf{a} \cdot \mathbf{b} = 0$ for two nonzero vectors $\mathbf{a}$ and $\mathbf{b}$, then $\cos \theta = 0$ and $\theta = 90°$. Thus, $\mathbf{a}$ and $\mathbf{b}$ are perpendicular or **orthogonal vectors.** See Figure 36.

Now try Exercises 87 and 89.

7.4 Exercises

*Concept Check Exercises 1–4 refer to the vectors **m**–**t** at the right.*

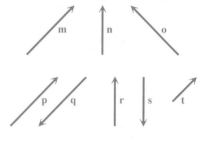

1. Name all pairs of vectors that appear to be equal.

2. Name all pairs of vectors that are opposites.

3. Name all pairs of vectors where the first is a scalar multiple of the other, with the scalar positive.

4. Name all pairs of vectors where the first is a scalar multiple of the other, with the scalar negative.

*Concept Check Refer to vectors **a**–**h** below. Make a copy or a sketch of each vector, and then draw a sketch to represent each vector in Exercises 5–16. For example, find **a** + **e** by placing **a** and **e** so that their initial points coincide. Then use the parallelogram rule to find the resultant, as shown in the figure on the right.*

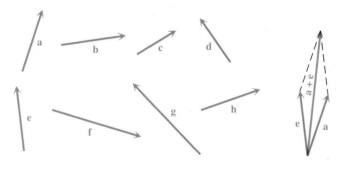

5. −**b**	**6.** −**g**	**7.** 3**a**	**8.** 2**h**
9. **a** + **b**	**10.** **h** + **g**	**11.** **a** − **c**	**12.** **d** − **e**
13. **a** + (**b** + **c**)	**14.** (**a** + **b**) + **c**	**15.** **c** + **d**	**16.** **d** + **c**

17. From the results of Exercises 13 and 14, does it appear that vector addition is associative?

18. From the results of Exercises 15 and 16, does it appear that vector addition is commutative?

*In Exercises 19–24, use the figure to find each vector: (a) **a** + **b** (b) **a** − **b** (c) −**a**. Use ⟨x, y⟩ notation as in Example 3.*

22.

23.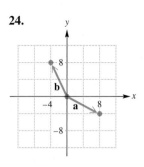

24.

Given vectors **a** *and* **b**, *find:* (a) 2**a** (b) 2**a** + 3**b** (c) **b** − 3**a**.

25. **a** = 2**i**, **b** = **i** + **j**   **26.** **a** = −**i** + 2**j**, **b** = **i** − **j**

27. **a** = ⟨−1, 2⟩, **b** = ⟨3, 0⟩   **28.** **a** = ⟨−2, −1⟩, **b** = ⟨−3, 2⟩

For each pair of vectors **u** *and* **w** *with angle* θ *between them, sketch the resultant.*

29. |**u**| = 12, |**w**| = 20, θ = 27°   **30.** |**u**| = 8, |**w**| = 12, θ = 20°

31. |**u**| = 20, |**w**| = 30, θ = 30°   **32.** |**u**| = 50, |**w**| = 70, θ = 40°

Find the magnitude and direction angle for **u**. *See Example 1.*

33. ⟨15, −8⟩   **34.** ⟨−7, 24⟩   **35.** $\left\langle -4, 4\sqrt{3} \right\rangle$ **36.** $\left\langle 8\sqrt{2}, -8\sqrt{2} \right\rangle$

For each of the following, vector **v** *has the given magnitude and direction. Find the magnitudes of the horizontal and vertical components of* **v**, *if* α *is the angle of inclination of* **v** *from the horizontal. See Example 2.*

37. α = 20°, |**v**| = 50   **38.** α = 50°, |**v**| = 26

39. α = 35° 50′, |**v**| = 47.8   **40.** α = 27° 30′, |**v**| = 15.4

41. α = 128.5°, |**v**| = 198   **42.** α = 146.3°, |**v**| = 238

Write each vector in the form ⟨a, b⟩. *See Example 3.*

43.

44.

45.

46.

47.

48.

Two forces act at a point in the plane. The angle between the two forces is given. Find the magnitude of the resultant force. See Example 4.

49. forces of 250 and 450 newtons, forming an angle of 85°

50. forces of 19 and 32 newtons, forming an angle of 118°

51. forces of 116 and 139 lb, forming an angle of 140° 50′

52. forces of 37.8 and 53.7 lb, forming an angle of 68.5°

Use the parallelogram rule to find the magnitude of the resultant force for the two forces shown in each figure. Round answers to the nearest tenth.

53.

54.

55.

56.

57. *Concept Check*   If $\mathbf{u} = \langle a, b \rangle$ and $\mathbf{v} = \langle c, d \rangle$, what is $\mathbf{u} + \mathbf{v}$?

58. Explain how addition of vectors is similar to addition of complex numbers.

Given $\mathbf{u} = \langle -2, 5 \rangle$ and $\mathbf{v} = \langle 4, 3 \rangle$, find the following. See Example 5.

59. $\mathbf{u} + \mathbf{v}$	**60.** $\mathbf{u} - \mathbf{v}$	**61.** $-4\mathbf{u}$	**62.** $-5\mathbf{v}$
63. $3\mathbf{u} - 6\mathbf{v}$	**64.** $-2\mathbf{u} + 4\mathbf{v}$	**65.** $\mathbf{u} + \mathbf{v} - 3\mathbf{u}$	**66.** $2\mathbf{u} + \mathbf{v} - 6\mathbf{v}$

Write each vector in the form $a\mathbf{i} + b\mathbf{j}$. See Figure 34(b).

67. $\langle -5, 8 \rangle$	**68.** $\langle 6, -3 \rangle$	**69.** $\langle 2, 0 \rangle$	**70.** $\langle 0, -4 \rangle$

Find the dot product for each pair of vectors. See Example 6.

71. $\langle 6, -1 \rangle, \langle 2, 5 \rangle$	**72.** $\langle -3, 8 \rangle, \langle 7, -5 \rangle$	**73.** $\langle 2, -3 \rangle, \langle 6, 5 \rangle$
74. $\langle 1, 2 \rangle, \langle 3, -1 \rangle$	**75.** $4\mathbf{i}, 5\mathbf{i} - 9\mathbf{j}$	**76.** $2\mathbf{i} + 4\mathbf{j}, -\mathbf{j}$

Find the angle between each pair of vectors. See Example 7.

77. $\langle 2, 1 \rangle, \langle -3, 1 \rangle$	**78.** $\langle 1, 7 \rangle, \langle 1, 1 \rangle$	**79.** $\langle 1, 2 \rangle, \langle -6, 3 \rangle$
80. $\langle 4, 0 \rangle, \langle 2, 2 \rangle$	**81.** $3\mathbf{i} + 4\mathbf{j}, \mathbf{j}$	**82.** $-5\mathbf{i} + 12\mathbf{j}, 3\mathbf{i} + 2\mathbf{j}$

Let $\mathbf{u} = \langle -2, 1 \rangle$, $\mathbf{v} = \langle 3, 4 \rangle$, and $\mathbf{w} = \langle -5, 12 \rangle$. Evaluate each expression.

83. $(3\mathbf{u}) \cdot \mathbf{v}$	**84.** $\mathbf{u} \cdot (\mathbf{v} - \mathbf{w})$	**85.** $\mathbf{u} \cdot \mathbf{v} - \mathbf{u} \cdot \mathbf{w}$	**86.** $\mathbf{u} \cdot (3\mathbf{v})$

Determine whether each pair of vectors is orthogonal. See Figure 36.

87. $\langle 1, 2 \rangle, \langle -6, 3 \rangle$	**88.** $\langle 3, 4 \rangle, \langle 6, 8 \rangle$
89. $\langle 1, 0 \rangle, \langle \sqrt{2}, 0 \rangle$	**90.** $\langle 1, 1 \rangle, \langle 1, -1 \rangle$
91. $\sqrt{5}\mathbf{i} - 2\mathbf{j}, -5\mathbf{i} + 2\sqrt{5}\mathbf{j}$	**92.** $-4\mathbf{i} + 3\mathbf{j}, 8\mathbf{i} - 6\mathbf{j}$

Relating Concepts

For individual or collaborative investigation
(Exercises 93–98)

 Consider the two vectors **u** *and* **v** *shown. Assume all values are exact.* **Work Exercises 93–98 in order.**

93. Use trigonometry alone (without using vector notation) to find the magnitude and direction angle of **u** + **v**. Use the law of cosines and the law of sines in your work.

94. Find the horizontal and vertical components of **u**, using your calculator.

95. Find the horizontal and vertical components of **v**, using your calculator.

96. Find the horizontal and vertical components of **u** + **v** by adding the results you obtained in Exercises 94 and 95.

97. Use your calculator to find the magnitude and direction angle of the vector **u** + **v**.

98. Compare your answers in Exercises 93 and 97. What do you notice? Which method of solution do you prefer?

7.5 Applications of Vectors

The Equilibrant ■ **Incline Applications** ■ **Navigation Applications**

The Equilibrant The previous section covered methods for finding the resultant of two vectors. Sometimes it is necessary to find a vector that will counterbalance the resultant. This opposite vector is called the **equilibrant;** that is, the equilibrant of vector **u** is the vector −**u**.

EXAMPLE 1 Finding the Magnitude and Direction of an Equilibrant

Find the magnitude of the equilibrant of forces of 48 newtons and 60 newtons acting on a point *A*, if the angle between the forces is 50°. Then find the angle between the equilibrant and the 48-newton force.

Solution

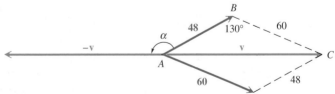

Figure 37

In Figure 37, the equilibrant is −**v**. The magnitude of **v**, and hence of −**v**, is found by using triangle *ABC* and the law of cosines.

$$|\mathbf{v}|^2 = 48^2 + 60^2 - 2(48)(60)\cos 130° \quad \text{Law of cosines (Section 7.3)}$$
$$|\mathbf{v}|^2 \approx 9606.5$$
$$|\mathbf{v}| \approx 98 \text{ newtons} \qquad \text{Two significant digits (Section 2.4)}$$

The required angle, labeled α in Figure 37, can be found by subtracting angle CAB from 180°. Use the law of sines to find angle CAB.

$$\frac{98}{\sin 130°} = \frac{60}{\sin CAB} \qquad \text{Law of sines (Section 7.1)}$$

$$\sin CAB \approx .46900680$$

$$CAB \approx 28° \qquad \text{Use the inverse sine function. (Section 6.1)}$$

Finally, $\alpha \approx 180° - 28° = 152°$.

> **Now try Exercise 1.**

Incline Applications   We can use vectors to solve incline problems.

EXAMPLE 2   Finding a Required Force

Find the force required to keep a 50-lb wagon from sliding down a ramp inclined at 20° to the horizontal. (Assume there is no friction.)

Solution   In Figure 38, the vertical 50-lb force **BA** represents the force of gravity. It is the sum of vectors **BC** and $-$**AC**. The vector **BC** represents the force with which the weight pushes against the ramp. Vector **BF** represents the force that would pull the weight up the ramp. Since vectors **BF** and **AC** are equal, $|$**AC**$|$ gives the magnitude of the required force.

Vectors **BF** and **AC** are parallel, so angle EBD equals angle A. Since angle BDE and angle C are right angles, triangles CBA and DEB have two corresponding angles equal and so are similar triangles. Therefore, angle ABC equals angle E, which is 20°. From right triangle ABC,

$$\sin 20° = \frac{|\mathbf{AC}|}{50} \qquad \text{(Section 2.1)}$$

$$|\mathbf{AC}| = 50 \sin 20° \approx 17.$$

Approximately a 17-lb force will keep the wagon from sliding down the ramp.

> **Now try Exercise 9.**

Figure 38

EXAMPLE 3   Finding an Incline Angle

A force of 16.0 lb is required to hold a 40.0 lb lawn mower on an incline. What angle does the incline make with the horizontal?

Solution   Figure 39 illustrates the situation. Consider right triangle ABC. Angle B equals angle θ, the magnitude of vector **BA** represents the weight of the mower, and vector **AC** equals vector **BE**, which represents the force required to hold the mower on the incline. From the figure,

$$\sin B = \frac{16}{40} = .4$$

$$B \approx 23.5782°. \quad \text{Use the inverse sine function.}$$

Therefore, the hill makes an angle of about 23.6° with the horizontal.

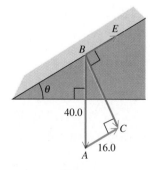

Figure 39

> **Now try Exercise 11.**

Navigation Applications Problems involving bearing (defined in Section 2.5) can also be worked with vectors.

EXAMPLE 4 Applying Vectors to a Navigation Problem

A ship leaves port on a bearing of 28.0° and travels 8.20 mi. The ship then turns due east and travels 4.30 mi. How far is the ship from port? What is its bearing from port?

Solution In Figure 40, vectors **PA** and **AE** represent the ship's path. The magnitude and bearing of the resultant **PE** can be found as follows. Triangle *PNA* is a right triangle, so angle *NAP* = 90° − 28.0° = 62.0°. Then angle *PAE* = 180° − 62.0° = 118.0°. Use the law of cosines to find $|\mathbf{PE}|$, the magnitude of vector **PE**.

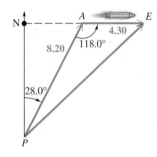

Figure 40

$$|\mathbf{PE}|^2 = 8.20^2 + 4.30^2 - 2(8.20)(4.30)\cos 118.0°$$
$$|\mathbf{PE}|^2 \approx 118.84$$
$$|\mathbf{PE}| \approx 10.9 \qquad \text{Square root property (Appendix A)}$$

The ship is about 10.9 mi from port.

To find the bearing of the ship from port, first find angle *APE*. Use the law of sines.

$$\frac{\sin APE}{4.30} = \frac{\sin 118.0°}{10.9}$$

$$\sin APE = \frac{4.30 \sin 118.0°}{10.9}$$

$$APE \approx 20.4° \qquad \text{Use the inverse sine function.}$$

Now add 20.4° to 28.0° to find that the bearing is 48.4°.

Now try Exercise 15.

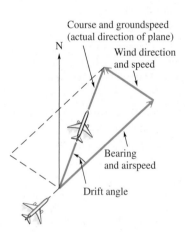

Course and groundspeed (actual direction of plane)

Wind direction and speed

Bearing and airspeed

Drift angle

Figure 41

In air navigation, the **airspeed** of a plane is its speed relative to the air, while the **groundspeed** is its speed relative to the ground. Because of wind, these two speeds are usually different. The groundspeed of the plane is represented by the vector sum of the airspeed and windspeed vectors. See Figure 41.

EXAMPLE 5 Applying Vectors to a Navigation Problem

A plane with an airspeed of 192 mph is headed on a bearing of 121°. A north wind is blowing (from north to south) at 15.9 mph. Find the groundspeed and the actual bearing of the plane.

Solution In Figure 42, the groundspeed is represented by $|\mathbf{x}|$. We must find angle α to determine the bearing, which will be 121° + α. From Figure 42, angle *BCO* equals angle *AOC*, which equals 121°. Find $|\mathbf{x}|$ by the law of cosines.

Figure 42

$$|\mathbf{x}|^2 = 192^2 + 15.9^2 - 2(192)(15.9)\cos 121°$$
$$|\mathbf{x}|^2 \approx 40{,}261$$
$$|\mathbf{x}| \approx 200.7 \quad \text{or} \quad \text{about 201 mph}$$

Now find α by using the law of sines. Use the value of $|\mathbf{x}|$ before rounding.

$$\frac{\sin \alpha}{15.9} = \frac{\sin 121°}{200.7}$$

$$\sin \alpha \approx .0679$$

$$\alpha \approx 3.89°$$

To the nearest degree, α is 4°. The groundspeed is about 201 mph on a bearing of $121° + 4° = 125°$.

Now try Exercise 25.

7.5 Exercises

Solve each problem. See Examples 1–5.

1. *Direction and Magnitude of an Equilibrant*   Two tugboats are pulling a disabled speedboat into port with forces of 1240 lb and 1480 lb. The angle between these forces is 28.2°. Find the direction and magnitude of the equilibrant.

2. *Direction and Magnitude of an Equilibrant*   Two rescue vessels are pulling a broken-down motorboat toward a boathouse with forces of 840 lb and 960 lb. The angle between these forces is 24.5°. Find the direction and magnitude of the equilibrant.

3. *Angle Between Forces*   Two forces of 692 newtons and 423 newtons act at a point. The resultant force is 786 newtons. Find the angle between the forces.

4. *Angle Between Forces*   Two forces of 128 lb and 253 lb act at a point. The equilibrant force is 320 lb. Find the angle between the forces.

5. *Magnitudes of Forces*   A force of 176 lb makes an angle of 78° 50′ with a second force. The resultant of the two forces makes an angle of 41° 10′ with the first force. Find the magnitudes of the second force and of the resultant.

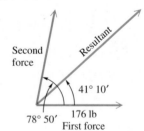

6. *Magnitudes of Forces*   A force of 28.7 lb makes an angle of 42° 10′ with a second force. The resultant of the two forces makes an angle of 32° 40′ with the first force. Find the magnitudes of the second force and of the resultant.

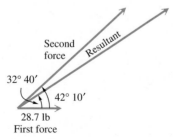

7. *Angle of a Hill Slope*   A force of 25 lb is required to hold an 80-lb crate on a hill. What angle does the hill make with the horizontal?

8. *Force Needed to Keep a Car Parked*   Find the force required to keep a 3000-lb car parked on a hill that makes an angle of 15° with the horizontal.

9. *Force Needed for a Monolith* To build the pyramids in Egypt, it is believed that giant causeways were constructed to transport the building materials to the site. One such causeway is said to have been 3000 ft long, with a slope of about 2.3°. How much force would be required to hold a 60-ton monolith on this causeway?

10. *Weight of a Box* Two people are carrying a box. One person exerts a force of 150 lb at an angle of 62.4° with the horizontal. The other person exerts a force of 114 lb at an angle of 54.9°. Find the weight of the box.

11. *Incline Angle* A force of 18 lb is required to hold a 60-lb stump grinder on an incline. What angle does the incline make with the horizontal?

12. *Incline Angle* A force of 30 lb is required to hold an 80-lb pressure washer on an incline. What angle does the incline make with the horizontal?

13. *Weight of a Crate and Tension of a Rope* A crate is supported by two ropes. One rope makes an angle of 46° 20′ with the horizontal and has a tension of 89.6 lb on it. The other rope is horizontal. Find the weight of the crate and the tension in the horizontal rope.

14. *Angles Between Forces* Three forces acting at a point are in equilibrium. The forces are 980 lb, 760 lb, and 1220 lb. Find the angles between the directions of the forces. (*Hint:* Arrange the forces to form the sides of a triangle.)

15. *Distance and Bearing of a Ship* A ship leaves port on a bearing of 34.0° and travels 10.4 mi. The ship then turns due east and travels 4.6 mi. How far is the ship from port, and what is its bearing from port?

16. *Distance and Bearing of a Luxury Liner* A luxury liner leaves port on a bearing of 110.0° and travels 8.8 mi. It then turns due west and travels 2.4 mi. How far is the liner from port, and what is its bearing from port?

17. *Distance of a Ship from Its Starting Point* Starting at point A, a ship sails 18.5 km on a bearing of 189°, then turns and sails 47.8 km on a bearing of 317°. Find the distance of the ship from point A.

18. *Distance of a Ship from Its Starting Point* Starting at point X, a ship sails 15.5 km on a bearing of 200°, then turns and sails 2.4 km on a bearing of 320°. Find the distance of the ship from point X.

19. *Distance and Direction of a Motorboat* A motorboat sets out in the direction N 80° E. The speed of the boat in still water is 20.0 mph. If the current is flowing directly south, and the actual direction of the motorboat is due east, find the speed of the current and the actual speed of the motorboat.

20. *Path Traveled by a Plane* The aircraft carrier *Tallahassee* is traveling at sea on a steady course with a bearing of 30° at 32 mph. Patrol planes on the carrier have enough fuel for 2.6 hr of flight when traveling at a speed of 520 mph. One of the pilots takes off on a bearing of 338° and then turns and heads in a straight line, so as to be able to catch the carrier and land on the deck at the exact instant that his fuel runs out. If the pilot left at 2 P.M., at what time did he turn to head for the carrier?

21. *Bearing and Groundspeed of a Plane* An airline route from San Francisco to Honolulu is on a bearing of 233.0°. A jet flying at 450 mph on that bearing runs into a wind blowing at 39.0 mph from a direction of 114.0°. Find the resulting bearing and groundspeed of the plane.

22. *Movement of a Motorboat* Suppose you would like to cross a 132-ft-wide river in a motorboat. Assume that the motorboat can travel at 7 mph relative to the water and that the current is flowing west at the rate of 3 mph. The bearing θ is chosen so that the motorboat will land at a point exactly across from the starting point.

   (a) At what speed will the motorboat be traveling relative to the banks?
   (b) How long will it take for the motorboat to make the crossing?
   (c) What is the measure of angle θ?

23. *Airspeed and Groundspeed* A pilot wants to fly on a bearing of 74.9°. By flying due east, he finds that a 42-mph wind, blowing from the south, puts him on course. Find the airspeed and the groundspeed.

24. *Bearing of a Plane* A plane flies 650 mph on a bearing of 175.3°. A 25-mph wind, from a direction of 266.6°, blows against the plane. Find the resulting bearing of the plane.

25. *Bearing and Groundspeed of a Plane* A pilot is flying at 190 mph. He wants his flight path to be on a bearing of 64° 30′. A wind is blowing from the south at 35.0 mph. Find the bearing he should fly, and find the plane's groundspeed.

26. *Bearing and Groundspeed of a Plane* A pilot is flying at 168 mph. She wants her flight path to be on a bearing of 57° 40′. A wind is blowing from the south at 27.1 mph. Find the bearing the pilot should fly, and find the plane's groundspeed.

27. *Bearing and Airspeed of a Plane*   What bearing and airspeed are required for a plane to fly 400 mi due north in 2.5 hr if the wind is blowing from a direction of 328° at 11 mph?

28. *Groundspeed and Bearing of a Plane*   A plane is headed due south with an airspeed of 192 mph. A wind from a direction of 78° is blowing at 23 mph. Find the groundspeed and resulting bearing of the plane.

29. *Groundspeed and Bearing of a Plane*   An airplane is headed on a bearing of 174° at an airspeed of 240 km per hr. A 30 km per hr wind is blowing from a direction of 245°. Find the groundspeed and resulting bearing of the plane.

30. *Velocity of a Star*   The space velocity **v** of a star relative to the sun can be expressed as the resultant vector of two perpendicular vectors—the radial velocity $\mathbf{v}_r$ and the tangential velocity $\mathbf{v}_t$, where $\mathbf{v} = \mathbf{v}_r + \mathbf{v}_t$. If a star is located near the sun and its space velocity is large, then its motion across the sky will also be large. Barnard's Star is a relatively close star with a distance of 35 trillion mi from the sun. It moves across the sky through an angle of 10.34″ per year, which is the largest motion of any known star. Its radial velocity is $\mathbf{v}_r = 67$ mi per sec toward the sun. (*Source:* Zeilik, M., S. Gregory, and E. Smith,

Not to scale

Introductory Astronomy and Astrophysics, Second Edition, Saunders College Publishing, 1998; Acker, A. and C. Jaschek, *Astronomical Methods and Calculations,* John Wiley & Sons, 1986.)

(a) Approximate the tangential velocity $\mathbf{v}_t$ of Barnard's Star. (*Hint:* Use the arc length formula $s = r\theta$.)

(b) Compute the magnitude of **v**.

31. *(Modeling) Measuring Rainfall*   Suppose that vector **R** models the amount of rainfall in inches and the direction it falls, and vector **A** models the area in square inches and orientation of the opening of a rain gauge, as illustrated in the figure. The total volume V of water collected in the rain gauge is given by $V = |\mathbf{R} \cdot \mathbf{A}|$. This formula calculates the volume of water collected even if the wind is blowing the rain in a slanted direction or the rain gauge is not exactly vertical. Let $\mathbf{R} = \mathbf{i} - 2\mathbf{j}$ and $\mathbf{A} = .5\mathbf{i} + \mathbf{j}$.

(a) Find $|\mathbf{R}|$ and $|\mathbf{A}|$. Interpret your results.
(b) Calculate V and interpret this result.
(c) For the rain gauge to collect the maximum amount of water, what should be true about vectors **R** and **A**?

32. *The Dot Product*   In the figure below, $\mathbf{a} = \langle a_1, a_2 \rangle$, $\mathbf{b} = \langle b_1, b_2 \rangle$, and $\mathbf{a} - \mathbf{b} = \langle a_1 - b_1, a_2 - b_2 \rangle$. Apply the law of cosines to the triangle and derive the equation $\mathbf{a} \cdot \mathbf{b} = |\mathbf{a}||\mathbf{b}| \cos \theta$.

Chapter 7 Summary

KEY TERMS

7.1 Side-Angle-Side
(SAS)
Angle-Side-Angle
(ASA)
Side-Side-Side (SSS)
oblique triangle
Side-Angle-Angle
(SAA)

7.2 ambiguous case
7.4 scalar
vector quantity
vector
magnitude
initial point
terminal point
parallelogram rule

resultant
opposite (of a vector)
zero vector
scalar product
position vector
horizontal component
vertical component
direction angle

unit vector
dot product
angle between two
vectors
orthogonal vectors
7.5 equilibrant
airspeed
groundspeed

NEW SYMBOLS

OP or $\overrightarrow{\textbf{OP}}$ vector **OP**

$|\textbf{OP}|$ magnitude of vector **OP**

$\langle a,b \rangle$ position vector

i, j   unit vectors

QUICK REVIEW

CONCEPTS

EXAMPLES

7.1 Oblique Triangles and the Law of Sines

Law of Sines
In any triangle ABC, with sides a, b, and c,

$$\frac{a}{\sin A} = \frac{b}{\sin B}, \quad \frac{a}{\sin A} = \frac{c}{\sin C}, \quad \text{and} \quad \frac{b}{\sin B} = \frac{c}{\sin C}.$$

In triangle ABC, find c if $A = 44°$, $C = 62°$, and $a = 12.00$ units. Then find its area.

$$\frac{a}{\sin A} = \frac{c}{\sin C}$$

$$\frac{12.00}{\sin 44°} = \frac{c}{\sin 62°}$$

$$c = \frac{12.00 \sin 62°}{\sin 44°} \approx 15.25 \text{ units}$$

Area of a Triangle
In any triangle ABC, the area is half the product of the lengths of two sides and the sine of the angle between them.

$$\mathcal{A} = \frac{1}{2}bc \sin A, \quad \mathcal{A} = \frac{1}{2}ab \sin C, \quad \mathcal{A} = \frac{1}{2}ac \sin B$$

For triangle ABC above,

$$\mathcal{A} = \frac{1}{2}ac \sin B$$

$$= \frac{1}{2}(12.00)(15.25) \sin 74° \quad B = 180° - 44° - 62°,$$

$$\approx 87.96 \text{ sq units.}$$

CONCEPTS	EXAMPLES

7.2 The Ambiguous Case of the Law of Sines

Ambiguous Case

If we are given the lengths of two sides and the angle opposite one of them, for example, A, a, and b in triangle ABC, then it is possible that zero, one, or two such triangles exist. If A is acute, h is the altitude from C, and

1. $a < h < b$, then there is no triangle.
2. $a = h < b$, then there is one triangle (a right triangle).
3. $a \geq b$, then there is one triangle.
4. $h < a < b$, then there are two triangles.

If A is obtuse and

1. $a \leq b$, then there is no triangle.
2. $a > b$, then there is one triangle.

See the table on page 287 that illustrates the possible outcomes.

Solve triangle ABC, given $A = 44.5°$, $a = 11.0$ in., and $c = 7.0$ in.

Find angle C.

$$\frac{\sin C}{7.0} = \frac{\sin 44.5°}{11.0}$$

$$\sin C \approx .4460$$

$$C \approx 26.5°$$

Another angle with this sine value is

$$180° - 26.5° = 153.5°.$$

However, $153.5° + 44.5° > 180°$, so there is only one triangle.

$$B = 180° - 44.5° - 26.5°$$

$$B = 109°$$

Using the law of sines again,

$$b \approx 14.8 \text{ in.}$$

7.3 The Law of Cosines

Law of Cosines

In any triangle ABC, with sides a, b, and c,

$$a^2 = b^2 + c^2 - 2bc \cos A$$
$$b^2 = a^2 + c^2 - 2ac \cos B$$
$$c^2 = a^2 + b^2 - 2ab \cos C.$$

In triangle ABC, find C if $a = 11$ units, $b = 13$ units, and $c = 20$ units. Then find its area.

$$c^2 = a^2 + b^2 - 2ab \cos C$$
$$20^2 = 11^2 + 13^2 - 2(11)(13) \cos C$$
$$400 = 121 + 169 - 286 \cos C$$
$$\frac{400 - 121 - 169}{-286} = \cos C$$
$$C = \cos^{-1}\left(\frac{400 - 121 - 169}{-286}\right)$$
$$C \approx 112.6°$$

Heron's Area Formula

If a triangle has sides of lengths a, b, and c, with semiperimeter

$$s = \frac{1}{2}(a + b + c),$$

then the area of the triangle is

$$\mathcal{A} = \sqrt{s(s - a)(s - b)(s - c)}.$$

The semiperimeter s is

$$s = \frac{1}{2}(11 + 13 + 20) = 22,$$

so

$$\mathcal{A} = \sqrt{22(22 - 11)(22 - 13)(22 - 20)} = 66 \text{ sq units.}$$

| CONCEPTS | EXAMPLES |

7.4 Vectors, Operations, and the Dot Product

Magnitude and Direction Angle of a Vector

The magnitude (length) of vector $\mathbf{u} = \langle a, b \rangle$ is given by

$$|\mathbf{u}| = \sqrt{a^2 + b^2}.$$

$$|\mathbf{u}| = \sqrt{(2\sqrt{3})^2 + 2^2} = \sqrt{16} = 4$$

The direction angle θ satisfies $\tan \theta = \frac{b}{a}$, where $a \neq 0$.

Since $\tan \theta = \frac{2}{2\sqrt{3}} = \frac{1}{\sqrt{3}} \cdot \frac{\sqrt{3}}{\sqrt{3}} = \frac{\sqrt{3}}{3}$, it follows that $\theta = 30°$.

Vector Operations

For any real numbers a, b, c, d, and k,

$$\langle a, b \rangle + \langle c, d \rangle = \langle a + c, b + d \rangle$$

$$k \cdot \langle a, b \rangle = \langle ka, kb \rangle.$$

If $\mathbf{a} = \langle a_1, a_2 \rangle$, then $-\mathbf{a} = \langle -a_1, -a_2 \rangle$.

$$\langle a, b \rangle - \langle c, d \rangle = \langle a, b \rangle + (-\langle c, d \rangle) = \langle a - c, b - d \rangle$$

$$\langle 4, 6 \rangle + \langle -8, 3 \rangle = \langle -4, 9 \rangle$$

$$5\langle -2, 1 \rangle = \langle -10, 5 \rangle$$

$$-\langle -9, 6 \rangle = \langle 9, -6 \rangle$$

$$\langle 4, 6 \rangle - \langle -8, 3 \rangle = \langle 12, 3 \rangle$$

If $\mathbf{u} = \langle x, y \rangle$ has direction angle θ, then

$$\mathbf{u} = \langle |\mathbf{u}| \cos \theta, |\mathbf{u}| \sin \theta \rangle.$$

For $\mathbf{u}$ defined above,

$$\mathbf{u} = \langle 4 \cos 30°, 4 \sin 30° \rangle$$

$$= \langle 2\sqrt{3}, 2 \rangle$$

i, j Form for Vectors

If $\mathbf{v} = \langle a, b \rangle$, then $\mathbf{v} = a\mathbf{i} + b\mathbf{j}$, where $\mathbf{i} = \langle 1, 0 \rangle$ and $\mathbf{j} = \langle 0, 1 \rangle$.

and

$$\mathbf{u} = 2\sqrt{3}\mathbf{i} + 2\mathbf{j}.$$

Dot Product

The dot product of the two vectors $\mathbf{u} = \langle a, b \rangle$ and $\mathbf{v} = \langle c, d \rangle$, denoted $\mathbf{u} \cdot \mathbf{v}$, is given by

$$\mathbf{u} \cdot \mathbf{v} = ac + bd.$$

$$\langle 2, 1 \rangle \cdot \langle 5, -2 \rangle = 2 \cdot 5 + 1(-2) = 8$$

If θ is the angle between $\mathbf{u}$ and $\mathbf{v}$, where $0° \leq \theta \leq 180°$, then

Find the angle θ between $\mathbf{u} = \langle 3, 1 \rangle$ and $\mathbf{v} = \langle 2, -3 \rangle$.

$$\mathbf{u} \cdot \mathbf{v} = |\mathbf{u}||\mathbf{v}| \cos \theta \quad \text{or} \quad \cos \theta = \frac{\mathbf{u} \cdot \mathbf{v}}{|\mathbf{u}||\mathbf{v}|}.$$

$$\cos \theta = \frac{\langle 3, 1 \rangle \cdot \langle 2, -3 \rangle}{\sqrt{3^2 + 1^2} \cdot \sqrt{2^2 + (-3)^2}}$$

$$\cos \theta = \frac{6 + (-3)}{\sqrt{10} \cdot \sqrt{13}}$$

$$\cos \theta = \frac{3}{\sqrt{130}}$$

$$\theta = \cos^{-1} \frac{3}{\sqrt{130}} \approx 74.7°$$

Chapter 7 Review Exercises

Use the law of sines to find the indicated part of each triangle ABC.

1. $C = 74.2°$, $c = 96.3$ m, $B = 39.5°$;   find b

2. $A = 129.7°$, $a = 127$ ft, $b = 69.8$ ft;   find B

3. $C = 51.3°$, $c = 68.3$ m, $b = 58.2$ m;   find B

4. $a = 165$ m, $A = 100.2°$, $B = 25.0°$;   find b

5. $B = 39° 50'$, $b = 268$ m, $a = 340$ m;   find A

6. $C = 79° 20'$, $c = 97.4$ mm, $a = 75.3$ mm;   find A

7. If we are given a, A, and C in a triangle ABC, does the possibility of the ambiguous case exist? If not, explain why.

8. Can triangle ABC exist if $a = 4.7$, $b = 2.3$, and $c = 7.0$? If not, explain why. Answer this question without using trigonometry.

9. Given $a = 10$ and $B = 30°$, determine the values of b for which A has

(a) exactly one value   (b) two possible values   (c) no value.

10. Explain why there can be no triangle ABC satisfying $A = 140°$, $a = 5$, and $b = 7$.

Use the law of cosines to find the indicated part of each triangle ABC.

11. $a = 86.14$ in., $b = 253.2$ in., $c = 241.9$ in.;   find A

12. $B = 120.7°$, $a = 127$ ft, $c = 69.8$ ft;   find b

13. $A = 51° 20'$, $c = 68.3$ m, $b = 58.2$ m;   find a

14. $a = 14.8$ m, $b = 19.7$ m, $c = 31.8$ m;   find B

15. $A = 60°$, $b = 5$ cm, $c = 21$ cm;   find a

16. $a = 13$ ft, $b = 17$ ft, $c = 8$ ft;   find A

Solve each triangle ABC having the given information.

17. $A = 25.2°$, $a = 6.92$ yd, $b = 4.82$ yd

18. $A = 61.7°$, $a = 78.9$ m, $b = 86.4$ m

19. $a = 27.6$ cm, $b = 19.8$ cm, $C = 42° 30'$

20. $a = 94.6$ yd, $b = 123$ yd, $c = 109$ yd

Find the area of each triangle ABC with the given information.

21. $b = 840.6$ m, $c = 715.9$ m, $A = 149.3°$

22. $a = 6.90$ ft, $b = 10.2$ ft, $C = 35° 10'$

23. $a = .913$ km, $b = .816$ km, $c = .582$ km

24. $a = 43$ m, $b = 32$ m, $c = 51$ m

Solve each problem.

25. *Distance Across a Canyon*   To measure the distance AB across a canyon for a power line, a surveyor measures angles B and C and the distance BC, as shown in the figure. What is the distance from A to B?

26. *Length of a Brace* A banner on an 8.0-ft pole is to be mounted on a building at an angle of 115°, as shown in the figure. Find the length of the brace.

27. *Height of a Tree* A tree leans at an angle of 8.0° from the vertical. From a point 7.0 m from the bottom of the tree, the angle of elevation to the top of the tree is 68°. How tall is the tree?

28. *Hanging Sculpture* A hanging sculpture in an art gallery is to be hung with two wires of lengths 15.0 ft and 12.2 ft so that the angle between them is 70.3°. How far apart should the ends of the wire be placed on the ceiling?

29. *Height of a Tree* A hill makes an angle of 14.3° with the horizontal. From the base of the hill, the angle of elevation to the top of a tree on top of the hill is 27.2°. The distance along the hill from the base to the tree is 212 ft. Find the height of the tree.

30. *Pipeline Position* A pipeline is to run between points A and B, which are separated by a protected wetlands area. To avoid the wetlands, the pipe will run from point A to C and then to B. The distances involved are $AB = 150$ km, $AC = 102$ km, and $BC = 135$ km. What angle should be used at point C?

31. *Distance Between Two Boats* Two boats leave a dock together. Each travels in a straight line. The angle between their courses measures 54° 10′. One boat travels 36.2 km per hr, and the other travels 45.6 km per hr. How far apart will they be after 3 hr?

32. *Distance from a Ship to a Lighthouse* A ship sailing parallel to shore sights a lighthouse at an angle of 30° from its direction of travel. After the ship travels 2.0 mi farther, the angle has increased to 55°. At that time, how far is the ship from the lighthouse?

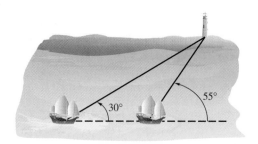

33. *Area of a Triangle* Find the area of the triangle shown in the figure using Heron's area formula.

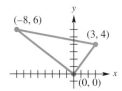

34. *Area of a Quadrilateral*   A lot has the shape of the quadrilateral in the figure. What is its area?

In Exercises 35 and 36, use the given vectors to sketch the following.

35. a − b

36. a + 3c

37. *Concept Check*   Decide whether each statement is true or false.

(a) Opposite angles of a parallelogram are equal.

(b) A diagonal of a parallelogram must bisect two angles of the parallelogram.

Given two forces and the angle between them, find the magnitude of the resultant force.

38. forces of 142 and 215 newtons, forming an angle of 112°

39.

*Vector **v** has the given magnitude and direction angle. Find the magnitudes of the horizontal and vertical components of **v**.*

40. $|\mathbf{v}| = 50$, $\theta = 45°$
   (Give exact values.)

41. $|\mathbf{v}| = 964$, $\theta = 154° \, 20'$

*Find the magnitude and direction angle for **u** rounded to the nearest tenth.*

42. $\mathbf{u} = \langle 21, -20 \rangle$

43. $\mathbf{u} = \langle -9, 12 \rangle$

44. Let $\mathbf{v} = 2\mathbf{i} - \mathbf{j}$ and $\mathbf{u} = -3\mathbf{i} + 2\mathbf{j}$. Express each in terms of **i** and **j**.

(a) $2\mathbf{v} + \mathbf{u}$

(b) $2\mathbf{v}$

(c) $\mathbf{v} - 3\mathbf{u}$

45. Let $\mathbf{a} = \langle 3, -2 \rangle$ and $\mathbf{b} = \langle -1, 3 \rangle$. Find $\mathbf{a} \cdot \mathbf{b}$ and the angle between **a** and **b**, rounded to the nearest tenth of a degree.

Find the vector of magnitude 1 having the same direction angle as the given vector.

46. $\mathbf{u} = \langle -4, 3 \rangle$

47. $\mathbf{u} = \langle 5, 12 \rangle$

Solve each problem.

48. *Force Placed on a Barge*   One rope pulls a barge directly east with a force of 100 newtons. Another rope pulls the barge to the northeast with a force of 200 newtons. Find the resultant force acting on the barge and the angle between the resultant and the first rope.

49. *Weight of a Sled and Passenger* Paula and Steve are pulling their daughter Jessie on a sled. Steve pulls with a force of 18 lb at an angle of 10°. Paula pulls with a force of 12 lb at an angle of 15°. Find the magnitude of the resultant force on Jessie and the sled.

50. *Angle of a Hill* A 186-lb force just keeps a 2800-lb car from rolling down a hill. What angle does the hill make with the horizontal?

51. *Direction and Speed of a Plane* A plane has an airspeed of 520 mph. The pilot wishes to fly on a bearing of 310°. A wind of 37 mph is blowing from a bearing of 212°. What direction should the pilot fly, and what will be her actual speed?

52. *Speed and Direction of a Boat* A boat travels 15 km per hr in still water. The boat is traveling across a large river, on a bearing of 130°. The current in the river, coming from the west, has a speed of 7 km per hr. Find the resulting speed of the boat and its resulting direction of travel.

53. *Control Points* To obtain accurate aerial photographs, ground control must determine the coordinates of *control points* located on the ground that can be identified in the photographs. Using these known control points, the orientation and scale of each photograph can be found. Then, unknown positions and distances can easily be determined. The figure shows three consecutive control points *A*, *B*, and *C*.

A surveyor measures a baseline distance of 92.13 ft from *B* to an arbitrary point *P*. Angles *BAP* and *BCP* are found to be 2° 22′ 47″ and 5° 13′ 11″, respectively. Then, angles *APB* and *CPB* are determined to be 63° 4′ 25″ and 74° 19′ 49″, respectively. Determine the distance between control points *A* and *B* and between *B* and *C*. (*Source:* Moffitt, F. and E. Mikhail, *Photogrammetry,* Third Edition, Harper & Row, 1980.)

The following identities involve all six parts of a triangle ABC and are thus useful for checking answers.

$$\frac{a + b}{c} = \frac{\cos \frac{1}{2}(A - B)}{\sin \frac{1}{2} C} \quad \text{Newton's formula}$$

$$\frac{a - b}{c} = \frac{\sin \frac{1}{2}(A - B)}{\cos \frac{1}{2} C} \quad \text{Mollweide's formula}$$

54. Apply Newton's formula to the given triangle to verify the accuracy of the information.

55. Apply Mollweide's formula to the given triangle to verify the accuracy of the information.

Chapter 7 Test

Find the indicated part of each triangle ABC.

1. $A = 25.2°$, $a = 6.92$ yd, $b = 4.82$ yd;   find C

2. $C = 118°$, $b = 130$ km, $a = 75$ km;   find c

3. $a = 17.3$ ft, $b = 22.6$ ft, $c = 29.8$ ft;   find B

4. Find the area of triangle ABC in Exercise 2.

5. Given $a = 10$ and $B = 150°$ in triangle ABC, determine the values of b for which A has

    (a) exactly one value   **(b)** two possible values   **(c)** no value.

6. Find the area of the triangle having sides of lengths 22, 26, and 40.

7. Find the magnitude and the direction angle for the vector shown in the figure.

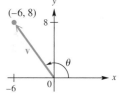

8. Use the given vectors to sketch $\mathbf{a} + \mathbf{b}$.

9. For the vectors $\mathbf{u} = \langle -1, 3 \rangle$ and $\mathbf{v} = \langle 2, -6 \rangle$, find each of the following.

    (a) $\mathbf{u} + \mathbf{v}$ **(b)** $-3\mathbf{v}$ **(c)** $\mathbf{u} \cdot \mathbf{v}$ **(d)** $|\mathbf{u}|$

Solve each problem.

10. *Height of a Balloon*   The angles of elevation of a balloon from two points A and B on level ground are $24° 50'$ and $47° 20'$, respectively. As shown in the figure, points A, B, and C are in the same vertical plane and points A and B are 8.4 mi apart. Approximate the height of the balloon above the ground to the nearest tenth of a mile.

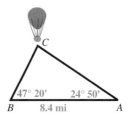

11. *Horizontal and Vertical Components*   Find the horizontal and vertical components of the vector with magnitude 569 that is inclined $127.5°$ from the horizontal. Give your answer in the form $\langle a, b \rangle$.

12. *Radio Direction Finders*   Radio direction finders are placed at points A and B, which are 3.46 mi apart on an east-west line, with A west of B. From A, the bearing of a certain illegal pirate radio transmitter is $48°$, and from B the bearing is $302°$. Find the distance between the transmitter and A to the nearest hundredth of a mile.

13. *Wires Supporting a Flagpole* A flagpole stands vertically on a hillside that makes an angle of 20° with the horizontal. Two supporting wires are attached as shown in the figure. What are the lengths of the supporting wires?

30 ft

15 ft 15 ft

20°

14. *Walking Dogs on Leashes* While Michael is walking his two dogs, Duke and Prince, they reach a corner and must wait for a WALK sign. Michael is holding the two leashes in the same hand, and the dogs are pulling on their leashes at the angles and forces shown in the figure. Find the magnitude of the force Michael must apply to restrain the dogs.

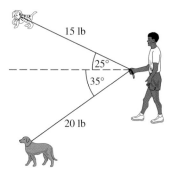

15 lb

25°

35°

20 lb

Chapter 7 Quantitative Reasoning

Just how much does the U.S. flag "show its colors"?

The flag of the United States includes the colors red, white, and blue. Which color is predominant? Clearly the answer is either red or white. (It can be shown that only 18.73% of the total area is blue.) (*Source: Slicing Pizzas, Racing Turtles, and Further Adventures in Applied Mathematics,* Banks, R., Princeton University Press, 1999.)

1. Let R denote the radius of the circumscribing circle of a five-pointed star appearing on the American flag. The star can be decomposed into ten congruent triangles. In the figure, r is the radius of the circumscribing circle of the pentagon in the interior of the star. Show that the area of a star is

$$A = \left[5 \frac{\sin A \sin B}{\sin(A + B)} \right] R^2.$$

(*Hint:* $\sin C = \sin[180° - (A + B)] = \sin(A + B)$.)

R r

2. Angles A and B have values 18° and 36°, respectively. Express the area of a star in terms of its radius, R.

3. To determine whether red or white is predominant, we must know the measurements of the flag. Consider a flag of width 10 in., length 19 in., length of each upper stripe 11.4 in., and radius R of the circumscribing circle of each star .308 in. The thirteen stripes consist of six matching pairs of red and white stripes and one additional red, upper stripe. Therefore, we must compare the area of a red, upper stripe with the total area of the 50 white stars.

(a) Compute the area of the red, upper stripe.
(b) Compute the total area of the 50 white stars.
(c) Which color occupies the greatest area on the flag?

A

R

C

r

B

8

Complex Numbers, Polar Equations, and Parametric Equations

Higig-resolution computer graphics and *complex numbers* make it possible to produce beautiful shapes called *fractals*. Benoit B. Mandelbrot first used the term *fractal* in 1975. At its basic level, a fractal is a unique, enchanting geometric figure with an endless self-similarity property. A fractal image repeats itself infinitely with ever-decreasing dimensions. If you look at smaller and smaller portions of a fractal image, you will continue to see the whole—much like looking into two parallel mirrors that are facing each other.

The above fractal, called *Newton's basins of attraction for the cube roots of unity,* is discussed in Exercise 50, Section 8.4. (*Source:* Crownover, R., *Introduction to Fractals and Chaos,* Jones and Bartlett, 1995; Lauwerier, H., *Fractals,* Princeton University Press, 1991. Reprinted with permission.)

8.1 | Complex Numbers

Basic Concepts of Complex Numbers ▪ Complex Solutions of Equations ▪ Operations on Complex Numbers

Basic Concepts of Complex Numbers The set of real numbers does not include all numbers needed in mathematics. For example, there is no real number solution of the equation

$$x^2 = -1,$$

since -1 has no real square root. Square roots of negative numbers were not incorporated into an integrated number system until the 16th century. They were then used as solutions of equations and, later in the 18th century, in surveying. Today, such numbers are used extensively in science and engineering.

To extend the real number system to include numbers such as $\sqrt{-1}$, the number i is defined to have the following property.

Looking Ahead to Calculus

The letters j and k are also used to represent $\sqrt{-1}$ in calculus and some applications (electronics, for example).

$$i^2 = -1$$

Thus, $i = \sqrt{-1}$. The number i is called the **imaginary unit.** Numbers of the form $a + bi$, where a and b are real numbers, are called **complex numbers.** In the complex number $a + bi$, a is the **real part** and b is the **imaginary part.***

The relationships among the various sets of numbers are shown in Figure 1.

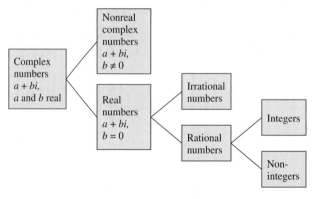

Figure 1

Two complex numbers $a + bi$ and $c + di$ are equal provided that their real parts are equal and their imaginary parts are equal; that is,

$$a + bi = c + di \quad \text{if and only if} \quad a = c \quad \text{and} \quad b = d.$$

For a complex number $a + bi$, if $b = 0$, then $a + bi = a$, which is a real number. Thus, the set of real numbers is a subset of the set of complex numbers. If $a = 0$ and $b \neq 0$, the complex number is said to be a **pure imaginary number.** For example, $3i$ is a pure imaginary number. A number such as $7 + 2i$ is a nonreal complex number. A complex number written in the form $a + bi$ (or $a + ib$) is in **standard form.** $\big($The form $a + ib$ is used to write expressions such as $i\sqrt{5}$, since $\sqrt{5}i$ could be mistaken for $\sqrt{5i}$.$\big)$

Now try Exercises 9, 11, and 13.

---

*In some texts, the term bi is defined to be the imaginary part.

The calculator is in complex number mode.

Figure 2

Some graphing calculators, such as the TI-83 Plus, are capable of working with complex numbers, as seen in Figure 2. The top screen supports the definition of i. The bottom screen shows how the calculator returns the real and imaginary parts of $7 + 2i$. ■

For a positive real number a, $\sqrt{-a}$ is defined as follows.

$\sqrt{-a}$

If $a > 0$, then $\qquad\qquad \sqrt{-a} = i\sqrt{a}.$

EXAMPLE 1   Writing $\sqrt{-a}$ as $i\sqrt{a}$

Write as the product of a real number and i, using the definition of $\sqrt{-a}$.

(a) $\sqrt{-16}$ **(b)** $\sqrt{-70}$ **(c)** $\sqrt{-48}$

Solution

(a) $\sqrt{-16} = i\sqrt{16} = 4i$ **(b)** $\sqrt{-70} = i\sqrt{70}$

(c) $\sqrt{-48} = i\sqrt{48} = i\sqrt{16 \cdot 3} = 4i\sqrt{3}$ Product rule for radicals; $\sqrt[n]{ab} = \sqrt[n]{a} \cdot \sqrt[n]{b}$

Now try Exercises 17, 19, and 21.

Complex Solutions of Equations

EXAMPLE 2   Solving Quadratic Equations for Complex Solutions

Solve each equation.

(a) $x^2 = -9$ **(b)** $x^2 + 24 = 0$

Solution

(a) Take the square root on both sides, remembering that we must find both roots, indicated by the $\pm$ sign.

$$x^2 = -9$$
$$x = \pm\sqrt{-9} \quad \text{Square root property (Appendix A)}$$
$$x = \pm i\sqrt{9} \quad \sqrt{-a} = i\sqrt{a}$$
$$x = \pm 3i \quad \sqrt{9} = 3$$

The solution set is $\{\pm 3i\}$.

(b) $x^2 + 24 = 0$

$$x^2 = -24 \qquad \text{Subtract 24.}$$
$$x = \pm\sqrt{-24} \quad \text{Square root property}$$
$$x = \pm i\sqrt{24} = \pm i\sqrt{4 \cdot 6} = \pm 2i\sqrt{6}$$

The solution set is $\left\{\pm 2i\sqrt{6}\right\}$.

Now try Exercises 25 and 27.

EXAMPLE 3 Solving a Quadratic Equation for Complex Solutions

Solve $9x^2 + 5 = 6x$.

Solution Write the equation in standard form, $9x^2 - 6x + 5 = 0$. Then use the quadratic formula.

$$x = \frac{-b \pm \sqrt{b^2 - 4ac}}{2a} \qquad \text{Quadratic formula (Appendix A)}$$

$$= \frac{-(-6) \pm \sqrt{(-6)^2 - 4(9)(5)}}{2(9)} \qquad a = 9, b = -6, c = 5$$

$$= \frac{6 \pm \sqrt{-144}}{18}$$

$$= \frac{6 \pm 12i}{18} \qquad \sqrt{-144} = 12i$$

$$x = \frac{6(1 \pm 2i)}{6 \cdot 3} = \frac{1 \pm 2i}{3} \qquad \text{Factor; write in lowest terms.}$$

The solution set is $\left\{ \frac{1}{3} \pm \frac{2}{3}i \right\}$.

Now try Exercise 29.

Operations on Complex Numbers Products or quotients with negative radicands are simplified by first rewriting $\sqrt{-a}$ as $i\sqrt{a}$ for a positive number a. Then the properties of real numbers are applied, together with the fact that $i^2 = -1$.

CAUTION When working with negative radicands, *use the definition* $\sqrt{-a} = i\sqrt{a}$ *before using any of the other rules for radicals.* In particular, the rule $\sqrt{c} \cdot \sqrt{d} = \sqrt{cd}$ is valid only when c and d are *not* both negative. For example,

$$\sqrt{(-4)(-9)} = \sqrt{36} = 6,$$

while
$$\sqrt{-4} \cdot \sqrt{-9} = 2i(3i) = 6i^2 = -6,$$

so
$$\sqrt{-4} \cdot \sqrt{-9} \neq \sqrt{(-4)(-9)}.$$

EXAMPLE 4 Finding Products and Quotients Involving Negative Radicands

Multiply or divide as indicated. Simplify each answer.

(a) $\sqrt{-7} \cdot \sqrt{-7}$ **(b)** $\sqrt{-6} \cdot \sqrt{-10}$ **(c)** $\dfrac{\sqrt{-20}}{\sqrt{-2}}$ **(d)** $\dfrac{\sqrt{-48}}{\sqrt{24}}$

Solution

(a) $\sqrt{-7} \cdot \sqrt{-7} = i\sqrt{7} \cdot i\sqrt{7}$

$\qquad = i^2 \cdot (\sqrt{7})^2$

$\qquad = -1 \cdot 7$

$\qquad\qquad\qquad i^2 = -1$

$\qquad = -7$

(b) $\sqrt{-6} \cdot \sqrt{-10} = i\sqrt{6} \cdot i\sqrt{10}$

$\qquad = i^2 \cdot \sqrt{60}$

$\qquad = -1\sqrt{4 \cdot 15}$

$\qquad = -1 \cdot 2\sqrt{15}$

$\qquad = -2\sqrt{15}$

(c) $\dfrac{\sqrt{-20}}{\sqrt{-2}} = \dfrac{i\sqrt{20}}{i\sqrt{2}} = \sqrt{\dfrac{20}{2}} = \sqrt{10}$ Quotient rule for radicals; $\frac{\sqrt[n]{a}}{\sqrt[n]{b}} = \sqrt[n]{\frac{a}{b}}$

(d) $\dfrac{\sqrt{-48}}{\sqrt{24}} = \dfrac{i\sqrt{48}}{\sqrt{24}} = i\sqrt{\dfrac{48}{24}} = i\sqrt{2}$

Now try Exercises 37, 39, 41, and 43.

The operations of addition and subtraction of complex numbers are defined in a manner similar to those for binomials.

Addition and Subtraction of Complex Numbers

For complex numbers $a + bi$ and $c + di$,

$$(a + bi) + (c + di) = (a + c) + (b + d)i$$

and $\qquad (a + bi) - (c + di) = (a - c) + (b - d)i.$

That is, to add or subtract complex numbers, add or subtract the real parts and add or subtract the imaginary parts.

EXAMPLE 5 Adding and Subtracting Complex Numbers

Find each sum or difference.

(a) $(3 - 4i) + (-2 + 6i)$ **(b)** $(-9 + 7i) + (3 - 15i)$

(c) $(-4 + 3i) - (6 - 7i)$ **(d)** $(12 - 5i) - (8 - 3i)$

Solution

(a) $(3 - 4i) + (-2 + 6i) = [3 + (-2)] + [-4 + 6]i$

Commutative, associative, distributive properties

$$= 1 + 2i$$

(b) $(-9 + 7i) + (3 - 15i) = -6 - 8i$

(c) $(-4 + 3i) - (6 - 7i) = (-4 - 6) + [3 - (-7)]i$

$$= -10 + 10i$$

(d) $(12 - 5i) - (8 - 3i) = 4 - 2i$

Now try Exercises 49 and 51.

The product of two complex numbers is found by multiplying as if the numbers were binomials and using the fact that $i^2 = -1$, as follows.

$(a + bi)(c + di) = ac + adi + bic + bidi$ FOIL (multiply First, Outer, Inner, Last terms)

$$= ac + adi + bci + bdi^2$$

$$= ac + (ad + bc)i + bd(-1) \quad \text{Distributive property; } i^2 = -1$$

$$= (ac - bd) + (ad + bc)i$$

> **Multiplication of Complex Numbers**
>
> For complex numbers $a + bi$ and $c + di$,
>
> $$(a + bi)(c + di) = (ac - bd) + (ad + bc)i.$$

This definition is not practical in routine calculations. To find a given product, it is easier just to multiply as with binomials.

EXAMPLE 6   Multiplying Complex Numbers

Find each product.

(a) $(2 - 3i)(3 + 4i)$ **(b)** $(4 + 3i)^2$

(c) $(2 + i)(-2 - i)$ **(d)** $(6 + 5i)(6 - 5i)$

Solution

```
(2-3i)(3+4i)
            18-i
(4+3i)²
           7+24i
(6+5i)(6-5i)
              61
```

This screen shows how the
TI–83 Plus displays the results
found in Example 6 (a), (b), and (d).

(a) $(2 - 3i)(3 + 4i) = 2(3) + 2(4i) - 3i(3) - 3i(4i)$ FOIL

$\qquad\qquad\qquad\qquad = 6 + 8i - 9i - 12i^2$

$\qquad\qquad\qquad\qquad = 6 - i - 12(-1)$ $i^2 = -1$

$\qquad\qquad\qquad\qquad = 18 - i$

(b) $(4 + 3i)^2 = 4^2 + 2(4)(3i) + (3i)^2$ Square of a binomial;
$\qquad\qquad\qquad\qquad\qquad\qquad\qquad\qquad (x + y)^2 = x^2 + 2xy + y^2$

$\qquad\qquad\quad = 16 + 24i + 9i^2$

$\qquad\qquad\quad = 16 + 24i + 9(-1)$ $i^2 = -1$

$\qquad\qquad\quad = 7 + 24i$

(c) $(2 + i)(-2 - i) = -4 - 2i - 2i - i^2$ FOIL

$\qquad\qquad\qquad\qquad = -4 - 4i - (-1)$ Combine terms; $i^2 = -1$

$\qquad\qquad\qquad\qquad = -4 - 4i + 1$

$\qquad\qquad\qquad\qquad = -3 - 4i$ Standard form

(d) $(6 + 5i)(6 - 5i) = 6^2 - (5i)^2$ Product of the sum and difference of two
$\qquad\qquad\qquad\qquad\qquad\qquad\qquad$ terms; $(x + y)(x - y) = x^2 - y^2$

$\qquad\qquad\qquad\qquad\quad = 36 - 25(-1)$ $i^2 = -1$

$\qquad\qquad\qquad\qquad\quad = 36 + 25$

$\qquad\qquad\qquad\qquad\quad = 61$ or $61 + 0i$ Standard form

---

Now try Exercises 57, 61, 63, and 65.

Powers of i can be simplified using the facts

$$i^2 = -1 \qquad \text{and} \qquad i^4 = (i^2)^2 = (-1)^2 = 1.$$

EXAMPLE 7 Simplifying Powers of i

Simplify each power of i.

(a) i^{15} **(b)** i^{-3}

Solution

(a) Since $i^2 = -1$ and $i^4 = 1$, write the given power as a product involving i^2 or i^4. For example,

$$i^3 = i^2 \cdot i = (-1) \cdot i = -i.$$

To simplify i^{15}, use i^4 and i^3 to rewrite i^{15}, giving

$$i^{15} = i^{12} \cdot i^3 = (i^4)^3 \cdot i^3 = 1^3(-i) = -i.$$

(b) $i^{-3} = i^{-4} \cdot i = (i^4)^{-1} \cdot i = (1)^{-1} \cdot i = i$

Now try Exercises 75 and 83.

We can use the method of Example 7 to construct the following list of powers of i.

Powers of i on the TI–83 Plus calculator

Powers of i

$i^1 = i$	$i^5 = i$	$i^9 = i$
$i^2 = -1$	$i^6 = -1$	$i^{10} = -1$
$i^3 = -i$	$i^7 = -i$	$i^{11} = -i$
$i^4 = 1$	$i^8 = 1$	$i^{12} = 1,$ and so on.

Example 6(d) showed that $(6 + 5i)(6 - 5i) = 61$. The numbers $6 + 5i$ and $6 - 5i$ differ only in the sign of their imaginary parts, and are called **complex conjugates.** The product of a complex number and its conjugate is always a real number.

Property of Complex Conjugates

For real numbers a and b,

$$(a + bi)(a - bi) = a^2 + b^2.$$

The complex conjugate of the divisor is used to find the quotient of two complex numbers in standard form. Multiplying both the numerator and the denominator by the complex conjugate of the denominator gives the quotient, as shown in the next example.

EXAMPLE 8   Dividing Complex Numbers

Find each quotient.

(a) $\dfrac{3 + 2i}{5 - i}$

(b) $\dfrac{3}{i}$

Solution

(a) $\dfrac{3 + 2i}{5 - i} = \dfrac{(3 + 2i)(5 + i)}{(5 - i)(5 + i)}$ Multiply by the complex conjugate of the denominator in both the numerator and the denominator.

$= \dfrac{15 + 3i + 10i + 2i^2}{25 - i^2}$ Multiply.

$= \dfrac{13 + 13i}{26}$ $i^2 = -1$

$= \dfrac{13}{26} + \dfrac{13i}{26}$ $\dfrac{a + bi}{c} = \dfrac{a}{c} + \dfrac{bi}{c}$

$= \dfrac{1}{2} + \dfrac{1}{2}i$ Lowest terms; standard form

To *check* this answer, show that

$$\left(\dfrac{1}{2} + \dfrac{1}{2}i\right)(5 - i) = 3 + 2i. \quad \text{Quotient} \times \text{Divisor} = \text{Dividend}$$

(b) $\dfrac{3}{i} = \dfrac{3(-i)}{i(-i)}$ $-i$ is the conjugate of i.

$= \dfrac{-3i}{-i^2}$

$= \dfrac{-3i}{1}$ $-i^2 = -(-1) = 1$

$= -3i \quad \text{or} \quad 0 - 3i$ Standard form

This screen supports the results in Example 8. Notice that the answer to part (a) must be interpreted $\frac{1}{2} + \frac{1}{2}i$, *not* $\frac{1}{2} + \frac{1}{2i}$.

> Now try Exercises 89 and 95.

8.1 Exercises

Concept Check   Determine whether each statement is true *or* false. *If it is false, tell why.*

1. Every real number is a complex number.

2. No real number is a pure imaginary number.

3. Every pure imaginary number is a complex number.

4. A number can be both real and complex.

5. There is no real number that is a complex number.

6. A complex number might not be a pure imaginary number.

Identify each number as real, complex, *or* pure imaginary. *(More than one of these descriptions may apply.)*

7. -6 **8.** 0 **9.** $10i$ **10.** $-8i$ **11.** $2 + i$

12. $-5 - 2i$ **13.** π **14.** $\sqrt{8}$ **15.** $\sqrt{-9}$ **16.** $\sqrt{-16}$

Write each number as the product of a real number and i. See Example 1.

17. $\sqrt{-25}$ **18.** $\sqrt{-36}$ **19.** $\sqrt{-10}$ **20.** $\sqrt{-15}$

21. $\sqrt{-288}$ **22.** $\sqrt{-500}$ **23.** $-\sqrt{-18}$ **24.** $-\sqrt{-80}$

Solve each quadratic equation and express all nonreal complex solutions in terms of i. See Examples 2 and 3.

25. $x^2 = -16$ **26.** $x^2 = -36$ **27.** $x^2 + 12 = 0$

28. $x^2 + 48 = 0$ **29.** $3x^2 + 2 = -4x$ **30.** $2x^2 + 3x = -2$

31. $x^2 - 6x + 14 = 0$ **32.** $x^2 + 4x + 11 = 0$ **33.** $4(x^2 - x) = -7$

34. $3(3x^2 - 2x) = -7$ **35.** $x^2 + 1 = -x$ **36.** $x^2 + 2 = 2x$

Multiply or divide as indicated. Simplify each answer. See Example 4.

37. $\sqrt{-13} \cdot \sqrt{-13}$ **38.** $\sqrt{-17} \cdot \sqrt{-17}$ **39.** $\sqrt{-3} \cdot \sqrt{-8}$

40. $\sqrt{-5} \cdot \sqrt{-15}$ **41.** $\dfrac{\sqrt{-30}}{\sqrt{-10}}$ **42.** $\dfrac{\sqrt{-70}}{\sqrt{-7}}$

43. $\dfrac{\sqrt{-24}}{\sqrt{8}}$ **44.** $\dfrac{\sqrt{-54}}{\sqrt{27}}$ **45.** $\dfrac{\sqrt{-10}}{\sqrt{-40}}$

46. $\dfrac{\sqrt{-40}}{\sqrt{20}}$ **47.** $\dfrac{\sqrt{-6} \cdot \sqrt{-2}}{\sqrt{3}}$ **48.** $\dfrac{\sqrt{-12} \cdot \sqrt{-6}}{\sqrt{8}}$

Find each sum or difference. Write the answer in standard form. See Example 5.

49. $(3 + 2i) + (4 - 3i)$ **50.** $(4 - i) + (2 + 5i)$

51. $(-2 + 3i) - (-4 + 3i)$ **52.** $(-3 + 5i) - (-4 + 5i)$

53. $(2 - 5i) - (3 + 4i) - (-2 + i)$ **54.** $(-4 - i) - (2 + 3i) + (-4 + 5i)$

55. $-i - 2 - (6 - 4i) - (5 - 2i)$ **56.** $3 - (4 - i) - 4i + (-2 + 5i)$

Find each product. Write the answer in standard form. See Example 6.

57. $(2 + i)(3 - 2i)$ **58.** $(-2 + 3i)(4 - 2i)$ **59.** $(2 + 4i)(-1 + 3i)$

60. $(1 + 3i)(2 - 5i)$ **61.** $(-3 + 2i)^2$ **62.** $(2 + i)^2$

63. $(3 + i)(-3 - i)$ **64.** $(-5 - i)(5 + i)$ **65.** $(2 + 3i)(2 - 3i)$

66. $(6 - 4i)(6 + 4i)$ **67.** $\left(\sqrt{6} + i\right)\left(\sqrt{6} - i\right)$ **68.** $\left(\sqrt{2} - 4i\right)\left(\sqrt{2} + 4i\right)$

69. $i(3 - 4i)(3 + 4i)$ **70.** $i(2 + 7i)(2 - 7i)$ **71.** $3i(2 - i)^2$

72. $-5i(4 - 3i)^2$ **73.** $(2 + i)(2 - i)(4 + 3i)$ **74.** $(3 - i)(3 + i)(2 - 6i)$

Simplify each power of i. See Example 7.

75. i^{21} **76.** i^{25} **77.** i^{22} **78.** i^{26}

79. i^{23} **80.** i^{27} **81.** i^{24} **82.** i^{32}

83. i^{-9} **84.** i^{-10} **85.** $\dfrac{1}{i^{-11}}$ **86.** $\dfrac{1}{i^{12}}$

87. Suppose that your friend, Ceci Fleming, tells you that she has discovered a method of simplifying a positive power of i. "Just divide the exponent by 4," she says, "and then look at the remainder. Then refer to the table of powers of i in this section. The large power of i is equal to the power indicated by the remainder. And if the remainder is 0, the result is $i^0 = 1$." Explain why her method works.

88. Explain why the following method of simplifying i^{-46} works.

$$i^{-46} = i^{-46} \cdot i^{48} = i^{-46+48} = i^2 = -1$$

Find each quotient. Write the answer in standard form. See Example 8.

89. $\dfrac{6 + 2i}{1 + 2i}$

90. $\dfrac{14 + 5i}{3 + 2i}$

91. $\dfrac{2 - i}{2 + i}$

92. $\dfrac{4 - 3i}{4 + 3i}$

93. $\dfrac{1 - 3i}{1 + i}$

94. $\dfrac{-3 + 4i}{2 - i}$

95. $\dfrac{5}{i}$

96. $\dfrac{6}{i}$

97. $\dfrac{-8}{-i}$

98. $\dfrac{-12}{-i}$

99. $\dfrac{2}{3i}$

100. $\dfrac{5}{9i}$

101. Show that $\dfrac{\sqrt{2}}{2} + \dfrac{\sqrt{2}}{2}i$ is a square root of i.

102. Show that $\dfrac{\sqrt{3}}{2} + \dfrac{1}{2}i$ is a cube root of i.

(Modeling) Alternating Current   Complex numbers are used to describe current, I, voltage, E, and impedance, Z (the opposition to current). These three quantities are related by the equation E = IZ. Thus, if any two of these quantities are known, the third can be found. In each problem, solve the equation E = IZ for the missing variable.

103. $I = 8 + 6i$, $Z = 6 + 3i$

104. $I = 10 + 6i$, $Z = 8 + 5i$

105. $I = 7 + 5i$, $E = 28 + 54i$

106. $E = 35 + 55i$, $Z = 6 + 4i$

(Modeling) Impedance   Impedance is a measure of the opposition to the flow of alternating electrical current found in common electrical outlets. It consists of two parts called resistance and reactance. Resistance occurs when a lightbulb is turned on, while reactance is produced when electricity passes through a coil of wire like that found in electric motors. Impedance Z in ohms (Ω) can be expressed as a complex number, where the real part represents resistance and the imaginary part represents reactance.

For example, if the resistive part is 3 ohms and the reactive part is 4 ohms, then the impedance could be described by the complex number Z = 3 + 4i. In the series circuit shown in the figure, the total impedance will be the sum of the individual impedances. (Source: Wilcox, G. and C. Hesselberth, Electricity for Engineering Technology, Allyn & Bacon, 1970.)

107. The circuit contains two lightbulbs and two electric motors. Assuming that the lightbulbs are pure resistive and the motors are pure reactive, find the total impedance in this circuit and express it in the form $Z = a + bi$.

108. The phase angle θ measures the phase difference between the voltage and the current in an electrical circuit. θ (in degrees) can be determined by the equation $\tan \theta = \frac{b}{a}$. Find θ for this circuit.

8.2 | Trigonometric (Polar) Form of Complex Numbers

The Complex Plane and Vector Representation ▪ Trigonometric (Polar) Form ▪ Converting Between Trigonometric and Polar Forms ▪ An Application of Complex Numbers to Fractals

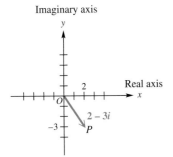

Figure 3

The Complex Plane and Vector Representation Unlike real numbers, complex numbers cannot be ordered. One way to organize and illustrate them is by using a graph. To graph a complex number such as $2 - 3i$, we modify the familiar coordinate system by calling the horizontal axis the **real axis** and the vertical axis the **imaginary axis.** Then complex numbers can be graphed in this **complex plane,** as shown in Figure 3. Each complex number $a + bi$ determines a unique position vector with initial point $(0, 0)$ and terminal point (a, b). This shows the close connection between vectors and complex numbers.

N O T E This geometric representation is the reason that $a + bi$ is called the **rectangular form** of a complex number. (*Rectangular form* is also called *standard form.*)

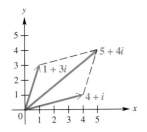

Figure 4

Recall that the sum of the two complex numbers $4 + i$ and $1 + 3i$ is

$$(4 + i) + (1 + 3i) = 5 + 4i. \quad \text{(Section 8.1)}$$

Graphically, the sum of two complex numbers is represented by the vector that is the resultant of the vectors corresponding to the two numbers, as shown in Figure 4.

EXAMPLE 1 Expressing the Sum of Complex Numbers Graphically

Find the sum of $6 - 2i$ and $-4 - 3i$. Graph both complex numbers and their resultant.

Solution The sum is found by adding the two numbers.

$$(6 - 2i) + (-4 - 3i) = 2 - 5i$$

The graphs are shown in Figure 5.

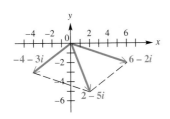

Figure 5

Now try Exercises 3 and 13.

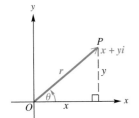

Figure 6

Trigonometric (Polar) Form Figure 6 shows the complex number $x + yi$ that corresponds to a vector **OP** with direction angle θ and magnitude r. The following relationships among x, y, r, and θ can be verified from Figure 6.

Relationships Among x, y, r, and θ
$x = r \cos \theta \qquad y = r \sin \theta$
$r = \sqrt{x^2 + y^2} \qquad \tan \theta = \dfrac{y}{x}, \quad \text{if } x \neq 0$

Substituting $x = r \cos \theta$ and $y = r \sin \theta$ from these relationships into $x + yi$ gives

$$x + yi = r \cos \theta + (r \sin \theta)i$$
$$= r(\cos \theta + i \sin \theta). \quad \text{Factor out } r.$$

Trigonometric (Polar) Form of a Complex Number

The expression

$$r(\cos \theta + i \sin \theta)$$

is called the **trigonometric form** (or **polar form**) of the complex number $x + yi$. The expression $\cos \theta + i \sin \theta$ is sometimes abbreviated cis θ. Using this notation,

$$r(\cos \theta + i \sin \theta) \text{ is written } r \text{ cis } \theta.$$

The number r is the **absolute value** (or **modulus**) of $x + yi$, and θ is the **argument** of $x + yi$. In this section we choose the value of θ in the interval $[0°, 360°)$. However, any angle coterminal with θ also could serve as the argument.

EXAMPLE 2 Converting from Trigonometric Form to Rectangular Form

Express $2(\cos 300° + i \sin 300°)$ in rectangular form.

Algebraic Solution

From Chapter 2, $\cos 300° = \frac{1}{2}$ and $\sin 300° = -\frac{\sqrt{3}}{2}$, so

$$2(\cos 300° + i \sin 300°) = 2\left(\frac{1}{2} - i\frac{\sqrt{3}}{2}\right)$$
$$= 1 - i\sqrt{3}.$$

Notice that the real part is positive and the imaginary part is negative; this is consistent with $300°$ being a quadrant IV angle.

Graphing Calculator Solution

Figure 7 confirms the algebraic solution.

The imaginary part is an approximation for $-\sqrt{3}$.

Figure 7

Now try Exercise 27.

Converting Between Trigonometric and Polar Forms

To convert from rectangular form to trigonometric form, we use the following procedure.

Converting from Rectangular Form to Trigonometric Form

Step 1 Sketch a graph of the number $x + yi$ in the complex plane.

Step 2 Find r by using the equation $r = \sqrt{x^2 + y^2}$.

Step 3 Find θ by using the equation $\tan \theta = \frac{y}{x}$, $x \neq 0$, choosing the quadrant indicated in Step 1.

CAUTION   Errors often occur in Step 3. Be sure to choose the correct quadrant for θ by referring to the graph sketched in Step 1.

EXAMPLE 3   Converting from Rectangular Form to Trigonometric Form

Write each complex number in trigonometric form.

(a) $-\sqrt{3} + i$ **(b)** $-3i$

Solution

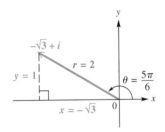

Figure 8

(a) We start by sketching the graph of $-\sqrt{3} + i$ in the complex plane, as shown in Figure 8. Since $x = -\sqrt{3}$ and $y = 1$,

$$r = \sqrt{x^2 + y^2} = \sqrt{\left(-\sqrt{3}\right)^2 + 1^2} = \sqrt{3 + 1} = 2,$$

and $\tan \theta = \dfrac{y}{x} = \dfrac{1}{-\sqrt{3}} = -\dfrac{1}{\sqrt{3}} \cdot \dfrac{\sqrt{3}}{\sqrt{3}} = -\dfrac{\sqrt{3}}{3}.$

Rationalize the denominator.

Since $\tan \theta = -\dfrac{\sqrt{3}}{3}$, the reference angle for θ in radians is $\dfrac{\pi}{6}$. From the graph, we see that θ is in quadrant II, so $\theta = \pi - \dfrac{\pi}{6} = \dfrac{5\pi}{6}$. Therefore,

$$-\sqrt{3} + i = 2\left(\cos \frac{5\pi}{6} + i \sin \frac{5\pi}{6}\right) = 2 \operatorname{cis} \frac{5\pi}{6}.$$

(b) The sketch of $-3i$ is shown in Figure 9.

Figure 9

Choices 5 and 6 in the top screen show how to convert from rectangular (x, y) form to trigonometric form. The calculator is in radian mode. The results agree with our algebraic results in Example 3(a).

Since $-3i = 0 - 3i$, we have $x = 0$ and $y = -3$ and
$$r = \sqrt{0^2 + (-3)^2} = \sqrt{0 + 9} = \sqrt{9} = 3.$$

We cannot find θ by using $\tan \theta = \dfrac{y}{x}$, because $x = 0$. From the graph, a value for θ is 270°. In trigonometric form,

$$-3i = 3(\cos 270° + i \sin 270°) = 3 \operatorname{cis} 270°.$$

Now try Exercises 41 and 47.

NOTE   In Example 3, we gave answers in both forms: $r(\cos \theta + i \sin \theta)$ and $r \operatorname{cis} \theta$. These forms will be used interchangeably from now on.

EXAMPLE 4 Converting Between Trigonometric and Rectangular Forms Using Calculator Approximations

Write each complex number in its alternative form, using calculator approximations as necessary.

(a) $6(\cos 115° + i \sin 115°)$ **(b)** $5 - 4i$

Solution

(a) Since 115° does not have a special angle as a reference angle, we cannot find exact values for cos 115° and sin 115°. Use a calculator set in degree mode to find $\cos 115° \approx -.4226182617$ and $\sin 115° \approx .906307787$. Therefore, in rectangular form,

$$6(\cos 115° + i \sin 115°) \approx 6(-.4226182617 + .906307787i)$$

$$\approx -2.5357 + 5.4378i.$$

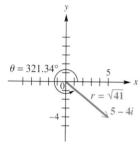

$\theta = 321.34°$

$r = \sqrt{41}$

$5 - 4i$

Figure 10

(b) A sketch of $5 - 4i$ shows that θ must be in quadrant IV. See Figure 10. Here $r = \sqrt{5^2 + (-4)^2} = \sqrt{41}$ and $\tan \theta = -\frac{4}{5}$. Use a calculator to find that one measure of θ is $-38.66°$. In order to express θ in the interval $[0, 360°)$, we find $\theta = 360° - 38.66° = 321.34°$. Use these results to get

$$5 - 4i = \sqrt{41} \text{ cis } 321.34°.$$

Now try Exercises 53 and 57.

Source: Figure from Crownover, R., *Introduction to Fractals and Chaos.* Copyright © 1995. Boston: Jones and Bartlett Publishers. Reprinted with permission.

Figure 11

An Application of Complex Numbers to Fractals   We can apply complex numbers to the study of *fractals*. An example of a fractal is the *Mandelbrot set* shown in Figure 11.

EXAMPLE 5 Deciding Whether a Complex Number Is in the Julia Set

The fractal called the *Julia set* is shown in Figure 12. To determine if a complex number $z = a + bi$ is in this Julia set, perform the following sequence of calculations. Repeatedly compute the values of $z^2 - 1$, $(z^2 - 1)^2 - 1$, $[(z^2 - 1)^2 - 1]^2 - 1, \ldots$. If the absolute values of any of the resulting complex numbers exceed 2, then the complex number z is not in the Julia set. Otherwise z is part of this set and the point (a, b) should be shaded in the graph.

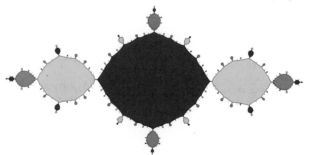

Source: Figure from Crownover, R., *Introduction to Fractals and Chaos.* Copyright © 1995. Boston: Jones and Bartlett Publishers. Reprinted with permission.

Figure 12

Determine whether each number belongs to the Julia set.

(a) $z = 0 + 0i$ **(b)** $z = 1 + 1i$

66. Use vectors to show that $-z = r[\cos(\theta + \pi) + i\sin(\theta + \pi)]$.

Concept Check   *In Exercises 67–69, identify the geometric condition (A, B, or C) that implies the situation.*

    A. *The corresponding vectors have opposite directions.*

    B. *The terminal points of the vectors corresponding to $a + bi$ and $c + di$ lie on a horizontal line.*

    C. *The corresponding vectors have the same direction.*

67. The difference between two nonreal complex numbers $a + bi$ and $c + di$ is a real number.

68. The absolute value of the sum of two complex numbers $a + bi$ and $c + di$ is equal to the sum of their absolute values.

69. The absolute value of the difference of two complex numbers $a + bi$ and $c + di$ is equal to the sum of their absolute values.

70. Show that z and iz have the same absolute value. How are the graphs of these two numbers related?

8.3 | The Product and Quotient Theorems

Products of Complex Numbers in Trigonometric Form   ▪   Quotients of Complex Numbers in Trigonometric Form

Products of Complex Numbers in Trigonometric Form   Using the FOIL method to multiply complex numbers in rectangular form, we find the product of $1 + i\sqrt{3}$ and $-2\sqrt{3} + 2i$ as follows.

$$\left(1 + i\sqrt{3}\right)\left(-2\sqrt{3} + 2i\right) = -2\sqrt{3} + 2i - 2i(3) + 2i^2\sqrt{3} \quad \text{(Section 8.1)}$$
$$= -2\sqrt{3} + 2i - 6i - 2\sqrt{3}$$
$$= -4\sqrt{3} - 4i$$

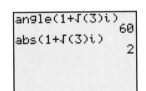

With the calculator in complex and degree modes, the MATH menu can be used to find the angle and the magnitude (absolute value) of the vector that corresponds to a given complex number.

We can also find this same product by first converting the complex numbers $1 + i\sqrt{3}$ and $-2\sqrt{3} + 2i$ to trigonometric form. Using the method explained in the preceding section,

$$1 + i\sqrt{3} = 2(\cos 60° + i\sin 60°) \quad \text{(Section 8.2)}$$

and $$-2\sqrt{3} + 2i = 4(\cos 150° + i\sin 150°).$$

    If we multiply the trigonometric forms, and if we use identities for the cosine and the sine of the sum of two angles, then the result is

$$[2(\cos 60° + i\sin 60°)][4(\cos 150° + i\sin 150°)]$$
$$= 2 \cdot 4(\cos 60° \cdot \cos 150° + i\sin 60° \cdot \cos 150°$$
$$+ i\cos 60° \cdot \sin 150° + i^2\sin 60° \cdot \sin 150°) \quad \text{(Section 5.3)}$$
$$= 8[(\cos 60° \cdot \cos 150° - \sin 60° \cdot \sin 150°)$$
$$+ i(\sin 60° \cdot \cos 150° + \cos 60° \cdot \sin 150°)]$$
$$= 8[\cos(60° + 150°) + i\sin(60° + 150°)]$$
$$= 8(\cos 210° + i\sin 210°).$$

The absolute value of the product, 8, is equal to the product of the absolute values of the factors, 2 · 4, and the argument of the product, 210°, is equal to the sum of the arguments of the factors, 60° + 150°.

As we would expect, the product obtained when multiplying by the first method is the rectangular form of the product obtained when multiplying by the second method.

$$8(\cos 210° + i \sin 210°) = 8\left(-\frac{\sqrt{3}}{2} - \frac{1}{2}i\right)$$

$$= -4\sqrt{3} - 4i$$

We can generalize this work in the following *product theorem*.

Product Theorem

If $r_1(\cos \theta_1 + i \sin \theta_1)$ and $r_2(\cos \theta_2 + i \sin \theta_2)$ are any two complex numbers, then

$$[r_1(\cos \theta_1 + i \sin \theta_1)] \cdot [r_2(\cos \theta_2 + i \sin \theta_2)]$$
$$= r_1 r_2 [\cos(\theta_1 + \theta_2) + i \sin(\theta_1 + \theta_2)].$$

In compact form, this is written

$$(r_1 \operatorname{cis} \theta_1)(r_2 \operatorname{cis} \theta_2) = r_1 r_2 \operatorname{cis}(\theta_1 + \theta_2).$$

That is, to multiply complex numbers in trigonometric form, multiply their absolute values and add their arguments.

EXAMPLE 1 Using the Product Theorem

Find the product of $3(\cos 45° + i \sin 45°)$ and $2(\cos 135° + i \sin 135°)$.

Solution

$$[3(\cos 45° + i \sin 45°)][2(\cos 135° + i \sin 135°)]$$
$$= 3 \cdot 2[\cos(45° + 135°) + i \sin(45° + 135°)] \quad \text{Product theorem}$$
$$= 6(\cos 180° + i \sin 180°) \quad \text{Multiply and add.}$$
$$= 6(-1 + i \cdot 0) = 6(-1) = -6$$

Now try Exercise 5.

Quotients of Complex Numbers in Trigonometric Form
The rectangular form of the quotient of the complex numbers $1 + i\sqrt{3}$ and $-2\sqrt{3} + 2i$ is

$$\frac{1 + i\sqrt{3}}{-2\sqrt{3} + 2i} = \frac{\left(1 + i\sqrt{3}\right)\left(-2\sqrt{3} - 2i\right)}{\left(-2\sqrt{3} + 2i\right)\left(-2\sqrt{3} - 2i\right)} \quad \begin{array}{l}\text{Multiply by the conjugate of the}\\ \text{denominator. (Section 8.1)}\end{array}$$

$$= \frac{-2\sqrt{3} - 2i - 6i - 2i^2\sqrt{3}}{12 - 4i^2} \quad \begin{array}{l}\text{FOIL;}\\ (x + y)(x - y) = x^2 - y^2\end{array}$$

$$= \frac{-8i}{16} = -\frac{1}{2}i. \quad \text{Simplify.}$$

Writing $1 + i\sqrt{3}$, $-2\sqrt{3} + 2i$, and $-\frac{1}{2}i$ in trigonometric form gives

$$1 + i\sqrt{3} = 2(\cos 60° + i \sin 60°),$$
$$-2\sqrt{3} + 2i = 4(\cos 150° + i \sin 150°),$$

and
$$-\frac{1}{2}i = \frac{1}{2}[\cos(-90°) + i \sin(-90°)].$$

The absolute value of the quotient, $\frac{1}{2}$, is the quotient of the two absolute values, $\frac{2}{4} = \frac{1}{2}$. The argument of the quotient, $-90°$, is the difference of the two arguments, $60° - 150° = -90°$. Generalizing from this example leads to the *quotient theorem.*

Quotient Theorem

If $r_1(\cos \theta_1 + i \sin \theta_1)$ and $r_2(\cos \theta_2 + i \sin \theta_2)$ are any two complex numbers, where $r_2(\cos \theta_2 + i \sin \theta_2) \neq 0$, then

$$\frac{r_1(\cos \theta_1 + i \sin \theta_1)}{r_2(\cos \theta_2 + i \sin \theta_2)} = \frac{r_1}{r_2}[\cos(\theta_1 - \theta_2) + i \sin(\theta_1 - \theta_2)].$$

In compact form, this is written

$$\frac{r_1 \text{ cis } \theta_1}{r_2 \text{ cis } \theta_2} = \frac{r_1}{r_2} \text{cis}(\theta_1 - \theta_2).$$

That is, to divide complex numbers in trigonometric form, divide their absolute values and subtract their arguments.

EXAMPLE 2 Using the Quotient Theorem

Find the quotient $\dfrac{10 \text{ cis}(-60°)}{5 \text{ cis } 150°}$. Write the result in rectangular form.

Solution

$$\frac{10 \text{ cis}(-60°)}{5 \text{ cis } 150°} = \frac{10}{5} \text{cis}(-60° - 150°) \qquad \text{Quotient theorem}$$

$$= 2 \text{ cis}(-210°) \qquad \text{Divide and subtract.}$$

$$= 2[\cos(-210°) + i \sin(-210°)] \qquad \text{Rewrite.}$$

$$= 2\left[-\frac{\sqrt{3}}{2} + i\left(\frac{1}{2}\right)\right] \qquad \begin{array}{l} \cos(-210°) = -\frac{\sqrt{3}}{2}; \\ \sin(-210°) = \frac{1}{2} \text{ (Section 2.2)} \end{array}$$

$$= -\sqrt{3} + i \qquad \text{Rectangular form}$$

Now try Exercise 17.

EXAMPLE 3 Using the Product and Quotient Theorems with a Calculator

Use a calculator to find the following. Write the results in rectangular form.

(a) $(9.3 \text{ cis } 125.2°)(2.7 \text{ cis } 49.8°)$

(b) $\dfrac{10.42\left(\cos \frac{3\pi}{4} + i \sin \frac{3\pi}{4}\right)}{5.21\left(\cos \frac{\pi}{5} + i \sin \frac{\pi}{5}\right)}$

Solution

(a) $(9.3 \text{ cis } 125.2°)(2.7 \text{ cis } 49.8°) = 9.3(2.7) \text{ cis}(125.2° + 49.8°)$

Product theorem

$$= 25.11 \text{ cis } 175°$$
$$= 25.11(\cos 175° + i \sin 175°)$$
$$\approx 25.11[-.99619470 + i(.08715574)]$$
$$\approx -25.0144 + 2.1885i$$

(b) $\dfrac{10.42\left(\cos \frac{3\pi}{4} + i \sin \frac{3\pi}{4}\right)}{5.21\left(\cos \frac{\pi}{5} + i \sin \frac{\pi}{5}\right)} = \dfrac{10.42}{5.21}\left[\cos\left(\dfrac{3\pi}{4} - \dfrac{\pi}{5}\right) + i \sin\left(\dfrac{3\pi}{4} - \dfrac{\pi}{5}\right)\right]$

Quotient theorem

$$= 2\left(\cos \dfrac{11\pi}{20} + i \sin \dfrac{11\pi}{20}\right)$$
$$\approx -.3129 + 1.9754$$

Now try Exercises 25 and 29.

8.3 Exercises

Concept Check   *Fill in the blanks with the correct responses.*

1. When multiplying two complex numbers in trigonometric form, we _____ their absolute values and _____ their arguments.

2. When dividing two complex numbers in trigonometric form, we _____ their absolute values and _____ their arguments.

Find each product and write it in rectangular form. See Example 1.

3. $[3(\cos 60° + i \sin 60°)][2(\cos 90° + i \sin 90°)]$

4. $[4(\cos 30° + i \sin 30°)][5(\cos 120° + i \sin 120°)]$

5. $[2(\cos 45° + i \sin 45°)][2(\cos 225° + i \sin 225°)]$

6. $[8(\cos 300° + i \sin 300°)][5(\cos 120° + i \sin 120°)]$

7. $[4(\cos 60° + i \sin 60°)][6(\cos 330° + i \sin 330°)]$

8. $[8(\cos 210° + i \sin 210°)][2(\cos 330° + i \sin 330°)]$

9. $(5 \text{ cis } 90°)(3 \text{ cis } 45°)$ 10. $(6 \text{ cis } 120°)[5 \text{ cis}(-30°)]$

11. $\left(\sqrt{3} \text{ cis } 45°\right)\left(\sqrt{3} \text{ cis } 225°\right)$ 12. $\left(\sqrt{2} \text{ cis } 300°\right)\left(\sqrt{2} \text{ cis } 270°\right)$

Find each quotient and write it in rectangular form. In Exercises 19–24, first convert the numerator and the denominator to trigonometric form. See Example 2.

13. $\dfrac{4(\cos 120° + i \sin 120°)}{2(\cos 150° + i \sin 150°)}$ 14. $\dfrac{24(\cos 150° + i \sin 150°)}{2(\cos 30° + i \sin 30°)}$

15. $\dfrac{10(\cos 225° + i \sin 225°)}{5(\cos 45° + i \sin 45°)}$ 16. $\dfrac{16(\cos 300° + i \sin 300°)}{8(\cos 60° + i \sin 60°)}$

17. $\dfrac{3 \text{ cis } 305°}{9 \text{ cis } 65°}$ 18. $\dfrac{12 \text{ cis } 293°}{6 \text{ cis } 23°}$

19. $\dfrac{8}{\sqrt{3} + i}$

20. $\dfrac{2i}{-1 - i\sqrt{3}}$

21. $\dfrac{-i}{1 + i}$

22. $\dfrac{1}{2 - 2i}$

23. $\dfrac{2\sqrt{6} - 2i\sqrt{2}}{\sqrt{2} - i\sqrt{6}}$

24. $\dfrac{4 + 4i}{2 - 2i}$

Use a calculator to perform the indicated operations. Give answers in rectangular form. See Example 3.

25. $[2.5(\cos 35° + i \sin 35°)][3.0(\cos 50° + i \sin 50°)]$

26. $[4.6(\cos 12° + i \sin 12°)][2.0(\cos 13° + i \sin 13°)]$

27. $(12 \operatorname{cis} 18.5°)(3 \operatorname{cis} 12.5°)$

28. $(4 \operatorname{cis} 19.25°)(7 \operatorname{cis} 41.75°)$

29. $\dfrac{45(\cos 127° + i \sin 127°)}{22.5(\cos 43° + i \sin 43°)}$

30. $\dfrac{30(\cos 130° + i \sin 130°)}{10(\cos 21° + i \sin 21°)}$

31. $\left[2 \operatorname{cis} \dfrac{5\pi}{9} \right]^2$

32. $\left[24.3 \operatorname{cis} \dfrac{7\pi}{12} \right]^2$

Relating Concepts

For individual or collaborative investigation
(Exercises 33–39)

Consider the complex numbers $w = -1 + i$ and $z = -1 - i$. **Work Exercises 33–39 in order.**

33. Multiply w and z using their rectangular forms and the FOIL method from Section 8.1. Leave the product in rectangular form.

34. Find the trigonometric forms of w and z.

35. Multiply w and z using their trigonometric forms and the method described in this section.

36. Use the result of Exercise 35 to find the rectangular form of wz. How does this compare to your result in Exercise 33?

37. Find the quotient $\dfrac{w}{z}$ using their rectangular forms and multiplying both the numerator and the denominator by the conjugate of the denominator. Leave the quotient in rectangular form.

38. Use the trigonometric forms of w and z, found in Exercise 34, to divide w by z using the method described in this section.

39. Use the result of Exercise 38 to find the rectangular form of $\dfrac{w}{z}$. How does this compare to your result in Exercise 37?

---

40. Notice that $(r \operatorname{cis} \theta)^2 = (r \operatorname{cis} \theta)(r \operatorname{cis} \theta) = r^2 \operatorname{cis}(\theta + \theta) = r^2 \operatorname{cis} 2\theta$. State in your own words how we can square a complex number in trigonometric form. (In the next section, we will develop this idea more fully.)

41. Without actually performing the operations, state why the following products are the same.

$$[2(\cos 45° + i \sin 45°)] \cdot [5(\cos 90° + i \sin 90°)]$$

and $\quad \{2[\cos(-315°) + i \sin(-315°)]\} \cdot \{5[\cos(-270°) + i \sin(-270°)]\}$

42. Show that $\dfrac{1}{z} = \dfrac{1}{r}(\cos \theta - i \sin \theta)$, where $z = r(\cos \theta + i \sin \theta)$.

(Modeling) Solve each problem.

43. *Electrical Current* The alternating current in an electric inductor is $I = \frac{E}{Z}$ amperes, where E is voltage and $Z = R + X_L i$ is impedance. If $E = 8(\cos 20° + i \sin 20°)$, $R = 6$, and $X_L = 3$, find the current. Give the answer in rectangular form, with real and imaginary parts to the nearest hundredth.

44. *Electrical Current* The current I in a circuit with voltage E, resistance R, capacitive reactance X_c, and inductive reactance X_L is

$$I = \frac{E}{R + (X_L - X_c)i}.$$

Find I if $E = 12(\cos 25° + i \sin 25°)$, $R = 3$, $X_L = 4$, and $X_c = 6$. Give the answer in rectangular form, with real and imaginary parts to the nearest tenth.

(Modeling) *Impedance* In the parallel electrical circuit shown in the figure, the impedance Z can be calculated using the equation

$$Z = \frac{1}{\dfrac{1}{Z_1} + \dfrac{1}{Z_2}},$$

where Z_1 and Z_2 are the impedances for the branches of the circuit.

45. If $Z_1 = 50 + 25i$ and $Z_2 = 60 + 20i$, calculate Z.

46. Determine the phase angle θ for the value of Z found in Exercise 45.

8.4 De Moivre's Theorem; Powers and Roots of Complex Numbers

Powers of Complex Numbers (De Moivre's Theorem) ▪ **Roots of Complex Numbers**

Powers of Complex Numbers (De Moivre's Theorem) In the previous section, we studied the product theorem for complex numbers in trigonometric form. Because raising a number to a positive integer power is a repeated application of the product rule, it would seem likely that a theorem for finding powers of complex numbers exists. This is indeed the case. For example, the square of the complex number $r(\cos \theta + i \sin \theta)$ is

$$[r(\cos \theta + i \sin \theta)]^2 = [r(\cos \theta + i \sin \theta)][r(\cos \theta + i \sin \theta)]$$
$$= r \cdot r[\cos(\theta + \theta) + i \sin(\theta + \theta)]$$
$$= r^2(\cos 2\theta + i \sin 2\theta).$$

In the same way,

$$[r(\cos \theta + i \sin \theta)]^3 = r^3(\cos 3\theta + i \sin 3\theta).$$

These results suggest the following theorem for positive integer values of n. Although the theorem is stated and can be proved for all n, we use it only for positive integer values of n and their reciprocals.

**Abraham De Moivre
(1667–1754)**

De Moivre's Theorem

If $r(\cos \theta + i \sin \theta)$ is a complex number, and if n is any real number, then

$$[r(\cos \theta + i \sin \theta)]^n = r^n(\cos n\theta + i \sin n\theta).$$

In compact form, this is written

$$[r \text{ cis } \theta]^n = r^n(\text{cis } n\theta).$$

This theorem is named after the French expatriate friend of Isaac Newton, Abraham De Moivre, although he never explicitly stated it.

EXAMPLE 1   Finding a Power of a Complex Number

Find $\left(1 + i\sqrt{3}\right)^8$ and express the result in rectangular form.

Solution   First write $1 + i\sqrt{3}$ in trigonometric form.

$$1 + i\sqrt{3} = 2(\cos 60° + i \sin 60°) \quad \text{(Section 8.2)}$$

Now, apply De Moivre's theorem.

$$\begin{aligned}
\left(1 + i\sqrt{3}\right)^8 &= [2(\cos 60° + i \sin 60°)]^8 \\
&= 2^8[\cos(8 \cdot 60°) + i \sin(8 \cdot 60°)] \quad \text{De Moivre's theorem} \\
&= 256(\cos 480° + i \sin 480°) \\
&= 256(\cos 120° + i \sin 120°) \quad \text{\small 480° and 120° are coterminal.} \\
&\qquad\qquad\qquad\qquad\qquad\qquad \text{\small (Section 1.1)} \\
&= 256\left(-\frac{1}{2} + i\frac{\sqrt{3}}{2}\right) \quad \text{\small $\cos 120° = -\frac{1}{2}$; $\sin 120° = \frac{\sqrt{3}}{2}$} \\
&\qquad\qquad\qquad\qquad\qquad\qquad \text{\small (Section 2.2)} \\
&= -128 + 128i\sqrt{3} \quad \text{\small Rectangular form}
\end{aligned}$$

Now try Exercise 7.

Roots of Complex Numbers   Every nonzero complex number has exactly n distinct complex nth roots. De Moivre's theorem can be extended to find all nth roots of a complex number.

nth Root

For a positive integer n, the complex number $a + bi$ is an ***n*th root** of the complex number $x + yi$ if

$$(a + bi)^n = x + yi.$$

To find the three complex cube roots of $8(\cos 135° + i \sin 135°)$, for example, look for a complex number, say $r(\cos \alpha + i \sin \alpha)$, that will satisfy

$$[r(\cos \alpha + i \sin \alpha)]^3 = 8(\cos 135° + i \sin 135°).$$

By De Moivre's theorem, this equation becomes

$$r^3(\cos 3\alpha + i \sin 3\alpha) = 8(\cos 135° + i \sin 135°).$$

Set $r^3 = 8$ and $\cos 3\alpha + i \sin 3\alpha = \cos 135° + i \sin 135°$, to satisfy this equation. The first of these conditions implies that $r = 2$, and the second implies that

$$\cos 3\alpha = \cos 135° \quad \text{and} \quad \sin 3\alpha = \sin 135°.$$

For these equations to be satisfied, 3α must represent an angle that is coterminal with $135°$. Therefore, we must have

$$3\alpha = 135° + 360° \cdot k, \quad k \text{ any integer}$$

or

$$a = \frac{135° + 360° \cdot k}{3}, \quad k \text{ any integer.}$$

Now, let k take on the integer values 0, 1, and 2.

$$\text{If } k = 0, \text{ then} \quad \alpha = \frac{135° + 0°}{3} = 45°.$$

$$\text{If } k = 1, \text{ then} \quad \alpha = \frac{135° + 360°}{3} = \frac{495°}{3} = 165°.$$

$$\text{If } k = 2, \text{ then} \quad \alpha = \frac{135° + 720°}{3} = \frac{855°}{3} = 285°.$$

In the same way, $\alpha = 405°$ when $k = 3$. But note that $405° = 45° + 360°$ so $\sin 405° = \sin 45°$ and $\cos 405° = \cos 45°$. Similarly, if $k = 4$, $\alpha = 525°$, which has the same sine and cosine values as $165°$. To continue with larger values of k would just be repeating solutions already found. Therefore, all of the cube roots (three of them) can be found by letting $k = 0$, 1, or 2.

When $k = 0$, the root is $\quad 2(\cos 45° + i \sin 45°)$.
When $k = 1$, the root is $\quad 2(\cos 165° + i \sin 165°)$.
When $k = 2$, the root is $\quad 2(\cos 285° + i \sin 285°)$.

In summary, we see that $2(\cos 45° + i \sin 45°)$, $2(\cos 165° + i \sin 165°)$, and $2(\cos 285° + i \sin 285°)$ are the three cube roots of $8(\cos 135° + i \sin 135°)$.

Generalizing our results, we state the following theorem.

*n*th Root Theorem

If n is any positive integer, r is a positive real number, and θ is in degrees, then the nonzero complex number $r(\cos \theta + i \sin \theta)$ has exactly n distinct nth roots, given by

$$\sqrt[n]{r}(\cos \alpha + i \sin \alpha) \quad \text{or} \quad \sqrt[n]{r} \text{ cis } \alpha,$$

where

$$\alpha = \frac{\theta + 360° \cdot k}{n} \quad \text{or} \quad \alpha = \frac{\theta}{n} + \frac{360° \cdot k}{n}, \quad k = 0, 1, 2, \ldots, n - 1.$$

NOTE In the statement of the nth root theorem, if θ is in radians, then

$$\alpha = \frac{\theta + 2\pi k}{n} \quad \text{or} \quad \alpha = \frac{\theta}{n} + \frac{2\pi k}{n}.$$

EXAMPLE 2 Finding Complex Roots

Find the two square roots of $4i$. Write the roots in rectangular form.

Solution First write $4i$ in trigonometric form as

$$4i = 4\left(\cos \frac{\pi}{2} + i \sin \frac{\pi}{2}\right).$$

Here $r = 4$ and $\theta = \frac{\pi}{2}$. The square roots have absolute value $\sqrt{4} = 2$ and arguments as follows.

$$\alpha = \frac{\frac{\pi}{2}}{2} + \frac{2\pi k}{2} = \frac{\pi}{4} + \pi k$$

Since there are two square roots, let $k = 0$ and 1.

> If $k = 0$, then $\quad \alpha = \frac{\pi}{4} + \pi \cdot 0 = \frac{\pi}{4}$.
>
> If $k = 1$, then $\quad \alpha = \frac{\pi}{4} + \pi \cdot 1 = \frac{5\pi}{4}$.

Using these values for α, the square roots are $2 \operatorname{cis} \frac{\pi}{4}$ and $2 \operatorname{cis} \frac{5\pi}{4}$, which can be written in rectangular form as $\sqrt{2} + i\sqrt{2}$ and $-\sqrt{2} - i\sqrt{2}$.

This screen confirms the result of Example 2.

Now try Exercise 17(a).

EXAMPLE 3 Finding Complex Roots

Find all fourth roots of $-8 + 8i\sqrt{3}$. Write the roots in rectangular form.

Solution $-8 + 8i\sqrt{3} = 16 \operatorname{cis} 120°$ Write in trigonometric form.

Here $r = 16$ and $\theta = 120°$. The fourth roots of this number have absolute value $\sqrt[4]{16} = 2$ and arguments as follows.

$$\alpha = \frac{120°}{4} + \frac{360° \cdot k}{4} = 30° + 90° \cdot k$$

Degree mode
This screen shows how a calculator finds r and θ for the number in Example 3.

Since there are four fourth roots, let $k = 0, 1, 2,$ and 3.

> If $k = 0$, then $\quad \alpha = 30° + 90° \cdot 0 = 30°$.
> If $k = 1$, then $\quad \alpha = 30° + 90° \cdot 1 = 120°$.
> If $k = 2$, then $\quad \alpha = 30° + 90° \cdot 2 = 210°$.
> If $k = 3$, then $\quad \alpha = 30° + 90° \cdot 3 = 300°$.

Using these angles, the fourth roots are

$$2 \operatorname{cis} 30°, \quad 2 \operatorname{cis} 120°, \quad 2 \operatorname{cis} 210°, \quad \text{and} \quad 2 \operatorname{cis} 300°.$$

These four roots can be written in rectangular form as

$$\sqrt{3} + i, \quad -1 + i\sqrt{3}, \quad -\sqrt{3} - i, \quad \text{and} \quad 1 - i\sqrt{3}.$$

The graphs of these roots lie on a circle with center at the origin and radius 2. See Figure 13. The roots are equally spaced about the circle, 90° apart.

Figure 13

Now try Exercises 23(a) and (b).

EXAMPLE 4 Solving an Equation by Finding Complex Roots

Find all complex number solutions of $x^5 - 1 = 0$. Graph them as vectors in the complex plane.

Solution Write the equation as

$$x^5 - 1 = 0 \quad \text{or} \quad x^5 = 1.$$

While there is only one real number solution, 1, there are five complex number solutions. To find these solutions, first write 1 in trigonometric form as

$$1 = 1 + 0i = 1(\cos 0° + i \sin 0°).$$

The absolute value of the fifth roots is $\sqrt[5]{1} = 1$, and the arguments are given by

$$0° + 72° \cdot k, \quad k = 0, 1, 2, 3, \text{ and } 4.$$

By using these arguments, the fifth roots are

$$1(\cos 0° + i \sin 0°), \qquad k = 0$$
$$1(\cos 72° + i \sin 72°), \qquad k = 1$$
$$1(\cos 144° + i \sin 144°), \quad k = 2$$
$$1(\cos 216° + i \sin 216°), \quad k = 3$$

and $\qquad\qquad 1(\cos 288° + i \sin 288°). \quad k = 4$

The solution set of the equation can be written as $\{\text{cis } 0°, \text{cis } 72°, \text{cis } 144°, \text{cis } 216°, \text{cis } 288°\}$. The first of these roots equals 1; the others cannot easily be expressed in rectangular form but can be approximated with a calculator. The tips of the arrows representing the five fifth roots all lie on a unit circle and are equally spaced around it every 72°, as shown in Figure 14.

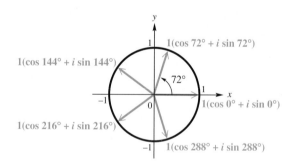

Figure 14

Now try Exercise 35.

8.4 Exercises

Find each power. Write each answer in rectangular form. See Example 1.

1. $[3(\cos 30° + i \sin 30°)]^3$

2. $[2(\cos 135° + i \sin 135°)]^4$

3. $(\cos 45° + i \sin 45°)^8$

4. $[2(\cos 120° + i \sin 120°)]^3$

5. $[3 \text{ cis } 100°]^3$

6. $[3 \text{ cis } 40°]^3$

7. $\left(\sqrt{3} + i\right)^5$

8. $\left(2\sqrt{2} - 2i\sqrt{2}\right)^6$

9. $\left(2 - 2i\sqrt{3}\right)^4$

10. $\left(\dfrac{\sqrt{2}}{2} - \dfrac{\sqrt{2}}{2}i\right)^8$

11. $(-2 - 2i)^5$

12. $(-1 + i)^7$

*In Exercises 13–24, **(a)** find all cube roots of each complex number. Leave answers in trigonometric form. **(b)** Graph each cube root as a vector in the complex plane. See Examples 2 and 3.*

13. $\cos 0° + i \sin 0°$ **14.** $\cos 90° + i \sin 90°$ **15.** $8 \text{ cis } 60°$

16. $27 \text{ cis } 300°$ **17.** $-8i$ **18.** $27i$

19. -64 **20.** 27 **21.** $1 + i\sqrt{3}$

22. $2 - 2i\sqrt{3}$ **23.** $-2\sqrt{3} + 2i$ **24.** $\sqrt{3} - i$

Find and graph all specified roots of 1.

25. second (square)   **26.** fourth   **27.** sixth   **28.** eighth

Find and graph all specified roots of i.

29. second (square)   **30.** fourth

Find all complex number solutions of each equation. Leave answers in trigonometric form. See Example 4.

31. $x^3 - 1 = 0$ **32.** $x^3 + 1 = 0$ **33.** $x^3 + i = 0$

34. $x^4 + i = 0$ **35.** $x^3 - 8 = 0$ **36.** $x^3 + 27 = 0$

37. $x^4 + 1 = 0$ **38.** $x^4 + 16 = 0$ **39.** $x^4 - i = 0$

40. $x^5 - i = 0$ **41.** $x^3 - \left(4 + 4i\sqrt{3}\right) = 0$ **42.** $x^4 - \left(8 + 8i\sqrt{3}\right) = 0$

43. Solve the equation $x^3 - 1 = 0$ by factoring the left side as the difference of two cubes and setting each factor equal to 0. Apply the quadratic formula as needed. Then compare your solutions to those of Exercise 31.

44. Solve the equation $x^3 + 27 = 0$ by factoring the left side as the sum of two cubes and setting each factor equal to 0. Apply the quadratic formula as needed. Then compare your solutions to those of Exercise 36.

Relating Concepts

For individual or collaborative investigation
(Exercises 45–48)

*In Chapter 5 we derived identities, or formulas, for $\cos 2\theta$ and $\sin 2\theta$. These identities can also be derived using De Moivre's theorem. **Work Exercises 45–48 in order,** to see how this is done.*

45. De Moivre's theorem states that $(\cos \theta + i \sin \theta)^2 = $ _____ .

46. Expand the left side of the equation in Exercise 45 as a binomial and collect terms to write the left side in the form $a + bi$.

47. Use the result of Exercise 46 to obtain the double-angle formula for cosine.

48. Repeat Exercise 47, but find the double-angle formula for sine.

Solve each problem.

49. *Mandelbrot Set* The fractal called the *Mandelbrot set* is shown in the figure. To determine if a complex number $z = a + bi$ is in this set, perform the following sequence of calculations. Repeatedly compute

$$z, \quad z^2 + z, \quad (z^2 + z)^2 + z,$$
$$[(z^2 + z)^2 + z]^2 + z, \ldots$$

In a manner analogous to the Julia set, the complex number z does not belong to the Mandelbrot set if any of the resulting absolute values exceed 2. Otherwise z is in the set and the point (a, b) should be shaded in the graph. Determine whether or not the following numbers belong to the Mandelbrot set. (*Source:* Lauwerier, H., *Fractals,* Princeton University Press, 1991.)

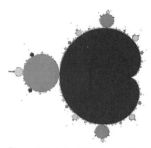

Source: Figure from Crownover, R., *Introduction to Fractals and Chaos.* Copyright © 1995. Boston: Jones and Bartlett Publishers. Reprinted with permission.

(a) $z = 0 + 0i$ **(b)** $z = 1 - 1i$ **(c)** $z = -.5i$

50. *Basins of Attraction* The fractal shown in the figure is the solution to Cayley's problem of determining the basins of attraction for the cube roots of unity. The three cube roots of unity are

$$w_1 = 1, \quad w_2 = -\frac{1}{2} + \frac{\sqrt{3}}{2}i,$$

and

$$w_3 = -\frac{1}{2} - \frac{\sqrt{3}}{2}i.$$

This fractal can be generated by repeatedly evaluating the function defined by $f(z) = \dfrac{2z^3 + 1}{3z^2}$, where z is a complex number. One begins by picking $z_1 = a + bi$ and then successively computing $z_2 = f(z_1)$, $z_3 = f(z_2)$, $z_4 = f(z_3), \ldots$. If the resulting values of $f(z)$ approach w_1, color the pixel at (a, b) red. If it approaches w_2, color it blue, and if it approaches w_3, color it yellow. If this process continues for a large number of different z_1, the fractal in the figure will appear. Determine the appropriate color of the pixel for each value of z_1. (*Source:* Crownover, R., *Introduction to Fractals and Chaos,* Jones and Bartlett Publishers, 1995.)

(a) $z_1 = i$ **(b)** $z_1 = 2 + i$ **(c)** $z_1 = -1 - i$

51. The screens here illustrate how a pentagon can be graphed using a graphing calculator. Note that a pentagon has five sides, and the T-step is $\frac{360}{5} = 72$. The display at the bottom of the graph screen indicates that one fifth root of 1 is $1 + 0i = 1$. Use this technique to find all fifth roots of 1, and express the real and imaginary parts in decimal form.

This is a continuation of the previous screen.

The calculator is in parametric, degree, and connected graph modes.

52. Use the method of Exercise 51 to find the first three of the ten 10th roots of 1.

53. One of the three cube roots of a complex number is $2 + 2\sqrt{3}i$. Determine the rectangular form of its other two cube roots.

Use a calculator to find all solutions of each equation in rectangular form.

54. $x^3 + 4 - 5i = 0$ **55.** $x^5 + 2 + 3i = 0$

56. *Concept Check*   How many complex 64th roots does 1 have? How many are real? How many are not?

57. *Concept Check*   True or false: Every real number must have two distinct real square roots.

58. *Concept Check*   True or false: Some real numbers have three real cube roots.

59. Show that if z is an nth root of 1, then so is $\frac{1}{z}$.

60. Explain why a real number can have only one real cube root.

61. Explain why the n nth roots of 1 are equally spaced around the unit circle.

62. Refer to Figure 14. A regular pentagon can be created by joining the tips of the arrows. Explain how you can use this principle to create a regular octagon.

8.5 Polar Equations and Graphs

Polar Coordinate System ▪ **Graphs of Polar Equations** ▪ **Converting from Polar to Rectangular Equations** ▪ **Classifying Polar Equations**

Figure 15

Figure 16

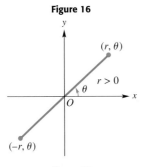

Figure 17

Polar Coordinate System   We have been using the rectangular coordinate system to graph equations. The **polar coordinate system** is based on a point, called the **pole,** and a ray, called the **polar axis.** The polar axis is usually drawn in the direction of the positive x-axis, as shown in Figure 15.

In Figure 16, the pole has been placed at the origin of a rectangular coordinate system so that the polar axis coincides with the positive x-axis. Point P has rectangular coordinates (x, y). Point P can also be located by giving the directed angle θ from the positive x-axis to ray OP and the directed distance r from the pole to point P. The ordered pair (r, θ) gives the **polar coordinates** of point P. If $r > 0$, then point P lies on the terminal side of θ and if $r < 0$, then point P lies on the ray pointing in the opposite direction of the terminal side of θ, a distance $|r|$ from the origin. See Figure 17.

Using Figure 16 and trigonometry, we can establish the following relationships between rectangular and polar coordinates.

Rectangular and Polar Coordinates

If a point has rectangular coordinates (x, y) and polar coordinates (r, θ), then these coordinates are related as follows.

$$x = r \cos \theta \qquad\qquad y = r \sin \theta$$

$$r^2 = x^2 + y^2 \qquad \tan \theta = \frac{y}{x}, \quad \text{if } x \neq 0$$

EXAMPLE 1 Plotting Points with Polar Coordinates

Plot each point by hand in the polar coordinate system. Then, determine the rectangular coordinates of each point.

(a) $P(2, 30°)$ **(b)** $Q\left(-4, \frac{2\pi}{3}\right)$ **(c)** $R\left(5, -\frac{\pi}{4}\right)$

Solution

(a) In this case, $r = 2$ and $\theta = 30°$, so the point P is located 2 units from the origin in the positive direction on a ray making a $30°$ angle with the polar axis, as shown in Figure 18.

Using the conversion equations, we find the rectangular coordinates as follows.

$$
\begin{array}{ll}
x = r \cos \theta & y = r \sin \theta \\
x = 2 \cos 30° & y = 2 \sin 30° \\
x = 2\left(\dfrac{\sqrt{3}}{2}\right) = \sqrt{3} & y = 2\left(\dfrac{1}{2}\right) = 1 \quad \text{(Section 2.1)}
\end{array}
$$

The rectangular coordinates are $\left(\sqrt{3}, 1\right)$.

Figure 18

(b) Since r is negative, Q is 4 units in the opposite direction from the pole on an extension of the $\frac{2\pi}{3}$ ray. See Figure 19. The rectangular coordinates are

$$
x = -4 \cos \frac{2\pi}{3} = -4\left(-\frac{1}{2}\right) = 2 \quad \text{and} \quad y = -4 \sin \frac{2\pi}{3} = -4\left(\frac{\sqrt{3}}{2}\right) = -2\sqrt{3}.
$$

Figure 19

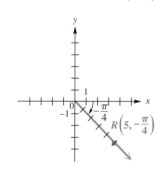

Figure 20

Looking Ahead to Calculus

Techniques studied in calculus associated with derivatives and integrals provide methods of finding slopes of tangent lines to polar curves, areas bounded by such curves, and lengths of their arcs.

(c) Point R is shown in Figure 20. Since θ is negative, the angle is measured in the clockwise direction. Furthermore, we have

$$
x = 5 \cos\left(-\frac{\pi}{4}\right) = \frac{5\sqrt{2}}{2} \quad \text{and} \quad y = 5 \sin\left(-\frac{\pi}{4}\right) = -\frac{5\sqrt{2}}{2}.
$$

Now try Exercises 3(a), (c), 5(a), (c), and 11(a), (c).

One important difference between rectangular coordinates and polar coordinates is that while a given point in the plane can have only one pair of rectangular coordinates, this same point can have an infinite number of pairs of polar coordinates. For example, $(2, 30°)$ locates the same point as $(2, 390°)$ or $(2, -330°)$ or $(-2, 210°)$.

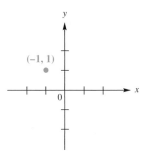

Figure 21

Figure 22

EXAMPLE 2   Giving Alternative Forms for Coordinates of a Point

(a) Give three other pairs of polar coordinates for the point $P(3, 140°)$.

(b) Determine two pairs of polar coordinates for the point with rectangular coordinates $(-1, 1)$.

Solution

(a) Three pairs that could be used for the point are $(3, -220°)$, $(-3, 320°)$, and $(-3, -40°)$. See Figure 21.

(b) As shown in Figure 22, the point $(-1, 1)$ lies in the second quadrant. Since $\tan \theta = \frac{1}{-1} = -1$, one possible value for θ is $135°$. Also,

$$r = \sqrt{x^2 + y^2} = \sqrt{(-1)^2 + 1^2} = \sqrt{2}.$$

Two pairs of polar coordinates are $\left(\sqrt{2}, 135°\right)$ and $\left(-\sqrt{2}, 315°\right)$.

Now try Exercises 3(b), 5(b), 11(b), and 13.

Graphs of Polar Equations   The graphs that we have studied so far have used rectangular coordinates. Equations in x and y are called **rectangular** (or **Cartesian) equations.** We now examine how equations in polar coordinates are graphed. An equation where r and θ are the variables is a **polar equation.**

$$r = 3 \sin \theta, \qquad r = 2 + \cos \theta, \qquad r = \theta \quad \text{Polar equations}$$

Lines and circles are two of the most common types of graphs studied in rectangular coordinates. While their rectangular forms are the ones most often encountered, they can also be defined in terms of polar coordinates. The polar equation of the line $ax + by = c$ can be derived as follows.

$$ax + by = c \quad \text{Rectangular equation}$$

$$a(r \cos \theta) + b(r \sin \theta) = c \quad \text{Convert from rectangular to polar coordinates.}$$

$$r(a \cos \theta + b \sin \theta) = c \quad \text{Factor out } r.$$

$$r = \frac{c}{a \cos \theta + b \sin \theta}$$

For the circle $x^2 + y^2 = a^2$, we have

$$x^2 + y^2 = a^2$$

$$\left(\sqrt{x^2 + y^2}\right)^2 = a^2 \quad \left(\sqrt{k}\right)^2 = k$$

$$r^2 = a^2 \quad \sqrt{x^2 + y^2} = r$$

$$r = \pm a. \quad r \text{ can be negative in polar coordinates.}$$

EXAMPLE 3   Examining Polar and Rectangular Equations of Lines and Circles

For each rectangular equation, give its equivalent polar equation and sketch its graph.

(a) $y = x - 3$

(b) $x^2 + y^2 = 4$

Function graphing mode

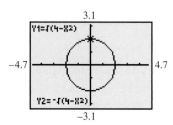

Polar graphing mode
Compare with Figure 23.

Function graphing mode

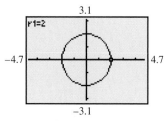

Polar graphing mode
Compare with Figure 24.

Solution

(a) This is the equation of a line. Rewrite $y = x - 3$ as

$$x - y = 3. \quad \text{Standard form } ax + by = c$$

Using the general form for the polar equation of a line,

$$r = \frac{c}{a \cos \theta + b \sin \theta}$$

$$r = \frac{3}{1 \cdot \cos \theta + (-1) \cdot \sin \theta} \quad a = 1, b = -1, c = 3$$

$$r = \frac{3}{\cos \theta - \sin \theta}.$$

Its graph is shown in Figure 23.

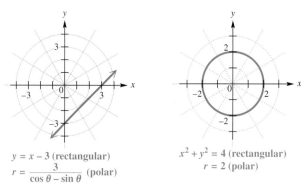

$y = x - 3$ (rectangular)
$r = \dfrac{3}{\cos \theta - \sin \theta}$ (polar)

Figure 23

$x^2 + y^2 = 4$ (rectangular)
$r = 2$ (polar)

Figure 24

(b) The graph of $x^2 + y^2 = 4$ is a circle with center at the origin and radius 2.

$$x^2 + y^2 = 4 \quad \text{(Appendix B)}$$

$$\left(\sqrt{x^2 + y^2}\right)^2 = (\pm 2)^2$$

$$\sqrt{x^2 + y^2} = 2 \quad \text{or} \quad \sqrt{x^2 + y^2} = -2$$

$$r = 2 \quad \text{or} \quad r = -2$$

(In polar coordinates, we may have $r < 0$.) The graphs of $r = 2$ and $r = -2$ coincide. The resulting graph is shown in Figure 24.

Now try Exercises 23 and 25.

To graph polar equations, evaluate r for various values of θ until a pattern appears, and then join the points with a smooth curve.

A graphing calculator can be used to graph an equation in the form $r = f(\theta)$. (See the margin screens next to Example 3.) Refer to your owner's manual to see how your model handles polar graphs. As always, you must set an appropriate window and choose the correct angle mode (radians or degrees). You will need to decide on maximum and minimum values of θ. Keep in mind the periods of the functions, so a complete set of ordered pairs is generated. ■

EXAMPLE 4 Graphing a Polar Equation (Cardioid)

Graph $r = 1 + \cos \theta$.

Algebraic Solution

To graph this equation, find some ordered pairs (as in the table) and then connect the points in order—from $(2, 0°)$ to $(1.9, 30°)$ to $(1.7, 45°)$ and so on. The graph is shown in Figure 25. This curve is called a **cardioid** because of its heart shape.

θ	$\cos \theta$	$r = 1 + \cos \theta$	θ	$\cos \theta$	$r = 1 + \cos \theta$
0°	1	2	135°	−.7	.3
30°	.9	1.9	150°	−.9	.1
45°	.7	1.7	180°	−1	0
60°	.5	1.5	270°	0	1
90°	0	1	315°	.7	1.7
120°	−.5	.5	330°	.9	1.9

Once the pattern of values of r becomes clear, it is not necessary to find more ordered pairs. That is why we stopped with the ordered pair $(1.9, 330°)$ in the table. Notice that the curve has been graphed on a **polar grid.**

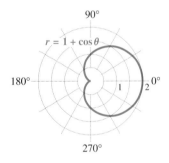

Figure 25

Graphing Calculator Solution

We choose degree mode and graph values of θ in the interval $[0°, 360°]$. The screens in Figure 26(a) show the choices needed to generate the graph in Figure 26(b).

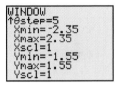

This is a continuation of the previous screen.

(a)

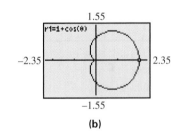

(b)

Figure 26

Now try Exercise 41.

EXAMPLE 5 Graphing a Polar Equation (Rose)

Graph $r = 3 \cos 2\theta$.

Solution Because of the argument 2θ, the graph requires a larger number of points than when the argument is just θ. A few ordered pairs are given in the table. You should complete the table similarly through the first 180° so that 2θ has values up to 360°.

θ	0°	15°	30°	45°	60°	75°	90°
2θ	0°	30°	60°	90°	120°	150°	180°
$\cos 2\theta$	1	.9	.5	0	−.5	−.9	−1
$r = 3 \cos 2\theta$	3	2.6	1.5	0	−1.5	−2.6	−3

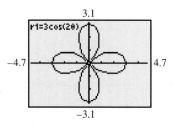

This is a calculator graph of the rose in Example 5.

Plotting the points from the table on the previous page in order gives the graph, called a **four-leaved rose.** Notice in Figure 27 how the graph is developed with a continuous curve, beginning with the upper half of the right horizontal leaf and ending with the lower half of that leaf. As the graph is traced, the curve goes through the pole four times.

Figure 27

Now try Exercise 45.

The equation $r = 3 \cos 2\theta$ in Example 5 has a graph that belongs to a family of curves called **roses.** The graphs of $r = a \sin n\theta$ and $r = a \cos n\theta$ are roses, with n leaves if n is odd, and $2n$ leaves if n is even. The value of a determines the length of the leaves.

EXAMPLE 6 Graphing a Polar Equation (Lemniscate)

Graph $r^2 = \cos 2\theta$.

Algebraic Solution

Complete a table of ordered pairs, and sketch the graph, as in Figure 28. The point $(-1, 0°)$, with r negative, may be plotted as $(1, 180°)$. Also, $(-.7, 30°)$ may be plotted as $(.7, 210°)$, and so on. This curve is called a **lemniscate.**

Values of θ for $45° < \theta < 135°$ are not included in the table because the corresponding values of $\cos 2\theta$ are negative (quadrants II and III) and so do not have real square roots. Values of θ larger than $180°$ give 2θ larger than $360°$ and would repeat the points already found.

θ	$0°$	$30°$	$45°$	$135°$	$150°$	$180°$
2θ	$0°$	$60°$	$90°$	$270°$	$300°$	$360°$
$\cos 2\theta$	1	.5	0	0	.5	1
$r = \pm\sqrt{\cos 2\theta}$	± 1	$\pm .7$	0	0	$\pm .7$	± 1

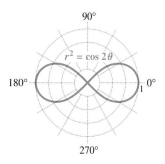

Figure 28

Graphing Calculator Solution

To graph $r^2 = \cos 2\theta$ with a graphing calculator, let

$$r_1 = \sqrt{\cos 2\theta}$$

and $\qquad r_2 = -\sqrt{\cos 2\theta}.$

See Figures 29(a) and (b).

(a)

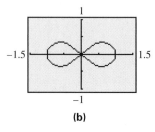

(b)

Figure 29

Now try Exercise 47.

EXAMPLE 7 Graphing a Polar Equation (Spiral of Archimedes)

Graph $r = 2\theta$ (θ measured in radians).

Solution Some ordered pairs are shown in the table. Since $r = 2\theta$, rather than a trigonometric function of θ, we must also consider negative values of θ. Radian measures have been rounded. These and other ordered pairs lead to the graph in Figure 30, called a **spiral of Archimedes.**

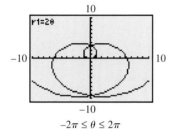

$-2\pi \leq \theta \leq 2\pi$

More of the spiral can be seen in this calculator graph of the spiral in Example 7 than is shown in Figure 30.

θ (radians)	$r = 2\theta$	θ (radians)	$r = 2\theta$
$-\pi$	-6.3	$\frac{\pi}{3}$	2.1
$-\frac{\pi}{2}$	-3.1	$\frac{\pi}{2}$	3.1
$-\frac{\pi}{4}$	-1.6	π	6.3
0	0	$\frac{3\pi}{2}$	9.4
$\frac{\pi}{6}$	1	2π	12.6

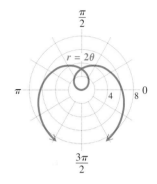

Figure 30

Now try Exercise 63.

Converting from Polar to Rectangular Equations We can transform polar equations to rectangular equations, as shown in the following example.

EXAMPLE 8 Converting a Polar Equation to a Rectangular Equation

Convert the equation to rectangular coordinates, and graph.

$$r = \frac{4}{1 + \sin \theta}$$

Solution

$$r = \frac{4}{1 + \sin \theta} \qquad \text{Polar equation}$$

$$r + r \sin \theta = 4 \qquad \text{Multiply by } 1 + \sin \theta.$$

$$\sqrt{x^2 + y^2} + y = 4 \qquad \text{Let } r = \sqrt{x^2 + y^2} \text{ and } r \sin \theta = y.$$

$$\sqrt{x^2 + y^2} = 4 - y \qquad \text{Subtract } y.$$

$$x^2 + y^2 = (4 - y)^2 \qquad \text{Square each side.}$$

$$x^2 + y^2 = 16 - 8y + y^2 \qquad \text{Expand the right side.}$$

$$x^2 = -8y + 16 \qquad \text{Subtract } y^2.$$

$$x^2 = -8(y - 2) \qquad \text{Rectangular equation}$$

The final equation represents a parabola and can be graphed using rectangular coordinates. See Figure 31.

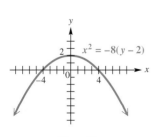

Figure 31

Now try Exercise 55.

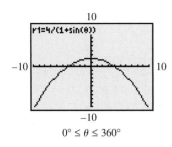

$$0° \le \theta \le 360°$$

Figure 32

The conversion in Example 8 is not necessary when using a graphing calculator. Figure 32 shows the graph of $r = \frac{4}{1 + \sin \theta}$, graphed directly using polar mode. ■

Classifying Polar Equations The following table summarizes some common polar graphs and forms of their equations. (In addition to circles, lemniscates, and roses, we include **limaçons.** Cardioids are a special case of limaçons, where $\left|\frac{a}{b}\right| = 1$.)

Circles and Lemniscates

Circles		Lemniscates	
$r = a \cos \theta$	$r = a \sin \theta$	$r^2 = a^2 \sin 2\theta$	$r^2 = a^2 \cos 2\theta$

Limaçons

$$r = a \pm b \sin \theta \quad \text{or} \quad r = a \pm b \cos \theta$$

			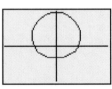
$\frac{a}{b} < 1$	$\frac{a}{b} = 1$	$1 < \frac{a}{b} < 2$	$\frac{a}{b} \ge 2$

Rose Curves

$2n$ leaves if n is even, $n \ge 2$		n leaves if n is odd	
$n = 2$	$n = 4$	$n = 3$	$n = 5$
$r = a \sin n\theta$	$r = a \cos n\theta$	$r = a \cos n\theta$	$r = a \sin n\theta$

8.5 Exercises

1. *Concept Check* For each point given in polar coordinates, state the quadrant in which the point lies if it is graphed in a rectangular coordinate system.

    (a) $(5, 135°)$ **(b)** $(2, 60°)$ **(c)** $(6, -30°)$ **(d)** $(4.6, 213°)$

2. *Concept Check* For each point given in polar coordinates, state the axis on which the point lies if it is graphed in a rectangular coordinate system. Also state whether it is on the positive portion or the negative portion of the axis. (For example, $(5, 0°)$ lies on the positive x-axis.)

    (a) $(7, 360°)$ **(b)** $(4, 180°)$ **(c)** $(2, -90°)$ **(d)** $(8, 450°)$

*For each pair of polar coordinates, **(a)** plot the point, **(b)** give two other pairs of polar coordinates for the point, and **(c)** give the rectangular coordinates for the point. See Examples 1 and 2.*

3. $(1, 45°)$ 4. $(3, 120°)$ 5. $(-2, 135°)$ 6. $(-4, 30°)$

7. $(5, -60°)$ 8. $(2, -45°)$ 9. $(-3, -210°)$ 10. $(-1, -120°)$

11. $\left(3, \dfrac{5\pi}{3}\right)$ 12. $\left(4, \dfrac{3\pi}{2}\right)$

*For each pair of rectangular coordinates, **(a)** plot the point and **(b)** give two pairs of polar coordinates for the point, where $0° \leq \theta < 360°$. See Example 2(b).*

13. $(1, -1)$ 14. $(1, 1)$ 15. $(0, 3)$ 16. $(0, -3)$

17. $\left(\sqrt{2}, \sqrt{2}\right)$ 18. $\left(-\sqrt{2}, \sqrt{2}\right)$ 19. $\left(\dfrac{\sqrt{3}}{2}, \dfrac{3}{2}\right)$ 20. $\left(-\dfrac{\sqrt{3}}{2}, -\dfrac{1}{2}\right)$

21. $(3, 0)$ 22. $(-2, 0)$

For each rectangular equation, give its equivalent polar equation and sketch its graph. See Example 3.

23. $x - y = 4$ 24. $x + y = -7$ 25. $x^2 + y^2 = 16$

26. $x^2 + y^2 = 9$ 27. $2x + y = 5$ 28. $3x - 2y = 6$

Relating Concepts

For individual or collaborative investigation
(Exercises 29–36)

In rectangular coordinates, the graph of $ax + by = c$ is a horizontal line if $a = 0$ or a vertical line if $b = 0$. **Work Exercises 29–36 in order,** *to determine the general forms of polar equations for horizontal and vertical lines.*

29. Begin with the equation $y = k$, whose graph is a horizontal line. Make a trigonometric substitution for y using r and θ.

30. Solve the equation in Exercise 29 for r.

31. Rewrite the equation in Exercise 30 using the appropriate reciprocal function.

32. Sketch the graph of $r = 3 \csc \theta$. What is the corresponding rectangular equation?

33. Begin with the equation $x = k$, whose graph is a vertical line. Make a trigonometric substitution for x using r and θ.

(continued)

34. Solve the equation in Exercise 33 for r.

35. Rewrite the equation in Exercise 34 using the appropriate reciprocal function.

36. Sketch the graph of $r = 3 \sec \theta$. What is the corresponding rectangular equation?

Concept Check *Match the polar graphs in Column II with their corresponding equations in Column I.*

I

37. $r = 3$

38. $r = \cos 3\theta$

39. $r = \cos 2\theta$

40. $r = \dfrac{2}{\cos \theta + \sin \theta}$

II

A.

B.

C.

D.

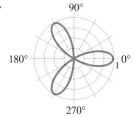

Give a complete graph of each polar equation. In Exercises 41–50, also identify the type of polar graph. See Examples 4–6.

41. $r = 2 + 2 \cos \theta$

42. $r = 8 + 6 \cos \theta$

43. $r = 3 + \cos \theta$

44. $r = 2 - \cos \theta$

45. $r = 4 \cos 2\theta$

46. $r = 3 \cos 5\theta$

47. $r^2 = 4 \cos 2\theta$

48. $r^2 = 4 \sin 2\theta$

49. $r = 4 - 4 \cos \theta$

50. $r = 6 - 3 \cos \theta$

51. $r = 2 \sin \theta \tan \theta$
(This is a *cissoid*.)

52. $r = \dfrac{\cos 2\theta}{\cos \theta}$
(This is a *cissoid* with a loop.)

For each equation, find an equivalent equation in rectangular coordinates and graph. See Example 8.

53. $r = 2 \sin \theta$

54. $r = 2 \cos \theta$

55. $r = \dfrac{2}{1 - \cos \theta}$

56. $r = \dfrac{3}{1 - \sin \theta}$

57. $r = -2 \cos \theta - 2 \sin \theta$

58. $r = \dfrac{3}{4 \cos \theta - \sin \theta}$

59. $r = 2 \sec \theta$

60. $r = -5 \csc \theta$

61. $r = \dfrac{2}{\cos \theta + \sin \theta}$

62. $r = \dfrac{2}{2 \cos \theta + \sin \theta}$

63. Graph $r = \theta$, a spiral of Archimedes. (See Example 7.) Use both positive and non-positive values for θ.

64. Use a graphing calculator window of $[-1250, 1250]$ by $[-1250, 1250]$, in degree mode, to graph more of $r = 2\theta$ (a spiral of Archimedes) than what is shown in Figure 30 and the accompanying margin screen. Use $-1250° \leq \theta \leq 1250°$.

65. Find the polar equation of the line that passes through the points $(1, 0°)$ and $(2, 90°)$.

66. Explain how to plot a point (r, θ) in polar coordinates, if $r < 0$.

Concept Check *The polar graphs in this section exhibit symmetry. (See Appendix D.) Visualize an xy-plane superimposed on the polar coordinate system, with the pole at the origin and the polar axis on the positive x-axis. Then a polar graph may be symmetric with respect to the x-axis (the polar axis), the y-axis $\left(\text{the line } \theta = \frac{\pi}{2}\right)$, or the origin (the pole). Use this information to work Exercises 67 and 68.*

67. Complete the missing ordered pairs in the graphs below.

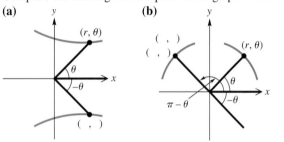

68. Based on your results in Exercise 67, fill in the blanks with the correct responses.

(a) The graph of $r = f(\theta)$ is symmetric with respect to the polar axis if substitution of _____ for θ leads to an equivalent equation.

(b) The graph of $r = f(\theta)$ is symmetric with respect to the vertical line $\theta = \frac{\pi}{2}$ if substitution of _____ for θ leads to an equivalent equation.

(c) Alternatively, the graph of $r = f(\theta)$ is symmetric with respect to the vertical line $\theta = \frac{\pi}{2}$ if substitution of _____ for r and _____ for θ leads to an equivalent equation.

(d) The graph of $r = f(\theta)$ is symmetric with respect to the pole if substitution of _____ for r leads to an equivalent equation.

(e) Alternatively, the graph of $r = f(\theta)$ is symmetric with respect to the pole if substitution of _____ for θ leads to an equivalent equation.

(f) In general, the completed statements in parts (a)–(e) mean that the graphs of polar equations of the form $r = a \pm b \cos \theta$ (where a may be 0) are symmetric with respect to _____ .

(g) In general, the completed statements in parts (a)–(e) mean that the graphs of polar equations of the form $r = a \pm b \sin \theta$ (where a may be 0) are symmetric with respect to _____ .

The graph of $r = a\theta$ in polar coordinates is an example of the spiral of Archimedes. With your calculator set to radian mode, use the given value of a and interval of θ to graph the spiral in the window specified.

69. $a = 1, 0 \leq \theta \leq 4\pi, [-15, 15]$ by $[-15, 15]$

70. $a = 2, -4\pi \leq \theta \leq 4\pi, [-30, 30]$ by $[-30, 30]$

71. $a = 1.5, -4\pi \leq \theta \leq 4\pi, [-20, 20]$ by $[-20, 20]$

72. $a = -1, 0 \leq \theta \leq 12\pi, [-40, 40]$ by $[-40, 40]$

Find the polar coordinates of the points of intersection of the given curves for the specified interval of θ.

73. $r = 4 \sin \theta$, $r = 1 + 2 \sin \theta$; $0 \le \theta < 2\pi$

74. $r = 3$, $r = 2 + 2 \cos \theta$; $0° \le \theta < 360°$

75. $r = 2 + \sin \theta$, $r = 2 + \cos \theta$; $0 \le \theta < 2\pi$

76. $r = \sin 2\theta$, $r = \sqrt{2} \cos \theta$; $0 \le \theta < \pi$

(Modeling)   Solve each problem.

77. *Planetary Orbits*   The polar equation

$$r = \frac{a(1 - e^2)}{1 + e \cos \theta}$$

can be used to graph the orbits of the planets, where a is the average distance in astronomical units from the sun and e is a constant called the *eccentricity*. The sun will be located at the pole. The table lists a and e for the planets.

Planet	a	e
Mercury	.39	.206
Venus	.78	.007
Earth	1.00	.017
Mars	1.52	.093
Jupiter	5.20	.048
Saturn	9.54	.056
Uranus	19.20	.047
Neptune	30.10	.009
Pluto	39.40	.249

Source: Karttunen, H., P. Kröger, H. Oja, M. Putannen, and K. Donners (Editors), *Fundamental Astronomy, 4th edition,* Springer-Verlag, 2003; Zeilik, M., S. Gregory, and E. Smith, *Introductory Astronomy and Astrophysics,* Saunders College Publishers, 1992.

(a) Graph the orbits of the four closest planets on the same polar grid. Choose a viewing window that results in a graph with nearly circular orbits.

(b) Plot the orbits of Earth, Jupiter, Uranus, and Pluto on the same polar grid. How does Earth's distance from the sun compare to the other planets' distances from the sun?

(c) Use graphing to determine whether or not Pluto is always the farthest planet from the sun.

78. *Radio Towers and Broadcasting Patterns*   Many times radio stations do not broadcast in all directions with the same intensity. To avoid interference with an existing station to the north, a new station may be licensed to broadcast only east and west. To create an east-west signal, two radio towers are sometimes used, as illustrated in the figure. Locations where the radio signal is received correspond to the interior of the curve defined by $r^2 = 40{,}000 \cos 2\theta$, where the polar axis (or positive x-axis) points east.

(a) Graph $r^2 = 40{,}000 \cos 2\theta$ for $0° \le \theta \le 360°$, where units are in miles. Assuming the radio towers are located near the pole, use the graph to describe the regions where the signal can be received and where the signal cannot be received.

(b) Suppose a radio signal pattern is given by $r^2 = 22{,}500 \sin 2\theta$. Graph this pattern and interpret the results.

8.6 | Parametric Equations, Graphs, and Applications

Basic Concepts ■ Parametric Graphs and Their Rectangular Equivalents ■ The Cycloid ■ Applications of Parametric Equations

Basic Concepts   Throughout this text, we have graphed sets of ordered pairs of real numbers that correspond to a function of the form $y = f(x)$ or $r = g(\theta)$. Another way to determine a set of ordered pairs involves two functions f and g defined by $x = f(t)$ and $y = g(t)$, where t is a real number in some interval I. Each value of t leads to a corresponding x-value and a corresponding y-value, and thus to an ordered pair (x, y).

Parametric Equations of a Plane Curve

A **plane curve** is a set of points (x, y) such that $x = f(t)$, $y = g(t)$, and f and g are both defined on an interval I. The equations $x = f(t)$ and $y = g(t)$ are **parametric equations** with **parameter t.**

Graphing calculators are capable of graphing plane curves defined by parametric equations. The calculator must be set in parametric mode, and the window requires intervals for the parameter t, as well as for x and y. ■

Parametric Graphs and Their Rectangular Equivalents

This is a continuation of the previous screen.

For the equation in Example 1, we make the choices seen in the first two screens to obtain the graph shown in the final screen.

EXAMPLE 1   Graphing a Plane Curve Defined Parametrically

Let $x = t^2$ and $y = 2t + 3$, for t in $[-3, 3]$. Graph the set of ordered pairs (x, y).

Solution   Make a table of corresponding values of t, x, and y over the domain of t. Then plot the points as shown in Figure 33. The graph is a portion of a parabola with horizontal axis $y = 3$. The arrowheads indicate the direction the curve traces as t increases.

t	x	y
-3	9	-3
-2	4	-1
-1	1	1
0	0	3
1	1	5
2	4	7
3	9	9

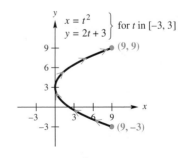

Figure 33

Now try Exercise 5(a).

EXAMPLE 2 Finding an Equivalent Rectangular Equation

Find a rectangular equation for the plane curve of Example 1 defined as follows.

$$x = t^2, \quad y = 2t + 3, \qquad \text{for } t \text{ in } [-3, 3]$$

Solution To eliminate the parameter t, solve either equation for t. Here, only the second equation, $y = 2t + 3$, leads to a unique solution for t, so we choose it.

$$y = 2t + 3 \qquad \text{Solve for } t. \textbf{ (Appendix A)}$$
$$2t = y - 3$$
$$t = \frac{y - 3}{2}$$

Now substitute this result into the first equation to get

$$x = t^2 = \left(\frac{y - 3}{2}\right)^2 = \frac{(y - 3)^2}{4} \qquad \text{or} \qquad 4x = (y - 3)^2.$$

This is the equation of a horizontal parabola opening to the right, which agrees with the graph given in Figure 33. Because t is in $[-3, 3]$, x is in $[0, 9]$ and y is in $[-3, 9]$. The rectangular equation must be given with its restricted domain as

$$4x = (y - 3)^2, \qquad \text{for } x \text{ in } [0, 9].$$

> **Now try Exercise 5(b).**

Trigonometric functions are often used to define a plane curve parametrically.

$x = 2 \sin t$ $\Big\}$ for
$y = 3 \cos t$ $\Big\}$ t in $[0, 2\pi]$

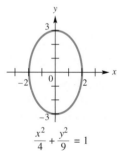

$$\frac{x^2}{4} + \frac{y^2}{9} = 1$$

Figure 34

EXAMPLE 3 Graphing a Plane Curve Defined Parametrically

Graph the plane curve defined by $x = 2 \sin t$, $y = 3 \cos t$, for t in $[0, 2\pi]$.

Solution To convert to a rectangular equation, it is not productive here to solve either equation for t. Instead, we use the fact that $\sin^2 t + \cos^2 t = 1$ to apply another approach. Square both sides of each equation; solve one for $\sin^2 t$, the other for $\cos^2 t$.

$x = 2 \sin t$	$y = 3 \cos t$ Given equations
$x^2 = 4 \sin^2 t$	$y^2 = 9 \cos^2 t$ Square both sides.
$\dfrac{x^2}{4} = \sin^2 t$	$\dfrac{y^2}{9} = \cos^2 t$ Divide.

Now add corresponding sides of the two equations.

$$\frac{x^2}{4} + \frac{y^2}{9} = \sin^2 t + \cos^2 t$$

$$\frac{x^2}{4} + \frac{y^2}{9} = 1 \qquad \sin^2 t + \cos^2 t = 1 \textbf{ (Section 5.1)}$$

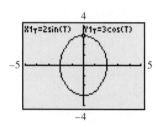

This is a calculator graph of the plane curve in Example 3.

This is the equation of an *ellipse*, as shown in Figure 34. (Ellipses are covered in more detail in college algebra courses.)

> **Now try Exercise 27.**

Parametric representations of a curve are not unique. In fact, there are infinitely many parametric representations of a given curve. If the curve can be described by a rectangular equation $y = f(x)$, with domain X, then one simple parametric representation is

$$x = t, \quad y = f(t), \qquad \text{for } t \text{ in } X.$$

EXAMPLE 4 Finding Alternative Parametric Equation Forms

Give two parametric representations for the equation of the parabola

$$y = (x - 2)^2 + 1.$$

Solution   The simplest choice is to let

$$x = t, \quad y = (t - 2)^2 + 1, \qquad \text{for } t \text{ in } (-\infty, \infty).$$

Another choice, which leads to a simpler equation for y, is

$$x = t + 2, \quad y = t^2 + 1, \qquad \text{for } t \text{ in } (-\infty, \infty).$$

Now try Exercise 29.

N O T E   Sometimes trigonometric functions are desirable. One choice in Example 4 might be $x = 2 + \tan t$, $y = \sec^2 t$, for t in $\left(-\frac{\pi}{2}, \frac{\pi}{2}\right)$.

The Cycloid   The path traced by a fixed point on the circumference of a circle rolling along a line is called a *cycloid*. A **cycloid** is defined by

$$x = at - a \sin t, \quad y = a - a \cos t, \qquad \text{for } t \text{ in } (-\infty, \infty).$$

Looking Ahead to Calculus

The cycloid is a special case of a curve traced out by a point at a given distance from the center of a circle as the circle rolls along a straight line. Such a curve is called a *trochoid*. It is just one of several parametrically defined curves studied in calculus. *Bezier curves* are used in manufacturing, and *Conchoids of Nicodemes* are so named because the shape of their outer branches resembles a conch shell. Other examples are *hypocycloids, epicycloids, the witch of Agnesi, swallowtail catastrophe curves,* and *Lissajou figures.* (*Source:* Stewart, J., *Calculus*, Fifth Edition, Brooks/Cole Publishing Co., 2003.)

EXAMPLE 5 Graphing a Cycloid

Graph the cycloid $x = t - \sin t$, $y = 1 - \cos t$, for t in $[0, 2\pi]$.

Algebraic Solution

There is no simple way to find a rectangular equation for the cycloid from its parametric equations. Instead, begin with a table of values.

t	0	$\frac{\pi}{4}$	$\frac{\pi}{2}$	π	$\frac{3\pi}{2}$	2π
x	0	.08	.6	π	5.7	2π
y	0	.3	1	2	1	0

Figure 35

Plotting the ordered pairs (x, y) from the table of values leads to the portion of the graph in Figure 35 from 0 to 2π.

Graphing Calculator Solution

It is easier to graph a cycloid with a graphing calculator in parametric mode than with traditional methods. See Figure 36.

Figure 36

Now try Exercise 33.

Figure 37

Looking Ahead to Calculus

At any time t, the velocity of an object is given by the vector $\mathbf{v} = \langle f'(t), g'(t) \rangle$. The object's speed at time t is

$$|\mathbf{v}| = \sqrt{(f'(t))^2 + (g'(t))^2}.$$

The cycloid has an interesting physical property. If a flexible cord or wire goes through points P and Q as in Figure 37, and a bead is allowed to slide due to the force of gravity without friction along this path from P to Q, the path that requires the shortest time takes the shape of the graph of an inverted cycloid.

Applications of Parametric Equations Parametric equations are used to simulate motion. If a ball is thrown with a velocity of v feet per second at an angle θ with the horizontal, its flight can be modeled by the parametric equations

$$x = (v \cos \theta)t \qquad \text{and} \qquad y = (v \sin \theta)t - 16t^2 + h,$$

where t is in seconds and h is the ball's initial height in feet above the ground. The term $-16t^2$ occurs because gravity is pulling downward. See Figure 38. These equations ignore air resistance.

Figure 38

(a)

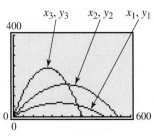

(b)

Figure 39

⌐⌐ **EXAMPLE 6** Simulating Motion with Parametric Equations

Three golf balls are hit simultaneously into the air at 132 ft per sec (90 mph) at angles of 30°, 50°, and 70° with the horizontal.

(a) Assuming the ground is level, determine graphically which ball travels the farthest. Estimate this distance.

(b) Which ball reaches the greatest height? Estimate this height.

Solution

(a) The three sets of parametric equations determined by the three golf balls are as follows since $h = 0$.

$$x_1 = (132 \cos 30°)t, \qquad y_1 = (132 \sin 30°)t - 16t^2$$
$$x_2 = (132 \cos 50°)t, \qquad y_2 = (132 \sin 50°)t - 16t^2$$
$$x_3 = (132 \cos 70°)t, \qquad y_3 = (132 \sin 70°)t - 16t^2$$

The graphs of the three sets of parametric equations are shown in Figure 39(a), where $0 \leq t \leq 9$. We used a graphing calculator in simultaneous mode so that we can view all three balls in flight at the same time. From the graph in Figure 39(b), we can see that the ball hit at 50° travels the farthest distance. Using the TRACE feature, we estimate this distance to be about 540 ft.

(b) Again, use the TRACE feature to find that the ball hit at 70° reaches the greatest height, about 240 ft.

Now try Exercise 39.

EXAMPLE 7 Examining Parametric Equations of Flight

Jack Lukas launches a small rocket from a table that is 3.36 ft above the ground. Its initial velocity is 64 ft per sec, and it is launched at an angle of 30° with respect to the ground. Find the rectangular equation that models its path. What type of path does the rocket follow?

Solution The path of the rocket is defined by the parametric equations

$$x = (64 \cos 30°)t \quad \text{and} \quad y = (64 \sin 30°)t - 16t^2 + 3.36$$

or, equivalently,

$$x = 32\sqrt{3}t \quad \text{and} \quad y = -16t^2 + 32t + 3.36.$$

From $x = 32\sqrt{3}t$, we obtain

$$t = \frac{x}{32\sqrt{3}}.$$

Substituting into the other parametric equation for t yields

$$y = -16\left(\frac{x}{32\sqrt{3}}\right)^2 + 32\left(\frac{x}{32\sqrt{3}}\right) + 3.36.$$

Simplifying, we find that the rectangular equation is

$$y = -\frac{1}{192}x^2 + \frac{\sqrt{3}}{3}x + 3.36.$$

Because this equation defines a parabola, the rocket follows a parabolic path.

Now try Exercise 43(a).

EXAMPLE 8 Analyzing the Path of a Projectile

Determine the total flight time and the horizontal distance traveled by the rocket in Example 7.

Algebraic Solution

The equation $y = -16t^2 + 32t + 3.36$ tells the vertical position of the rocket at time t. We need to determine those values of t for which $y = 0$ since these values correspond to the rocket at ground level. This yields

$$0 = -16t^2 + 32t + 3.36.$$

Using the quadratic formula, the solutions are $t = -.1$ or $t = 2.1$. Since t represents time, $t = -.1$ is an unacceptable answer. Therefore, the flight time is 2.1 sec.

The rocket was in the air for 2.1 sec, so we can use $t = 2.1$ and the parametric equation that models the horizontal position, $x = 32\sqrt{3}t$, to obtain

$$x = 32\sqrt{3}(2.1) \approx 116.4 \text{ ft.}$$

Graphing Calculator Solution

Figure 40 shows that when T = 2.1, the horizontal distance X covered is approximately 116.4 ft, which agrees with the algebraic solution.

Figure 40

Now try Exercise 43(b).

8.6 Exercises

Concept Check *Match the ordered pair from Column II with the pair of parametric equations in Column I on whose graph the point lies. In each case, consider the given value of t.*

<table>
<tr><td align="center">**I**</td><td align="center">**II**</td></tr>
</table>

1. $x = 3t + 6, y = -2t + 4;$ $t = 2$ **A.** $(5, 25)$

2. $x = \cos t, y = \sin t;$ $t = \dfrac{\pi}{4}$ **B.** $(7, 2)$

     C. $(12, 0)$

3. $x = t, y = t^2;$ $t = 5$

4. $x = t^2 + 3, y = t^2 - 2;$ $t = 2$ **D.** $\left(\dfrac{\sqrt{2}}{2}, \dfrac{\sqrt{2}}{2}\right)$

For each plane curve (a) graph the curve, and (b) find a rectangular equation for the curve. See Examples 1 and 2.

5. $x = t + 2, y = t^2,$ for t in $[-1, 1]$ **6.** $x = 2t, y = t + 1,$ for t in $[-2, 3]$

7. $x = \sqrt{t}, y = 3t - 4,$ for t in $[0, 4]$ **8.** $x = t^2, y = \sqrt{t},$ for t in $[0, 4]$

9. $x = t^3 + 1, y = t^3 - 1,$ for t in $(-\infty, \infty)$

10. $x = 2t - 1, y = t^2 + 2,$ for t in $(-\infty, \infty)$

11. $x = 2 \sin t, y = 2 \cos t,$ for t in $[0, 2\pi]$

12. $x = \sqrt{5} \sin t, y = \sqrt{3} \cos t,$ for t in $[0, 2\pi]$

13. $x = 3 \tan t, y = 2 \sec t,$ for t in $\left(-\dfrac{\pi}{2}, \dfrac{\pi}{2}\right)$

14. $x = \cot t, y = \csc t,$ for t in $(0, \pi)$

15. $x = \sin t, y = \csc t,$ for t in $(0, \pi)$ **16.** $x = \tan t, y = \cot t,$ for t in $\left(0, \dfrac{\pi}{2}\right)$

17. $x = t, y = \sqrt{t^2 + 2},$ for t in $(-\infty, \infty)$ **18.** $x = \sqrt{t}, y = t^2 - 1,$ for t in $[0, \infty)$

19. $x = 2 + \sin t, y = 1 + \cos t,$ for t in $[0, 2\pi]$

20. $x = 1 + 2 \sin t, y = 2 + 3 \cos t,$ for t in $[0, 2\pi]$

21. $x = t + 2, y = \dfrac{1}{t + 2},$ for $t \neq -2$ **22.** $x = t - 3, y = \dfrac{2}{t - 3},$ for $t \neq 3$

23. $x = t + 2, y = t - 4,$ for t in $(-\infty, \infty)$

24. $x = t^2 + 2, y = t^2 - 4,$ for t in $(-\infty, \infty)$

Graph each plane curve defined by the parametric equations for t in $[0, 2\pi]$. Then find a rectangular equation for the plane curve. See Example 3.

25. $x = 3 \cos t, y = 3 \sin t$ **26.** $x = 2 \cos t, y = 2 \sin t$

27. $x = 3 \sin t, y = 2 \cos t$ **28.** $x = 4 \sin t, y = 3 \cos t$

Give two parametric representations for the equation of each parabola. See Example 4.

29. $y = (x + 3)^2 - 1$ **30.** $y = (x + 4)^2 + 2$

31. $y = x^2 - 2x + 3$ **32.** $y = x^2 - 4x + 6$

Graph each cycloid defined by the given equations for t in the specified interval. See Example 5.

33. $x = 2t - 2 \sin t, y = 2 - 2 \cos t,$ for t in $[0, 8\pi]$

34. $x = t - \sin t, y = 1 - \cos t,$ for t in $[0, 4\pi]$

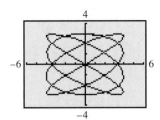

Lissajous Figures *The screen shown here is an example of a Lissajous figure. Lissajous figures occur in electronics and may be used to find the frequency of an unknown voltage. Graph each Lissajous figure for t in $[0, 6.5]$ in the window $[-6, 6]$ by $[-4, 4]$.*

35. $x = 2 \cos t,\ y = 3 \sin 2t$

36. $x = 3 \cos 2t,\ y = 3 \sin 3t$

37. $x = 3 \sin 4t,\ y = 3 \cos 3t$

38. $x = 4 \sin 4t,\ y = 3 \sin 5t$

(Modeling) *In Exercises 39–42, do the following.*

(a) *Determine the parametric equations that model the path of the projectile.*

(b) *Determine the rectangular equation that models the path of the projectile.*

(c) *Determine how long the projectile is in flight and the horizontal distance covered.*

See Examples 6–8.

39. *Flight of a Model Rocket* A model rocket is launched from the ground with velocity 48 ft per sec at an angle of 60° with respect to the ground.

40. *Flight of a Golf Ball* Tiger is playing golf. He hits a golf ball from the ground at an angle of 60° with respect to the ground at velocity 150 ft per sec.

41. *Flight of a Softball* Sally hits a softball when it is 2 ft above the ground. The ball leaves her bat at an angle of 20° with respect to the ground at velocity 88 ft per sec.

42. *Flight of a Baseball* Luis hits a baseball when it is 2.5 ft above the ground. The ball leaves his bat at an angle of 29° from the horizontal with velocity 136 ft per sec.

(Modeling) *Solve each problem. See Examples 7 and 8.*

43. *Path of a Rocket* A small rocket is launched from the top of an 8-ft ladder. Its initial velocity is 128 ft per sec, and it is launched at an angle of 60° with respect to the ground.

(a) Find the rectangular equation that models its path. What type of path does the rocket follow?

(b) Determine the total flight time and the horizontal distance the rocket travels.

44. *Simulating Gravity on the Moon* If an object is thrown on the moon, then the parametric equations of flight are

$$x = (v \cos \theta)t \quad \text{and} \quad y = (v \sin \theta)t - 2.66t^2 + h.$$

Estimate the distance that a golf ball hit at 88 ft per sec (60 mph) at an angle of 45° with the horizontal travels on the moon if the moon's surface is level.

45. *Flight of a Baseball* A baseball is hit from a height of 3 ft at a 60° angle above the horizontal. Its initial velocity is 64 ft per sec.

   (a) Write parametric equations that model the flight of the baseball.

   (b) Determine the horizontal distance traveled by the ball in the air. Assume that the ground is level.

   (c) What is the maximum height of the baseball? At that time, how far has the ball traveled horizontally?

   (d) Would the ball clear a 5-ft-high fence that is 100 ft from the batter?

(Modeling) Path of a Projectile In Exercises 46 and 47, a projectile has been launched from the ground with initial velocity 88 ft per sec. You are supplied with the parametric equations modeling the path of the projectile.

(a) *Graph the parametric equations.*

(b) *Approximate θ, the angle the projectile makes with the horizontal at launch, to the nearest tenth of a degree.*

(c) *Based on your answer to part (b), write parametric equations for the projectile using the cosine and sine functions.*

46. $x = 82.69295063t,\ y = -16t^2 + 30.09777261t$

47. $x = 56.56530965t,\ y = -16t^2 + 67.41191099t$

48. Give two parametric representations of the line through the point (x_1, y_1) with slope m.

49. Give two parametric representations of the parabola $y = a(x - h)^2 + k$.

50. Give a parametric representation of the rectangular equation $\frac{x^2}{a^2} - \frac{y^2}{b^2} = 1$.

51. Give a parametric representation of the rectangular equation $\frac{x^2}{a^2} + \frac{y^2}{b^2} = 1$.

52. The spiral of Archimedes has polar equation $r = a\theta$, where $r^2 = x^2 + y^2$. Show that a parametric representation of the spiral of Archimedes is

$$x = a\theta \cos \theta, \quad y = a\theta \sin \theta, \qquad \text{for } \theta \text{ in } (-\infty, \infty).$$

53. Show that the hyperbolic spiral $r\theta = a$, where $r^2 = x^2 + y^2$, is given parametrically by

$$x = \frac{a \cos \theta}{\theta}, \quad y = \frac{a \sin \theta}{\theta}, \qquad \text{for } \theta \text{ in } (-\infty, 0) \cup (0, \infty).$$

54. The parametric equations $x = \cos t,\ y = \sin t$, for t in $[0, 2\pi]$ and the parametric equations $x = \cos t,\ y = -\sin t$, for t in $[0, 2\pi]$ both have the unit circle as their graph. However, in one case the circle is traced out clockwise (as t moves from 0 to 2π) and in the other case the circle is traced out counterclockwise. For which pair of equations is the circle traced out in the clockwise direction?

Concept Check *Consider the parametric equations* $x = f(t),\ y = g(t)$, *for t in* $[a, b]$, *with* $c > 0, d > 0$.

55. How is the graph affected if the equation $x = f(t)$ is replaced by $x = c + f(t)$?

56. How is the graph affected if the equation $y = g(t)$ is replaced by $y = d + g(t)$?

Chapter 8 Summary

KEY TERMS

8.1 imaginary unit
complex number
real part
imaginary part
pure imaginary number
standard form
complex conjugates

8.2 real axis
imaginary axis
complex plane

rectangular form of a
complex number
trigonometric (polar)
form of a complex
number
absolute value
(modulus)
argument

8.4 nth root of a complex
number

8.5 polar coordinate
system
pole
polar axis
polar coordinates
rectangular (Cartesian)
equation
polar equation
cardioid
polar grid

rose curve (four-
leaved rose)
lemniscate
spiral of Archimedes
limaçon

8.6 plane curve
parametric equations
of a plane curve
parameter
cycloid

NEW SYMBOLS

i	imaginary unit	$a + bi$	complex number

QUICK REVIEW

CONCEPTS	EXAMPLES

8.1 Complex Numbers

Definition of i
$$i^2 = -1 \quad \text{or} \quad i = \sqrt{-1}$$

Definition of $\sqrt{-a}$
For $a > 0$,
$$\sqrt{-a} = i\sqrt{a}.$$

$$\sqrt{-4} = 2i$$
$$\sqrt{-12} = i\sqrt{12} = 2i\sqrt{3}$$

Adding and Subtracting Complex Numbers
Add or subtract the real parts and add or subtract the imaginary parts.

$$(2 + 3i) + (3 + i) - (2 - i)$$
$$= (2 + 3 - 2) + (3 + 1 + 1)i$$
$$= 3 + 5i$$

Multiplying and Dividing Complex Numbers
Multiply complex numbers as with binomials, and use the fact that $i^2 = -1$.

$$(6 + i)(3 - 2i) = 18 - 12i + 3i - 2i^2 \quad \text{FOIL}$$
$$= (18 + 2) + (-12 + 3)i \quad i^2 = -1$$
$$= 20 - 9i$$

Divide complex numbers by multiplying the numerator and denominator by the complex conjugate of the denominator.

$$\frac{3 + i}{1 + i} = \frac{(3 + i)(1 - i)}{(1 + i)(1 - i)} = \frac{3 - 3i + i - i^2}{1 - i^2}$$
$$= \frac{4 - 2i}{2} = \frac{2(2 - i)}{2} = 2 - i$$

CONCEPTS	EXAMPLES

8.2 Trigonometric (Polar) Form of Complex Numbers

Trigonometric (Polar) Form of Complex Numbers
If the complex number $x + yi$ corresponds to the vector with direction angle θ and magnitude r, then

$$x = r \cos \theta \qquad y = r \sin \theta$$

$$r = \sqrt{x^2 + y^2} \qquad \tan \theta = \frac{y}{x}, \quad \text{if } x \neq 0.$$

The expression

$$r(\cos \theta + i \sin \theta) \qquad \text{or} \qquad r \text{ cis } \theta$$

is the trigonometric form (or polar form) of $x + yi$.

Write $2(\cos 60° + i \sin 60°)$ in rectangular form.

$$2(\cos 60° + i \sin 60°) = 2\left(\frac{1}{2} + i \cdot \frac{\sqrt{3}}{2}\right) = 1 + \sqrt{3}i$$

Write $-\sqrt{2} + i\sqrt{2}$ in trigonometric form.

$$r = \sqrt{\left(-\sqrt{2}\right)^2 + \left(\sqrt{2}\right)^2} = 2$$

$\tan \theta = -1$ and θ is in quadrant II, so $\theta = 180° - 45° = 135°$. Therefore,

$$-\sqrt{2} + i\sqrt{2} = 2 \text{ cis } 135°.$$

8.3 The Product and Quotient Theorems

Product and Quotient Theorems
For any two complex numbers $r_1(\cos \theta_1 + i \sin \theta_1)$ and $r_2(\cos \theta_2 + i \sin \theta_2)$,

$$[r_1(\cos \theta_1 + i \sin \theta_1)] \cdot [r_2(\cos \theta_2 + i \sin \theta_2)]$$
$$= r_1 r_2[\cos(\theta_1 + \theta_2) + i \sin(\theta_1 + \theta_2)]$$

and

$$\frac{r_1(\cos \theta_1 + i \sin \theta_1)}{r_2(\cos \theta_2 + i \sin \theta_2)} = \frac{r_1}{r_2}[\cos(\theta_1 - \theta_2) + i \sin(\theta_1 - \theta_2)],$$

where $r_2 \text{ cis } \theta_2 \neq 0$.

If $\quad z_1 = 4(\cos 135° + i \sin 135°)$

and $\quad z_2 = 2(\cos 45° + i \sin 45°)$, then

$$z_1 z_2 = 8(\cos 180° + i \sin 180°)$$
$$= 8(-1 + i \cdot 0) = -8,$$

and

$$\frac{z_1}{z_2} = 2(\cos 90° + i \sin 90°)$$
$$= 2(0 + i \cdot 1) = 2i.$$

8.4 De Moivre's Theorem; Powers and Roots of Complex Numbers

De Moivre's Theorem

$$[r(\cos \theta + i \sin \theta)]^n = r^n(\cos n\theta + i \sin n\theta)$$

nth Root Theorem
If n is any positive integer, r is a positive real number, and θ is in degrees, then the nonzero complex number $r(\cos \theta + i \sin \theta)$ has exactly n distinct nth roots, given by

$$\sqrt[n]{r} (\cos \alpha + i \sin \alpha), \qquad \text{or} \qquad \sqrt[n]{r} \text{ cis } \alpha,$$

where

$$\alpha = \frac{\theta + 360° \cdot k}{n} \qquad \text{or} \qquad \alpha = \frac{\theta}{n} + \frac{360° \cdot k}{n},$$

$k = 0, 1, 2, \ldots, n - 1$.

Let $z = 4(\cos 180° + i \sin 180°)$. Find z^3 and the square roots of z.

$$z^3 = 4^3(\cos 3 \cdot 180° + i \sin 3 \cdot 180°)$$
$$= 64(\cos 540° + i \sin 540°)$$
$$= 64(-1 + i \cdot 0)$$
$$= -64$$

For the given z, $r = 4$ and $\theta = 180°$. Its square roots are

$$\sqrt{4}\left(\cos \frac{180°}{2} + i \sin \frac{180°}{2}\right) = 2(0 + i \cdot 1) = 2i$$

and

$$\sqrt{4}\left(\cos \frac{180° + 360°}{2} + i \sin \frac{180° + 360°}{2}\right)$$
$$= 2(0 + i(-1)) = -2i.$$

CONCEPTS	EXAMPLES

8.5 Polar Equations and Graphs

Rectangular and Polar Coordinates

The following relationships hold between the point (x, y) in the rectangular coordinate plane and the same point (r, θ) in the polar coordinate plane.

$$x = r \cos \theta \qquad y = r \sin \theta$$

$$r^2 = x^2 + y^2 \qquad \tan \theta = \frac{y}{x}, \quad \text{if } x \neq 0$$

Find the rectangular coordinates for the point $(5, 60°)$ in polar coordinates.

$$x = 5 \cos 60° = 5\left(\frac{1}{2}\right) = \frac{5}{2}$$

$$y = 5 \sin 60° = 5\left(\frac{\sqrt{3}}{2}\right) = \frac{5\sqrt{3}}{2}$$

The rectangular coordinates are $\left(\frac{5}{2}, \frac{5\sqrt{3}}{2}\right)$.

Find polar coordinates for $(-1, -1)$ in rectangular coordinates.

$$r = \sqrt{(-1)^2 + (-1)^2} = \sqrt{2}$$

$\tan \theta = 1$ and θ is in quadrant III, so $\theta = 225°$.

One pair of polar coordinates for $(-1, -1)$ is $\left(\sqrt{2}, 225°\right)$.

Polar Graphs

Examples of polar graphs include lines, circles, limaçons, lemniscates, roses, and spirals. (See page 366.)

Graph $r = 4 \cos 2\theta$.

$r = 4 \cos 2\theta$

8.6 Parametric Equations, Graphs, and Applications

Plane Curve

A **plane curve** is a set of points (x, y) such that $x = f(t)$, $y = g(t)$, and f and g are both defined on an interval I. The equations $x = f(t)$ and $y = g(t)$ are **parametric equations** with **parameter t.**

Graph $x = 2 - \sin t$, $y = \cos t - 1$, for $0 \leq t \leq 2\pi$.

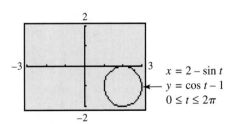

$x = 2 - \sin t$
$y = \cos t - 1$
$0 \leq t \leq 2\pi$

(continued)

CONCEPTS	EXAMPLES
Flight of an Object If an object has an initial velocity v, initial height h, and travels so that its initial angle of elevation is θ, then its flight after t seconds is modeled by the parametric equations $$x = (v \cos \theta)t \quad \text{and} \quad y = (v \sin \theta)t - 16t^2 + h.$$	Joe kicks a football from the ground at an angle of 45° with a velocity of 48 ft per sec. Give the parametric equations that model its path and the distance it travels before hitting the ground. $$x = (48 \cos 45°)t = 24\sqrt{2}\,t$$ $$y = (48 \sin 45°)t - 16t^2 = 24\sqrt{2}\,t - 16t^2$$ When the ball hits the ground, $y = 0$. $$24\sqrt{2}\,t - 16t^2 = 0$$ $$8t(3\sqrt{2} - 2t) = 0$$ $$t = 0 \quad \text{or} \quad t = \frac{3\sqrt{2}}{2}$$ $$\text{(Reject)}$$ The distance it travels is $x = 24\sqrt{2}\left(\dfrac{3\sqrt{2}}{2}\right) = 72$ ft.

Chapter 8 Review Exercises

Write as a multiple of i.

1. $\sqrt{-9}$

2. $\sqrt{-12}$

Solve each quadratic equation.

3. $x^2 = -81$

4. $x(2x + 3) = -4$

Perform each operation. Write answers in rectangular form.

5. $(1 - i) - (3 + 4i) + 2i$

6. $(2 - 5i) + (9 - 10i) - 3$

7. $(6 - 5i) + (2 + 7i) - (3 - 2i)$

8. $(4 - 2i) - (6 + 5i) - (3 - i)$

9. $(3 + 5i)(8 - i)$

10. $(4 - i)(5 + 2i)$

11. $(2 + 6i)^2$

12. $(6 - 3i)^2$

13. $(1 - i)^3$

14. $(2 + i)^3$

15. $\dfrac{6 + 2i}{3 - i}$

16. $\dfrac{2 - 5i}{1 + i}$

17. $\dfrac{2 + i}{1 - 5i}$

18. $\dfrac{3 + 2i}{i}$

19. i^{53}

20. i^{-41}

Perform each operation. Write answers in rectangular form.

21. $[5(\cos 90° + i \sin 90°)][6(\cos 180° + i \sin 180°)]$

22. $[3 \operatorname{cis} 135°][2 \operatorname{cis} 105°]$

23. $\dfrac{2(\cos 60° + i \sin 60°)}{8(\cos 300° + i \sin 300°)}$

24. $\dfrac{4 \text{ cis } 270°}{2 \text{ cis } 90°}$

25. $\left(\sqrt{3} + i\right)^3$

26. $(2 - 2i)^5$

27. $(\cos 100° + i \sin 100°)^6$

28. *Concept Check*   The vector representing a real number will lie on the _____-axis in the complex plane.

Graph each complex number as a vector.

29. $5i$

30. $-4 + 2i$

31. $3 - 3i\sqrt{3}$

32. Find and graph the resultant of $7 + 3i$ and $-2 + i$.

In Exercises 33–39, perform each conversion.

Rectangular Form	Trigonometric Form
33. $-2 + 2i$	_____
34. _____	$3(\cos 90° + i \sin 90°)$
35. _____	$2(\cos 225° + i \sin 225°)$
36. $-4 + 4i\sqrt{3}$	_____
37. $1 - i$	_____
38. _____	$4 \text{ cis } 240°$
39. $-4i$	_____

Concept Check   *The complex number z, where $z = x + yi$, can be graphed in the plane as (x, y). Describe the graphs of all complex numbers z satisfying the conditions in Exercises 40 and 41.*

40. The absolute value of z is 2.

41. The imaginary part of z is the negative of the real part of z.

Find all roots as indicated. Express them in trigonometric form.

42. the fifth roots of $-2 + 2i$

43. the cube roots of $1 - i$

44. *Concept Check*   How many real fifth roots does -32 have?

45. *Concept Check*   How many real sixth roots does -64 have?

Solve each equation. Leave answers in trigonometric form.

46. $x^3 + 125 = 0$

47. $x^4 + 16 = 0$

48. $x^2 + i = 0$

49. Convert $\left(-1, \sqrt{3}\right)$ to polar coordinates, with $0° \le \theta < 360°$ and $r > 0$.

50. Convert $(5, 315°)$ to rectangular coordinates.

51. *Concept Check*   If a point lies on an axis in the rectangular plane, then what kind of angle must θ be if (r, θ) represents the point in polar coordinates?

52. *Concept Check*   What will the graph of $r = k$ be, for $k > 0$?

Identify and graph each polar equation for θ in $[0°, 360°)$.

53. $r = 4 \cos \theta$

54. $r = -1 + \cos \theta$

55. $r = 2 \sin 4\theta$

56. $r = \dfrac{2}{2 \cos \theta - \sin \theta}$

Find an equivalent equation in rectangular coordinates.

57. $r = \dfrac{3}{1 + \cos \theta}$ **58.** $r = \sin \theta + \cos \theta$ **59.** $r = 2$

Find an equivalent equation in polar coordinates.

60. $y = x$ **61.** $y = x^2$

In Exercises 62–65, identify the geometric symmetry (A, B, or C) that the graph will possess.

 A. *symmetry with respect to the origin*
 B. *symmetry with respect to the y-axis*
 C. *symmetry with respect to the x-axis*

62. Whenever (r, θ) is on the graph, then so is $(-r, -\theta)$.

63. Whenever (r, θ) is on the graph, then so is $(-r, \theta)$.

64. Whenever (r, θ) is on the graph, then so is $(r, -\theta)$.

65. Whenever (r, θ) is on the graph, then so is $(r, \pi - \theta)$.

In Exercises 66–69, find a polar equation having the given graph.

66.

67.

68.

69.

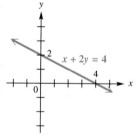

70. Show that the distance between (r_1, θ_1) and (r_2, θ_2) in polar coordinates is given by

$$d = \sqrt{r_1^2 + r_2^2 - 2r_1 r_2 \cos(\theta_1 - \theta_2)}.$$

71. Graph the plane curve defined by the parametric equations $x = t + \cos t$, $y = \sin t$, for t in $[0, 2\pi]$.

Find a rectangular equation for each plane curve with the given parametric equations.

72. $x = 3t + 2, y = t - 1$, for t in $[-5, 5]$

73. $x = \sqrt{t - 1}, y = \sqrt{t}$, for t in $[1, \infty)$

74. $x = t^2 + 5, y = \dfrac{1}{t^2 + 1}$, for t in $(-\infty, \infty)$

75. $x = 5 \tan t, y = 3 \sec t$, for t in $\left(-\dfrac{\pi}{2}, \dfrac{\pi}{2} \right)$

76. $x = \cos 2t,\ y = \sin t,$ for t in $(-\pi, \pi)$

77. Find a pair of parametric equations whose graph is the circle having center $(3, 4)$ and passing through the origin.

78. *Mandelbrot Set* Follow the steps in Exercise 64 of Section 8.2 to show that the graph of the Mandelbrot set in Exercise 49 of Section 8.4 is symmetric with respect to the x-axis.

79. *Flight of a Baseball* Manny hits a baseball when it is 3.2 ft above the ground. It leaves the bat with velocity 118 ft per sec at an angle of 27° with respect to the ground.

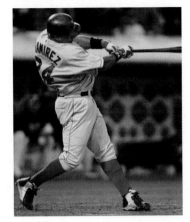

(a) Determine the parametric equations that model the path of the baseball.

(b) Determine the rectangular equation that models the path of the baseball.

(c) Determine how long the projectile is in flight and the horizontal distance covered.

Chapter 8 Test

1. For the complex numbers $w = 2 - 4i$ and $z = 5 + i$, find each of the following in rectangular form.

(a) $w + z$ (and give a geometric representation) (b) $w - z$ (c) wz (d) $\dfrac{w}{z}$

2. Express each of the following in rectangular form.

(a) i^{15} (b) $(1 + i)^2$

3. Solve $2x^2 - x + 4 = 0$ over the complex number system.

4. Write each complex number in trigonometric (polar) form, where $0° \le \theta < 360°$.

(a) $3i$ (b) $1 + 2i$ (c) $-1 - \sqrt{3}i$

5. Write each complex number in rectangular form.

(a) $3(\cos 30° + i \sin 30°)$ (b) $4 \text{ cis } 40°$ (c) $3(\cos 90° + i \sin 90°)$

6. For the complex numbers $w = 8(\cos 40° + i \sin 40°)$ and $z = 2(\cos 10° + i \sin 10°)$, find each of the following in the form specified.

(a) wz (trigonometric form) (b) $\dfrac{w}{z}$ (rectangular form) (c) z^3 (rectangular form)

7. Find the four complex fourth roots of $-16i$. Express them in trigonometric form.

8. Convert the given rectangular coordinates to polar coordinates. Give two pairs of polar coordinates for each point.

(a) $(0, 5)$ (b) $(-2, -2)$

9. Convert the given polar coordinates to rectangular coordinates.

(a) $(3, 315°)$ (b) $(-4, 90°)$

Identify and graph each polar equation for θ in $[0°, 360°)$.

10. $r = 1 - \cos \theta$

11. $r = 3 \cos 3\theta$

12. Convert the polar equation to a rectangular equation, and sketch its graph.

(a) $r = \dfrac{4}{2 \sin \theta - \cos \theta}$

(b) $r = 6$

Graph each pair of parametric equations.

13. $x = 4t - 3, y = t^2,$ for t in $[-3, 4]$

14. $x = 2 \cos 2t, y = 2 \sin 2t,$ for t in $[0, 2\pi]$

15. *Julia Set* Consider the complex number $z = -1 + i$. Compute the value of $z^2 - 1$, and show that its absolute value exceeds 2, indicating that $-1 + i$ is not in the Julia set.

Chapter 8 Quantitative Reasoning

How Rugged Is Your Coastline?*

An interesting feature of coastlines is that their ruggedness is independent of the distance from which they are viewed. From an airplane, we see irregularities as bays, peninsulas, river mouths, and so on. On foot, we see each rock outcropping and creek that makes the coastline appear more rugged. An ant sees every pebble as a mountain to be scaled.

An interesting result of this phenomenon is that the total distance you travel along a coastline is dependent on the size of the steps you take. The closer (or smaller) you are, the smaller your steps will be. This means you will have more obstacles in your way, which results in a longer distance to travel. The more rugged the coastline, the longer it will be. In theory, this means that if you could take small enough steps on a rugged enough coastline, the length of the coastline would approach infinity. This is related to the study of fractals.

From a mathematical perspective, we can say that the number of steps needed (y) varies inversely with some power of the size of the steps taken (x):

$$y = \frac{k}{x^n}.$$

Lake Tahoe

Each coastline will have different values for k and n, depending on its ruggedness. For a particular map of Lake Tahoe, these values are $k = 23.5$ and $n = 1.153$. Use the equation to determine how long the coastline would be if you "walked" with the given step sizes. Remember that the equation gives you the number of steps, so multiply that value by the step size to get the total length.

1. 6 in.

2. .1 in.

3. .01 in.

*This material is based on an idea presented by Lori Lambertson, of the Nueva School and the Exploratorium in San Francisco.

9

Exponential and Logarithmic Functions

In 1990 the International Panel on Climate Change (IPCC) reported that if current trends of burning fossil fuels and deforestation continue, then future amounts of atmospheric carbon dioxide in parts per million (ppm) would increase exponentially. Atmospheric carbon dioxide accumulation is irreversible and its effect requires hundreds of years to disappear.

In Example 11 of Section 9.1 and other examples and exercises in this chapter, we learn about exponential growth.

9.1 | Exponential Functions

Exponents and Properties ▪ **Exponential Functions** ▪ **Exponential Equations** ▪ **Compound Interest** ▪ **The Number e and Continuous Compounding** ▪ **Exponential Models and Curve Fitting**

Exponents and Properties Recall from algebra the definition of a^r, where r is a rational number: if $r = \frac{m}{n}$, then for appropriate values of m and n,

$$a^{m/n} = \left(\sqrt[n]{a}\right)^m.$$

For example,
$$16^{3/4} = \left(\sqrt[4]{16}\right)^3 = 2^3 = 8,$$

$$27^{-1/3} = \frac{1}{27^{1/3}} = \frac{1}{\sqrt[3]{27}} = \frac{1}{3}, \quad \text{and} \quad 64^{-1/2} = \frac{1}{64^{1/2}} = \frac{1}{\sqrt{64}} = \frac{1}{8}.$$

In this section we extend the definition of a^r to include all real (not just rational) values of the exponent r. For example, $2^{\sqrt{3}}$ might be evaluated by approximating the exponent $\sqrt{3}$ with the rational numbers 1.7, 1.73, 1.732, and so on. Since these decimals approach the value of $\sqrt{3}$ more and more closely, it seems reasonable that $2^{\sqrt{3}}$ should be approximated more and more closely by the numbers $2^{1.7}$, $2^{1.73}$, $2^{1.732}$, and so on. (Recall, for example, that $2^{1.7} = 2^{17/10} = \left(\sqrt[10]{2}\right)^{17}$. In fact, this is exactly how $2^{\sqrt{3}}$ is defined (in a more advanced course). To show that this assumption is reasonable, Figure 1 gives graphs of the function $f(x) = 2^x$ with three different domains.

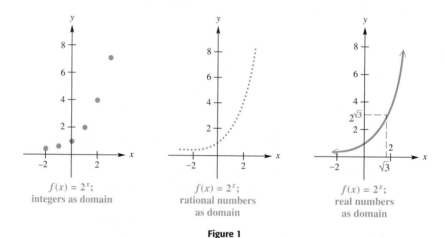

$f(x) = 2^x;$
integers as domain

$f(x) = 2^x;$
rational numbers
as domain

$f(x) = 2^x;$
real numbers
as domain

Figure 1

Using this interpretation of real exponents, all rules and theorems for exponents are valid for all real number exponents, not just rational ones. In addition to the rules for exponents, we use several new properties in this chapter. For example, if $y = 2^x$, then each real value of x leads to exactly one value of y, and therefore, $y = 2^x$ defines a function. Furthermore,

$$\text{if} \qquad 2^x = 2^4, \qquad \text{then} \qquad x = 4,$$

and $\qquad$ if $\qquad x = 4, \qquad$ then $\qquad 2^x = 2^4.$

Also, $\qquad\qquad 4^2 < 4^3 \qquad$ but $\qquad \left(\frac{1}{2}\right)^2 > \left(\frac{1}{2}\right)^3.$

In general, when $a > 1$, increasing the exponent on a leads to a *larger* number, but when $0 < a < 1$, increasing the exponent on a leads to a *smaller* number.

These properties are generalized below. Proofs of the properties are not given here, as they require more advanced mathematics.

Additional Properties of Exponents

For any real number $a > 0$, $a \neq 1$, the following statements are true.

(a) a^x **is a unique real number for all real numbers** x.

(b) $a^b = a^c$ **if and only if** $b = c$.

(c) If $a > 1$ **and** $m < n$, **then** $a^m < a^n$.

(d) If $0 < a < 1$ **and** $m < n$, **then** $a^m > a^n$.

Properties (a) and (b) require $a > 0$ so that a^x is always defined. For example, $(-6)^x$ is not a real number if $x = \frac{1}{2}$. This means that a^x will always be positive, since a must be positive. In property (a), a cannot equal 1 because $1^x = 1$ for every real number value of x, so each value of x leads to the same real number, 1. For property (b) to hold, a must not equal 1 since, for example, $1^4 = 1^5$, even though $4 \neq 5$.

EXAMPLE 1 Evaluating an Exponential Expression

If $f(x) = 2^x$, find each of the following.

(a) $f(-1)$ **(b)** $f(3)$ **(c)** $f\left(\dfrac{5}{2}\right)$ **(d)** $f(4.92)$

Solution

(a) $f(-1) = 2^{-1} = \dfrac{1}{2}$ Replace x with -1. **(Appendix C)**

(b) $f(3) = 2^3 = 8$

(c) $f\left(\dfrac{5}{2}\right) = 2^{5/2} = (2^5)^{1/2} = 32^{1/2} = \sqrt{32} = \sqrt{16} \cdot \sqrt{2} = 4\sqrt{2}$

(d) $f(4.92) = 2^{4.92} \approx 30.2738447$ Use a calculator.

Now try Exercises 3, 9, and 11.

Exponential Functions We can now define a function $f(x) = a^x$ whose domain is the set of all real numbers.

Exponential Function

If $a > 0$ and $a \neq 1$, then

$$f(x) = a^x$$

defines the **exponential function** with base a.

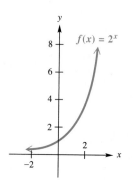

NOTE If $a = 1$, the function becomes the constant function defined by $f(x) = 1$, not an exponential function.

Figure 1 showed the graph of $f(x) = 2^x$ with three different domains. We repeat the final graph (with real numbers as domain) in the margin. The y-intercept is $y = 2^0 = 1$. Since $2^x > 0$ for all x and $2^x \to 0$ as $x \to -\infty$, the x-axis is a horizontal asymptote. As the graph suggests, the domain of the function is $(-\infty, \infty)$ and the range is $(0, \infty)$. The function is increasing on its entire domain, and therefore is one-to-one. Our observations from Figure 1 lead to the following generalizations about the graphs of exponential functions.

EXPONENTIAL FUNCTION $\quad f(x) = a^x$

Domain: $(-\infty, \infty)$ Range: $(0, \infty)$

For $f(x) = 2^x$:

x	$f(x)$
-2	$\frac{1}{4}$
-1	$\frac{1}{2}$
0	1
1	2
2	4
3	8

Figure 2

- $f(x) = a^x$, $a > 1$, is increasing and continuous on its entire domain, $(-\infty, \infty)$.
- The x-axis is a horizontal asymptote as $x \to -\infty$.
- The graph goes through the points $\left(-1, \frac{1}{a}\right)$, $(0, 1)$, and $(1, a)$.

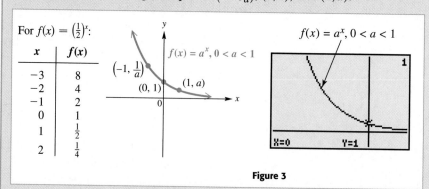

For $f(x) = \left(\frac{1}{2}\right)^x$:

x	$f(x)$
-3	8
-2	4
-1	2
0	1
1	$\frac{1}{2}$
2	$\frac{1}{4}$

Figure 3

- $f(x) = a^x$, $0 < a < 1$, is decreasing and continuous on its entire domain, $(-\infty, \infty)$.
- The x-axis is a horizontal asymptote as $x \to \infty$.
- The graph goes through the points $\left(-1, \frac{1}{a}\right)$, $(0, 1)$, and $(1, a)$.

Starting with $f(x) = 2^x$ and replacing x with $-x$ gives $f(-x) = 2^{-x} = (2^{-1})^x = \left(\frac{1}{2}\right)^x$. For this reason, the graphs of $f(x) = 2^x$ and $f(x) = \left(\frac{1}{2}\right)^x$ are reflections of each other across the y-axis. This is supported by the graphs in Figures 2 and 3.

The graph of $f(x) = 2^x$ is typical of graphs of $f(x) = a^x$ where $a > 1$. For larger values of a, the graphs rise more steeply, but the general shape is similar to the graph in Figure 2. When $0 < a < 1$, the graph decreases in a manner similar to the graph of $f(x) = \left(\frac{1}{2}\right)^x$. In Figure 4, the graphs of several typical exponential functions illustrate these facts.

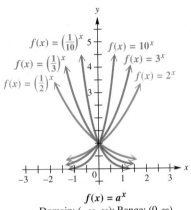

$f(x) = a^x$
Domain: $(-\infty, \infty)$; Range: $(0, \infty)$
When $a > 1$, the function is increasing.
When $0 < a < 1$, the function is decreasing.

Figure 4

In summary, the graph of a function of the form $f(x) = a^x$ has the following features.

Characteristics of the Graph of $f(x) = a^x$

1. The points $\left(-1, \frac{1}{a}\right)$, $(0, 1)$, and $(1, a)$ are on the graph.

2. If $a > 1$, then f is an increasing function; if $0 < a < 1$, then f is a decreasing function.

3. The x-axis is a horizontal asymptote.

4. The domain is $(-\infty, \infty)$, and the range is $(0, \infty)$.

EXAMPLE 2 Graphing an Exponential Function

Graph $f(x) = 5^x$.

Solution The y-intercept is 1, and the x-axis is a horizontal asymptote. Plot a few ordered pairs, and draw a smooth curve through them as shown in Figure 5 on the next page. Like the function $f(x) = 2^x$, this function also has domain $(-\infty, \infty)$ and range $(0, \infty)$ and is one-to-one. The graph is increasing on its entire domain.

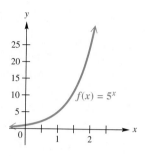

Figure 5

Now try Exercise 13.

EXAMPLE 3 Graphing Reflections and Translations

Graph each function.

(a) $f(x) = -2^x$ **(b)** $f(x) = 2^{x+3}$ **(c)** $f(x) = 2^x + 3$

Solution

(a) The graph of $f(x) = -2^x$ is that of $f(x) = 2^x$ reflected across the x-axis. The domain is $(-\infty, \infty)$, and the range is $(-\infty, 0)$. See Figure 6.

(b) The graph of $f(x) = 2^{x+3}$ is the graph of $f(x) = 2^x$ translated 3 units to the left, as shown in Figure 7.

(c) The graph of $f(x) = 2^x + 3$ is that of $f(x) = 2^x$ translated 3 units up. See Figure 8.

The screen shows how a graphing calculator can be directed to graph the three functions in Example 3. Y_1 is defined as 2^X, and Y_2, Y_3, and Y_4 are defined as reflections and/or translations of Y_1.

Figure 6

Figure 7

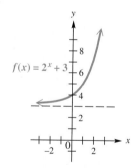

Figure 8

Now try Exercises 25 and 27.

Exponential Equations Property (b) given at the beginning of this section is used to solve **exponential equations,** equations with variables as exponents.

$$2^x = 32, \qquad 4^{3x-1} = 16^{x+2}, \qquad \left(\frac{1}{27}\right)^{x-1} = 9^{2x} \quad \text{Exponential equations}$$

$$Y_1 = \left(\tfrac{1}{3}\right)^X - 81$$

The screen supports the algebraic result in Example 4, using the *x*-intercept method of solution.

EXAMPLE 4 Using a Property of Exponents to Solve an Equation

Solve $\left(\dfrac{1}{3}\right)^x = 81$.

Solution   Write each side of the equation using a common base.

$$\left(\frac{1}{3}\right)^x = 81$$

$\quad (3^{-1})^x = 81 \qquad$ Definition of negative exponent; $\frac{1}{a^n} = a^{-n}$

$\qquad\ 3^{-x} = 81 \qquad (a^m)^n = a^{mn}$

$\qquad\ 3^{-x} = 3^4 \qquad$ Write 81 as a power of 3.

$\qquad\ -x = 4 \qquad\ $ Property (b)

$\qquad\ \ x = -4 \quad$ Multiply by -1.

The solution set of the original equation is $\{-4\}$.

Now try Exercise 45.

EXAMPLE 5 Using a Property of Exponents to Solve an Equation

Solve $2^{x+4} = 8^{x-6}$.

Solution   Write each side of the equation using a common base.

$$2^{x+4} = 8^{x-6}$$

$\qquad 2^{x+4} = (2^3)^{x-6} \qquad$ Write 8 as a power of 2.

$\qquad 2^{x+4} = 2^{3x-18} \qquad (a^m)^n = a^{mn}$

$\qquad x + 4 = 3x - 18 \quad$ Property (b)

$\qquad\ -2x = -22 \qquad$ Subtract $3x$ and 4. **(Appendix A)**

$\qquad\quad\ x = 11 \qquad\ $ Divide by -2.

Check by substituting 11 for x in the original equation. The solution set is $\{11\}$.

Now try Exercise 53.

EXAMPLE 6 Using a Property of Exponents to Solve an Equation

Solve $81 = b^{4/3}$.

Solution $\qquad\quad 81 = b^{4/3}$

$\qquad\qquad\qquad 81 = \left(\sqrt[3]{b}\right)^4 \qquad a^{m/n} = \left(\sqrt[n]{a}\right)^m$

$\qquad\qquad\quad \pm 3 = \sqrt[3]{b} \qquad$ Take fourth roots on both sides.

$\qquad\qquad\ \pm 27 = b \qquad$ Cube both sides.

Check *both* solutions in the original equation. Both check, so the solution set is $\{-27, 27\}$.

Now try Exercise 61.

Compound Interest Recall the formula for simple interest, $I = Prt$, where P is principal (amount deposited), r is annual rate of interest expressed as a decimal, and t is time in years that the principal earns interest. Suppose $t = 1$ yr. Then at the end of the year the amount has grown to

$$P + Pr = P(1 + r),$$

the original principal plus interest. If this balance earns interest at the same interest rate for another year, the balance at the end of that year will be

$$[P(1 + r)] + [P(1 + r)]r = [P(1 + r)](1 + r)$$
$$= P(1 + r)^2.$$

After the third year, this will grow to

$$[P(1 + r)^2] + [P(1 + r)^2]r = [P(1 + r)^2](1 + r)$$
$$= P(1 + r)^3.$$

Continuing in this way produces the formula

$$A = P(1 + r)^t$$

for interest compounded annually. The following general formula for *compound interest* can be derived in the same way as the formula given above.

Compound Interest

If P dollars are deposited in an account paying an annual rate of interest r compounded (paid) m times per year, then after t years the account will contain A dollars, where

$$A = P\left(1 + \frac{r}{m}\right)^{tm}.$$

In the formula for compound interest, A is sometimes called the **future value** and P the **present value**. A is also called the **compound amount** and is the balance *after* interest has been earned.

EXAMPLE 7 Using the Compound Interest Formula

Suppose $1000 is deposited in an account paying 4% interest per year compounded quarterly (four times per year).

(a) Find the amount in the account after 10 yr with no withdrawals.

(b) How much interest is earned over the 10-yr period?

Solution

(a) $A = P\left(1 + \dfrac{r}{m}\right)^{tm}$　　　　Compound interest formula

$A = 1000\left(1 + \dfrac{.04}{4}\right)^{10(4)}$　　　　Let $P = 1000$, $r = .04$, $m = 4$, and $t = 10$.

$A = 1000(1 + .01)^{40} = 1488.86$　　　Round to the nearest cent.

Thus, $1488.86 is in the account after 10 yr.

(b) The interest earned for that period is

$$\$1488.86 - \$1000 = \$488.86.$$

Now try Exercise 63(a).

EXAMPLE 8 Finding Present Value

Becky Anderson must pay a lump sum of $6000 in 5 yr.

(a) What amount deposited today at 3.1% compounded annually will grow to $6000 in 5 yr?

(b) If only $5000 is available to deposit now, what annual interest rate is necessary for the money to increase to $6000 in 5 yr?

Solution

(a)
$$A = P\left(1 + \frac{r}{m}\right)^{tm} \qquad \text{Compound interest formula}$$

$$6000 = P\left(1 + \frac{.031}{1}\right)^{5(1)} \qquad \text{Let } A = 6000, r = .031, m = 1, \text{ and } t = 5.$$

$$6000 = P(1.031)^5$$

$$P = \frac{6000}{(1.031)^5} \qquad \text{Solve for } P; \text{ divide by } (1.031)^5.$$

$$P \approx 5150.60 \qquad \text{Use a calculator.}$$

If Becky leaves $5150.60 for 5 yr in an account paying 3.1% compounded annually, she will have $6000 when she needs it. We say that $5150.60 is the *present value* of $6000 if interest of 3.1% is compounded annually for 5 yr.

(b)
$$A = P\left(1 + \frac{r}{m}\right)^{tm}$$

$$6000 = 5000(1 + r)^5 \qquad \text{Let } A = 6000, P = 5000, m = 1, \text{ and } t = 5.$$

$$\frac{6}{5} = (1 + r)^5 \qquad \text{Divide by 5000.}$$

$$\left(\frac{6}{5}\right)^{1/5} = 1 + r \qquad \text{Take the fifth root.}$$

$$\left(\frac{6}{5}\right)^{1/5} - 1 = r \qquad \text{Subtract 1.}$$

$$r \approx .0371 \qquad \text{Use a calculator.}$$

An interest rate of 3.71% will produce enough interest to increase the $5000 to $6000 by the end of 5 yr.

Now try Exercises 65 and 69.

m	$\left(1 + \dfrac{1}{m}\right)^m$ (rounded)
1	2
2	2.25
5	2.48832
10	2.59374
25	2.66584
50	2.69159
100	2.70481
500	2.71557
1000	2.71692
10,000	2.71815
1,000,000	2.71828

The Number e and Continuous Compounding The more often interest is compounded within a given time period, the more interest will be earned. Surprisingly, however, there is a limit on the amount of interest, no matter how often it is compounded. To see this, suppose that $1 is invested at 100% interest per year, compounded m times per year. Then the interest rate (in decimal form) is 1.00 and the interest rate per period is $\frac{1}{m}$. According to the formula (with $P = 1$), the compound amount at the end of 1 yr will be

$$A = \left(1 + \frac{1}{m}\right)^m.$$

A calculator gives the results in the margin for various values of m. The table suggests that as m increases, the value of $\left(1 + \frac{1}{m}\right)^m$ gets closer and closer to some fixed number. This is indeed the case. This fixed number is called e.

Value of e

To nine decimal places, $e \approx 2.718281828.$

Figure 9

Figure 9 shows the functions defined by $y = 2^x$, $y = 3^x$, and $y = e^x$. Notice that because $2 < e < 3$, the graph of $y = e^x$ lies "between" the other two graphs.

As mentioned above, the amount of interest earned increases with the frequency of compounding, but the value of the expression $\left(1 + \frac{1}{m}\right)^m$ approaches e as m gets larger. Consequently, the formula for compound interest approaches a limit as well, called the compound amount from *continuous compounding*.

Continuous Compounding

If P dollars are deposited at a rate of interest r compounded continuously for t years, the compound amount in dollars on deposit is

$$A = Pe^{rt}.$$

EXAMPLE 9   Solving a Continuous Compounding Problem

Suppose $5000 is deposited in an account paying 3% interest compounded continuously for 5 yr. Find the total amount on deposit at the end of 5 yr.

Solution

$$
\begin{aligned}
A &= Pe^{rt} && \text{Continuous compounding formula} \\
&= 5000e^{.03(5)} && \text{Let } P = 5000, t = 5, \text{ and } r = .03. \\
&= 5000e^{.15} \\
&\approx 5000(1.161834) && \text{Use a calculator.} \\
&= 5809.17
\end{aligned}
$$

or $5809.17. Check that daily compounding would have produced a compound amount about 3¢ less.

Looking Ahead to Calculus

In calculus, the derivative allows us to determine the slope of a tangent line to the graph of a function. For the function $f(x) = e^x$, the derivative is the function f itself: $f'(x) = e^x$. Therefore, in calculus the exponential function with base e is much easier to work with than exponential functions having other bases.

Now try Exercise 63(b).

EXAMPLE 10   Comparing Interest Earned as Compounding Is More Frequent

In Example 7, we found that $1000 invested at 4% compounded quarterly for 10 yr grew to $1488.86. Compare this same investment compounded annually, semiannually, monthly, daily, and continuously.

Solution   Substituting into the compound interest formula and the formula for continuous compounding gives the following results for amounts of $1 and $1000.

Compounded	$1	$1000
annually	$(1 + .04)^{10} \approx 1.48024$	$1480.24
semiannually	$\left(1 + \dfrac{.04}{2}\right)^{10(2)} \approx 1.48595$	$1485.95
quarterly	$\left(1 + \dfrac{.04}{4}\right)^{10(4)} \approx 1.48886$	$1488.86
monthly	$\left(1 + \dfrac{.04}{12}\right)^{10(12)} \approx 1.49083$	$1490.83
daily	$\left(1 + \dfrac{.04}{365}\right)^{10(365)} \approx 1.49179$	$1491.79
continuously	$e^{10(.04)} \approx 1.49182$	$1491.82

Compounding semiannually rather than annually increases the value of the account after 10 yr by $5.71. Quarterly compounding grows to $2.91 more than semiannual compounding after 10 yr. Each increase in the frequency of compounding earns less and less additional interest, until going from daily to continuous compounding increases the value by only $.03.

Now try Exercise 71.

Exponential Models and Curve Fitting   The number e is important as the base of an exponential function in many practical applications. For example, in situations involving growth or decay of a quantity, the amount or number present at time t often can be closely modeled by a function defined by

$$y = y_0 e^{kt},$$

where y_0 is the amount or number present at time $t = 0$ and k is a constant.

The next example, which refers to the problem stated at the beginning of this chapter, illustrates exponential growth.

EXAMPLE 11   Using Data to Model Exponential Growth

If current trends of burning fossil fuels and deforestation continue, then future amounts of atmospheric carbon dioxide in parts per million (ppm) will increase as shown in the table on the next page.

Year	Carbon Dioxide (ppm)
1990	353
2000	375
2075	590
2175	1090
2275	2000

Source: International Panel on Climate Change (IPCC), 1990.

(a) Make a scatter diagram of the data. Do the carbon dioxide levels appear to grow exponentially?

(b) The function defined by

$$y = 353e^{.0060857(t-1990)}$$

is a good model for the data. Use a graph of this model to estimate when future levels of carbon dioxide will double and triple over the preindustrial level of 280 ppm.

Solution

(a) We show a calculator graph for the data in Figure 10(a). The data appear to have the shape of the graph of an increasing exponential function.

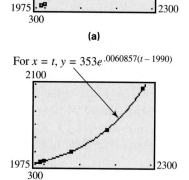

(a)

For $x = t$, $y = 353e^{.0060857(t-1990)}$

(b)

Figure 10

(b) A graph of $y = 353e^{.0060857(t-1990)}$ in Figure 10(b) shows that it is very close to the data points. We graph $y = 2 \cdot 280 = 560$ in Figure 11(a) and $y = 3 \cdot 280 = 840$ in Figure 11(b) on the same coordinate axes as the given function, and use the calculator to find the intersection points.

(a)

(b)

Figure 11

The graph of the function intersects the horizontal lines at approximately 2065.8 and 2132.5. According to this model, carbon dioxide levels will double by 2065 and triple by 2132.

Now try Exercise 73.

Graphing calculators are capable of fitting exponential curves to scatter diagrams like the one found in Example 11. Figure 12(a) shows how the TI-83 Plus displays another (different) equation for the atmospheric carbon dioxide example: $y = .0019 \cdot 1.0061^x$. (Coefficients are rounded here.) Notice that this calculator form differs from the model in Example 11. Figure 12(b) shows the data points and the graph of this exponential regression equation. ∎

(a)

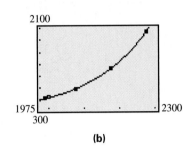

(b)

Figure 12

9.1 Exercises

If $f(x) = 3^x$ and $g(x) = \left(\frac{1}{4}\right)^x$, find each of the following. If a result is irrational, round the answer to the nearest thousandth. See Example 1.

1. $f(2)$ **2.** $f(3)$ **3.** $f(-2)$ **4.** $f(-3)$

5. $g(2)$ **6.** $g(3)$ **7.** $g(-2)$ **8.** $g(-3)$

9. $f\left(\frac{3}{2}\right)$ **10.** $g\left(\frac{3}{2}\right)$ **11.** $g(2.34)$ **12.** $f(1.68)$

Graph each function. See Example 2.

13. $f(x) = 3^x$ **14.** $f(x) = 4^x$ **15.** $f(x) = \left(\frac{1}{3}\right)^x$ **16.** $f(x) = \left(\frac{1}{4}\right)^x$

17. $f(x) = \left(\frac{3}{2}\right)^x$ **18.** $f(x) = \left(\frac{2}{3}\right)^x$ **19.** $f(x) = 10^x$ **20.** $f(x) = 10^{-x}$

21. $f(x) = 4^{-x}$ **22.** $f(x) = 6^{-x}$ **23.** $f(x) = 2^{|x|}$ **24.** $f(x) = 2^{-|x|}$

Sketch the graph of $f(x) = 2^x$. Then refer to it and use graphing techniques to graph each function as defined. See Example 3.

25. $f(x) = 2^x + 1$ **26.** $f(x) = 2^x - 4$ **27.** $f(x) = 2^{x+1}$ **28.** $f(x) = 2^{x-4}$

Sketch the graph of $f(x) = \left(\frac{1}{3}\right)^x$. Then refer to it and use graphing techniques to graph each function as defined. See Example 3.

29. $f(x) = \left(\frac{1}{3}\right)^x - 2$ **30.** $f(x) = \left(\frac{1}{3}\right)^x + 4$

31. $f(x) = \left(\frac{1}{3}\right)^{x+2}$ **32.** $f(x) = \left(\frac{1}{3}\right)^{x-4}$

Concept Check *The graphs of $y = a^x$ for $a = 1.8, 2.3, 3.2, .4, .75,$ and $.31$ are given in the figure. They are identified by letter, but not necessarily in the same order as the values just given. Use your knowledge of how the exponential function behaves for various values of the base to identify each lettered graph.*

33. A

34. B

35. C

36. D

37. E

38. F

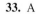 *Use a graphing calculator to graph each function as defined.*

39. $f(x) = \dfrac{e^x - e^{-x}}{2}$

40. $f(x) = \dfrac{e^x + e^{-x}}{2}$

41. $f(x) = x \cdot 2^x$

42. $f(x) = x^2 \cdot 2^{-x}$

Solve each equation. See Examples 4–6.

43. $4^x = 2$

44. $125^r = 5$

45. $\left(\dfrac{1}{2}\right)^k = 4$

46. $\left(\dfrac{2}{3}\right)^x = \dfrac{9}{4}$

47. $2^{3-y} = 8$

48. $5^{2p+1} = 25$

49. $e^{4x-1} = (e^2)^x$

50. $e^{3-x} = (e^3)^{-x}$

51. $27^{4z} = 9^{z+1}$

52. $32^t = 16^{1-t}$

53. $4^{x-2} = 2^{3x+3}$

54. $\left(\dfrac{1}{2}\right)^{3x-6} = 8^{x+1}$

55. $\left(\dfrac{1}{e}\right)^{-x} = \left(\dfrac{1}{e^2}\right)^{x+1}$

56. $e^{k-1} = \left(\dfrac{1}{e^4}\right)^{k+1}$

57. $\left(\sqrt{2}\right)^{x+4} = 4^x$

58. $\left(\sqrt[3]{5}\right)^{-x} = \left(\dfrac{1}{5}\right)^{x+2}$

59. $\dfrac{1}{27} = b^{-3}$

60. $\dfrac{1}{81} = k^{-4}$

61. $4 = r^{2/3}$

62. $z^{5/2} = 32$

Solve each problem involving compound interest. See Examples 7–9.

63. *Future Value*   Find the future value and interest earned if $8906.54 is invested for 9 yr at 5% compounded
(a) semiannually (b) continuously.

64. *Future Value*   Find the future value and interest earned if $56,780 is invested at 5.3% compounded

(a) quarterly for 23 quarters (b) continuously for 15 yr.

65. *Present Value*   Find the present value of $25,000 if interest is 6% compounded quarterly for 11 quarters.

66. *Present Value*   Find the present value of $45,000 if interest is 3.6% compounded monthly for 1 yr.

67. *Present Value*   Find the present value of $5000 if interest is 3.5% compounded quarterly for 10 yr.

68. *Interest Rate*   Find the required annual interest rate to the nearest tenth of a percent for $65,000 to grow to $65,325 if interest is compounded monthly for 6 months.

69. *Interest Rate*   Find the required annual interest rate to the nearest tenth of a percent for $1200 to grow to $1500 if interest is compounded quarterly for 5 yr.

70. *Interest Rate*   Find the required annual interest rate to the nearest tenth of a percent for $5000 to grow to $8400 if interest is compounded quarterly for 8 yr.

Solve each problem. See Example 10.

71. *Comparing Loans*   Bank A is lending money at 6.4% interest compounded annually. The rate at Bank B is 6.3% compounded monthly, and the rate at Bank C is 6.35% compounded quarterly. Which bank will you pay the *least* interest?

72. *Future Value*   Suppose $10,000 is invested at an annual rate of 5% for 10 yr. Find the future value if interest is compounded as follows.

(a) annually   (b) quarterly   (c) monthly   (d) daily (365 days)

(Modeling)   Solve each problem. See Example 11.

73. *Atmospheric Pressure*   The atmospheric pressure (in millibars) at a given altitude (in meters) is shown in the table.

Altitude	Pressure	Altitude	Pressure
0	1013	6000	472
1000	899	7000	411
2000	795	8000	357
3000	701	9000	308
4000	617	10,000	265
5000	541		

Source: Miller, A. and J. Thompson, *Elements of Meteorology,* Fourth Edition, Charles E. Merrill Publishing Company, Columbus, Ohio, 1993.

(a) Use a graphing calculator to make a scatter diagram of the data for atmospheric pressure P at altitude x.

(b) Would a linear or exponential function fit the data better?

(c) The function defined by

$$P(x) = 1013e^{-.0001341x}$$

approximates the data. Use a graphing calculator to graph P and the data on the same coordinate axes.

(d) Use P to predict the pressures at 1500 m and 11,000 m, and compare them to the actual values of 846 millibars and 227 millibars, respectively.

74. *World Population Growth*   Since 1990, world population in millions closely fits the exponential function defined by

$$y = 6073e^{.0137x},$$

where x is the number of years since 1990.

(a) The world population was about 6079 million in 2000. How closely does the function approximate this value?

(b) Use this model to approximate the population in 2005.

(c) Use this model to predict the population in 2015.

(d) Explain why this model may not be accurate for 2015.

75. *Deer Population*   The exponential growth of the deer population in Massachusetts can be calculated using the model

$$T = 50{,}000(1 + .06)^n,$$

where 50,000 is the initial deer population and .06 is the rate of growth. T is the total population after n years have passed.

(a) Predict the total population after 4 yr.
(b) If the initial population was 30,000 and the growth rate was .12, approximately how many deer would be present after 3 yr?
(c) How many additional deer can we expect in 5 yr if the initial population is 45,000 and the current growth rate is .08?

76. *Employee Training*   A person learning certain skills involving repetition tends to learn quickly at first. Then learning tapers off and approaches some upper limit. Suppose the number of symbols per minute that a person using a word processor can type is given by

$$p(t) = 250 - 120(2.8)^{-.5t},$$

where t is the number of months the operator has been in training. Find each value.

(a) $p(2)$ **(b)** $p(4)$ **(c)** $p(10)$
(d) What happens to the number of symbols per minute after several months of training?

9.2 Logarithmic Functions

Logarithms ▪ Logarithmic Equations ▪ Logarithmic Functions ▪ Properties of Logarithms

Logarithms   The previous section dealt with exponential functions of the form $y = a^x$ for all positive values of a, where $a \neq 1$. The horizontal line test shows that exponential functions are one-to-one, and thus have inverse functions. The equation defining the inverse of a function is found by interchanging x and y in the equation that defines the function. Doing so with

$$y = a^x \qquad \text{gives} \qquad x = a^y \quad \text{(Section 6.1)}$$

as the equation of the inverse function of the exponential function defined by $y = a^x$. This equation can be solved for y by using the following definition.

> **Logarithm**
>
> For all real numbers y, and all positive numbers a and x, where $a \neq 1$:
>
> $$y = \log_a x \qquad \text{if and only if} \qquad x = a^y.$$

The "log" in the definition above is an abbreviation for **logarithm.** Read $\log_a x$ as "the logarithm to the base a of x."

By the definition of logarithm, if $y = \log_a x$, then the power to which a must be raised to obtain x is y, or $x = a^y$. To remember the location of the base and the exponent in each form, refer to the diagrams in the margin.

Logarithmic form: $y = \log_a x$ (Exponent ↓, Base ↑)

Exponential form: $a^y = x$ (Exponent ↓, Base ↑)

> **Meaning of $\log_a x$**
>
> A logarithm is an exponent. The expression $\log_a x$ is the exponent to which the base a must be raised to obtain x.

The table shows several pairs of equivalent statements, written in both logarithmic and exponential forms.

Logarithmic Form	Exponential Form
$\log_2 8 = 3$	$2^3 = 8$
$\log_{1/2} 16 = -4$	$\left(\dfrac{1}{2}\right)^{-4} = 16$
$\log_{10} 100{,}000 = 5$	$10^5 = 100{,}000$
$\log_3 \dfrac{1}{81} = -4$	$3^{-4} = \dfrac{1}{81}$
$\log_5 5 = 1$	$5^1 = 5$
$\log_{3/4} 1 = 0$	$\left(\dfrac{3}{4}\right)^0 = 1$

Now try Exercises 3, 5, 7, and 9.

Logarithmic Equations   The definition of logarithm can be used to solve a **logarithmic equation,** an equation with a logarithm in at least one term, by first writing the equation in exponential form.

EXAMPLE 1   Solving Logarithmic Equations

Solve each equation.

(a) $\log_x \dfrac{8}{27} = 3$ **(b)** $\log_4 x = \dfrac{5}{2}$ **(c)** $\log_{49} \sqrt[3]{7} = x$

Solution

(a) $\log_x \dfrac{8}{27} = 3$

$x^3 = \dfrac{8}{27}$ Write in exponential form.

$x^3 = \left(\dfrac{2}{3}\right)^3$ $\frac{8}{27} = \left(\frac{2}{3}\right)^3$

$x = \dfrac{2}{3}$ Take cube roots.

The solution set is $\left\{\dfrac{2}{3}\right\}$.

(b) $\log_4 x = \dfrac{5}{2}$

$4^{5/2} = x$ Write in exponential form.

$\left(4^{1/2}\right)^5 = x$ $a^{m/n} = \left(a^{1/n}\right)^m$

$2^5 = x$ $4^{1/2} = (2^2)^{1/2} = 2$

$32 = x$

The solution set is $\{32\}$.

(c)

$$\log_{49} \sqrt[3]{7} = x$$

$$49^x = \sqrt[3]{7} \qquad \text{Write in exponential form.}$$

$$(7^2)^x = 7^{1/3} \qquad \text{Write with the same base.}$$

$$7^{2x} = 7^{1/3} \qquad (a^m)^n = a^{mn}$$

$$2x = \frac{1}{3} \qquad \text{Set exponents equal.}$$

$$x = \frac{1}{6} \qquad \text{Divide by 2. (Appendix A)}$$

The solution set is $\left\{\frac{1}{6}\right\}$.

Now try Exercises 13, 25, and 27.

Logarithmic Functions We define the logarithmic function with base a as follows.

Logarithmic Function

If $a > 0$, $a \neq 1$, and $x > 0$, then

$$f(x) = \log_a x$$

defines the **logarithmic function** with base a.

Exponential and logarithmic functions are inverses of each other. The graph of $y = 2^x$ is shown in red in Figure 13(a). The graph of its inverse is found by reflecting the graph of $y = 2^x$ across the line $y = x$. The graph of the inverse function, defined by $y = \log_2 x$, shown in blue, has the y-axis as a vertical asymptote. Figure 13(b) shows a calculator graph of the two functions.

x	2^x
-2	.25
-1	.5
0	1
1	2
2	4

x	$\log_2 x$
.25	-2
.5	-1
1	0
2	1
4	2

(a)

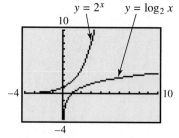

The graph of $y = \log_2 x$ can be obtained by drawing the inverse of $y = 2^x$.

(b)

Figure 13

Since the domain of an exponential function is the set of all real numbers, the range of a logarithmic function also will be the set of all real numbers. In the same way, both the range of an exponential function and the domain of a logarithmic function are the set of all positive real numbers, so *logarithms can be found for positive numbers only.*

We now summarize information about the graphs of logarithmic functions.

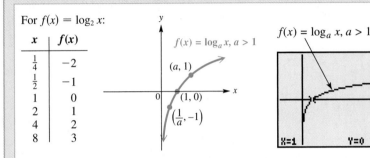

LOGARITHMIC FUNCTION $f(x) = \log_a x$

Domain: $(0, \infty)$ Range: $(-\infty, \infty)$

For $f(x) = \log_2 x$:

x	$f(x)$
$\frac{1}{4}$	-2
$\frac{1}{2}$	-1
1	0
2	1
4	2
8	3

Figure 14

- $f(x) = \log_a x$, $a > 1$, is increasing and continuous on its entire domain, $(0, \infty)$.
- The y-axis is a vertical asymptote as $x \to 0$ from the right.
- The graph goes through the points $\left(\frac{1}{a}, -1\right)$, $(1, 0)$, and $(a, 1)$.

For $f(x) = \log_{1/2} x$:

x	$f(x)$
$\frac{1}{4}$	2
$\frac{1}{2}$	1
1	0
2	-1
4	-2
8	-3

Figure 15

- $f(x) = \log_a x$, $0 < a < 1$, is decreasing and continuous on its entire domain, $(0, \infty)$.
- The y-axis is a vertical asymptote as $x \to 0$ from the right.
- The graph goes through the points $\left(\frac{1}{a}, -1\right)$, $(1, 0)$, and $(a, 1)$.

Calculator graphs of logarithmic functions do not, in general, give an accurate picture of the behavior of the graphs near the vertical asymptotes. While it may seem as if the graph has an endpoint, this is not the case. The resolution of the calculator screen is not precise enough to indicate that the graph approaches the vertical asymptote as the value of x gets closer to it. Do not draw incorrect conclusions just because the calculator does not show this behavior. ∎

The graphs in Figures 14 and 15 and the information with them suggest the following generalizations about the graphs of logarithmic functions of the form $f(x) = \log_a x$.

Characteristics of the Graph of $f(x) = \log_a x$

1. The points $\left(\frac{1}{a}, -1\right)$, $(1, 0)$, and $(a, 1)$ are on the graph.
2. If $a > 1$, then f is an increasing function; if $0 < a < 1$, then f is a decreasing function.
3. The y-axis is a vertical asymptote.
4. The domain is $(0, \infty)$, and the range is $(-\infty, \infty)$.

EXAMPLE 2 Graphing Logarithmic Functions

Graph each function.

(a) $f(x) = \log_{1/2} x$ **(b)** $f(x) = \log_3 x$

Solution

(a) First graph $y = \left(\frac{1}{2}\right)^x$, which defines the inverse function of f, by plotting points. Some ordered pairs are given in the table with the graph shown in red in Figure 16. The graph of $f(x) = \log_{1/2} x$ is the reflection of the graph of $y = \left(\frac{1}{2}\right)^x$ across the line $y = x$. The ordered pairs for $y = \log_{1/2} x$ are found by interchanging the x- and y-values in the ordered pairs for $y = \left(\frac{1}{2}\right)^x$. See the graph in blue in Figure 16.

x	$y = \left(\frac{1}{2}\right)^x$	$f(x) = y = \log_{1/2} x$
-2	4	undefined
-1	2	undefined
0	1	undefined
1	$\frac{1}{2}$	0
2	$\frac{1}{4}$	-1
4	$\frac{1}{16}$	-2

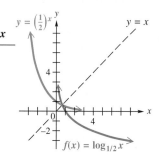

Figure 16

x	$y = \log_3 x$
$\frac{1}{3}$	-1
1	0
3	1
9	2

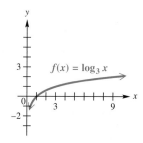

Figure 17

(b) Another way to graph a logarithmic function is to write $f(x) = y = \log_3 x$ in exponential form as $x = 3^y$. Now find some ordered pairs; several are shown in the table with the graph in Figure 17. Be careful to get the ordered pairs in the correct order.

Now try Exercise 45.

EXAMPLE 3 Graphing Translated Logarithmic Functions

Graph each function.

(a) $f(x) = \log_2(x - 1)$ **(b)** $f(x) = (\log_3 x) - 1$

Solution

(a) The graph of $f(x) = \log_2(x - 1)$ is the graph of $f(x) = \log_2 x$ translated 1 unit to the right. The vertical asymptote is $x = 1$. The domain of this function is $(1, \infty)$ since logarithms can be found only for positive numbers.

Figure 18

Figure 19

To find some ordered pairs to plot, use the equivalent exponential form of the equation $y = \log_2(x - 1)$.

$$y = \log_2(x - 1)$$
$$x - 1 = 2^y \qquad \text{Write in exponential form.}$$
$$x = 2^y + 1 \qquad \text{Add 1.}$$

We choose values for y and then calculate each of the corresponding x-values. See Figure 18.

(b) The function defined by $f(x) = (\log_3 x) - 1$ has the same graph as $g(x) = \log_3 x$ translated 1 unit down. We find ordered pairs to plot by writing $y = (\log_3 x) - 1$ in exponential form.

$$y = (\log_3 x) - 1$$
$$y + 1 = \log_3 x \qquad \text{Add 1.}$$
$$x = 3^{y+1} \qquad \text{Write in exponential form.}$$

Again, choose y-values and calculate the corresponding x-values. The graph is shown in Figure 19.

Now try Exercises 33 and 37.

Properties of Logarithms Since a logarithmic statement can be written as an exponential statement, it is not surprising that the properties of logarithms are based on the properties of exponents. The properties of logarithms allow us to change the form of logarithmic statements so that products can be converted to sums, quotients can be converted to differences, and powers can be converted to products.

Looking Ahead to Calculus

A technique called *logarithmic differentiation*, which uses the properties of logarithms, can often be used to differentiate complicated functions.

Properties of Logarithms

For $x > 0$, $y > 0$, $a > 0$, $a \neq 1$, and any real number r:

Property	Description
Product Property $\log_a xy = \log_a x + \log_a y$	The logarithm of the product of two numbers is equal to the sum of the logarithms of the numbers.
Quotient Property $\log_a \dfrac{x}{y} = \log_a x - \log_a y$	The logarithm of the quotient of two numbers is equal to the difference between the logarithms of the numbers.
Power Property $\log_a x^r = r \log_a x$	The logarithm of a number raised to a power is equal to the exponent multiplied by the logarithm of the number.

Two additional properties of logarithms follow directly from the definition of $\log_a x$ since $a^0 = 1$ and $a^1 = a$.

$$\log_a 1 = 0 \qquad \text{and} \qquad \log_a a = 1$$

EXAMPLE 4 Using the Properties of Logarithms

Rewrite each expression. Assume all variables represent positive real numbers, with $a \neq 1$ and $b \neq 1$.

(a) $\log_6(7 \cdot 9)$ **(b)** $\log_9 \dfrac{15}{7}$ **(c)** $\log_5 \sqrt{8}$

(d) $\log_a \dfrac{mnq}{p^2}$ **(e)** $\log_a \sqrt[3]{m^2}$ **(f)** $\log_b \sqrt[n]{\dfrac{x^3 y^5}{z^m}}$

Solution

(a) $\log_6(7 \cdot 9) = \log_6 7 + \log_6 9$ Product property

(b) $\log_9 \dfrac{15}{7} = \log_9 15 - \log_9 7$ Quotient property

(c) $\log_5 \sqrt{8} = \log_5(8^{1/2}) = \dfrac{1}{2}\log_5 8$ Power property

(d) $\log_a \dfrac{mnq}{p^2} = \log_a m + \log_a n + \log_a q - \log_a p^2$

$= \log_a m + \log_a n + \log_a q - 2\log_a p$

(e) $\log_a \sqrt[3]{m^2} = \log_a m^{2/3} = \dfrac{2}{3}\log_a m$

(f) $\log_b \sqrt[n]{\dfrac{x^3 y^5}{z^m}} = \log_b \left(\dfrac{x^3 y^5}{z^m}\right)^{1/n}$ $\sqrt[n]{a} = a^{1/n}$

$= \dfrac{1}{n}\log_b \dfrac{x^3 y^5}{z^m}$ Power property

$= \dfrac{1}{n}(\log_b x^3 + \log_b y^5 - \log_b z^m)$ Product and quotient properties

$= \dfrac{1}{n}(3\log_b x + 5\log_b y - m\log_b z)$ Power property

$= \dfrac{3}{n}\log_b x + \dfrac{5}{n}\log_b y - \dfrac{m}{n}\log_b z$ Distributive property

Notice the use of parentheses in the second and third steps. The factor $\frac{1}{n}$ applies to each term.

Now try Exercises 53, 55, and 59.

EXAMPLE 5 Using the Properties of Logarithms

Write each expression as a single logarithm with coefficient 1. Assume all variables represent positive real numbers, with $a \neq 1$ and $b \neq 1$.

(a) $\log_3(x + 2) + \log_3 x - \log_3 2$ **(b)** $2\log_a m - 3\log_a n$

(c) $\dfrac{1}{2}\log_b m + \dfrac{3}{2}\log_b 2n - \log_b m^2 n$

Solution

(a) $\log_3(x + 2) + \log_3 x - \log_3 2 = \log_3 \dfrac{(x + 2)x}{2}$ Product and quotient properties

(b) $2 \log_a m - 3 \log_a n = \log_a m^2 - \log_a n^3$ Power property

$$= \log_a \dfrac{m^2}{n^3} \qquad\qquad\text{Quotient property}$$

(c) $\dfrac{1}{2} \log_b m + \dfrac{3}{2} \log_b 2n - \log_b m^2 n$

$$= \log_b m^{1/2} + \log_b(2n)^{3/2} - \log_b m^2 n \qquad \text{Power property}$$

$$= \log_b \dfrac{m^{1/2}(2n)^{3/2}}{m^2 n} \qquad\qquad \text{Product and quotient properties}$$

$$= \log_b \dfrac{2^{3/2} n^{1/2}}{m^{3/2}} \qquad\qquad \text{Rules for exponents}$$

$$= \log_b \left(\dfrac{2^3 n}{m^3}\right)^{1/2} \qquad\qquad \text{Rules for exponents}$$

$$= \log_b \sqrt{\dfrac{8n}{m^3}} \qquad\qquad\qquad \text{Definition of } a^{1/n}$$

Now try Exercises 61, 63, and 65.

CAUTION There is no property of logarithms to rewrite a logarithm of a *sum* or *difference*. That is why, in Example 5(a), $\log_3(x + 2)$ was not written as $\log_3 x + \log_3 2$. Remember, $\log_3 x + \log_3 2 = \log_3(x \cdot 2)$.

The distributive property does not apply in a situation like this because $\log_3(x + y)$ is one term; "log" is a function name, not a factor.

EXAMPLE 6 Using the Properties of Logarithms with Numerical Values

Assume that $\log_{10} 2 = .3010$. Find each logarithm.

(a) $\log_{10} 4$ **(b)** $\log_{10} 5$

Solution

(a) $\log_{10} 4 = \log_{10} 2^2 = 2 \log_{10} 2 = 2(.3010) = .6020$

(b) $\log_{10} 5 = \log_{10} \dfrac{10}{2} = \log_{10} 10 - \log_{10} 2 = 1 - .3010 = .6990.$

Now try Exercise 69.

Recall from algebra that if $f(x)$ and $g(x)$ are inverse functions, then

$$f[g(x)] = x \qquad \text{and} \qquad g[f(x)] = x.$$

Since $f(x) = a^x$ and $g(x) = \log_a x$ are inverses, the next theorem follows.

> ### Theorem on Inverses
>
> For $a > 0, a \neq 1$:
>
> $$a^{\log_a x} = x \qquad \text{and} \qquad \log_a a^x = x.$$

By the results of this theorem,

$$7^{\log_7 10} = 10, \qquad \log_5 5^3 = 3, \qquad \text{and} \qquad \log_r r^{k+1} = k + 1.$$

The second statement in the theorem will be useful in Section 9.3 when we solve other logarithmic and exponential equations.

9.2 Exercises

Concept Check *In Exercises 1 and 2, match the logarithm in Column I with its value in Column II. Remember that* $\log_a x$ *is the exponent to which a must be raised in order to obtain x.*

I	II	I	II
1. (a) $\log_2 16$	**A.** 0	**2. (a)** $\log_3 81$	**A.** -2
(b) $\log_3 1$	**B.** $\dfrac{1}{2}$	**(b)** $\log_3 \dfrac{1}{3}$	**B.** -1
(c) $\log_{10} .1$	**C.** 4	**(c)** $\log_{10} .01$	**C.** 0
(d) $\log_2 \sqrt{2}$	**D.** -3	**(d)** $\log_6 \sqrt{6}$	**D.** $\dfrac{1}{2}$
(e) $\log_e \dfrac{1}{e^2}$	**E.** -1	**(e)** $\log_e 1$	**E.** $\dfrac{9}{2}$
(f) $\log_{1/2} 8$	**F.** -2	**(f)** $\log_3 27^{3/2}$	**F.** 4

For each statement, write an equivalent statement in logarithmic form.

3. $3^4 = 81$ **4.** $2^5 = 32$ **5.** $\left(\dfrac{2}{3}\right)^{-3} = \dfrac{27}{8}$ **6.** $10^{-4} = .0001$

For each statement, write an equivalent statement in exponential form.

7. $\log_6 36 = 2$ **8.** $\log_5 5 = 1$ **9.** $\log_{\sqrt{3}} 81 = 8$ **10.** $\log_4 \dfrac{1}{64} = -3$

11. Explain why logarithms of negative numbers are not defined.

12. *Concept Check* Why does $\log_a 1$ always equal 0 for any valid base a?

Solve each logarithmic equation. See Example 1.

13. $x = \log_5 \dfrac{1}{625}$ **14.** $x = \log_3 \dfrac{1}{81}$ **15.** $x = \log_{10} .001$

16. $x = \log_6 \dfrac{1}{216}$ **17.** $x = \log_8 \sqrt[4]{8}$ **18.** $x = 8 \log_{100} 10$

19. $x = 3^{\log_3 8}$ **20.** $x = 12^{\log_{12} 5}$ **21.** $x = 2^{\log_2 9}$

22. $x = 8^{\log_8 11}$ **23.** $\log_x 25 = -2$ **24.** $\log_x \dfrac{1}{16} = -2$

25. $\log_4 x = 3$ **26.** $\log_2 x = -1$ **27.** $x = \log_4 \sqrt[4]{16}$

28. $x = \log_5 \sqrt[4]{25}$ **29.** $\log_x 3 = -1$ **30.** $\log_x 1 = 0$

31. Compare the summary of characteristics of the graph of $f(x) = \log_a x$ with the similar summary about the graph of $f(x) = a^x$ in Section 9.1. Make a list of characteristics that reinforce the idea that these are inverse functions.

32. The calculator graph of $y = \log_2 x$ shows the values of the ordered pair with $x = 5$. What does the value of y represent?

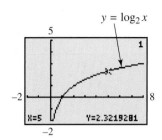

Sketch the graph of $f(x) = \log_2 x$. Then refer to it and use graphing techniques to graph each function. See Example 3.

33. $f(x) = (\log_2 x) + 3$ **34.** $f(x) = \log_2(x + 3)$ **35.** $f(x) = |\log_2(x + 3)|$

Sketch the graph of $f(x) = \log_{1/2} x$. Then refer to it and use graphing techniques to graph each function. See Example 3.

36. $f(x) = (\log_{1/2} x) - 2$ **37.** $f(x) = \log_{1/2}(x - 2)$ **38.** $f(x) = |\log_{1/2}(x - 2)|$

Concept Check In Exercises 39–44, match the function with its graph from choices A–F.

39. $f(x) = \log_2 x$ **40.** $f(x) = \log_2 2x$ **41.** $f(x) = \log_2 \dfrac{1}{x}$

42. $f(x) = \log_2 \dfrac{x}{2}$ **43.** $f(x) = \log_2(x - 1)$ **44.** $f(x) = \log_2(-x)$

A.

B.

C.

D.

E.

F.

Graph each function. See Examples 2 and 3.

45. $f(x) = \log_5 x$ **46.** $f(x) = \log_{10} x$ **47.** $f(x) = \log_{1/2}(1 - x)$

48. $f(x) = \log_{1/3}(3 - x)$ **49.** $f(x) = \log_3(x - 1)$ **50.** $f(x) = \log_2(x^2)$

We write log x as an abbreviation for $\log_{10} x$. Use the log key on your graphing calculator to graph each function.

51. $f(x) = x \log_{10} x$ **52.** $f(x) = x^2 \log_{10} x$

Use the properties of logarithms to rewrite each expression. Simplify the result if possible. Assume all variables represent positive real numbers. See Example 4.

53. $\log_2 \dfrac{6x}{y}$ **54.** $\log_3 \dfrac{4p}{q}$ **55.** $\log_5 \dfrac{5\sqrt{7}}{3}$ **56.** $\log_2 \dfrac{2\sqrt{3}}{5}$

57. $\log_4(2x + 5y)$ **58.** $\log_6(7m + 3q)$ **59.** $\log_m \sqrt{\dfrac{5r^3}{z^5}}$ **60.** $\log_p \sqrt[3]{\dfrac{m^5 n^4}{t^2}}$

Write each expression as a single logarithm with coefficient 1. Assume all variables represent positive real numbers. See Example 5.

61. $\log_a x + \log_a y - \log_a m$

62. $(\log_b k - \log_b m) - \log_b a$

63. $2 \log_m a - 3 \log_m b^2$

64. $\dfrac{1}{2} \log_y p^3 q^4 - \dfrac{2}{3} \log_y p^4 q^3$

65. $2 \log_a(z - 1) + \log_a(3z + 2), \quad z > 1$

66. $\log_b(2y + 5) - \dfrac{1}{2} \log_b(y + 3)$

67. $-\dfrac{2}{3} \log_5 5m^2 + \dfrac{1}{2} \log_5 25m^2$

68. $-\dfrac{3}{4} \log_3 16p^4 - \dfrac{2}{3} \log_3 8p^3$

Given $\log_{10} 2 = .3010$ and $\log_{10} 3 = .4771$, find each logarithm without using a calculator. See Example 6.

69. $\log_{10} 6$

70. $\log_{10} 12$

71. $\log_{10} \dfrac{9}{4}$

72. $\log_{10} \dfrac{20}{27}$

73. $\log_{10} \sqrt{30}$

74. $\log_{10} 36^{1/3}$

Solve each problem.

75. *(Modeling) Interest Rates of Treasury Securities* The table gives interest rates for various U.S. Treasury Securities on June 27, 2003.

   (a) Make a scatter diagram of the data.

   (b) Discuss which type of function will model these data best: linear, exponential, or logarithmic.

Time	Yield
3-month	.83%
6-month	.91%
2-year	1.35%
5-year	2.46%
10-year	3.54%
30-year	4.58%

Source: Charles Schwab.

76. *Concept Check* Use the graph to estimate each logarithm.

   (a) $\log_3 .3$ **(b)** $\log_3 .8$

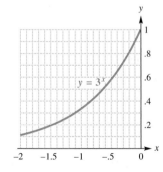

77. *Concept Check* Suppose $f(x) = \log_a x$ and $f(3) = 2$. Determine each function value.

   (a) $f\left(\dfrac{1}{9}\right)$ **(b)** $f(27)$

78. *Concept Check* If $(5, 4)$ is on the graph of the logarithmic function of base a, does $5 = \log_a 4$ or does $4 = \log_a 5$?

9.3 | Evaluating Logarithms; Equations and Applications

Common Logarithms ▪ Natural Logarithms ▪ Logarithms to Other Bases ▪ Exponential and Logarithmic Equations

Common Logarithms The two most important bases for logarithms are 10 and *e*. Base 10 logarithms are called **common logarithms.** The common logarithm of *x* is written log *x*, where the base is understood to be 10.

> ### Common Logarithm
>
> For all positive numbers *x*, $\quad \log x = \log_{10} x.$

   A calculator with a log key can be used to find the base 10 logarithm of any positive number. Figure 20 shows how a graphing calculator displays common logarithms. A common logarithm of a power of 10, such as 1000, is an integer (in this case, 3). Most common logarithms used in applications, such as log 142 and log .005832, are irrational numbers.

Figure 21 reinforces the concept presented in the previous section: log *x* is the exponent to which 10 must be raised in order to obtain *x*.

| Figure 20 | Figure 21 | ■ |

N O T E   Base *a*, *a* > 1, logarithms of numbers less than 1 are always negative, as suggested by the graphs in Section 9.2.

In chemistry, the **pH** of a solution is defined using logarithms as

$$\text{pH} = -\log[H_3O^+],$$

where $[H_3O^+]$ is the hydronium ion concentration in moles* per liter. The pH value is a measure of the acidity or alkalinity of a solution. Pure water has pH 7.0, substances with pH values greater than 7.0 are alkaline, and substances with pH values less than 7.0 are acidic. It is customary to round pH values to the nearest tenth.

---

*A *mole* is the amount of a substance that contains the same number of molecules as the number of atoms in exactly 12 grams of carbon 12.

EXAMPLE 1 Finding pH

(a) Find the pH of a solution with $[H_3O^+] = 2.5 \times 10^{-4}$.

(b) Find the hydronium ion concentration of a solution with pH $= 7.1$.

Solution

The screens show how a graphing calculator evaluates the pH and $[H_3O^+]$ in Example 1.

(a)
$$
\begin{aligned}
pH &= -\log[H_3O^+] \\
&= -\log(2.5 \times 10^{-4}) && \text{Substitute.} \\
&= -(\log 2.5 + \log 10^{-4}) && \text{Product property (Section 9.2)} \\
&= -(.3979 - 4) && \log 10^{-4} = -4 \text{ (Section 9.2)} \\
&= -.3979 + 4 && \text{Distributive property} \\
pH &\approx 3.6
\end{aligned}
$$

(b)
$$
\begin{aligned}
pH &= -\log[H_3O^+] \\
7.1 &= -\log[H_3O^+] && \text{Substitute.} \\
-7.1 &= \log[H_3O^+] && \text{Multiply by } -1. \\
[H_3O^+] &= 10^{-7.1} && \text{Write in exponential form. (Section 9.2)} \\
[H_3O^+] &\approx 7.9 \times 10^{-8} && \text{Evaluate } 10^{-7.1} \text{ with a calculator.}
\end{aligned}
$$

Now try Exercises 23 and 27.

EXAMPLE 2 Using pH in an Application

Wetlands are classified as *bogs*, *fens*, *marshes*, and *swamps* based on pH values. A pH value between 6.0 and 7.5 indicates that the wetland is a "rich fen." When the pH is between 4.0 and 6.0, it is a "poor fen," and if the pH falls to 3.0 or less, the wetland is a "bog." (*Source:* R. Mohlenbrock, "Summerby Swamp, Michigan," *Natural History,* March 1994.)

Suppose that the hydronium ion concentration of a sample of water from a wetland is 6.3×10^{-5}. How would this wetland be classified?

Solution
$$
\begin{aligned}
pH &= -\log[H_3O^+] && \text{Definition of pH} \\
&= -\log(6.3 \times 10^{-5}) && \text{Substitute.} \\
&= -(\log 6.3 + \log 10^{-5}) && \text{Product property} \\
&= -\log 6.3 - (-5) && \text{Distributive property; } \log 10^n = n \\
&= -\log 6.3 + 5 \\
pH &\approx 4.2 && \text{Use a calculator.}
\end{aligned}
$$

Since the pH is between 4.0 and 6.0, the wetland is a poor fen.

Now try Exercise 31.

EXAMPLE 3 Measuring the Loudness of Sound

The loudness of sounds is measured in a unit called a *decibel.* To measure with this unit, we assign intensity I_0 to a very faint sound, called the *threshold sound.* If a particular sound has intensity I, then the decibel rating of this louder sound is $d = 10 \log \frac{I}{I_0}$. Find the decibel rating of a sound with intensity $10{,}000 I_0$.

Solution

$$d = 10 \log \frac{10,000 I_0}{I_0} \qquad \text{Let } I = 10,000 I_0.$$

$$= 10 \log 10,000$$

$$= 10(4) \qquad \log 10,000 = \log 10^4 = 4 \text{ (Section 9.2)}$$

$$= 40$$

The sound has a decibel rating of 40.

Now try Exercise 47.

Looking Ahead to Calculus

The natural logarithmic function defined by $f(x) = \ln x$ and the reciprocal function defined by $g(x) = \dfrac{1}{x}$ have an important relationship in calculus. The derivative of the natural logarithmic function is the reciprocal function. Using *Leibniz notation* (named after one of the co-inventors of calculus), this fact is written $\dfrac{d}{dx}(\ln x) = \dfrac{1}{x}$.

Natural Logarithms In Section 9.1 we introduced the irrational number e. In most practical applications of logarithms, e is used as base. Logarithms to base e are called **natural logarithms,** since they occur in the life sciences and economics in natural situations that involve growth and decay. The base e logarithm of x is written $\ln x$ (read "el-en x").

Natural Logarithm

For all positive numbers x, $\qquad$ $\ln x = \log_e x.$

A graph of the natural logarithmic function defined by $f(x) = \ln x$ is given in Figure 22.

Figure 23

Figure 24

Figure 22

Natural logarithms can be found using a calculator. (Consult your owner's manual.) As in the case of common logarithms, when used in applications natural logarithms are usually irrational numbers. Figure 23 shows how three natural logarithms are evaluated with a graphing calculator. Figure 24 reinforces the fact that $\ln x$ is the exponent to which e must be raised in order to obtain x. ■

EXAMPLE 4 Measuring the Age of Rocks

Geologists sometimes measure the age of rocks by using "atomic clocks." By measuring the amounts of potassium 40 and argon 40 in a rock, the age t of the specimen in years is found with the formula

$$t = (1.26 \times 10^9) \frac{\ln\left[1 + 8.33\left(\frac{A}{K}\right)\right]}{\ln 2},$$

where A and K are the numbers of atoms of argon 40 and potassium 40, respectively, in the specimen.

(a) How old is a rock in which $A = 0$ and $K > 0$?

(b) The ratio $\frac{A}{K}$ for a sample of granite from New Hampshire is .212. How old is the sample?

Solution

(a) If $A = 0$, $\frac{A}{K} = 0$ and the equation becomes

$$t = (1.26 \times 10^9)\frac{\ln 1}{\ln 2} = (1.26 \times 10^9)(0) = 0.$$

The rock is new (0 yr old).

(b) Since $\frac{A}{K} = .212$, we have

$$t = (1.26 \times 10^9)\frac{\ln[1 + 8.33(.212)]}{\ln 2} \approx 1.85 \times 10^9.$$

The granite is about 1.85 billion yr old.

Now try Exercise 53.

Logarithms to Other Bases We can use a calculator to find the values of either natural logarithms (base e) or common logarithms (base 10). However, sometimes we must use logarithms to other bases. The following theorem can be used to convert logarithms from one base to another.

Looking Ahead to Calculus

In calculus, natural logarithms are more convenient to work with than logarithms to other bases. The change-of-base theorem enables us to convert any logarithmic function to a *natural* logarithmic function.

Change-of-Base Theorem

For any positive real numbers x, a, and b, where $a \neq 1$ and $b \neq 1$:

$$\log_a x = \frac{\log_b x}{\log_b a}.$$

Any positive number other than 1 can be used for base b in the change-of-base theorem, but usually the only practical bases are e and 10 since calculators give logarithms only for these two bases.

EXAMPLE 5 Using the Change-of-Base Theorem

Use the change-of-base theorem to find an approximation to four decimal places for each logarithm.

(a) $\log_5 17$ **(b)** $\log_2 .1$

Solution

```
log(17)/log(5)
            1.7604
ln(.1)/ln(2)
           -3.3219
```

The screen shows how the result of Example 5(a) can be found using *common* logarithms, and how the result of Example 5(b) can be found using *natural* logarithms. The results are the same as those in Example 5.

(a) $\log_5 17 = \frac{\ln 17}{\ln 5} \approx \frac{2.8332}{1.6094} \approx 1.7604$ Using natural logarithms

(b) $\log_2 .1 = \frac{\log .1}{\log 2} \approx \frac{-1.0000}{.3010} \approx -3.3219$ Using common logarithms

---

Now try Exercises 35 and 37.

N O T E In Example 5, we showed logarithms such as ln 17 and ln 5 evaluated in intermediate steps to four decimal places. However, the final answers were obtained *without* rounding these intermediate values, using all the digits obtained with the calculator. In general, wait until the final step to round off the answer; otherwise, a build-up of round-off errors may cause the final answer to have an incorrect final decimal place digit.

EXAMPLE 6 Modeling Diversity of Species

One measure of the diversity of the species in an ecological community is modeled by the formula

$$H = -[P_1 \log_2 P_1 + P_2 \log_2 P_2 + \cdots + P_n \log_2 P_n],$$

where $P_1, P_2, \ldots, P_n$ are the proportions of a sample that belong to each of n species found in the sample. (*Source:* Ludwig, J. and J. Reynolds, *Statistical Ecology: A Primer on Methods and Computing,* New York, Wiley, 1988, p. 92.)

Find the measure of diversity in a community with two species where there are 90 of one species and 10 of the other.

Solution Since there are 100 members in the community, $P_1 = \frac{90}{100} = .9$ and $P_2 = \frac{10}{100} = .1$, so

$$H = -[.9 \log_2 .9 + .1 \log_2 .1].$$

In Example 5(b), we found that $\log_2 .1 \approx -3.32$. Now we find $\log_2 .9$.

$$\log_2 .9 = \frac{\log .9}{\log 2} \approx \frac{-.0458}{.3010} \approx -.152$$

Therefore,

$$H = -[.9 \log_2 .9 + .1 \log_2 .1]$$
$$\approx -[.9(-.152) + .1(-3.32)] \approx .469.$$

Verify that $H \approx .971$ if there are 60 of one species and 40 of the other. As the proportions of n species get closer to $\frac{1}{n}$ each, the measure of diversity increases to a maximum of $\log_2 n$.

Now try Exercise 57.

Exponential and Logarithmic Equations We solved exponential equations in earlier sections. General methods for solving these equations depend on the property below, which follows from the fact that logarithmic functions are one-to-one.

Property of Logarithms

If $x > 0$, $y > 0$, $a > 0$, and $a \neq 1$, then

$$x = y \quad \text{if and only if} \quad \log_a x = \log_a y.$$

EXAMPLE 7   Solving an Exponential Equation

Solve $7^x = 12$. Give the solution to four decimal places.

Solution   The properties of exponents given in Section 9.1 cannot be used to solve this equation, so we apply the preceding property of logarithms. While any appropriate base b can be used, the best practical base is base 10 or base e. We choose base e (natural) logarithms here.

$$7^x = 12$$

$$\ln 7^x = \ln 12 \qquad \text{Property of logarithms}$$

$$x \ln 7 = \ln 12 \qquad \text{Power property (Section 9.2)}$$

$$x = \frac{\ln 12}{\ln 7} \qquad \text{Divide by } \ln 7.$$

$$x \approx 1.2770 \qquad \text{Use a calculator.}$$

The solution set is $\{1.2770\}$.

$Y_1 = 7^X$ $Y_2 = 12$

As seen in the display at the bottom of the screen, when rounded to four decimal places, the solution agrees with that found in Example 7.

Now try Exercise 63.

C A U T I O N   Be careful when evaluating a quotient like $\frac{\ln 12}{\ln 7}$ in Example 7. Do not confuse this quotient with $\ln \frac{12}{7}$, which can be written as $\ln 12 - \ln 7$. You *cannot* change the quotient of *two logarithms* to a difference of logarithms.

$$\frac{\ln 12}{\ln 7} \neq \ln \frac{12}{7}$$

EXAMPLE 8   Solving an Exponential Equation

Solve $3^{2x-1} = .4^{x+2}$. Give the solution to four decimal places.

Solution
$$3^{2x-1} = .4^{x+2}$$

$$\ln 3^{2x-1} = \ln .4^{x+2} \qquad \text{Take natural logarithms on both sides.}$$

$$(2x - 1) \ln 3 = (x + 2) \ln .4 \qquad \text{Power property}$$

$$2x \ln 3 - \ln 3 = x \ln .4 + 2 \ln .4 \qquad \text{Distributive property}$$

$$2x \ln 3 - x \ln .4 = 2 \ln .4 + \ln 3 \qquad \text{Write the terms with } x \text{ on one side.}$$

$$x(2 \ln 3 - \ln .4) = 2 \ln .4 + \ln 3 \qquad \text{Factor out } x.$$

$$x = \frac{2 \ln .4 + \ln 3}{2 \ln 3 - \ln .4} \qquad \text{Divide by } 2 \ln 3 - \ln .4.$$

$$x = \frac{\ln .16 + \ln 3}{\ln 9 - \ln .4} \qquad \text{Power property}$$

$$x = \frac{\ln .48}{\ln \frac{9}{.4}} \qquad \begin{array}{l} \text{Product property} \\ \text{Quotient property} \end{array} \text{(Section 9.2)}$$

$$x \approx -.2357 \qquad \text{Use a calculator.}$$

The solution set is $\{-.2357\}$.

$Y_2 = .4^{X+2}$ $Y_1 = 3^{2X-1}$

This screen supports the solution found in Example 8.

Now try Exercise 67.

EXAMPLE 9 Solving Base e Exponential Equations

Solve each equation. Give solutions to four decimal places.

(a) $e^{x^2} = 200$ **(b)** $e^{2x+1} \cdot e^{-4x} = 3e$

Solution

(a)
$$e^{x^2} = 200$$
$$\ln e^{x^2} = \ln 200 \qquad \text{Take natural logarithms on both sides.}$$
$$x^2 = \ln 200 \qquad \ln e^{x^2} = x^2 \text{ (Section 9.2)}$$
$$x = \pm\sqrt{\ln 200} \qquad \text{Square root property (Appendix A)}$$
$$x \approx \pm 2.3018 \qquad \text{Use a calculator.}$$

The solution set is $\{\pm 2.3018\}$.

(b)
$$e^{2x+1} \cdot e^{-4x} = 3e$$
$$e^{-2x+1} = 3e \qquad a^m \cdot a^n = a^{m+n}$$
$$e^{-2x} = 3 \qquad \text{Divide by } e; \frac{a^m}{a^n} = a^{m-n}.$$
$$\ln e^{-2x} = \ln 3 \qquad \text{Take natural logarithms on both sides.}$$
$$-2x \ln e = \ln 3 \qquad \text{Power property}$$
$$-2x = \ln 3 \qquad \ln e = 1$$
$$x = -\frac{1}{2} \ln 3 \qquad \text{Multiply by } -\tfrac{1}{2}.$$
$$x \approx -.5493$$

The solution set is $\{-.5493\}$.

Now try Exercises 69 and 75.

EXAMPLE 10 Solving a Logarithmic Equation

Solve $\log_a(x + 6) - \log_a(x + 2) = \log_a x$.

Solution
$$\log_a \frac{x + 6}{x + 2} = \log_a x \qquad \begin{array}{l}\text{Rewrite the equation using the}\\\text{quotient property.}\end{array}$$
$$\frac{x + 6}{x + 2} = x \qquad \text{Property of logarithms}$$
$$x + 6 = x(x + 2) \qquad \text{Multiply by } x + 2.$$
$$x + 6 = x^2 + 2x \qquad \text{Distributive property}$$
$$x^2 + x - 6 = 0 \qquad \text{Standard form}$$
$$(x + 3)(x - 2) = 0 \qquad \text{Factor.}$$
$$x = -3 \quad \text{or} \quad x = 2 \qquad \text{Zero-factor property (Appendix A)}$$

The negative solution ($x = -3$) is not in the domain of $\log_a x$ in the original equation, so the only valid solution is the number 2, giving the solution set $\{2\}$.

Now try Exercise 81.

CAUTION Recall that the domain of $y = \log_b x$ is $(0, \infty)$. For this reason, *it is always necessary to check that apparent solutions of a logarithmic equation result in logarithms of positive numbers in the original equation.*

EXAMPLE 11 Solving a Logarithmic Equation

Solve $\log(3x + 2) + \log(x - 1) = 1$.

Solution The notation $\log x$ is an abbreviation for $\log_{10} x$, and $1 = \log_{10} 10$.

$$\log(3x + 2) + \log(x - 1) = 1$$

$$\log(3x + 2) + \log(x - 1) = \log 10 \qquad \text{Substitute.}$$

$$\log[(3x + 2)(x - 1)] = \log 10 \qquad \text{Product property}$$

$$(3x + 2)(x - 1) = 10 \qquad \text{Property of logarithms}$$

$$3x^2 - x - 2 = 10 \qquad \text{Multiply.}$$

$$3x^2 - x - 12 = 0 \qquad \text{Subtract 10.}$$

$$x = \frac{1 \pm \sqrt{1 + 144}}{6} \qquad \text{Quadratic formula (\textbf{Appendix A})}$$

The number $\frac{1 - \sqrt{145}}{6}$ is negative, so $x - 1$ is negative. Therefore, $\log(x - 1)$ is not defined and this proposed solution must be discarded. Since $\frac{1 + \sqrt{145}}{6} > 1$, both $3x + 2$ and $x - 1$ are positive and the solution set is $\left\{ \frac{1 + \sqrt{145}}{6} \right\}$.

Now try Exercise 85.

EXAMPLE 12 Solving a Base *e* Logarithmic Equation

Solve $\ln e^{\ln x} - \ln(x - 3) = \ln 2$.

Solution

$$\ln e^{\ln x} - \ln(x - 3) = \ln 2$$

$$\ln x - \ln(x - 3) = \ln 2 \qquad e^{\ln x} = x \text{ (Section 9.2)}$$

$$\ln \frac{x}{x - 3} = \ln 2 \qquad \text{Quotient property}$$

$$\frac{x}{x - 3} = 2 \qquad \text{Property of logarithms}$$

$$x = 2(x - 3) \qquad \text{Multiply by } x - 3.$$

$$x = 2x - 6 \qquad \text{Distributive property}$$

$$6 = x \qquad \text{Solve for } x. \text{ (\textbf{Appendix A})}$$

Verify that the solution set is $\{6\}$.

Now try Exercise 91.

A summary of the methods used for solving equations in this section follows.

Solving Exponential or Logarithmic Equations

To solve an exponential or logarithmic equation, change the given equation into one of the following forms, where a and b are real numbers, $a > 0$, and $a \neq 1$.

1. $a^{f(x)} = b$
   Solve by taking logarithms on both sides.
2. $\log_a f(x) = b$
   Solve by changing to exponential form $a^b = f(x)$.
3. $\log_a f(x) = \log_a g(x)$
   The given equation is equivalent to the equation $f(x) = g(x)$. Solve algebraically.
4. In a more complicated equation, such as the one in Example 9(b), it may be necessary to first solve for $a^{f(x)}$ or $\log_a f(x)$ and then solve the resulting equation using one of the methods given above.

9.3 Exercises

Concept Check   *Answer each of the following.*

1. For the exponential function defined by $f(x) = a^x$, where $a > 1$, is the function increasing or is it decreasing over its entire domain?

2. For the logarithmic function defined by $g(x) = \log_a x$, where $a > 1$, is the function increasing or is it decreasing over its entire domain?

3. If $f(x) = 5^x$, what is the rule for $f^{-1}(x)$?

4. What is the name given to the exponent to which 4 must be raised in order to obtain 11?

5. A base e logarithm is called a(n) _____ logarithm; a base 10 logarithm is called a(n) _____ logarithm.

6. How is $\log_3 12$ written in terms of natural logarithms?

7. Why is $\log_2 0$ undefined?

8. Between what two consecutive integers must $\log_2 12$ lie?

9. The graph of $y = \log x$ shows a point on the graph. Write the logarithmic equation associated with that point.

10. The graph of $y = \ln x$ shows a point on the graph. Write the logarithmic equation associated with that point.

Use a calculator with logarithm keys to find an approximation to four decimal places for each expression. See Figures 20 and 23.

11. log 36   **12.** log 72   **13.** log .042   **14.** log .319

15. $\log(2 \times 10^4)$ **16.** $\log(2 \times 10^{-4})$ **17.** ln 36   **18.** ln 72

19. ln .042   **20.** ln .319   **21.** $\ln(2 \times e^4)$ **22.** $\ln(2 \times e^{-4})$

For each substance, find the pH from the given hydronium ion concentration. See Example 1(a).

23. grapefruit, 6.3×10^{-4} **24.** crackers, 3.9×10^{-9}

25. limes, 1.6×10^{-2} **26.** sodium hydroxide (lye), 3.2×10^{-14}

Find the $[H_3O^+]$ for each substance with the given pH. See Example 1(b).

27. soda pop, 2.7   **28.** wine, 3.4

29. beer, 4.8   **30.** drinking water, 6.5

In Exercises 31–33, suppose that water from a wetland area is sampled and found to have the given hydronium ion concentration. Determine whether the wetland is a rich fen, poor fen, or bog. See Example 2.

31. 2.49×10^{-5} **32.** 2.49×10^{-2} **33.** 2.49×10^{-7}

34. Use your calculator to find an approximation for each logarithm.

     (a) log 398.4   **(b)** log 39.84   **(c)** log 3.984

     (d) From your answers to parts (a)–(c), make a conjecture concerning the decimal values in the approximations of common logarithms of numbers greater than 1 that have the same digits.

Use the change-of-base theorem to find an approximation to four decimal places for each logarithm. See Example 5.

35. $\log_2 5$ **36.** $\log_2 9$ **37.** $\log_8 .59$ **38.** $\log_8 .71$

39. $\log_{\sqrt{13}} 12$ **40.** $\log_{\sqrt{19}} 5$ **41.** $\log_{.32} 5$ **42.** $\log_{.91} 8$

43. Given $g(x) = e^x$, evaluate **(a)** $g(\ln 3)$ **(b)** $g[\ln(5^2)]$ **(c)** $g\left[\ln\left(\dfrac{1}{e}\right)\right]$.

44. Given $f(x) = 3^x$, evaluate **(a)** $f(\log_3 7)$ **(b)** $f[\log_3(\ln 3)]$ **(c)** $f[\log_3(2 \ln 3)]$.

45. Given $f(x) = \ln x$, evaluate **(a)** $f(e^5)$ **(b)** $f(e^{\ln 3})$ **(c)** $f(e^{2\ln 3})$.

46. Given $f(x) = \log_2 x$, evaluate **(a)** $f(2^3)$ **(b)** $f(2^{\log_2 2})$ **(c)** $f(2^{2\log_2 2})$.

Solve each application of logarithms. See Examples 3 and 4.

47. *Decibel Levels*   Find the decibel ratings of sounds having the following intensities.

     (a) $100I_0$ **(b)** $1000I_0$ **(c)** $100,000I_0$ **(d)** $1,000,000I_0$

     (e) If the intensity of a sound is doubled, by how much is the decibel rating increased?

48. *Decibel Levels* Find the decibel ratings of the following sounds, having intensities as given. Round each answer to the nearest whole number.

(a) whisper, $115I_0$ **(b)** busy street, $9{,}500{,}000I_0$
(c) heavy truck, 20 m away, $1{,}200{,}000{,}000I_0$
(d) rock music, $895{,}000{,}000{,}000I_0$
(e) jetliner at takeoff, $109{,}000{,}000{,}000{,}000I_0$

49. *Earthquake Intensity* The magnitude of an earthquake, measured on the Richter scale, is $\log_{10} \frac{I}{I_0}$, where I is the amplitude registered on a seismograph 100 km from the epicenter of the earthquake, and I_0 is the amplitude of an earthquake of a certain (small) size. Find the Richter scale ratings for earthquakes having the following amplitudes.

(a) $1000I_0$ **(b)** $1{,}000{,}000I_0$ **(c)** $100{,}000{,}000I_0$

50. *Earthquake Intensity* On June 16, 1999, the city of Puebla in central Mexico was shaken by an earthquake that measured 6.7 on the Richter scale. Express this reading in terms of I_0. See Exercise 49. *(Source: Times Picayune.)*

51. *Earthquake Intensity* On September 19, 1985, Mexico's largest recent earthquake, measuring 8.1 on the Richter scale, killed about 9500 people. Express the magnitude of an 8.1 reading in terms of I_0. *(Source: Times Picayune.)*

52. Compare your answers to Exercises 50 and 51. How much greater was the force of the 1985 earthquake than the 1999 earthquake?

53. *(Modeling) Visitors to U.S. National Parks* The heights of the bars in the graph represent the number of visitors (in millions) to U.S. National Parks from 1950–2000. Suppose x represents the number of years since 1900—thus, 1950 is represented by 50, 1960 is represented by 60, and so on. The logarithmic function defined by

$$f(x) = -269 + 73 \ln x$$

closely models the data. Use this function to estimate the number of visitors in the year 2004. What

Visitors to National Parks (in millions)

Source: Statistical Abstract of the United States, 2000.

assumption must we make to estimate the number of visitors in years beyond 2000?

54. *(Modeling) Volunteerism among College Freshmen* The growth in the percentage of college freshmen who reported involvement in volunteer work during their last year of high school is shown in the bar graph. Connecting the tops of the bars with a continuous curve would give a graph that indicates logarithmic growth. The function defined by

$$f(t) = 74.61 + 3.84 \ln t, \quad t \geq 1,$$

Volunteerism among College Freshman

Source: Higher Education Research Institute, UCLA.

where t represents the number of years since 1994 and $f(t)$ is the percent, approximates the curve reasonably well.

(a) What does the function predict for the percent of freshmen entering college in 2002 who performed volunteer work during their last year of high school? How does this compare to the actual percent of 82.6?

(b) Explain why an exponential function would *not* provide a good model for these data.

55. *(Modeling) Diversity of Species*   The number of species in a sample is given by

$$S(n) = a \ln\left(1 + \frac{n}{a}\right).$$

Here n is the number of individuals in the sample, and a is a constant that indicates the diversity of species in the community. If $a = .36$, find $S(n)$ for each value of n. (*Hint:* $S(n)$ must be a whole number.)

(a) 100   (b) 200   (c) 150   (d) 10

56. *(Modeling) Diversity of Species*   In Exercise 55, find $S(n)$ if a changes to .88. Use the following values of n.

(a) 50   (b) 100   (c) 250

57. *(Modeling) Diversity of Species*   Suppose a sample of a small community shows two species with 50 individuals each. Find the measure of diversity H. (See Example 6.)

58. *(Modeling) Diversity of Species*   A virgin forest in northwestern Pennsylvania has 4 species of large trees with the following proportions of each: hemlock, .521; beech, .324; birch, .081; maple, .074. Find the measure of diversity H. (See Example 6.)

Concept Check   An exponential equation such as $5^x = 9$ can be solved for its exact solution using the meaning of logarithm and the change-of-base theorem. Since x is the exponent to which 5 must be raised in order to obtain 9, the exact solution is $\log_5 9$, or $\frac{\log 9}{\log 5}$ or $\frac{\ln 9}{\ln 5}$. For the following equations, give the exact solution in three forms similar to the forms explained here.

59. $7^x = 19$ **60.** $3^x = 10$ **61.** $\left(\frac{1}{2}\right)^x = 12$ **62.** $\left(\frac{1}{3}\right)^x = 4$

Solve each equation. When solutions are irrational, give them as decimals correct to four decimal places. See Examples 7–12.

63. $3^x = 6$ **64.** $4^x = 12$ **65.** $6^{1-2x} = 8$

66. $3^{2x-5} = 13$ **67.** $2^{x+3} = 5^x$ **68.** $6^{x+3} = 4^x$

69. $e^{x-1} = 4$ **70.** $e^{2-x} = 12$ **71.** $2e^{5x+2} = 8$

72. $10e^{3x-7} = 5$ **73.** $2^x = -3$ **74.** $3^x = -6$

75. $e^{8x} \cdot e^{2x} = e^{20}$ **76.** $e^{6x} \cdot e^x = e^{21}$

77. $\ln(6x + 1) = \ln 3$ **78.** $\ln(7 - x) = \ln 12$

79. $\log 4x - \log(x - 3) = \log 2$ **80.** $\ln(-x) + \ln 3 = \ln(2x - 15)$

81. $\log(2x - 1) + \log 10x = \log 10$ **82.** $\ln 5x - \ln(2x - 1) = \ln 4$

83. $\log(x + 25) = 1 + \log(2x - 7)$ **84.** $\ln(5 + 4x) - \ln(3 + x) = \ln 3$

85. $\log x + \log(3x - 13) = 1$ **86.** $\ln(2x + 5) + \ln x = \ln 7$

87. $\log_6 4x - \log_6(x - 3) = \log_6 12$ **88.** $\log_2 3x + \log_2 3 = \log_2(2x + 15)$

89. $5^{x+2} = 2^{2x-1}$ **90.** $6^{x-3} = 3^{4x+1}$

91. $\ln e^x - \ln e^3 = \ln e^5$ **92.** $\ln e^x - 2 \ln e = \ln e^4$

93. $\log_2(\log_2 x) = 1$ **94.** $\log x = \sqrt{\log x}$

95. $\log x^2 = (\log x)^2$ **96.** $\log_2 \sqrt{2x^2} = \frac{3}{2}$

(Modeling) Solve each application.

Year	Average Annual Cost
2000	$7,990
2001	$8,470
2002	$9,338
2003	$9,805
2004	$10,295
2005	$10,810
2006	$11,351
2007	$11,918
2008	$12,514

Source: www.princetonreview.com

97. *Average Annual Public University Costs* The table shows the cost of a year's tuition, room and board, and fees at a public university from 2000–2008. (*Note:* The amounts for 2004–2008 are projections.) Letting y represent the cost and x represent the number of years since 2000, we find that the function defined by

$$f(x) = 8160(1.06)^x$$

models the data quite well. According to this function, when will the cost in 2000 be doubled?

98. *Race Speed* At the World Championship races held at Rome's Olympic Stadium in 1987, American sprinter Carl Lewis ran the 100-m race in 9.86 sec. His speed in meters per second after t seconds is closely modeled by the function defined by

$$f(t) = 11.65(1 - e^{-t/1.27}).$$

(*Source:* Banks, R. B., *Towing Icebergs, Falling Dominoes, and Other Adventures in Applied Mathematics,* Princeton University Press, 1998.)

(a) How fast was he running as he crossed the finish line?

(b) After how many seconds was he running at the rate of 10 m per sec?

99. *Fatherless Children* The percent of U.S. children growing up without a father has increased rapidly since 1950. If x represents the number of years since 1900, the function defined by

$$f(x) = \frac{25}{1 + 1364.3e^{-x/9.316}}$$

models the percent fairly well. (*Source:* National Longitudinal Survey of Youth; U.S. Department of Commerce; U.S. Bureau of the Census.)

(a) What percent of U.S. children lived in a home without a father in 1997?

(b) In what year were 10% of these children living in a home without a father?

100. *Height of the Eiffel Tower* The right side of the Eiffel Tower in Paris, France has a shape that can be approximated by the graph of the function defined by

$$f(x) = -301 \ln \frac{x}{207}.$$

See the figure. (*Source:* Banks, R. B., *Towing Icebergs, Falling Dominoes, and Other Adventures in Applied Mathematics,* Princeton University Press, 1998.)

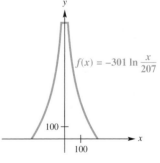

$f(x) = -301 \ln \dfrac{x}{207}$

(a) Explain why the shape of the left side of the Eiffel Tower has the formula given by $f(-x)$.

(b) The short horizontal line at the top of the figure has length 15.7488 ft. Approximately how tall is the Eiffel Tower?

(c) Approximately how far from the center of the tower is the point on the right side that is 500 ft above the ground?

Chapter 9 Summary

KEY TERMS

9.1 exponential function
exponential equation
compound interest
future value

present value
compound amount
continuous
   compounding

9.2 logarithm
logarithmic equation
logarithmic function

9.3 common logarithm
pH
natural logarithm

NEW SYMBOLS

e a constant, approximately 2.718281828

$\log_a x$ the logarithm of x to the base a

$\log x$ common (base 10) logarithm of x

$\ln x$ natural (base e) logarithm of x

QUICK REVIEW

CONCEPTS	EXAMPLES

9.1 Exponential Functions

Additional Properties of Exponents
For any real number $a > 0$, $a \neq 1$,

(a) a^x is a unique real number for all real numbers x.
(b) $a^b = a^c$ if and only if $b = c$.
(c) If $a > 1$ and $m < n$, then $a^m < a^n$.

(d) If $0 < a < 1$ and $m < n$, then $a^m > a^n$.

(a) 2^x is a unique real number for all real numbers x.
(b) $2^x = 2^3$ if and only if $x = 3$.
(c) $2^5 < 2^{10}$, because $2 > 1$ and $5 < 10$.

(d) $\left(\dfrac{1}{2}\right)^5 > \left(\dfrac{1}{2}\right)^{10}$ because $0 < \dfrac{1}{2} < 1$ and $5 < 10$.

Exponential Function
If $a > 0$ and $a \neq 1$, then $f(x) = a^x$ defines the exponential function with base a.

$f(x) = 3^x$ defines the exponential function with base 3.

Graph of $f(x) = a^x$
1. The graph contains the points $\left(-1, \frac{1}{a}\right)$, $(0, 1)$, and $(1, a)$.
2. If $a > 1$, then f is an increasing function; if $0 < a < 1$, then f is a decreasing function.
3. The x-axis is a horizontal asymptote.
4. The domain is $(-\infty, \infty)$; the range is $(0, \infty)$.

x	y
-1	$\frac{1}{3}$
0	1
1	3

CONCEPTS	EXAMPLES

9.2 Logarithmic Functions

Logarithmic Function

If $a > 0$, $a \neq 1$, and $x > 0$, then $f(x) = \log_a x$ defines the logarithmic function with base a.

$f(x) = \log_3 x$ defines the logarithmic function with base 3.

Graph of $f(x) = \log_a x$
1. The graph contains the points $\left(\frac{1}{a}, -1\right)$, $(1, 0)$, and $(a, 1)$.
2. If $a > 1$, then f is an increasing function; if $0 < a < 1$, then f is a decreasing function.
3. The y-axis is a vertical asymptote.
4. The domain is $(0, \infty)$; the range is $(-\infty, \infty)$.

x	y
$\frac{1}{3}$	-1
1	0
3	1

Properties of Logarithms

For any positive real numbers x and y, real number r, and positive real number a, $a \neq 1$:

$$\log_a xy = \log_a x + \log_a y \quad \text{Product property}$$

$$\log_a \frac{x}{y} = \log_a x - \log_a y \quad \text{Quotient property}$$

$$\log_a x^r = r \log_a x. \quad \text{Power property}$$

Also,

$$\log_a 1 = 0 \quad \text{and} \quad \log_a a = 1.$$

$$\log_2 (3 \cdot 5) = \log_2 3 + \log_2 5$$

$$\log_2 \frac{3}{5} = \log_2 3 - \log_2 5$$

$$\log_6 3^5 = 5 \log_6 3$$

$$\log_{10} 1 = 0 \quad \text{and} \quad \log_{10} 10 = 1$$

Theorem on Inverses

$$a^{\log_a x} = x \quad \text{and} \quad \log_a (a^x) = x$$

$$e^{\ln 4} = 4 \quad \text{and} \quad \ln e^2 = 2$$

9.3 Evaluating Logarithms; Equations and Applications

Common and Natural Logarithms

For all positive numbers x,

$$\log x = \log_{10} x \quad \text{Common logarithm}$$

$$\ln x = \log_e x. \quad \text{Natural logarithm}$$

$$\log .045 \approx -1.3468$$

$$\ln 247.1 \approx 5.5098$$

Change-of-Base Theorem

For any positive real numbers x, a, and b, where $a \neq 1$ and $b \neq 1$:

$$\log_a x = \frac{\log_b x}{\log_b a}.$$

$$\log_8 7 = \frac{\log 7}{\log 8} = \frac{\ln 7}{\ln 8} \approx .9358$$

Property of Logarithms

If $x > 0$, $y > 0$, $a > 0$, and $a \neq 1$, then

$$x = y \quad \text{if and only if} \quad \log_a x = \log_a y.$$

If $e^{5x} = 10$, then $\ln e^{5x} = \ln 10$.

If $\log_a (x + 1) = \log_a 7$, then $x + 1 = 7$.

Chapter 9 Review Exercises

Match each equation with the figure that most closely resembles its graph.

1. $y = \log_{.3} x$

2. $y = e^x$

3. $y = \ln x$

4. $y = (.3)^x$

A.

B.

C.

D.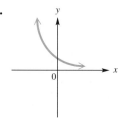

Write each equation in logarithmic form.

5. $2^5 = 32$

6. $100^{1/2} = 10$

7. $\left(\dfrac{3}{4}\right)^{-1} = \dfrac{4}{3}$

8. Graph $y = (1.5)^{x+2}$.

9. *Concept Check*   Is the logarithm to the base 3 of 4 written as $\log_4 3$ or $\log_3 4$?

Relating Concepts

For individual or collaborative investigation
(Exercises 10–15)

Work Exercises 10–15 in order.

10. What is the exact value of $\log_3 9$?

11. What is the exact value of $\log_3 27$?

12. Between what two consecutive integers must $\log_3 16$ lie?

13. Use the change-of-base theorem to support your answer for Exercise 12.

14. Repeat Exercises 10 and 11 for $\log_5 \frac{1}{5}$ and $\log_5 1$.

15. Repeat Exercises 12 and 13 for $\log_5 .68$.

Write each equation in exponential form.

16. $\log_9 27 = \dfrac{3}{2}$

17. $\log 3.45 \approx .5378$

18. $\ln 45 \approx 3.8067$

19. *Concept Check*   What is the base of the logarithmic function whose graph contains the point $(81, 4)$?

20. *Concept Check*   What is the base of the exponential function whose graph contains the point $\left(-4, \frac{1}{16}\right)$?

Use properties of logarithms to rewrite each expression. Simplify the result, if possible. Assume all variables represent positive numbers.

21. $\log_3 \dfrac{mn}{5r}$

22. $\log_5\left(x^2 y^4 \sqrt[5]{m^3 p}\right)$

23. $\log_7(7k + 5r^2)$

Find each logarithm. Round to four decimal places.

24. $\log 45.6$

25. $\log .0411$

26. $\ln 470$

27. $\ln 144{,}000$

28. $\log_3 769$

29. $\log_{2/3} \dfrac{5}{8}$

Solve each equation. Round irrational solutions to four decimal places.

30. $8^x = 32$

31. $x^{-3} = \dfrac{8}{27}$

32. $10^{2x-3} = 17$

33. $4^{x+3} = 5^{2-x}$

34. $e^{x+1} = 10$

35. $\log_{64} x = \dfrac{1}{3}$

36. $\ln(6x) - \ln(x + 1) = \ln 4$

37. $\log_{16}\sqrt{x + 1} = \dfrac{1}{4}$

38. $\ln x + 3 \ln 2 = \ln \dfrac{2}{x}$

39. $\log x + \log(x - 3) = 1$

Solve each problem.

40. *Earthquake Intensity* On July 14, 1991, Peshawar, Pakistan, was shaken by an earthquake that measured 6.6 on the Richter scale.

   (a) Express this reading in terms of I_0. (See Section 9.3 Exercises.)
   (b) In February of the same year a quake measuring 6.5 on the Richter scale killed about 900 people in the mountains of Pakistan and Afghanistan. Express the magnitude of a 6.5 reading in terms of I_0.
   (c) How much greater was the force of the earthquake with measure 6.6?

41. *Earthquake Intensity*

   (a) The San Francisco earthquake of 1906 had a Richter scale rating of 8.3. Express the magnitude of this earthquake as a multiple of I_0.
   (b) In 1989, the San Francisco region experienced an earthquake with a Richter scale rating of 7.1. Express the magnitude of this earthquake as a multiple of I_0.
   (c) Compare the magnitudes of the two San Francisco earthquakes discussed in parts (a) and (b).

42. *(Modeling) Decibel Levels* Recall from Section 9.3 that the model for the decibel rating of the loudness of a sound is

$$d = 10 \log \dfrac{I}{I_0}.$$

A few years ago, there was a controversy about a proposed government limit on factory noise. One group wanted a maximum of 89 decibels, while another group wanted 86. This difference seemed very small to many people. Find the percent by which the 89-decibel intensity exceeds that for 86 decibels.

43. *Interest Rate* What annual interest rate, to the nearest tenth, will produce $5760 if $3500 is left at interest (compounded annually) for 10 yr?

44. *Growth of an Account* Find the number of years (to the nearest tenth) needed for $48,000 to become $58,344 at 5% interest compounded semiannually.

45. *Growth of an Account* Manuel deposits $10,000 for 12 yr in an account paying 8% compounded annually. He then puts this total amount on deposit in another account paying 10% compounded semiannually for another 9 yr. Find the total amount on deposit after the entire 21-yr period.

46. *Growth of an Account* Anne Kelly deposits $12,000 for 8 yr in an account paying 5% compounded annually. She then leaves the money alone with no further deposits at 6% compounded annually for an additional 6 yr. Find the total amount on deposit after the entire 14-yr period.

47. *Cost from Inflation* If the inflation rate were 4%, use the formula for continuous compounding to find the number of years, to the nearest tenth, for a $1 item to cost $2.

48. *(Modeling) Drug Level in the Bloodstream* After a medical drug is injected directly into the bloodstream, it is gradually eliminated from the body. Graph the following functions on the interval $[0, 10]$. Use $[0, 500]$ for the range of $A(t)$. Determine the function that best models the amount $A(t)$ (in milligrams) of a drug remaining in the body after t hours if 350 mg were initially injected.

(a) $A(t) = t^2 - t + 350$ 　　　　**(b)** $A(t) = 350 \log(t + 1)$

(c) $A(t) = 350(.75)^t$ 　　　　　　**(d)** $A(t) = 100(.95)^t$

49. *(Modeling) New York Yankees' Payroll* The table shows the total payroll (in millions of dollars) of the New York Yankees baseball team for the years 2000–2003.

Year	Total Payroll (millions of dollars)
2000	92.9
2001	112.3
2002	125.9
2003	152.7

Source: USA Today.

Letting y represent the total payroll and x represent the number of years since 2000, we find that the function defined by

$$f(x) = 93.54e^{.16x}$$

models the data quite well. According to this function, when would the total payroll double its 2003 value?

50. *(Modeling) Transistors on Computer Chips* Computing power of personal computers has increased dramatically as a result of the ability to place an increasing number of transistors on a single processor chip. The table lists the number of transistors on some popular computer chips made by Intel.

(a) Make a scatter diagram of the data. Let the x-axis represent the year, where $x = 0$ corresponds to 1971, and let the y-axis represent the number of transistors.

Year	Chip	Transistors
1971	4004	2300
1986	386DX	275,000
1989	486DX	1,200,000
1993	Pentium	3,300,000
1995	P6	5,500,000
1997	Pentium 3	9,500,000
2000	Pentium 4	42,000,000

Source: Intel.

(b) Decide whether a linear, a logarithmic, or an exponential function describes the data best.

(c) Determine a function f that approximates these data. Plot f and the data on the same coordinate axes.

(d) Assuming that the present trend continues, use f to predict the number of transistors on a chip in the year 2005.

Chapter 9 Test

1. Match each equation with its graph.

   (a) $y = \log_{1/3} x$ (b) $y = e^x$ (c) $y = \ln x$ (d) $y = \left(\dfrac{1}{3}\right)^x$

   A.   B.   C.   D.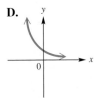

2. Solve $\left(\dfrac{1}{8}\right)^{2x-3} = 16^{x+1}$.

3. Write $4^{3/2} = 8$ in logarithmic form.

4. Write $\log_8 4 = \dfrac{2}{3}$ in exponential form.

5. Graph $f(x) = \left(\dfrac{1}{2}\right)^x$ and $g(x) = \log_{1/2} x$ on the same axes. What is their relationship?

6. Use properties of logarithms to write $\log_7 \dfrac{x^2 \sqrt[4]{y}}{z^3}$ as a sum, difference, or product of logarithms. Assume all variables represent positive numbers.

Use a calculator to find an approximation for each logarithm. Express answers to four decimal places.

7. $\log 237.4$

8. $\ln .0467$

9. $\log_9 13$

10. $\log(2.49 \times 10^{-3})$

Use properties of logarithms to solve each equation. Express the solutions as directed.

11. $\log_x 25 = 2$ (exact)

12. $\log_4 32 = x$ (exact)

13. $\log_2 x + \log_2(x + 2) = 3$ (exact)

14. $5^{x+1} = 7^x$ (approximate, to four decimal places)

15. $\ln x - 4 \ln 3 = \ln \dfrac{5}{x}$ (approximate, to four decimal places)

16. One of your friends is taking another mathematics course and tells you, "I have no idea what an expression like $\log_5 27$ really means." Write an explanation of what it means, and tell how you can find an approximation for it with a calculator.

Solve each problem.

17. *(Modeling) Skydiver Fall Speed* A skydiver in free fall travels at a speed modeled by

$$v(t) = 176(1 - e^{-.18t})$$

feet per second after t seconds. How long will it take for the skydiver to attain a speed of 147 ft per sec (100 mph)?

18. *Growth of an Account* How many years, to the nearest tenth, will be needed for $5000 to increase to $18,000 at 6.8% compounded **(a)** monthly **(b)** continuously?

Chapter 9 Quantitative Reasoning

Financial Planning for Retirement—What are your options?

The traditional IRA (Individual Retirement Account) is a common tax-deferred savings plan in the United States. Earned income deposited into an IRA is not taxed in the current year, and no taxes are incurred on the interest paid in subsequent years. However, when you withdraw the money from the account after age $59\frac{1}{2}$, you pay taxes on the entire amount.

Suppose you deposited $3000 of earned income into an IRA, you can earn an annual interest rate of 8%, and you are in a 40% tax bracket. (*Note:* Interest rates and tax brackets are subject to change over time, but some assumptions must be made to evaluate the investment.) Also, suppose you deposit the $3000 at age 25 and withdraw it at age 60.

1. How much money will remain after you pay the taxes at age 60?

2. Suppose that instead of depositing the money into an IRA, you pay taxes on the money and the annual interest. How much money will you have at age 60? (*Note:* You effectively start with $1800 (60% of $3000), and the money earns 4.8% (60% of 8%) interest after taxes.)

3. To the nearest dollar, how much additional money will you earn with the IRA?

4. Suppose you pay taxes on the original $3000, but are then able to earn 8% in a tax-free investment. Compare your balance at age 60 with the IRA balance.

Appendices

A | Equations and Inequalities

Equations ▪ Solving Linear Equations ▪ Solving Quadratic Equations ▪ Inequalities ▪ Solving Linear Inequalities ▪ Interval Notation ▪ Three-Part Inequalities

Equations Recall from algebra that an **equation** is a statement that two expressions are equal.

$$x + 2 = 9, \qquad 11x = 5x + 6x, \qquad x^2 - 2x - 1 = 0 \qquad \text{Equations}$$

To solve an equation means to find all numbers that make the equation a true statement. These numbers are called **solutions** or **roots** of the equation. A number that is a solution of an equation is said to *satisfy* the equation, and the solutions of an equation make up its **solution set.** Equations with the same solution set are **equivalent equations.** For example, $x = 4$, $x + 1 = 5$, and $6x + 3 = 27$ are equivalent equations because they have the same solution set, $\{4\}$. However, the equations $x^2 = 9$ and $x = 3$ are not equivalent, since the first has solution set $\{-3, 3\}$ while the solution set of the second is $\{3\}$.

To solve an equation, we use the *addition and multiplication properties of equality.* For real numbers a, b, and c:

$$a = b \text{ and } a + c = b + c \text{ are equivalent,}$$

$$a = b \text{ and } ac = bc \ (c \neq 0) \text{ are equivalent.}$$

That is, the same number may be added to (or subtracted from) both sides of an equation and both sides of an equation may be multiplied (or divided) by the same nonzero number, without changing the solution set.

Solving Linear Equations A **linear equation in one variable** is an equation that can be written in the form

$$ax + b = 0,$$

where a and b are real numbers with $a \neq 0$. A linear equation is also called a *first-degree* equation since the greatest degree of the variable is one.

$$3x + \sqrt{2} = 0, \qquad \frac{3}{4}x = 12, \qquad .5(x + 3) = 2x - 6 \qquad \text{Linear equations}$$

$$\sqrt{x} + 2 = 5, \qquad \frac{1}{x} = -8, \qquad x^2 + 3x + .2 = 0 \qquad \text{Nonlinear equations}$$

433

EXAMPLE 1 Solving a Linear Equation

Solve $3(2x - 4) = 7 - (x + 5)$.

Solution

$$3(2x - 4) = 7 - (x + 5)$$

$$3(2x) + 3(-4) = 7 - x - 5 \qquad \text{Distributive property}$$

$$6x - 12 = 7 - x - 5 \qquad \text{Multiply.}$$

$$6x - 12 = 2 - x \qquad \text{Combine terms.}$$

$$6x - 12 + x = 2 - x + x \qquad \text{Add } x \text{ to each side.}$$

$$7x - 12 = 2 \qquad \text{Combine terms.}$$

$$7x - 12 + 12 = 2 + 12 \qquad \text{Add 12 to each side.}$$

$$7x = 14 \qquad \text{Combine terms.}$$

$$\frac{7x}{7} = \frac{14}{7} \qquad \text{Divide each side by 7.}$$

$$x = 2$$

Check:
$$3(2x - 4) = 7 - (x + 5) \qquad \text{Original equation}$$

$$3(2 \cdot 2 - 4) = 7 - (2 + 5) \qquad ? \quad \text{Let } x = 2.$$

$$3(4 - 4) = 7 - (7) \qquad ?$$

$$0 = 0 \qquad \text{True}$$

Since replacing x with 2 results in a true statement, 2 is a solution of the given equation. The solution set is $\{2\}$.

Now try Exercise 9.

EXAMPLE 2 Clearing Fractions before Solving a Linear Equation

Solve $\dfrac{2t + 4}{3} + \dfrac{1}{2}t = \dfrac{1}{4}t - \dfrac{7}{3}$.

Solution

$$\frac{2t + 4}{3} + \frac{1}{2}t = \frac{1}{4}t - \frac{7}{3}$$

$$12\left(\frac{2t + 4}{3} + \frac{1}{2}t\right) = 12\left(\frac{1}{4}t - \frac{7}{3}\right) \qquad \begin{array}{l} \text{Multiply by 12, the least} \\ \text{common denominator} \\ \text{(LCD) of the fractions.} \end{array}$$

$$12\left(\frac{2t + 4}{3}\right) + 12\left(\frac{1}{2}t\right) = 12\left(\frac{1}{4}t\right) + 12\left(-\frac{7}{3}\right) \qquad \begin{array}{l} \text{Distributive} \\ \text{property} \end{array}$$

$$4(2t + 4) + 6t = 3t - 28 \qquad \text{Multiply.}$$

$$8t + 16 + 6t = 3t - 28 \qquad \text{Distributive property}$$

$$14t + 16 = 3t - 28 \qquad \text{Combine terms.}$$

$$11t = -44 \qquad \text{Subtract } 3t; \text{ subtract 16.}$$

$$t = -4 \qquad \text{Divide by 11.}$$

Check: $\dfrac{2(-4)+4}{3} + \dfrac{1}{2}(-4) = \dfrac{1}{4}(-4) - \dfrac{7}{3}$? Let $t = -4$.

$$\dfrac{-4}{3} + (-2) = -1 - \dfrac{7}{3}$$?

$$-\dfrac{10}{3} = -\dfrac{10}{3}$$ True

The solution set is $\{-4\}$.

Now try Exercise 17.

   An equation satisfied by every number that is a meaningful replacement for the variable is called an **identity.** The equation $3(x+1) = 3x + 3$ is an example of an identity. An equation that is satisfied by some numbers but not others, such as $2x = 4$, is called a **conditional equation.** The equations in Examples 1 and 2 are conditional equations. An equation that has no solution, such as $x = x + 1$, is called a **contradiction.**

EXAMPLE 3   Identifying Types of Equations

Decide whether each equation is an identity, a conditional equation, or a contradiction. Give the solution set.

(a) $-2(x+4) + 3x = x - 8$ **(b)** $5x - 4 = 11$ **(c)** $3(3x - 1) = 9x + 7$

Solution

(a) $-2(x+4) + 3x = x - 8$

$\qquad -2x - 8 + 3x = x - 8$ Distributive property

$\qquad\qquad\quad x - 8 = x - 8$ Combine terms.

$\qquad\qquad\qquad\quad 0 = 0$ Subtract x; add 8.

   When a *true* statement such as $0 = 0$ results, the equation is an identity, and the solution set is {all real numbers}.

(b) $5x - 4 = 11$

$\qquad\quad 5x = 15$ Add 4.

$\qquad\quad\; x = 3$ Divide by 5.

   This is a conditional equation, and its solution set is $\{3\}$.

(c) $3(3x - 1) = 9x + 7$

$\qquad 9x - 3 = 9x + 7$ Distributive property

$\qquad\qquad -3 = 7$ Subtract $9x$.

   When a *false* statement such as $-3 = 7$ results, the equation is a contradiction, and the solution set is the **empty set** or **null set,** symbolized $\emptyset$.

Now try Exercises 23, 25, and 27.

Solving Quadratic Equations An equation that can be written in the form

$$ax^2 + bx + c = 0,$$

where a, b, and c are real numbers with $a \neq 0$, is a **quadratic equation.** A quadratic equation is a *second-degree* equation—that is, an equation with a squared term and no terms of greater degree.

$$x^2 = 25, \qquad 4x^2 + 4x - 5 = 0, \ \backslash \ 3x^2 = 4x - 8 \qquad \text{Quadratic equations}$$

A quadratic equation written in the form $ax^2 + bx + c = 0$ is in *standard form.*

Factoring is the simplest method of solving a quadratic equation (but one that is not always easily applied). This method depends on the following **zero-factor property.**

> If a and b **are complex numbers with** $ab = 0,$ Zero-factor property
> **then** $a = 0$ **or** $b = 0$ **or both.**

EXAMPLE 4 Using the Zero-Factor Property

Solve $6x^2 + 7x = 3$.

Solution

$$6x^2 + 7x = 3$$
$$6x^2 + 7x - 3 = 0 \qquad \text{Standard form}$$
$$(3x - 1)(2x + 3) = 0 \qquad \text{Factor.}$$

$$3x - 1 = 0 \qquad \text{or} \qquad 2x + 3 = 0 \qquad \text{Zero-factor property}$$
$$3x = 1 \qquad \text{or} \qquad 2x = -3 \qquad \text{Solve each equation.}$$
$$x = \frac{1}{3} \qquad \text{or} \qquad x = -\frac{3}{2}$$

Check: $\qquad 6x^2 + 7x = 3 \qquad$ Original equation

$$6\left(\frac{1}{3}\right)^2 + 7\left(\frac{1}{3}\right) = 3 \quad \text{Let } x = \tfrac{1}{3}. \qquad 6\left(-\frac{3}{2}\right)^2 + 7\left(-\frac{3}{2}\right) = 3 \quad \text{Let } x = -\tfrac{3}{2}.$$

$$\frac{6}{9} + \frac{7}{3} = 3 \quad ? \qquad\qquad \frac{54}{4} - \frac{21}{2} = 3 \quad ?$$

$$3 = 3 \quad \text{True} \qquad\qquad\qquad 3 = 3 \quad \text{True}$$

Both values check, so the solution set is $\left\{\frac{1}{3}, -\frac{3}{2}\right\}$.

Now try Exercise 35.

A quadratic equation of the form $x^2 = k$ can be solved using the **square root property.**

> **If** $x^2 = k$**, then** $x = \sqrt{k}$ **or** $x = -\sqrt{k}$.   Square root property

That is, the solution set of $x^2 = k$ is $\left\{\sqrt{k}, -\sqrt{k}\right\}$, which may be abbreviated $\left\{\pm\sqrt{k}\right\}$.

EXAMPLE 5 Using the Square Root Property

Solve each quadratic equation.

(a) $x^2 = 17$

(b) $(x - 4)^2 = 12$

Solution

(a) By the square root property, the solution set of $x^2 = 17$ is $\left\{\pm\sqrt{17}\right\}$.

(b)
$$(x - 4)^2 = 12$$
$$x - 4 = \pm\sqrt{12} \qquad \text{Generalized square root property}$$
$$x = 4 \pm \sqrt{12} \qquad \text{Add 4.}$$
$$x = 4 \pm 2\sqrt{3} \qquad \sqrt{12} = \sqrt{4 \cdot 3} = 2\sqrt{3}$$

The solution set is $\left\{4 \pm 2\sqrt{3}\right\}$.

Now try Exercises 41 and 43.

Any quadratic equation can be solved using the **quadratic formula,** which says that the solutions of the quadratic equation $ax^2 + bx + c = 0$, where $a \neq 0$, are

$$x = \frac{-b \pm \sqrt{b^2 - 4ac}}{2a}. \qquad \text{Quadratic formula}$$

EXAMPLE 6 Using the Quadratic Formula

Solve $x^2 - 4x = -2$.

Solution $\qquad x^2 - 4x + 2 = 0 \qquad$ Write in standard form.

Here $a = 1$, $b = -4$, and $c = 2$. Substitute these values into the quadratic formula and solve for x.

$$x = \frac{-b \pm \sqrt{b^2 - 4ac}}{2a} \qquad \text{Quadratic formula}$$

$$= \frac{-(-4) \pm \sqrt{(-4)^2 - 4(1)2}}{2(1)} \qquad a = 1, b = -4, c = 2$$

$$= \frac{4 \pm \sqrt{16 - 8}}{2}$$

$$= \frac{4 \pm 2\sqrt{2}}{2} \qquad \sqrt{16 - 8} = \sqrt{8} = \sqrt{4 \cdot 2} = 2\sqrt{2}$$

$$= \frac{2(2 \pm \sqrt{2})}{2} \qquad \text{Factor out 2 in the numerator.}$$

$$x = 2 \pm \sqrt{2} \qquad \text{Lowest terms}$$

The solution set is $\left\{2 \pm \sqrt{2}\right\}$.

Now try Exercise 49.

Inequalities An **inequality** says that one expression is greater than, greater than or equal to, less than, or less than or equal to, another. As with equations, a value of the variable for which the inequality is true is a solution of the inequality; the set of all solutions is the solution set of the inequality. Two inequalities with the same solution set are equivalent. Inequalities are solved with the following *properties of inequality*. For real numbers a, b, and c:

1. If $a < b$, then $a + c < b + c$,

2. If $a < b$ and if $c > 0$, then $ac < bc$,

3. If $a < b$ and if $c < 0$, then $ac > bc$.

Replacing $<$ with $>$, $\leq$, or $\geq$ results in similar properties. (Restrictions on c remain the same.) Multiplication may be replaced by division in properties 2 and 3. *Always remember to reverse the direction of the inequality symbol when multiplying or dividing by a negative number.*

Solving Linear Inequalities A **linear inequality in one variable** is an inequality that can be written in the form

$$ax + b > 0,$$

where a and b are real numbers with $a \neq 0$. (Any of the symbols $\geq$, $<$, or $\leq$ may also be used.)

EXAMPLE 7 Solving a Linear Inequality

Solve $-3x + 5 > -7$.

Solution
$$-3x + 5 > -7$$
$$-3x + 5 - 5 > -7 - 5 \quad \text{Subtract 5.}$$
$$-3x > -12$$
$$\frac{-3x}{-3} < \frac{-12}{-3} \quad \text{Divide by } -3; \text{ reverse the direction of the inequality symbol.}$$
$$x < 4$$

Figure 1

The original inequality is satisfied by any real number less than 4. The solution set can be written using **set-builder notation** as $\{x \mid x < 4\}$ (read "the set of all x such that x is less than 4"). A graph of the solution set is shown in Figure 1, where the parenthesis is used to show that 4 itself does not belong to the solution set.

Interval Notation The solution set for the inequality in Example 7, $\{x \mid x < 4\}$, is an example of an **interval.** We use a simplified notation, called **interval notation,** to write intervals. With this notation, we write $\{x \mid x < 4\}$ as $(-\infty, 4)$. The symbol $-\infty$ does not represent an actual number; it is used to show that the interval includes all real numbers less than 4. The interval $(-\infty, 4)$ is an example of an **open interval,** since the endpoint, 4, is not part of the interval. A **closed interval** includes both endpoints. A square bracket is used to show that a number *is* part of the graph, and a parenthesis is used to indicate that a number *is not* part of the graph. In the table that follows, we assume that $a < b$.

Type of Interval	Set	Interval Notation	Graph
Open interval	$\{x \mid x > a\}$	(a, ∞)	
	$\{x \mid a < x < b\}$	(a, b)	
	$\{x \mid x < b\}$	$(-\infty, b)$	
Half-open interval	$\{x \mid x \geq a\}$	$[a, \infty)$	
	$\{x \mid a < x \leq b\}$	$(a, b]$	
	$\{x \mid a \leq x < b\}$	$[a, b)$	
	$\{x \mid x \leq b\}$	$(-\infty, b]$	
Closed interval	$\{x \mid a \leq x \leq b\}$	$[a, b]$	
All real numbers	$\{x \mid x \text{ is a real number}\}$	$(-\infty, \infty)$	

Now try Exercises 57 and 61.

Three-Part Inequalities The inequality $-2 < 5 + 3x \leq 20$ in the next example says that $5 + 3x$ is *between* -2 and 20, excluding -2 and including 20. This inequality is solved using an extension of the properties of inequality given on the preceding page, working with all three expressions at the same time.

EXAMPLE 8 Solving a Three-Part Inequality

Solve $-2 < 5 + 3x \leq 20$.

Solution

$$-2 < 5 + 3x \leq 20$$
$$-2 - 5 < 5 + 3x - 5 \leq 20 - 5 \quad \text{Subtract 5 from each part.}$$
$$-7 < 3x \leq 15$$
$$\frac{-7}{3} < \frac{3x}{3} \leq \frac{15}{3} \qquad \text{Divide each part by 3.}$$
$$-\frac{7}{3} < x \leq 5$$

Figure 2

The solution set, graphed in Figure 2, is the interval $\left(-\frac{7}{3}, 5\right]$.

Now try Exercise 69.

Appendix A Exercises

Concept Check In Exercises 1–3, decide whether each statement is true or false.

1. The solution set of $2x + 3 = x - 5$ is $\{-8\}$.

2. The equation $5(x - 9) = 5x - 45$ is an example of an identity.

3. It is possible for a linear equation to have exactly two solutions.

4. Explain the difference between an identity and a conditional equation.

5. *Concept Check* Which one is not a linear equation?

    A. $5x + 7(x - 1) = -3x$ **B.** $8x^2 - 4x + 3 = 0$

    C. $7x + 8x = 13x$ **D.** $.04x - .08x = .40$

6. In solving the equation $3(2x - 4) = 6x - 12$, a student obtains the result $0 = 0$ and gives the solution set $\{0\}$. Is this correct? Explain.

Solve each equation. See Examples 1 and 2.

7. $5x + 2 = 3x - 6$ **8.** $9x + 1 = 7x - 9$

9. $6(3x - 1) = 8 - (10x - 14)$ **10.** $4(-2x + 1) = 6 - (2x - 4)$

11. $\dfrac{5}{6}x - 2x + \dfrac{1}{3} = \dfrac{2}{3}$ **12.** $\dfrac{3}{4} + \dfrac{1}{5}x - \dfrac{1}{2} = \dfrac{4}{5}x$

13. $3x + 2 - 5(x + 1) = 6x + 4$ **14.** $5(x + 3) + 4x - 5 = -(2x - 4)$

15. $2[x - (4 + 2x) + 3] = 2x + 2$ **16.** $4[2x - (3 - x) + 5] = -7x - 2$

17. $\dfrac{1}{7}(3x - 2) = \dfrac{x + 10}{5}$ **18.** $\dfrac{1}{5}(2x + 5) = \dfrac{x + 2}{3}$

19. $.2x - .5 = .1x + 7$ **20.** $.01x + 3.1 = 2.03x - 2.96$

21. $-4(2x - 6) + 7x = 5x + 24$ **22.** $-8(3x + 4) + 2x = 4(x - 8)$

Decide whether each equation is an identity, *a* conditional equation, *or a* contradiction. *Give the solution set. See Example 3.*

23. $4(2x + 7) = 2x + 25 + 3(2x + 1)$ **24.** $\dfrac{1}{2}(6x + 14) = x + 1 + 2(x + 3)$

25. $2(x - 7) = 3x - 14$ **26.** $-8(x + 3) = -8x - 5(x + 1)$

27. $8(x + 7) = 4(x + 12) + 4(x + 1)$ **28.** $-6(2x + 1) - 3(x - 4) = -15x + 1$

29. *Concept Check* Match each equation (a)–(e) in Column I with its solution(s) A–E in Column II.

                      I   **II**

    (a) $x^2 = 4$ **(b)** $x^2 - 2 = 0$ **A.** $\pm 2\sqrt{2}$ **B.** 2

    (c) $x^2 = 8$ **(d)** $x - 2 = 0$ **C.** $\pm\sqrt{2}$ **D.** -2

    (e) $x + 2 = 0$ **E.** ± 2

Concept Check Answer each question.

30. Which one of the following equations is set up for direct use of the zero-factor property? Solve it.

    A. $3x^2 - 17x - 6 = 0$ **B.** $(2x + 5)^2 = 7$

    C. $x^2 + x = 12$ **D.** $(3x + 1)(x - 7) = 0$

31. Which one of the following equations is set up for direct use of the square root property? Solve it.

A. $3x^2 - 17x - 6 = 0$

B. $(2x + 5)^2 = 7$

C. $x^2 + x = 12$

D. $(3x + 1)(x - 7) = 0$

32. Only one of the following equations is set up so that the values of a, b, and c can be determined immediately. Which one is it? Solve it.

A. $3x^2 - 17x - 6 = 0$

B. $(2x + 5)^2 = 7$

C. $x^2 + x = 12$

D. $(3x + 1)(x - 7) = 0$

Solve each equation by the zero-factor property. See Example 4.

33. $x^2 - 5x + 6 = 0$

34. $x^2 + 2x - 8 = 0$

35. $5x^2 - 3x - 2 = 0$

36. $2x^2 - x - 15 = 0$

37. $-4x^2 + x = -3$

38. $-6x^2 + 7x = -10$

Solve each equation by the square root property. See Example 5.

39. $x^2 = 16$

40. $x^2 = 25$

41. $x^2 = 27$

42. $x^2 = 48$

43. $(x + 5)^2 = 40$

44. $(x - 7)^2 = 24$

45. $(3x - 1)^2 = 12$

46. $(4x + 1)^2 = 20$

Solve each equation using the quadratic formula. See Example 6.

47. $x^2 - x - 1 = 0$

48. $x^2 - 3x - 2 = 0$

49. $x^2 - 6x = -7$

50. $x^2 - 4x = -1$

51. $2x^2 - 4x - 3 = 0$

52. $-3x^2 + 6x + 5 = 0$

53. $\frac{1}{2}x^2 + \frac{1}{4}x - 3 = 0$

54. $\frac{2}{3}x^2 + \frac{1}{4}x = 3$

55. $.1x^2 - .1x = .3$

56. $.2x^2 + .4x - .3 = 0$

57. *Concept Check* Match each inequality (a)–(j) in Column I with its equivalent interval notation A–J in Column II.

I	II
(a) $x < -4$	**A.** $(-2, 6]$
(b) $x \leq 4$	**B.** $[-2, 6)$
(c) $-2 < x \leq 6$	**C.** $(-\infty, -4]$
(d) $0 \leq x \leq 8$	**D.** $[4, \infty)$
(e) $x \geq -3$	**E.** $(3, \infty)$
(f) $4 \leq x$	**F.** $(-\infty, -4)$
(g)	**G.** $(0, 8)$
	H. $[0, 8]$
(h)	**I.** $[-3, \infty)$
	J. $(-\infty, 4]$

(i) ++++++(+++++→
     0   3

(j) ←++[}++++++++→
      -4   0

58. *Concept Check* The three-part inequality $a < x < b$ means "a is less than x and x is less than b." Which one of the following inequalities is not satisfied by some real number x?

A. $-3 < x < 5$

B. $0 < x < 4$

C. $-3 < x < -2$

D. $-7 < x < -10$

Solve each inequality. Write each solution set in interval notation, and graph it. See Example 7.

59. $2x + 1 \leq 9$ **60.** $3x - 2 \leq 13$

61. $-3x - 2 \leq 1$ **62.** $-5x + 3 \geq -2$

63. $2(x + 5) + 1 \geq 5 + 3x$ **64.** $6x - (2x + 3) \geq 4x - 5$

65. $8x - 3x + 2 < 2(x + 7)$ **66.** $2 - 4x + 5(x - 1) < -6(x - 2)$

67. $\dfrac{4x + 7}{-3} \leq 2x + 5$ **68.** $\dfrac{2x - 5}{-8} \leq 1 - x$

Solve each inequality. Write each solution set in interval notation, and graph it. See Example 8.

69. $-3 < 7 + 2x < 13$ **70.** $-4 < 5 + 3x < 8$

71. $10 \leq 2x + 4 \leq 16$ **72.** $-6 \leq 6x + 3 \leq 21$

73. $-10 > -3x + 2 > -16$ **74.** $4 > -6x + 5 > -1$

75. $-4 \leq \dfrac{x + 1}{2} \leq 5$ **76.** $-5 \leq \dfrac{x - 3}{3} \leq 1$

B | Graphs of Equations

The Rectangular Coordinate System ▪ The Pythagorean Theorem and the Distance Formula ▪ The Midpoint Formula ▪ Graphing Equations ▪ Circles

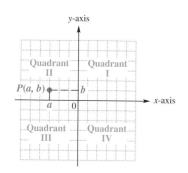

Figure 1

The Rectangular Coordinate System   Many ideas in trigonometry are best explained with a graph in a plane. Recall from algebra that each point in the plane corresponds to an **ordered pair,** two numbers written inside parentheses, such as $(3, 4)$, in which the order of the numbers is important. We graph ordered pairs of real numbers by using two perpendicular number lines, one horizontal and one vertical, that intersect at their zero-points. This point of intersection is called the **origin.** The horizontal line is called the **x-axis,** and the vertical line is called the **y-axis.**

The x-axis and y-axis together make up a **rectangular coordinate system,** or **Cartesian coordinate system** (named for one of its coinventors, René Descartes; the other coinventor was Pierre de Fermat). The plane into which the coordinate system is introduced is the **coordinate plane,** or **xy-plane.** The x-axis and y-axis divide the plane into four regions, or **quadrants,** labeled as shown in Figure 1. The points on the x-axis and y-axis belong to no quadrant.

Each point P in the xy-plane corresponds to a unique ordered pair (a, b) of real numbers. The numbers a and b are the **coordinates** of point P. (See Figure 1.) To locate on the xy-plane the point corresponding to the ordered pair $(3, 4)$, for example, start at the origin, move 3 units in the positive x-direction, and then move 4 units in the positive y-direction. See Figure 2, where point A corresponds to the ordered pair $(3, 4)$.

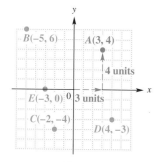

Figure 2

> Now try Exercises 1, 3, 5, and 9.

The Pythagorean Theorem and the Distance Formula   The distance between any two points in a plane can be found by using a formula derived from the **Pythagorean theorem.**

Pythagorean Theorem

In a right triangle, the sum of the squares of the lengths of the legs is equal to the square of the length of the hypotenuse.

$$a^2 + b^2 = c^2$$

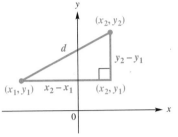

Figure 3

To find the distance between two points (x_1, y_1) and (x_2, y_2), draw the line segment connecting the points, as shown in Figure 3. Complete a right triangle by drawing a line through (x_1, y_1) parallel to the x-axis and a line through (x_2, y_2) parallel to the y-axis. The ordered pair at the right angle of this triangle is (x_2, y_1).

The horizontal side of the right triangle in Figure 3 has length $x_2 - x_1$, while the vertical side has length $y_2 - y_1$. If d represents the distance between the two original points, then by the Pythagorean theorem,

$$d^2 = (x_2 - x_1)^2 + (y_2 - y_1)^2.$$

Solving for d, we obtain the *distance formula*.

Distance Formula

Suppose that $P(x_1, y_1)$ and $R(x_2, y_2)$ are two points in a coordinate plane. Then the distance between P and R, written $d(P, R)$, is given by the **distance formula**,

$$d(P, R) = \sqrt{(x_2 - x_1)^2 + (y_2 - y_1)^2}.$$

That is, the distance between two points in a coordinate plane is the square root of the sum of the square of the difference between their x-coordinates and the square of the difference between their y-coordinates.

EXAMPLE 1   Using the Distance Formula

Find the distance between $P(-8, 4)$ and $Q(3, -2)$.

Solution   According to the distance formula,

$$d(P, Q) = \sqrt{[3 - (-8)]^2 + (-2 - 4)^2} \qquad {\scriptstyle x_1 = -8,\, y_1 = 4,\, x_2 = 3,\, y_2 = -2}$$
$$= \sqrt{11^2 + (-6)^2}$$
$$= \sqrt{121 + 36}$$
$$= \sqrt{157}.$$

Now try Exercise 27(a).

The Midpoint Formula   The *midpoint formula* is used to find the coordinates of the midpoint of a line segment. (Recall that the midpoint of a line segment is equidistant from the endpoints of the segment.) To develop the midpoint formula, let (x_1, y_1) and (x_2, y_2) be any two distinct points in a plane. (Although Figure 4 shows $x_1 < x_2$, no particular order is required.) Let (x, y) be the midpoint of the segment connecting (x_1, y_1) and (x_2, y_2). Draw vertical lines from each of the three points to the x-axis, as shown in Figure 4.

Figure 4

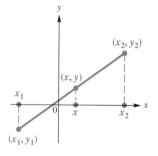

Figure 4 (repeated)

Since (x, y) is the midpoint of the line segment connecting (x_1, y_1) and (x_2, y_2), the distance between x and x_1 equals the distance between x and x_2, so

$$x_2 - x = x - x_1$$
$$x_2 + x_1 = 2x \qquad \text{Add } x; \text{ add } x_1. \text{ (Appendix A)}$$
$$x = \frac{x_1 + x_2}{2}. \qquad \text{Divide by 2; rewrite.}$$

Similarly, the y-coordinate of the midpoint is $\dfrac{y_1 + y_2}{2}$.

Midpoint Formula

The midpoint of the line segment with endpoints (x_1, y_1) and (x_2, y_2) is

$$\left(\frac{x_1 + x_2}{2}, \frac{y_1 + y_2}{2} \right).$$

That is, the x-coordinate of the midpoint of a line segment is the average of the x-coordinates of the segment's endpoints, and the y-coordinate is the average of the y-coordinates of the segment's endpoints.

EXAMPLE 2   Using the Midpoint Formula

Find the midpoint M of the segment with endpoints $(8, -4)$ and $(-6, 1)$.

Solution   Use the midpoint formula to find that the coordinates of M are

$$\left(\frac{8 + (-6)}{2}, \frac{-4 + 1}{2} \right) = \left(1, -\frac{3}{2} \right).$$

Now try Exercise 27(b).

Graphing Equations   Ordered pairs are used to express the solutions of equations in two variables. When an ordered pair represents the solution of an equation with the variables x and y, the x-value is written first. For example, we say that $(1, 2)$ is a solution of $2x - y = 0$, since substituting 1 for x and 2 for y in the equation gives a true statement.

$$2(1) - 2 = 0$$
$$0 = 0 \quad \text{True}$$

EXAMPLE 3   Finding Ordered Pairs That Are Solutions of Equations

For each equation, find three ordered pairs that are solutions.

(a) $y = 4x - 1$ \qquad\qquad **(b)** $x^2 + y^2 = 9$

Solution

(a) Choose any real number for x or y and substitute in the equation to get the corresponding value of the other variable. For example, let $x = -2$ and then let $y = 3$.

$$y = 4x - 1$$
$$y = 4(-2) - 1 \quad \text{Let } x = -2.$$
$$y = -8 - 1$$
$$y = -9$$

$$y = 4x - 1$$
$$3 = 4x - 1 \quad \text{Let } y = 3.$$
$$4 = 4x$$
$$1 = x$$

This gives the ordered pairs $(-2, -9)$ and $(1, 3)$. Verify that the ordered pair $(0, -1)$ is also a solution.

(b)
$$x^2 + y^2 = 9 \quad \text{Given equation}$$
$$2^2 + y^2 = 9 \quad \text{Let } x = 2.$$
$$y^2 = 5$$
$$y = \sqrt{5} \quad \text{or} \quad y = -\sqrt{5} \quad \text{Square root property (Appendix A)}$$

We get two ordered pairs here, $\left(2, \sqrt{5}\right)$ and $\left(2, -\sqrt{5}\right)$. Each choice of a real number for x such that $-3 < x < 3$ will give two ordered pairs. Verify that $(0, 3)$ and $(0, -3)$ are also solutions.

Now try Exercises 43(a) and 45(a).

We graph an equation by plotting ordered pairs that are solutions of the equation. An **x-intercept** is an x-value where the graph intersects the x-axis. A **y-intercept** is a y-value where the graph intersects the y-axis.* In other words, the x-intercept is the x-coordinate of an ordered pair where $y = 0$, and the y-intercept is the y-coordinate of an ordered pair where $x = 0$.

Graphing An Equation by Point Plotting

Step 1 Find the intercepts.

Step 2 Find as many additional ordered pairs as needed.

Step 3 Plot the ordered pairs from Steps 1 and 2.

Step 4 Connect the points from Step 3 with a smooth line or curve.

EXAMPLE 4 Graphing Equations

Graph each equation from Example 3.

(a) $y = 4x - 1$ **(b)** $x^2 + y^2 = 9$

Solution

(a) *Step 1* Let $y = 0$ to find the x-intercept, and let $x = 0$ to find the y-intercept.

$$y = 4x - 1$$
$$0 = 4x - 1$$
$$1 = 4x$$
$$\frac{1}{4} = x \qquad x\text{-intercept}$$

$$y = 4x - 1$$
$$y = 4(0) - 1$$
$$y = 0 - 1$$
$$y = -1 \qquad y\text{-intercept}$$

---

*The intercepts are sometimes defined as ordered pairs, such as $(3, 0)$ and $(0, -4)$, instead of numbers, like x-intercept 3 and y-intercept -4. At this level, however, they are usually defined as numbers.

Figure 5

Figure 6

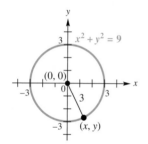

Figure 7

The intercepts $x = \frac{1}{4}$ and $y = -1$ lead to the ordered pairs $\left(\frac{1}{4}, 0\right)$ and $(0, -1)$. Note that the y-intercept yields one of the ordered pairs we found in Example 3(a).

Step 2 We use the other ordered pairs found in Example 3(a): $(-2, -9)$, $(1, 3)$.

Step 3 Plot the four ordered pairs from Steps 1 and 2 as shown in Figure 5.

Step 4 Connect the points plotted in Step 3 with a straight line. This line, also shown in Figure 5, is the graph of the equation $y = 4x - 1$.

(b) For $x^2 + y^2 = 9$, we found the y-intercepts 3 and -3 in Example 3(b). Verify that the x-intercepts are 3 and -3. We also found the solutions $\left(2, -\sqrt{5}\right)$ and $\left(2, \sqrt{5}\right)$ in Example 3(b). Verify that $\left(-2, -\sqrt{5}\right)$ and $\left(-2, \sqrt{5}\right)$ also lie on the graph. Plotting these ordered pairs and connecting the corresponding points with a smooth curve gives the circle in Figure 6.

Now try Exercises 43(b) and 45(b).

Circles In Examples 3(b) and 4(b) we worked with the equation $x^2 + y^2 = 9$, which we saw is the equation of a circle. By definition, a **circle** is the set of all points in a plane that lie a given distance from a given point. The given distance is the **radius** of the circle, and the given point is the **center.** We can find the equation of a circle from its definition by using the distance formula.

Figure 7 shows the graph of a circle of radius 3 with center at the origin. To find the equation of this circle, we let (x, y) be any point on the circle. The distance between (x, y) and the center of the circle, $(0, 0)$, is given by

$$\sqrt{(x - 0)^2 + (y - 0)^2}.$$

Since this distance equals the radius, 3,

$$\sqrt{(x - 0)^2 + (y - 0)^2} = 3$$
$$\sqrt{x^2 + y^2} = 3$$
$$\left(\sqrt{x^2 + y^2}\right)^2 = 3^2 \quad \text{Square both sides.}$$
$$x^2 + y^2 = 9.$$

EXAMPLE 5 Finding the Equation of a Circle

Find an equation for the circle having radius 6 and center at $(-3, 4)$.

Solution Use the distance formula. Let (x, y) be any point on the circle. The distance from (x, y) to $(-3, 4)$ is given by

$$\sqrt{[x - (-3)]^2 + (y - 4)^2} = \sqrt{(x + 3)^2 + (y - 4)^2}.$$

This distance equals the radius, 6. Therefore,

$$\sqrt{(x + 3)^2 + (y - 4)^2} = 6$$

or

$$(x + 3)^2 + (y - 4)^2 = 36. \quad \text{Square both sides.}$$

Generalizing from the work in Example 5 gives the following result.

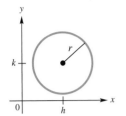

Figure 8

Center-Radius Form of the Equation of a Circle

The circle with center (h, k) and radius r has equation

$$(x - h)^2 + (y - k)^2 = r^2,$$

the **center-radius form** of the equation of a circle. (See Figure 8.) A circle with center $(0, 0)$ and radius r has equation

$$x^2 + y^2 = r^2.$$

Now try Exercise 53(a).

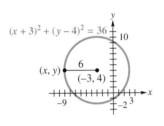

Figure 9

EXAMPLE 6   Graphing a Circle

Graph the circle with equation $(x + 3)^2 + (y - 4)^2 = 36$ from Example 5.

Solution   Writing the equation as

$$[x - (-3)]^2 + (y - 4)^2 = 6^2$$

gives $(-3, 4)$ as the center and 6 as the radius. The graph is shown in Figure 9.

Now try Exercise 53(b).

Appendix B Exercises

Graph the points on a coordinate system and identify the quadrant or axis for each point.

1. $(3, 2)$ **2.** $(-7, 6)$ **3.** $(-7, -4)$ **4.** $(8, -5)$

5. $(0, 5)$ **6.** $(-8, 0)$ **7.** $(4.5, 7)$ **8.** $(-7.5, 8)$

Give the quadrant in which each point lies.

9. $(-5, 1.25)$ **10.** $(\pi, -3)$ **11.** $(-1.4, -2.8)$ **12.** $\left(1 + \sqrt{3}, \dfrac{1}{2}\right)$

13. Suppose the point (a, b) lies in the first quadrant. Describe how you would move from the point (a, b) to the point $(a, -b)$.

14. Suppose the point (a, b) lies in the first quadrant. Describe how you would move from the point (a, b) to the point $(-a, b)$.

15. *Concept Check*   If (a, b) represents a point that lies in quadrant II, in which quadrant will each point lie?

(a) $(-a, b)$ (b) $(-a, -b)$ (c) $(a, -b)$ (d) (b, a)

16. *Concept Check*   If $xy = 1$ is graphed, in which quadrants will the points of the graph lie?

A triple of positive integers (a, b, c) is called a Pythagorean triple *if it satisfies the Pythagorean theorem, $a^2 + b^2 = c^2$. Determine whether each triple is a Pythagorean triple.*

17. $(9, 12, 15)$ **18.** $(6, 8, 10)$ **19.** $(5, 10, 15)$ **20.** $(7, 24, 25)$

The following exercises require the Pythagorean theorem.

21. *Slant Height of a Pyramid*   The Washington Monument is a marble obelisk with an aluminum pyramid at the top that acts as a lightning rod. See the figure. Find the length *x* of a slant edge of the pyramid to the nearest hundredth of a foot. (*Source:* National Park Service.)

22. *Square Corners*   Carpenters use the Pythagorean theorem to test for square corners. Suppose a rectangular floor has width 5 ft and length 12 ft. What should the diagonal of the rectangle measure to show that the floor has a square corner?

23. *Railroad Track Expansion*   A 1000-ft section of railroad track expands 6 in. because the day is very hot. This causes end *C* to break off and move to position *B*, forming right triangle *ABC*. (See the figure.) Find *BC*. *Note:* The surprising answer to this simple problem explains why railroad tracks, bridges, and similar structures must be designed to allow for expansion. (*Source: Trigonometry with Calculators* by Lawrence S. Levy. Reprinted by permission of the author.)

24. *Cutting on the Bias*   Clothing manufacturers sometimes cut their material "on the bias" (that is, at 45° to the direction the threads run) to give it more elasticity. A tie maker wants to cut twenty 8-in. strips of silk on the bias from a rectangular piece of material that costs $10 per (linear) yd of material 42 in. wide. (See the figure.) Find the total cost of the material. (*Hint:* First find length *AB*, using isosceles triangle *ABX*.) *Note:* This unappealing combination of units—inches, yards, and dollars—is typical of many practical problems, not just in the clothing industry. (*Source: Trigonometry with Calculators* by Lawrence S. Levy. Reprinted by permission of the author.)

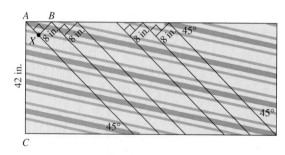

25. *Concept Check*   Determine the distance between $(5, -6)$ and the *x*-axis.

26. *Concept Check*   Determine the distance between $(5, -6)$ and the *y*-axis.

For the points P and Q, find **(a)** *the distance $d(P, Q)$ and* **(b)** *the coordinates of the midpoint of the segment PQ. See Examples 1 and 2.*

27. $P(-5, -7), Q(-13, 1)$ **28.** $P(-4, 3), Q(2, -5)$

29. $P(8, 2), Q(3, 5)$ **30.** $P(-6, -5), Q(6, 10)$

31. $P(-8, 4), Q(3, -5)$ **32.** $P(6, -2), Q(4, 6)$

33. $P(3\sqrt{2}, 4\sqrt{5}), Q(\sqrt{2}, -\sqrt{5})$ **34.** $P(-\sqrt{7}, 8\sqrt{3}), Q(5\sqrt{7}, -\sqrt{3})$

Find all values of x or y such that the distance between the given points is as indicated.

35. $(x, 7)$ and $(2, 3)$; 5   **36.** $(5, y)$ and $(8, -1)$; 5

37. $(3, y)$ and $(-2, 9)$; 12   **38.** $(x, 11)$ and $(5, -4)$; 17

39. Use the distance formula to write an equation for all points that are 5 units from $(0, 0)$. Sketch a graph showing these points.

40. Write an equation for all points 3 units from $(-5, 6)$. Sketch a graph showing these points.

Solve each problem.

41. *Aging of College Freshmen* The graph shows a straight line that approximates the results from an annual survey of college freshmen. Use the midpoint formula and the two given points to estimate the percent in 1992. Compare your answer with the actual percent of 75.3.

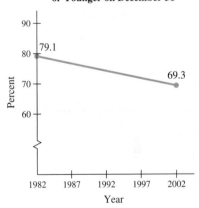

Percent of College Freshmen Age 18 or Younger on December 31

Source: Astin, A., L. Oseguera, L. Sax, and W. Korn, *The American Freshmen: Thirty-Five Year Trends;* Higher Education Research Institute, UCLA, 2002.

42. *Two-Year College Enrollment* Enrollments in two-year colleges for recent years are shown in the table. Assuming a linear relationship, estimate the enrollments for 1985 and 1995.

Year	Enrollment (in millions)
1980	4.5
1990	5.2
2000	5.8

Source: Statistical Abstract of the United States.

For each equation, **(a)** *give a table with three ordered pairs that are solutions, and* **(b)** *graph the equation. (Hint: You will need more than three points for Exercises 45–48.) See Examples 3 and 4.*

43. $6y = 3x - 12$ **44.** $6y = -6x + 18$ **45.** $y = x^2$

46. $y = x^2 + 2$ **47.** $y = x^3$ **48.** $y = -x^3$

In Exercises 49–54, (a) find the center-radius form of the equation of each circle, and (b) graph it. See Examples 5 and 6.

49. center $(0, 0)$, radius 6   **50.** center $(0, 0)$, radius 9   **51.** center $(2, 0)$, radius 6

52. center $(0, -3)$, radius 7   **53.** center $(-2, 5)$, radius 4   **54.** center $(4, 3)$, radius 5

55. Find the center-radius form of the equation of a circle with center $(3, 2)$ and tangent to the *x*-axis. (*Hint:* A line *tangent to* a circle means touching it at exactly one point.)

56. Find the equation of a circle with center at $(-4, 3)$, passing through the point $(5, 8)$. Write it in center-radius form.

C | Functions

Relations and Functions ▪ **Domain and Range** ▪ **Determining Whether a Relation Is a Function** ▪ **Function Notation** ▪ **Increasing, Decreasing, and Constant Functions**

Relations and Functions In algebra, we use ordered pairs to represent corresponding quantities. For example, $(3, \$5.25)$ might indicate that you pay $\$5.25$ for 3 gallons of gas. Since the amount you pay *depends* on the number of gallons pumped, the amount (in dollars) is called the *dependent variable,* and the number of gallons pumped is called the *independent variable.* Generalizing, if the value of the variable *y* depends on the value of the variable *x*, then *y* is the **dependent variable** and *x* is the **independent variable.**

Independent variable ⌐ ⌐ Dependent variable
$$(x, y)$$

Because we can write related quantities using ordered pairs, a set of ordered pairs such as $\{(3, 5.25), (8, 10), (10, 12.50)\}$ is called a **relation.** A special kind of relation called a *function* is very important in algebra and in trigonometry.

> **Function**
>
> A **function** is a relation in which, for each value of the first component of the ordered pairs, there is *exactly one* value of the second component.

EXAMPLE 1   Identifying Functions

Decide whether each relation defines a function.

(a) $F = \{(1, 2), (-2, 4), (3, -1)\}$ **(b)** $G = \{(-4, 1), (-2, 1), (-2, 0)\}$

Solution

(a) Relation F is a function, because for each different *x*-value there is exactly one *y*-value. We can show this correspondence as follows.

$$\{1, -2, 3\} \quad \text{\textit{x}-values of } F$$
$$\downarrow \quad \downarrow \quad \downarrow$$
$$\{2, \quad 4, -1\} \quad \text{\textit{y}-values of } F$$

(b) In relation G the last two ordered pairs have the same x-value paired with two different y-values (-2 is paired with both 1 and 0), so G is a relation but not a function. *In a function, no two ordered pairs can have the same first component and different second components.*

Different y-values

$$G = \{(-4, 1), (-2, 1), (-2, 0)\} \quad \text{Not a function}$$

Same x-value

Now try Exercises 5 and 7.

Since relations and functions are sets of ordered pairs, we can represent them using tables and graphs. A table and graph for function F in Example 1 is shown in Figure 1.

x	y
1	2
-2	4
3	-1

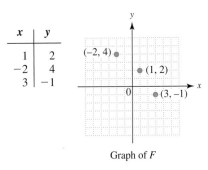

Graph of F

Figure 1

We can also describe a relation or function using a rule that tells how to determine the dependent variable for a specific value of the independent variable. The rule may be given in words: for instance, "the dependent variable is twice the independent variable." Usually the rule is an equation, such as

$$y = 2x.$$

Dependent   Independent
variable   variable

This is the most efficient way to define a relation or function.

Domain and Range For every relation there are two important sets of elements called the *domain* and *range*.

Domain and Range

In a relation, the set of all values of the independent variable (x) is the **domain.** The set of all values of the dependent variable (y) is the **range.**

EXAMPLE 2 Finding Domains and Ranges of Relations

Give the domain and range of each relation. Tell whether the relation defines a function.

(a) $\{(3, -1), (4, 2), (4, 5), (6, 8)\}$

(b)

x	y
-5	2
0	2
5	2

Solution

(a) The domain of $\{(3,-1), (4,2), (4,5), (6,8)\}$ is the set of x-values, $\{3,4,6\}$. The range is the set of y-values, $\{-1,2,5,8\}$. This relation is not a function because the same x-value, 4, is paired with two different y-values, 2 and 5.

x	y
-5	2
0	2
5	2

(b) This relation (repeated in the margin) is a set of ordered pairs, so the domain is the set of x-values $\{-5,0,5\}$ and the range is the set of y-values $\{2\}$. The table defines a function because each different x-value corresponds to exactly one y-value (even though it is the same y-value).

Now try Exercises 11 and 13.

As mentioned previously, the graph of a relation is the graph of its ordered pairs. The graph gives a picture of the relation, which can be used to determine its domain and range.

EXAMPLE 3 Finding Domains and Ranges from Graphs

Give the domain and range of each relation.

(a)

(b)

(c)

(d)

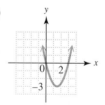

Solution

(a) The domain is the set of x-values, $\{-1,0,1,4\}$. The range is the set of y-values, $\{-3,-1,1,2\}$.

(b) The x-values of the points on the graph include all numbers between -4 and 4, inclusive. The y-values include all numbers between -6 and 6, inclusive. Using interval notation (Appendix A),

the domain is $[-4,4]$ and the range is $[-6,6]$.

(c) The arrowheads indicate that the line extends indefinitely left and right, as well as up and down. Therefore, both the domain and the range include all real numbers, written $(-\infty,\infty)$.

(d) The arrowheads indicate that the graph extends indefinitely left and right, as well as upward. The domain is $(-\infty, \infty)$. Because there is a least y-value, -3, the range includes all numbers greater than or equal to -3, written $[-3, \infty)$.

Now try Exercises 15, 17, and 19.

Function–each x-value corresponds to only one y-value.

Figure 2

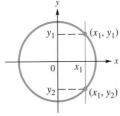

Not a function–the same x-value corresponds to two different y-values.

Figure 3

Determining Whether a Relation Is a Function Since each value of x leads to only one value of y in a function, any vertical line drawn through the graph of a function must intersect the graph in at most one point. This is the *vertical line test* for a function.

> ### Vertical Line Test
>
> If each vertical line intersects a graph in at most one point, then the graph is that of a function.

The graph in Figure 2 represents a function—each vertical line intersects the graph in at most one point. The graph in Figure 3 is not the graph of a function since a vertical line intersects the graph in more than one point.

Now try Exercise 3.

The vertical line test is a simple method for identifying a function defined by a graph. It is more difficult deciding whether a relation defined by an equation is a function, as well as determining the domain and range. The next example gives some hints that may help.

EXAMPLE 4 Identifying Functions, Domains, and Ranges from Equations

Decide whether each relation defines a function and give the domain and range.

(a) $y = x + 4$ **(b)** $y = \sqrt{2x - 1}$ **(c)** $y^2 = x$ **(d)** $y = \dfrac{5}{x - 1}$

Solution

(a) In the defining equation (or rule), $y = x + 4$, y is always found by adding 4 to x. Thus, each value of x corresponds to just one value of y and the relation defines a function; x can be any real number, so the domain is $\{x \mid x \text{ is a real number}\}$ or $(-\infty, \infty)$. Since y is always 4 more than x, y also may be any real number, and so the range is $(-\infty, \infty)$.

(b) For any choice of x in the domain of $y = \sqrt{2x - 1}$, there is exactly one corresponding value for y (the radical is a nonnegative number), so this equation defines a function. Since the equation involves a square root, the quantity under the radical sign cannot be negative. Thus,

$$2x - 1 \geq 0 \qquad \text{Solve the inequality. (Appendix A)}$$
$$2x \geq 1$$
$$x \geq \frac{1}{2},$$

Figure 4

Figure 5

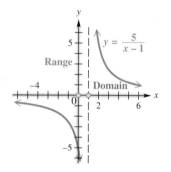

Figure 6

and the domain of $y = \sqrt{2x - 1}$ is $\left[\frac{1}{2}, \infty\right)$. Because the radical is a non-negative number, as x takes values greater than or equal to $\frac{1}{2}$, the range is $[0, \infty)$. See Figure 4.

(c) The ordered pairs $(16, 4)$ and $(16, -4)$ both satisfy the equation $y^2 = x$. Since one value of x, 16, corresponds to two values of y, 4 and -4, *this equation does not define a function.* Because x is equal to the square of y, the values of x must always be nonnegative. The domain of the relation is $[0, \infty)$. Any real number can be squared, so the range of the relation is $(-\infty, \infty)$. See Figure 5.

(d) Given any value of x in the domain of $y = \frac{5}{x - 1}$, we find y by subtracting 1, then dividing the result into 5. This process produces exactly one value of y for each value in the domain, so this equation defines a function. The domain includes all real numbers except those that make the denominator 0. We find these numbers by setting the denominator equal to 0 and solving for x.

$$x - 1 = 0$$
$$x = 1 \quad \text{Solve the equation. (Appendix A)}$$

Thus, the domain includes all real numbers except 1, written as the interval $(-\infty, 1) \cup (1, \infty)$. Values of y can be positive or negative, but never 0, because a fraction cannot equal 0 unless its numerator is 0. Therefore, the range is the interval $(-\infty, 0) \cup (0, \infty)$, as shown in Figure 6.

Now try Exercises 23, 25, 29, and 31.

NOTE   Some trigonometric functions are defined in such a way that restrictions are necessary since their denominators cannot equal 0. Example 4(d) illustrates this idea with an algebraic function.

Function Notation   When a function f is defined with a rule or an equation using x and y for the independent and dependent variables, we say "y is a function of x" to emphasize that y *depends on* x. We use the notation

$$y = f(x),$$

called **function notation,** to express this and read $f(x)$ as "f of x." (In this notation the parentheses do not indicate multiplication.) The letter f stands for *function.* For example, if $y = 9x - 5$, we can name this function f and write

$$f(x) = 9x - 5.$$

Note that $f(x)$ *is just another name for the dependent variable y.* For example, if $y = f(x) = 9x - 5$ and $x = 2$, then we find y, or $f(2)$, by replacing x with 2.

$$y = f(2) = 9 \cdot 2 - 5$$
$$= 18 - 5$$
$$= 13$$

The statement "for function f, if $x = 2$, then $y = 13$" represents the ordered pair $(2, 13)$ and is abbreviated with function notation as

$$f(2) = 13.$$

Read $f(2)$ as "f of 2" or "f at 2." Also,

$$f(0) = 9 \cdot 0 - 5 = -5 \qquad \text{and} \qquad f(-3) = 9(-3) - 5 = -32.$$

These ideas can be illustrated as follows.

$$\underset{\text{Value of the function}}{y} \quad = \quad \underset{}{f(x)} \quad = \quad \overset{\text{Defining expression}}{9x - 5}$$

Name of the function

Value of the function   Name of the independent variable

EXAMPLE 5   Using Function Notation

Let $f(x) = -x^2 + 5x - 3$. Find the following.

(a) $f(2)$ **(b)** $f(q)$

Solution

(a) $f(x) = -x^2 + 5x - 3$

$f(2) = -2^2 + 5 \cdot 2 - 3$ Replace x with 2.

$ = -4 + 10 - 3$

$ = 3$

Thus, $f(2) = 3$; the ordered pair $(2, 3)$ belongs to f.

(b) $f(x) = -x^2 + 5x - 3$

$f(q) = -q^2 + 5q - 3$ Replace x with q.

Now try Exercises 35 and 39.

Sometimes letters other than f, such as g, h, or capital letters F, G, and H, are used to name functions.

EXAMPLE 6   Using Function Notation

For each function, find $g(3)$.

(a) $g(x) = 3x - 7$ **(b)** $g = \{(-3, 5), (0, 3), (3, 1), (6, -1)\}$

(c)

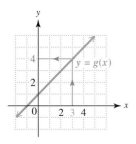

Figure 7

Solution

(a) $g(x) = 3x - 7$

$g(3) = 3(3) - 7 = 2$ Replace x with 3.

(b) For $g = \{(-3, 5), (0, 3), (3, 1), (6, -1)\}$, we want $g(3)$, the y-value of the ordered pair where $x = 3$. As indicated by the ordered pair $(3, 1)$, when $x = 3$, $y = 1$, so $g(3) = 1$.

(c) To evaluate $g(3)$, find 3 on the x-axis. See Figure 7. Then move up until the graph of g is reached. Moving horizontally to the y-axis gives 4 for the corresponding y-value. Thus, $g(3) = 4$.

Now try Exercises 45 and 47.

Increasing, Decreasing, and Constant Functions   Informally speaking, a function *increases* on an interval of its domain if its graph rises from left to right on the interval. It *decreases* on an interval of its domain if its graph falls from left to right on the interval. It is *constant* on an interval of its domain if its graph is horizontal on the interval. The formal definitions of these concepts follow.

Increasing, Decreasing, and Constant Functions

Suppose that a function f is defined over an interval I. If x_1 and x_2 are in I,

(a) f **increases** on I if, whenever $x_1 < x_2$, $f(x_1) < f(x_2)$;

(b) f **decreases** on I if, whenever $x_1 < x_2$, $f(x_1) > f(x_2)$;

(c) f is **constant** on I if, for every x_1 and x_2, $f(x_1) = f(x_2)$.

N O T E   To decide whether a function is increasing, decreasing, or constant on an interval, ask yourself **"What does y do as x goes from left to right?"**

EXAMPLE 7   Determining Intervals over Which a Function Is Increasing, Decreasing, or Constant

Figure 8 shows the graph of a function. Determine the intervals over which the function is increasing, decreasing, or constant.

Solution   Ask, "What is happening to the y-values as the x-values get larger?" Moving from left to right on the graph, we see that on the interval $(-\infty, 1)$, the y-values are *decreasing;* on the interval $[1, 3]$, the y-values are *increasing;* and on the interval $[3, \infty)$, the y-values are *constant* (and equal to 6). Therefore, the function is decreasing on $(-\infty, 1)$, increasing on $[1, 3]$, and constant on $[3, \infty)$.

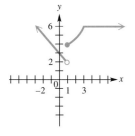

Figure 8

Now try Exercise 53.

C A U T I O N   When specifying intervals over which a function is increasing, decreasing, or constant, use *domain* values. Range values are not involved when writing the intervals.

Appendix C Exercises

 1. In your own words, define a function and give an example.

 2. In your own words, define the domain of a function and give an example.

3. *Concept Check*   Use the vertical line test to determine whether each relation graphed in Example 3 on page 452 is a function.

4. *Concept Check*   Give an example of a relation that is not a function, having domain $\{-3, 2, 6\}$ and range $\{4, 6\}$. (There are many possible correct answers.)

Decide whether each relation defines a function. See Example 1.

5. $\{(5, 1), (3, 2), (4, 9), (7, 6)\}$

6. $\{(8, 0), (5, 4), (9, 3), (3, 8)\}$

7. $\{(2, 4), (0, 2), (2, 5)\}$

8. $\{(9, -2), (-3, 5), (9, 2)\}$

9. $\{(-3, 1), (4, 1), (-2, 7)\}$

10. $\{(-12, 5), (-10, 3), (8, 3)\}$

Decide whether each relation defines a function and give the domain and range. See Examples 1–3.

11. $\{(1, 1), (1, -1), (0, 0), (2, 4), (2, -4)\}$

12. $\{(2, 5), (3, 7), (4, 9), (5, 11)\}$

13.

x	y
2	1
5	1
11	7
17	20

14.

x	y
1	10
2	15
2	20
3	25

15.

16.

17.

18.

19.

20.

Decide whether each relation defines y as a function of x. Give the domain and range. See Example 4.

21. $y = x^2$

22. $y = x^3$

23. $x = y^6$

24. $x = y^4$

25. $y = 2x - 6$

26. $y = -6x + 8$

27. $y = \sqrt{x}$

28. $y = -\sqrt{x}$

29. $y = \sqrt{4x + 2}$

30. $y = \sqrt{9 - 2x}$

31. $y = \dfrac{2}{x - 9}$

32. $y = \dfrac{-7}{x - 16}$

33. *Concept Check* Choose the correct response: The notation $f(3)$ means

   A. the variable f times 3 or $3f$.

   B. the value of the dependent variable when the independent variable is 3.

   C. the value of the independent variable when the dependent variable is 3.

   D. f equals 3.

34. *Concept Check* Give an example of a function from everyday life. (*Hint:* Fill in the blanks: _____ depends on _____, so _____ is a function of _____.)

Let $f(x) = -3x + 4$ and $g(x) = -x^2 + 4x + 1$. Find the following. See Example 5.

35. $f(0)$ **36.** $f(-3)$ **37.** $g(-2)$ **38.** $g(10)$

39. $f(p)$ **40.** $g(k)$ **41.** $f(-x)$ **42.** $g(-x)$

43. $f(a + 4)$ **44.** $f(2m - 3)$

For each function, find (a) $f(2)$ and (b) $f(-1)$. See Example 6.

45. $f = \{(-1, 3), (4, 7), (0, 6), (2, 2)\}$ **46.** $f = \{(2, 5), (3, 9), (-1, 11), (5, 3)\}$

47.

48.

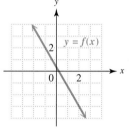

Concept Check Use the graph of $y = f(x)$ to find each function value: (a) $f(-2)$, (b) $f(0)$, (c) $f(1)$, and (d) $f(4)$.

49.

50.

51. *Concept Check* Let $f(x)$ be the function in the graph. Find each of the following.

   (a) $f(0)$ **(b)** $f(6)$

   (c) a negative number a for which $f(a) = 0$

   (d) three positive values of x for which $f(x) = 10$

   (e) the distance between $(8, f(8))$ and $(10, f(10))$

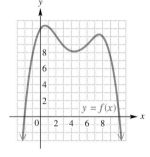

52. *Concept Check* Use the table to answer each question for a function $y = f(x)$.

   (a) What is $f(2)$?

   (b) If $f(x) = -2.4$, what is the value of x?

   (c) At what point does the graph of $y = f(x)$ intersect the y-axis?

   (d) At what point does the graph of $y = f(x)$ intersect the x-axis?

x	y
0	3.6
1	2.4
2	1.2
3	0
4	-1.2
5	-2.4
6	-3.6

Determine the intervals of the domain for which each function is (a) increasing, (b) decreasing, and (c) constant. See Example 7.

53. **54.** **55.**

56. **57.** **58.**

Concept Check   Solve each problem by obtaining information from the associated graph.

59. *Electricity Usage* The graph shows the daily megawatts of electricity used on a record-breaking summer day in Sacramento, California.

   (a) Is this the graph of a function?

   (b) What is the domain?

   (c) Estimate the number of megawatts used at 8 A.M.

   (d) At what time was the most electricity used? the least electricity?

   (e) Call this function f. What is $f(12)$? What does it mean?

   (f) During what time intervals is electricity usage increasing? decreasing?

Electricity Use

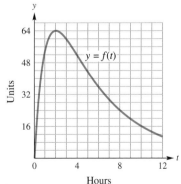

Source: Sacramento Municipal Utility District.

60. *Drug Levels in the Bloodstream* When a drug is taken orally, the amount of the drug in the bloodstream at t hours is given by the function defined by $y = f(t)$, as shown in the graph.

   (a) How many units of the drug are in the bloodstream at 8 hr?

   (b) During what time interval is the drug level in the bloodstream increasing? decreasing?

   (c) When does the level of the drug in the bloodstream reach its maximum value, and how many units are in the bloodstream at that time?

Drug Levels in the Bloodstream

   (d) When the drug reaches its maximum level in the bloodstream, how many additional hours are required for the level to drop to 16 units?

D Graphing Techniques

Stretching and Shrinking ▪ **Reflecting** ▪ **Symmetry** ▪ **Translations**

Graphing techniques presented here review how to graph functions that are defined by altering the equation of a basic function.

Stretching and Shrinking   We begin by considering how the graph of $y = a \cdot f(x)$ or $y = f(ax)$ compares to the graph of $y = f(x)$.

EXAMPLE 1   Stretching or Shrinking a Graph

Graph $g(x) = 2x^2$ and $h(x) = \frac{1}{2}x^2$.

Solution   Comparing the values for $f(x) = x^2$ and $g(x) = 2x^2$ in the table beside Figure 1, we see that for corresponding x-values, the y-values of g are each twice those of f. Thus the graph of $g(x)$, shown in blue in Figure 1, is narrower than that of $f(x)$, shown in red for comparison.

x	$f(x)$ x^2	$g(x)$ $2x^2$	$h(x)$ $\frac{1}{2}x^2$
-2	4	8	2
-1	1	2	$\frac{1}{2}$
0	0	0	0
1	1	2	$\frac{1}{2}$
2	4	8	2

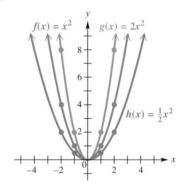

Figure 1

The graph of $h(x) = \frac{1}{2}x^2$ is also the same general shape as that of $f(x) = x^2$, but here the coefficient $\frac{1}{2}$ causes the graph of $h(x)$ to be wider than the graph of $f(x)$. See the values in the table. The graph of $h(x)$ is shown in green in Figure 1.

Now try Exercises 5 and 7.

$|a|$ is the absolute value of a number a.

$$|a| = \begin{cases} a \text{ if } a \text{ is positive or } 0 \\ -a \text{ if } a \text{ is negative} \end{cases}$$

Thus, $|2| = |2|$ and $|-2| = |2|$.

Graph of the absolute value function

NOTE   Recall that the graphs in Figure 1 are called *parabolas*. The lowest (or highest) point on a parabola is its *vertex*. The vertex of each of the parabolas in Figure 1 is $(0,0)$.

The graphs in Example 1 suggest the following generalizations.

Stretching and Shrinking

The graph of $g(x) = a \cdot f(x)$ has the same general shape as the graph of $f(x)$.

If $|a| > 1$, then the graph is stretched vertically (narrower) compared to the graph of $f(x)$.

If $0 < |a| < 1$, then the graph is shrunken vertically (wider) compared to the graph of $f(x)$.

In general, the larger the value of $|a|$, the greater the stretch. The smaller the value of $|a|$, the greater the shrink.

Reflecting   Forming the mirror image of a graph across a line is called *reflecting the graph across the line.*

EXAMPLE 2   Reflecting a Graph Across an Axis

Graph $g(x) = -\sqrt{x}$ and $h(x) = \sqrt{-x}$.

Solution   The table of values for $g(x) = -\sqrt{x}$ and $f(x) = \sqrt{x}$ are shown with their graphs in Figure 2. As the table suggests, every y-value of the graph of $g(x) = -\sqrt{x}$ is the negative of the corresponding y-value of $f(x) = \sqrt{x}$. This has the effect of reflecting the graph across the x-axis.

x	$f(x)$ $\sqrt{x}$	$g(x)$ $-\sqrt{x}$	$h(x)$ $\sqrt{-x}$
-4	undefined	undefined	2
-1	undefined	undefined	1
0	0	0	0
1	1	-1	undefined
4	2	-2	undefined

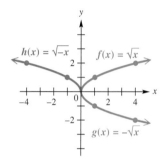

Figure 2

The domain of $h(x) = \sqrt{-x}$ is $(-\infty, 0]$, while the domain of $f(x) = \sqrt{x}$ is $[0, \infty)$. If we choose x-values for $h(x)$ that are the negatives of those we use for $f(x)$, we see that the corresponding y-values are the same. Thus, the graph of h is a reflection of the graph of f across the y-axis. See Figure 2.

Now try Exercises 9 and 11.

The graphs in Example 2 suggest the following generalizations.

Reflecting Across an Axis

The graph of $y = -f(x)$ is the same as the graph of $y = f(x)$ reflected across the x-axis.

The graph of $y = f(-x)$ is the same as the graph of $y = f(x)$ reflected across the y-axis.

(a)

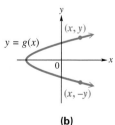

(b)

Figure 3

Symmetry   The graph of f shown in Figure 3(a) is cut in half by the y-axis with each half the mirror image of the other half. A graph with this property is said to be *symmetric with respect to the y-axis.* As this graph suggests, a graph is symmetric with respect to the y-axis if the point $(-x, y)$ is on the graph whenever the point (x, y) is on the graph.

Similarly, if the graph of g in Figure 3(b) were folded in half along the x-axis, the portion at the top would exactly match the portion at the bottom. Such a graph is *symmetric with respect to the x-axis:* the point $(x, -y)$ is on the graph whenever the point (x, y) is on the graph.

Symmetry with Respect to an Axis

The graph of an equation is **symmetric with respect to the y-axis** if the replacement of x with $-x$ results in an equivalent equation.

The graph of an equation is **symmetric with respect to the x-axis** if the replacement of y with $-y$ results in an equivalent equation.

EXAMPLE 3 Testing for Symmetry with Respect to an Axis

Test for symmetry with respect to the x-axis and the y-axis.

(a) $y = x^2 + 4$ **(b)** $x = y^2 - 3$ **(c)** $x^2 + y^2 = 16$ **(d)** $2x + y = 4$

Solution

(a) In $y = x^2 + 4$, replace x with $-x$.

$$y = x^2 + 4$$
$$y = (-x)^2 + 4 \quad \text{Equivalent}$$
$$y = x^2 + 4$$

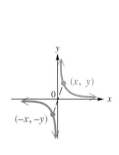

Figure 4

The result is the same as the original equation, so the graph, shown in Figure 4, is symmetric with respect to the y-axis. Check algebraically that the graph is *not* symmetric with respect to the x-axis.

(b) In $x = y^2 - 3$, replace y with $-y$ to get $x = (-y)^2 - 3 = y^2 - 3$, the same as the original equation. The graph is symmetric with respect to the x-axis, as shown in Figure 5. It is not symmetric with respect to the y-axis.

Figure 5

(c) Substituting $-x$ for x and $-y$ for y in $x^2 + y^2 = 16$, we get

$$(-x)^2 + y^2 = 16 \quad \text{and} \quad x^2 + (-y)^2 = 16.$$

Both simplify to $\qquad x^2 + y^2 = 16.$

Thus the graph, a circle of radius 4 centered at the origin, is symmetric with respect to both axes.

(d) In $2x + y = 4$, replace x with $-x$, and then replace y with $-y$; neither case produces an equivalent equation. This graph is not symmetric with respect to either axis.

Symmetry with respect to the origin occurs when a graph can be rotated 180° about the origin, with the result coinciding exactly with the original graph. A graph is symmetric with respect to the origin if the point $(-x, -y)$ is on the graph whenever the point (x, y) is on the graph. Figure 6 shows two graphs that are symmetric with respect to the origin.

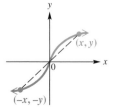

Figure 6

Symmetry with Respect to the Origin

The graph of an equation is **symmetric with respect to the origin** if the replacement of both x with $-x$ and y with $-y$ results in an equivalent equation.

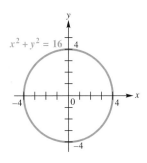

$x^2 + y^2 = 16$

Figure 7

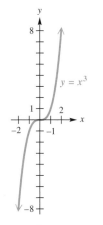

$y = x^3$

Figure 8

EXAMPLE 4 Testing for Symmetry with Respect to the Origin

Test for symmetry with respect to the origin.

(a) $x^2 + y^2 = 16$ **(b)** $y = x^3$

Solution In each case, replace x with $-x$ and y with $-y$.

(a)
$$x^2 + y^2 = 16$$
$$(-x)^2 + (-y)^2 = 16 \quad \text{Equivalent}$$
$$x^2 + y^2 = 16$$

(b)
$$y = x^3$$
$$-y = (-x)^3$$
$$-y = -x^3 \quad \text{Equivalent}$$
$$y = x^3$$

The graphs, both symmetric with respect to the origin, are shown in Figures 7 and 8.

Now try Exercise 23.

A graph symmetric with respect to both the x- and y-axes is automatically symmetric with respect to the origin. However, a graph symmetric with respect to the origin need not be symmetric with respect to either axis. (See Figure 8.) Of the three types of symmetry—with respect to the x-axis, the y-axis, and the origin—a graph possessing any two must also exhibit the third type.

Now try Exercises 19 and 21.

Translations The next examples show the results of horizontal and vertical shifts, called **translations,** of the graph of $f(x) = x^2$.

EXAMPLE 5 Translating Graphs

Graph each function.

(a) $g(x) = x^2 + 3$ **(b)** $g(x) = (x - 4)^2$

Solution

(a) By comparing the table of values for $g(x) = x^2 + 3$ and $f(x) = x^2$ shown with Figure 9, we see that for corresponding x-values, the y-values of g are each 3 more than those for f. Thus, the graph of $g(x) = x^2 + 3$ is the same as that of $f(x) = x^2$, but translated 3 units *up*. The vertex of the graph is at $(0, 3)$, and the graph is symmetric with respect to the y-axis.

x	$f(x)$ x^2	$g(x)$ $x^2 + 3$
-2	4	7
-1	1	4
0	0	3
1	1	4
2	4	7

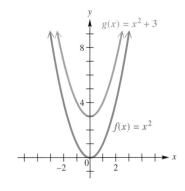

Figure 9

(b) The graph of $g(x) = (x - 4)^2$ in Figure 10 is the same as that of $f(x) = x^2$, but translated 4 units *to the right.* The vertex is at $(4, 0)$, and the graph is symmetric with respect to the line $x = 4$.

x	$f(x)$ x^2	$g(x)$ $(x - 4)^2$
1	1	9
2	4	4
3	9	1
4	16	0
5	25	1
6	36	4

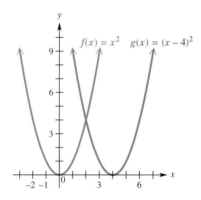

Figure 10

Now try Exercises 27 and 31.

The graphs in Example 5 and Figures 11 and 12 suggest the following generalizations.

Figure 11

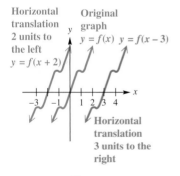

Figure 12

Vertical and Horizontal Translations

If a function g is defined by $g(x) = f(x) + c$, where c is a real number, then for every point (x, y) on the graph of f, there will be a corresponding point $(x, y + c)$ on the graph of g. The graph of g will be the same as the graph of f, but translated c units up if c is positive or $|c|$ units down if c is negative. The graph of g is called a **vertical translation** of the graph of f. See Figure 11.

If a function g is defined by $g(x) = f(x - c)$, where c is a real number, then for every point (x, y) on the graph of f, there will be a corresponding point $(x + c, y)$ on the graph of g. The graph of g will be the same as the graph of f, but translated c units to the right if c is positive or $|c|$ units to the left if c is negative. The graph of g is called a **horizontal translation** of the graph of f. See Figure 12.

Vertical and horizontal translations are summarized in the table, where f is a function, and c is a positive number.

To Graph:	Shift the Graph of $y = f(x)$ by c Units:
$g(x) = f(x) + c$	up
$g(x) = f(x) - c$	down
$g(x) = f(x + c)$	left
$g(x) = f(x - c)$	right

Now try Exercise 1.

EXAMPLE 6 Using More Than One Transformation on a Graph

Graph $f(x) = -\dfrac{1}{2}(x - 4)^2 + 3$.

Solution The graph of f will have the same shape as that of $y = x^2$, but is wider (that is, shrunken vertically) because $\left|-\frac{1}{2}\right| < 1$, and reflected across the x-axis because $-\frac{1}{2} < 0$. The vertex of f is translated 4 units to the right and 3 units up compared to the graph of $y = x^2$. Because of the reflection, the vertex is the highest point on the graph, as shown in Figure 13. The graph is symmetric with respect to the line $x = 4$.

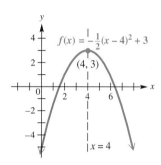

Figure 13

Now try Exercise 39.

Appendix D Exercises

1. *Concept Check* Match each equation in Column I with a description of its graph from Column II as it relates to the graph of $y = x^2$.

I	II
(a) $y = (x - 7)^2$	**A.** a translation 7 units to the left
(b) $y = x^2 - 7$	**B.** a translation 7 units to the right
(c) $y = 7x^2$	**C.** a vertical shrink
(d) $y = (x + 7)^2$	**D.** a translation 7 units up
(e) $y = x^2 + 7$	**E.** a translation 7 units down
(f) $y = \dfrac{1}{7}x^2$	**F.** a vertical stretch

2. *Concept Check* Match each equation in Column I with a description of its graph from Column II as it relates to the graph of $y = x^2$.

I	II
(a) $y = 4x^2$	**A.** a translation 4 units to the left
(b) $y = -x^2$	**B.** a translation 4 units up
(c) $y = (-x)^2$	**C.** a reflection across the x-axis
(d) $y = (x + 4)^2$	**D.** a reflection across the y-axis
(e) $y = x^2 + 4$	**E.** a vertical stretch

3. *Concept Check* Match each equation in parts (a)–(h) with the sketch of its graph.

(a) $y = x^2 + 2$

(b) $y = x^2 - 2$

(c) $y = (x + 2)^2$

(d) $y = (x - 2)^2$

(e) $y = 2x^2$

(f) $y = -x^2$

(g) $y = (x - 2)^2 + 1$

(h) $y = (x + 2)^2 + 1$

A.

B.

C.

D.

E.

F.

G.

H.

Graph each function. See Examples 1 and 2.

4. $y = 3x^2$

5. $y = 4x^2$

6. $y = \frac{1}{3}x^2$

7. $y = \frac{2}{3}x^2$

8. $y = -\frac{1}{2}x^2$

9. $y = -3x^2$

10. $y = 2\sqrt{-x}$

11. $y = \sqrt{-2x}$

Concept Check Suppose the point $(8, 12)$ is on the graph of $y = f(x)$. Find a point on the graph of each function.

12. (a) $y = f(x + 4)$

(b) $y = f(x) + 4$

13. (a) $y = \frac{1}{4}f(x)$

(b) $y = 4f(x)$

14. (a) the reflection of the graph of $y = f(x)$ across the x-axis

(b) the reflection of the graph of $y = f(x)$ across the y-axis

Concept Check Plot each point, and then plot the points that are symmetric to the given point with respect to the (a) x-axis, (b) y-axis, and (c) origin.

15. $(5, -3)$

16. $(-6, 1)$

17. $(-4, -2)$

18. $(-8, 0)$

Without graphing, determine whether each equation has a graph that is symmetric with respect to the x-axis, the y-axis, the origin, or none of these. See Examples 3 and 4.

19. $y = x^2 + 2$

20. $y = 2x^4 - 1$

21. $x^2 + y^2 = 10$

22. $y^2 = \frac{-5}{x^2}$

23. $y = -3x^3$

24. $y = x^3 - x$

25. $y = x^2 - x + 7$

26. $y = x + 12$

Graph each function. See Examples 5 and 6.

27. $y = x^2 - 1$　　　　**28.** $y = x^2 + 1$　　　　**29.** $y = x^2 + 2$

30. $y = x^2 - 2$　　　　**31.** $y = (x - 1)^2$　　　**32.** $y = (x - 2)^2$

33. $y = (x + 2)^2$　　　**34.** $y = (x + 3)^2$　　　**35.** $y = (x + 3)^2 - 4$

36. $y = (x - 5)^2 - 4$　　**37.** $y = 2x^2 - 1$　　　**38.** $y = \dfrac{2}{3}(x - 2)^2$

39. $f(x) = 2(x - 2)^2 - 4$　　**40.** $f(x) = -3(x - 2)^2 + 1$

Concept Check　Each of the following graphs is obtained from the graph of $f(x) = |x|$ or $g(x) = \sqrt{x}$ by applying several of the transformations discussed in this section. Describe the transformations and give the equation for the graph.

41.

42.

43.

44.

45. Find the function g whose graph can be obtained by translating the graph of $f(x) = 2x + 5$ up 2 units and to the left 3 units.

46. Find the function g whose graph can be obtained by translating the graph of $f(x) = 3 - x$ down 2 units and to the right 3 units.

47. Suppose the equation $y = F(x)$ is changed to $y = c \cdot F(x)$, for some constant c. What is the effect on the graph of $y = F(x)$? Discuss the effect depending on whether $c > 0$ or $c < 0$, and $|c| > 1$ or $|c| < 1$.

48. Suppose $y = F(x)$ is changed to $y = F(x + h)$. How are the graphs of these equations related? Is the graph of $y = F(x) + h$ the same as the graph of $y = F(x + h)$? If not, how do they differ?

Glossary

A

absolute value (modulus) of a complex number   When a complex number $x + yi$ is written in trigonometric (or polar) form as $r(\cos\theta + i\sin\theta)$, the number r is called the absolute value or modulus of the complex number. (Section 8.2)

acute angle   An acute angle is an angle measuring between $0°$ and $90°$. (Section 1.1)

addition of ordinates   Addition of ordinates is a method for graphing a function that is the sum of two other functions by adding the y-values of the two functions at selected x-values. (Section 4.3)

airspeed   In air navigation, the airspeed of a plane is its speed relative to the air. (Section 7.5)

ambiguous case   The situation in which the lengths of two sides of a triangle and the measure of the angle opposite one of them are given (SSA) is called the ambiguous case of the law of sines. Depending on the given measurements, this combination of given parts may result in 0, 1, or 2 possible triangles. (Section 7.2)

amplitude   The amplitude of a periodic function is half the difference between the maximum and minimum values of the function. (Section 4.1)

angle   An angle is formed by rotating a ray around its endpoint. (Section 1.1)

angle of depression   The angle of depression from point X to point Y (below X) is the acute angle formed by ray XY and a horizontal ray with endpoint at X. (Section 2.4)

angle of elevation   The angle of elevation from point X to point Y (above X) is the acute angle formed by ray XY and a horizontal ray with endpoint at X. (Section 2.4)

Angle-Side-Angle (ASA)   The Angle-Side-Angle (ASA) congruence axiom states that if two angles and the included side of one triangle are equal, respectively, to two angles and the included side of a second triangle, then the triangles are congruent. (Section 7.1)

angle in standard position   An angle is in standard position if its vertex is at the origin and its initial side is along the positive x-axis. (Section 1.1)

angle between two vectors   The angle between two vectors is defined to be the angle θ, for $0° \leq \theta \leq 180°$, having the two vectors as its sides. (Section 7.4)

angular speed ω   Angular speed ω (omega) measures the speed of rotation and is defined by $\omega = \frac{\theta}{t}$, where θ is the angle of rotation in radians and t is time. (Section 3.4)

argument of a complex number   When a complex number $x + yi$ is written in trigonometric (or polar) form as $r(\cos\theta + i\sin\theta)$, the angle θ is called the argument of the complex number. (Section 8.2)

argument of a function   The argument of a function is the expression containing the independent variable of the function. For example, in the function $y = f(x - d)$, the expression $x - d$ is the argument. (Section 4.2)

B

bearing   Bearing is used to identify angles in navigation. One method for expressing bearing uses a single angle, with bearing measured in a clockwise direction from due north. A second method for expressing bearing starts with a north-south line and uses an acute angle to show the direction, either east or west, from this line. (Section 2.5)

C

cardioid   A cardioid is a heart-shaped curve that is the graph of a polar equation of the form $r = a \pm b\sin\theta$ or $r = a \pm b\cos\theta$, where $\left|\frac{a}{b}\right| = 1$. (Section 8.5)

center of a circle   The center of a circle is the given point that is a given distance from all points on the circle. (Appendix B)

circle   A circle is the set of all points in a plane that lie a given distance from a given point. (Appendix B)

circular functions   The trigonometric functions of arc lengths, or real numbers, are called circular functions. (Section 3.3)

closed interval   A closed interval is an interval that includes both of its endpoints. (Appendix A)

cofunctions   The function pairs sine and cosine, tangent and cotangent, and secant and cosecant are called cofunctions. (Section 2.1)

common logarithm   A base 10 logarithm is called a common logarithm. (Section 9.3)

complementary angles   Two positive angles are complementary angles if the sum of their measures is $90°$. (Section 1.1)

complex conjugates   Two complex numbers that differ only in the sign of their imaginary parts are called complex conjugates. (Section 8.1)

complex number   A complex number is a number of the form $a + bi$, where a and b are real numbers and $i = \sqrt{-1}$. (Section 8.1)

complex plane   The complex plane is a two-dimensional representation of the complex numbers in which the horizontal axis is the real axis and the vertical axis is the imaginary axis. (Section 8.2)

compound amount   In an investment paying compound interest, the compound amount is the balance *after* interest has been earned. (The compound amount is sometimes called the *future value*.) (Section 9.1)

compound interest   In compound interest, interest is paid both on the principal and previously earned interest. (Section 9.1)

conditional equation   An equation that is satisfied by some numbers but not by others is called a conditional equation. (Appendix A)

congruent triangles   Triangles that are both the same size and the same shape are called congruent triangles. (Section 1.2)

constant function   A function f is constant on an interval I if, for every x_1 and x_2 in I, $f(x_1) = f(x_2)$. (Appendix C)

continuous compounding   As the frequency of compounding increases, compound interest approaches a limit, called continuous compounding. (Section 9.1)

contradiction   An equation that has no solution is called a contradiction. (Appendix A)

coordinate plane (*xy*-plane)   The plane into which the rectangular coordinate system is introduced is called the coordinate plane, or *xy*-plane. (Appendix B)

coordinates (in the *xy*-plane)   The coordinates of a point in the *xy*-plane are the numbers in the ordered pair that correspond to that point. (Appendix B)

cosecant   Let $P(x, y)$ be a point other than the origin on the terminal side of an angle θ in standard position. Let $r = \sqrt{x^2 + y^2}$ represent the distance from the origin to P. Then the cosecant function is defined by $\csc \theta = \frac{r}{y} \, (y \neq 0)$. (Section 1.3)

cosine   Let $P(x, y)$ be a point other than the origin on the terminal side of an angle θ in standard position. Let $r = \sqrt{x^2 + y^2}$ represent the distance from the origin to P. Then the cosine function is defined by $\cos \theta = \frac{x}{r}$. (Section 1.3)

cotangent   Let $P(x, y)$ be a point other than the origin on the terminal

side of an angle θ in standard position. Let $r = \sqrt{x^2 + y^2}$ represent the distance from the origin to P. Then the cotangent function is defined by $\cot \theta = \frac{x}{y} \, (y \neq 0)$. (Section 1.3)

coterminal angles   Two angles that have the same initial side and the same terminal side, but different amounts of rotation, are called coterminal angles. The measures of coterminal angles differ by a multiple of $360°$. (Section 1.1)

cycloid   A cycloid is a curve that represents the path traced by a fixed point on the circumference of a circle rolling along a line. (Section 8.6)

D

damped oscillatory motion
Damped oscillatory motion is oscillatory motion that has been slowed down (damped) by the force of friction. Friction causes the amplitude of the motion to diminish gradually until the weight comes to rest. (Section 4.4)

decreasing function   A function f is decreasing on an interval I if, whenever $x_1 < x_2$ in I, $f(x_1) > f(x_2)$. (Appendix C)

degree   The degree is the most common unit of measure for angles. One degree, written $1°$, represents $\frac{1}{360}$ of a rotation. (Section 1.1)

dependent variable   If the value of the variable y depends on the value of the variable x, then y is called the dependent variable. (Appendix C)

direction angle   The positive angle between the x-axis and a position vector is the direction angle for the vector. (Section 7.4)

domain   In a relation, the set of all values of the independent variable (x) is called the domain. (Appendix C)

dot product   The dot product of two vectors is the sum of the product of their first components and the product of their second components. The dot product of the two vectors $\mathbf{u} = \langle a, b \rangle$ and $\mathbf{v} = \langle c, d \rangle$ is denoted $\mathbf{u} \cdot \mathbf{v}$ and given by $\mathbf{u} \cdot \mathbf{v} = ac + bd$. (Section 7.4)

E

empty set (null set)   The empty set or null set, written $\emptyset$ or $\{ \ \}$, is the set containing no elements. (Appendix A)

equation   An equation is a statement that two expressions are equal. (Appendix A)

equilibrant   The opposite vector of the resultant of two vectors is called the equilibrant. (Section 7.5)

exact number   A number that represents the result of counting, or a number that results from theoretical work and is not the result of a measurement, is an exact number. (Section 2.4)

exponential equation   An exponential equation is an equation with a variable in an exponent. (Section 9.1)

exponential function   If $a > 0$ and $a \neq 1$, then $f(x) = a^x$ defines the exponential function with base a. (Section 9.1)

F

four-leaved rose   A four-leaved rose is a curve that is the graph of a polar equation of the form $r = a \sin 2\theta$ or $r = a \cos 2\theta$. (Section 8.5)

frequency   In simple harmonic motion, the frequency is the number of cycles per unit of time, or the reciprocal of the period. (Section 4.4)

function   A function is a relation (set of ordered pairs) in which, for each value of the first component of the ordered pairs, there is *exactly one* value of the second component. (Appendix C)

function notation   Function notation $f(x)$ (read "f of x") represents the y-value of the function f for the indicated x-value. (Appendix C)

future value   In an investment paying compound interest, the future value is the balance *after* interest has been earned. (The future value is sometimes called the *compound amount*.) (Section 9.1)

G

graph of an equation The graph of an equation is the set of all points that correspond to all of the ordered pairs that satisfy the equation. (Appendix B)

groundspeed In air navigation, the groundspeed of a plane is its speed relative to the ground. (Section 7.5)

H

horizontal component When a vector $\mathbf{u}$ is expressed as an ordered pair in the form $\mathbf{u} = \langle a, b \rangle$, the number a is the horizontal component of the vector. (Section 7.4)

I

identity An equation satisfied by every number that is a meaningful replacement for the variable is called an identity. (Appendix A)

imaginary axis In the complex plane, the vertical axis is called the imaginary axis. (Section 8.2)

imaginary part In the complex number $a + bi$, b is called the imaginary part. (Section 8.1)

imaginary unit The number i, defined by $i^2 = -1$ $\left(\text{so } i = \sqrt{-1} \right)$, is called the imaginary unit. (Section 8.1)

increasing function A function f is increasing on an interval I if, whenever $x_1 < x_2$ in I, $f(x_1) < f(x_2)$. (Appendix C)

independent variable If the value of the variable y depends on the value of the variable x, then x is called the independent variable. (Appendix C)

inequality An inequality says that one expression is greater than, greater than or equal to, less than, or less than or equal to, another. (Appendix A)

initial point When two letters are used to name a vector, the first letter indicates the initial (starting) point of the vector. (Section 7.4)

initial side When a ray is rotated around its endpoint to form an angle, the ray in its initial position is called the initial side of the angle. (Section 1.1)

interval An interval is a portion of the real number line, which may or may not include its endpoint(s). (Appendix A)

interval notation Interval notation is a simplified notation for writing intervals. It uses parentheses and brackets to show whether the endpoints are included. (Appendix A)

inverse function The inverse function of the one-to-one function f is defined as $\{(y,x) \,|\, (x,y) \text{ belongs to } f\}$. (Section 6.1)

L

lemniscate A lemniscate is a figure-eight-shaped curve that is the graph of a polar equation of the form $r^2 = a^2 \sin 2\theta$ or $r^2 = a^2 \cos 2\theta$. (Section 8.5)

line Two distinct points A and B determine a line called line AB. (Section 1.1)

line segment (segment) Line segment AB (or segment AB) is the portion of line AB between A and B, including A and B themselves. (Section 1.1)

linear equation (first-degree equation) in one variable A linear equation in one variable is an equation that can be written in the form $ax + b = 0$, where a and b are real numbers with $a \neq 0$. (Appendix A)

linear inequality in one variable A linear inequality in one variable is an inequality that can be written in the form $ax + b > 0$, where a and b are real numbers with $a \neq 0$. (Any of the symbols $<$, $\geq$, or $\leq$ may also be used.) (Appendix A)

linear speed v The linear speed v measures the distance traveled per unit of time. (Section 3.4)

logarithm A logarithm is an exponent. The expression $\log_a x$ is the power to which the base a must be raised to obtain x. (Section 9.2)

logarithmic equation A logarithmic equation is an equation with a logarithm in at least one term. (Section 9.2)

logarithmic function If $a > 0$, $a \neq 1$, and $x > 0$, then $f(x) = \log_a x$ defines the logarithmic function with base a. (Section 9.2)

M

magnitude The length of a vector represents the magnitude of the vector quantity. (Section 7.4)

minute One minute, written $1'$, is $\frac{1}{60}$ of a degree. (Section 1.1)

N

natural logarithm A logarithm to base e is called a natural logarithm. (Section 9.3)

negative angle A negative angle is an angle that is formed by clockwise rotation around its endpoint. (Section 1.1)

nth root of a complex number For a positive integer n, the complex number $a + bi$ is an nth root of the complex number $x + yi$ if $(a + bi)^n = x + yi$. (Section 8.4)

O

oblique triangle A triangle that is not a right triangle is called an oblique triangle. (Section 7.1)

obtuse angle An obtuse angle is an angle measuring more than $90°$ but less than $180°$. (Section 1.1)

one-to-one function If a function is defined so that each range element is used only once, then it is called a one-to-one function. (Section 6.1)

open interval An open interval is an interval that does not include its endpoint(s). (Appendix A)

opposite of a vector The opposite of a vector $\mathbf{v}$ is a vector $-\mathbf{v}$ that has the same magnitude as $\mathbf{v}$ but opposite direction. (Section 7.4)

ordered pair An ordered pair consists of two components, written inside parentheses, in which the order of the components is important. Ordered pairs are used to identify points in the coordinate plane. (Appendix B)

origin The point of intersection of the x-axis and the y-axis of a rectangular coordinate system is called the origin. (Appendix B)

orthogonal vectors Orthogonal vectors are vectors that are perpendicular, that is, the angle between the two vectors is $90°$. (Section 7.4)

P

parallel lines Parallel lines are lines that lie in the same plane and do not intersect. (Section 1.2)

parallelogram rule The parallelogram rule is a way to find the sum of two vectors. If the two vectors are placed so that their initial points coincide and a parallelogram is completed that has these two vectors as two of its sides, then the diagonal vector of the parallelogram that has the same initial point as the two vectors is their sum. (Section 7.4)

parameter A parameter is a variable in terms of which two or more other variables are expressed. In a pair of parametric equations $x = f(t)$ and $y = g(t)$, the variable t is the parameter. (Section 8.6)

parametric equations of a plane curve A pair of equations $x = f(t)$ and $y = g(t)$ are parametric equations of a plane curve. (Section 8.6)

period For a periodic function such that $f(x) = f(x + np)$, the smallest possible positive value of p is the period of the function. (Section 4.1)

periodic function A periodic function is a function f such that $f(x) = f(x + np)$, for every real number x in the domain of f, every integer n, and some positive real number p. (Section 4.1)

phase shift With trigonometric functions, a horizontal translation is called a phase shift. (Section 4.2)

plane curve A plane curve is a set of points (x, y) such that $x = f(t)$ and $y = g(t)$, and f and g are both defined on an interval I. (Section 8.6)

polar axis The polar axis is a specific ray in the polar coordinate system that has the pole as its endpoint. The polar axis is usually drawn in the direction of the positive x-axis. (Section 8.5)

polar coordinates In the polar coordinate system, the ordered pair (r, θ) gives polar coordinates of point P, where r is the directed distance from the pole to P and θ is the directed angle from the positive x-axis to ray OP. (Section 8.5)

polar coordinate system The polar coordinate system is a coordinate system based on a point (the pole) and a ray (the polar axis). (Section 8.5)

polar equation A polar equation is an equation that uses polar coordinates. The variables are r and θ. (Section 8.5)

pole The pole is the single fixed point in the polar coordinate system that is the endpoint of the polar axis. The pole is usually placed at the origin of a rectangular coordinate system. (Section 8.5)

position vector A vector with its initial point at the origin is called a position vector. (Section 7.4)

positive angle A positive angle is an angle that is formed by counterclockwise rotation around its endpoint. (Section 1.1)

present value In an investment paying compound interest, the principal is sometimes called the present value. (Section 9.1)

pure imaginary number A complex number $a + bi$ in which $a = 0$ and $b \neq 0$ is called a pure imaginary number. (Section 8.1)

Pythagorean theorem The Pythagorean theorem states that in a right triangle, the sum of the squares of the lengths of the legs is equal to the square of the length of the hypotenuse. (Appendix B)

Q

quadrantal angle A quadrantal angle is an angle that, when placed in standard position, has its terminal side along the x-axis or the y-axis. (Section 1.1)

quadrants The quadrants are the four regions into which the x-axis and y-axis divide the coordinate plane. (Appendix B)

quadratic equation An equation that can be written in the form $ax^2 + bx + c = 0$, where a, b, and c are real numbers with $a \neq 0$, is a quadratic equation. (Appendix A)

quadratic formula The quadratic formula $x = \frac{-b \pm \sqrt{b^2 - 4ac}}{2a}$ is a general formula that can be used to solve any quadratic equation. (Appendix A)

R

radian A radian is a unit of measure for angles. An angle with its vertex at the center of a circle that intercepts an arc on the circle equal in length to the radius of the circle has a measure of 1 radian. (Section 3.1)

radius The radius of a circle is the given distance between the center and any point on the circle. (Appendix B)

range In a relation, the set of all values of the dependent variable (y) is called the range. (Appendix C)

ray The portion of line AB that starts at A and continues through B, and on past B, is called ray AB. (Section 1.1)

real axis In the complex plane, the horizontal axis is called the real axis. (Section 8.2)

real part In the complex number $a + bi$, a is called the real part. (Section 8.1)

rectangular (Cartesian) coordinate system The x-axis and y-axis together make up a rectangular coordinate system. (Appendix B)

rectangular (Cartesian) equation A rectangular or Cartesian equation is an equation that uses rectangular coordinates. If it is an equation in two variables, the variables are x and y. (Section 8.5)

rectangular form (standard form) of a complex number The rectangular form or standard form of a complex number is $a + bi$, where a and b are real numbers. (Section 8.2)

reference angle The reference angle for an angle θ, written θ', is the positive acute angle made by the terminal side of angle θ and the x-axis. (Section 2,2)

relation A relation is a set of ordered pairs. (Appendix C)

resultant If **A** and **B** are vectors, the vector sum **A** + **B** is called the resultant of vectors **A** and **B**. (Section 7.4)

right angle A right angle is an angle measuring exactly 90°. (Section 1.1)

rose curve A rose curve is a member of a family of curves that resemble flowers and that is the graph of an equation of the form $r = a \sin n\theta$ or $r = a \cos n\theta$. (Section 8.5)

S

scalar A scalar is a quantity that involves a magnitude and can be represented by a real number. (Section 7.4)

scalar product The scalar product of a real number (or scalar) k and a vector **u** is the vector $k \cdot$ **u**, which has magnitude $|k|$ times the magnitude of **u**. (Section 7.4)

secant Let $P(x, y)$ be a point other than the origin on the terminal side of an angle θ in standard position. Let $r = \sqrt{x^2 + y^2}$ represent the distance from the origin to P. Then the secant function is defined by $\sec \theta = \frac{r}{x} (x \neq 0)$. (Section 1.3)

second One second, written $1''$, is $\frac{1}{60}$ of a minute. (Section 1.1)

sector of a circle A sector of a circle is the portion of the interior of a circle intercepted by a central angle. (Section 3.2)

Side-Angle-Side (SAS) The Side-Angle-Side (SAS) congruence axiom states that if two sides and the included angle of one triangle are equal, respectively, to two sides and the included angle of a second triangle, then the triangles are congruent. (Section 7.1)

Side-Side-Side (SSS) The Side-Side-Side (SSS) congruence axiom states that if three sides of one triangle are equal, respectively, to three sides of a second triangle, then the triangles are congruent. (Section 7.1)

significant digit A significant digit is a digit obtained by actual measurement. (Section 2.4)

similar triangles Triangles that are the same shape, but not necessarily the same size, are called similar triangles. (Section 1.2)

simple harmonic motion Simple harmonic motion is oscillatory motion about an equilibrium position. If friction is neglected, this motion can be described by a sinusoid. (Section 4.4)

sine Let $P(x, y)$ be a point other than the origin on the terminal side of an angle θ in standard position. Let $r = \sqrt{x^2 + y^2}$ represent the distance from the origin to P. Then the sine function is defined by $\sin \theta = \frac{y}{r}$. (Section 1.3)

sine wave (sinusoid) The graph of a sine function is called a sine wave or sinusoid. (Section 4.1)

solution (root) A solution or root of an equation is a number that makes the equation a true statement. (Appendix A)

solution set The solution set of an equation is the set of all numbers that satisfy the equation. (Appendix A)

spiral of Archimedes A spiral of Archimedes is an infinite curve that is the graph of a polar equation of the form $r = n\theta$. (Section 8.5)

standard form of a complex number A complex number written in the form $a + bi$ (or $a + ib$) is in standard form. (Section 8.1)

straight angle A straight angle is an angle measuring exactly 180°. (Section 1.1)

supplementary angles Two positive angles are supplementary angles if the sum of their measures is 180°. (Section 1.1)

T

tangent Let $P(x, y)$ be a point other than the origin on the terminal side of an angle θ in standard position. Let $r = \sqrt{x^2 + y^2}$ represent the distance from the origin to P. Then the tangent function is defined by $\tan \theta = \frac{y}{x} (x \neq 0)$. (Section 1.3)

terminal point When two letters are used to name a vector, the second letter indicates the terminal (ending) point of the vector. (Section 7.4)

terminal side When a ray is rotated around its endpoint to form an angle, the ray in its location after rotation is called the terminal side of the angle. (Section 1.1)

translation A translation is a horizontal or vertical shift of a graph. (Appendix D)

transversal A line that intersects two or more other lines, which may be parallel, is called a transversal. (Section 1.2)

trigonometric (polar) form of a complex number The expression $r(\cos \theta + i \sin \theta)$ is called the trigonometric form (or polar form) of the complex number $x + yi$. The expression $\cos \theta + i \sin \theta$ is sometimes abbreviated as cis θ. (Section 8.2)

U

unit circle The unit circle is the circle with center at the origin and radius 1. (Section 3.3)

unit vector A unit vector is a vector that has magnitude 1. Two useful unit vectors are **i** = $\langle 1, 0 \rangle$ and **j** = $\langle 0, 1 \rangle$. (Section 7.4)

V

vector A vector is a directed line segment that represents a vector quantity. (Section 7.4)

vector quantities Quantities that involve both magnitude and direction are called vector quantities. (Section 7.4)

vertex of an angle The vertex of an angle is the endpoint of the ray that is rotated to form the angle. (Section 1.1)

vertical angles Vertical angles are opposite angles formed by intersecting lines. (Section 1.2)

vertical asymptote A vertical line that a graph approaches, but never touches or intersects, is called a vertical asymptote. The line $x = a$ is a vertical asymptote if $|f(x)|$ gets larger and larger as x approaches a. (Section 4.3)

vertical component When a vector **u** is expressed as an ordered pair in the form **u** = $\langle a, b \rangle$, the number b is the vertical component of the vector. (Section 7.4)

X

***x*-axis**   The horizontal number line in a rectangular coordinate system is called the *x*-axis. (Appendix B)

***x*-intercept**   An *x*-intercept is the *x*-value of a point where the graph of an equation intersects the *x*-axis. (Appendix B)

Y

***y*-axis**   The vertical number line in a rectangular coordinate system is called the *y*-axis. (Appendix B)

***y*-intercept**   A *y*-intercept is the *y*-value of a point where the graph of an equation intersects the *y*-axis. (Appendix B)

Z

zero-factor property   The zero-factor property states that if the product of two (or more) complex numbers is 0, then at least one of the numbers must be 0. (Appendix A)

zero vector   The zero vector is the vector with magnitude 0. (Section 7.4)

Solutions to Selected Exercises

CHAPTER 1   TRIGONOMETRIC FUNCTIONS

1.1 Exercises *(page 6)*

29. $90° - 72° 58' 11''$

$\qquad 89° 59' 60''$ Write $90°$ as $89° 59' 60''$.

$\qquad \underline{-72° 58' 11''}$

$\qquad 17° 01' 49''$

Thus, $90° - 72° 58' 11'' = 17° 1' 49''$.

77. 600 rotations per min

$\qquad = \dfrac{600}{60}$ rotations per sec

$\qquad = 10$ rotations per sec

$\qquad = 5$ rotations per $\frac{1}{2}$ sec

$\qquad = 5(360°)$ per $\frac{1}{2}$ sec

$\qquad = 1800°$ per $\frac{1}{2}$ sec

A point on the edge of the tire will move $1800°$ in $\frac{1}{2}$ sec.

1.2 Exercises *(page 14)*

23. Angle 1 and the $55°$ angle are vertical angles, which are equal, so angle $1 = 55°$. Angle 5 and the $120°$ angle are interior angles on the same side of the transversal, which means they are supplements, so

$\qquad$ angle $5 + 120° = 180°$

$\qquad\qquad$ angle $5 = 60°$.   Subtract $120°$.

Since angles 3 and 5 are vertical angles, angle $3 = 60°$.

$\qquad$ angle 1 + angle 2 + angle 3 = $180°$

$\qquad\qquad 55°$ + angle 2 + $60°$ = $180°$

$\qquad\qquad\qquad$ angle 2 = $65°$ Subtract $115°$.

Since angles 2 and 4 are vertical angles, angle $4 = 65°$. Angle 6 and the $120°$ angle are vertical angles, so angle $6 = 120°$. Angles 6 and 8 are supplements.

$\qquad$ angle 6 + angle 8 = $180°$

$\qquad\quad 120°$ + angle 8 = $180°$

$\qquad\qquad\qquad$ angle 8 = $60°$ Subtract $120°$.

Since angles 7 and 8 are vertical angles, angle $7 = 60°$.

angle 7 + angle 4 + angle 10 = $180°$

$\qquad$ The sum of the measures of the angles in a triangle is $180°$.

$\qquad 60° + 65° +$ angle 10 = $180°$

$\qquad\qquad$ angle 10 = $55°$ Subtract $125°$.

Since angles 9 and 10 are vertical angles, angle $9 = 55°$. Thus, the measures of the angles are 1: $55°$, 2: $65°$, 3: $60°$, 4: $65°$, 5: $60°$, 6: $120°$, 7: $60°$, 8: $60°$, 9: $55°$, 10: $55°$.

33. The triangle is obtuse because it has an angle of $96°$, which is between $90°$ and $180°$. It is a scalene triangle because no two sides are equal.

1.3 Exercises *(page 24)*

37. Evaluate $\tan 360° + 4 \sin 180° + 5 \cos^2 180°$.

$\qquad \tan 360° = \tan 0° = \dfrac{y}{x} = \dfrac{0}{1} = 0$

$\qquad \sin 180° = \dfrac{y}{r} = \dfrac{0}{1} = 0$

$\qquad \cos 180° = \dfrac{x}{r} = \dfrac{-1}{1} = -1$

$\tan 360° + 4 \sin 180° + 5 \cos^2 180° = 0 + 4(0) + 5(-1)^2$

$\qquad$ Substitute; $\cos^2 x = (\cos x)^2$.

$\qquad\qquad\qquad\qquad = 5$

53. Since $\cos \theta = .8$ occurs when $x = .8$, $r = 1$, and $\theta \approx 36.87°$, $\cos \theta = -.8$ occurs when $x = -.8$ and $r = 1$. As a result, $\theta = 180° - 36.87° = 143.13°$ and $\theta = 180° + 36.87° = 216.87°$.

1.4 Exercises *(page 33)*

21. $\tan(3\theta - 4°) = \dfrac{1}{\cot(5\theta - 8°)}$ Given equation

$\tan(3\theta - 4°) = \tan(5\theta - 8°)$ Reciprocal identity

The second equation above will be true if $3\theta - 4° = 5\theta - 8°$, so solving this equation will give a value (but not the only value) for which the given equation is true.

$\qquad 3\theta - 4° = 5\theta - 8°$

$\qquad\qquad 4° = 2\theta$

$\qquad\qquad \theta = 2°$

41. $\tan 30° = \dfrac{\sin 30°}{\cos 30°}$ Quotient identity

$\cos 30° < 1$, so $\tan 30° = \dfrac{\sin 30°}{\cos 30°} > \sin 30°$; that is, $\tan 30°$ is greater than $\sin 30°$.

65. Given $\tan \theta = -\dfrac{15}{8}$, with θ in quadrant II

Draw θ in standard position in quadrant II. Because $\tan \theta = \dfrac{y}{x}$ and θ is in quadrant II, we can use the values $y = 15$ and $x = -8$ for a point on its terminal side.

$r = \sqrt{x^2 + y^2} = \sqrt{(-8)^2 + 15^2} = \sqrt{64 + 225}$
$\quad = \sqrt{289} = 17$

Use the values of x, y, and r and the definitions of the trigonometric functions to find the six trigonometric function values for θ.

$\sin \theta = \dfrac{y}{r} = \dfrac{15}{17}$ $\csc \theta = \dfrac{r}{y} = \dfrac{17}{15}$

$\cos \theta = \dfrac{x}{r} = \dfrac{-8}{17} = -\dfrac{8}{17}$ $\sec \theta = \dfrac{r}{x} = \dfrac{17}{-8} = -\dfrac{17}{8}$

$\tan \theta = \dfrac{y}{x} = \dfrac{15}{-8} = -\dfrac{15}{8}$ $\cot \theta = \dfrac{x}{y} = \dfrac{-8}{15} = -\dfrac{8}{15}$

CHAPTER 2   ACUTE ANGLES AND RIGHT TRIANGLES

2.1 Exercises (page 51)

57. One point on the line $y = \sqrt{3}x$ is the origin, $(0, 0)$. Let (x, y) be any other point on this line. Then, by the definition of slope, $m = \dfrac{y - 0}{x - 0} = \dfrac{y}{x} = \sqrt{3}$, but also, by the definition of tangent, $\tan \theta = \dfrac{y}{x}$. Thus, $\tan \theta = \sqrt{3}$. Because $\tan 60° = \sqrt{3}$, the line $y = \sqrt{3}x$ makes a $60°$ angle with the positive x-axis. (See Exercise 54.)

61. Apply the relationships between the lengths of the sides of a $30°$–$60°$ right triangle first to the triangle on the left to find the values of x and y, and then to the triangle on the right to find the values of z and w. In a $30°$–$60°$ right triangle, the side opposite the $30°$ angle is $\frac{1}{2}$ the length of the hypotenuse. The longer leg is $\sqrt{3}$ times the shorter leg.

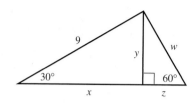

Thus,

$y = \dfrac{1}{2}(9) = \dfrac{9}{2}$ and $x = y\sqrt{3} = \dfrac{9\sqrt{3}}{2}$.

Since $y = z\sqrt{3}$,

$z = \dfrac{y}{\sqrt{3}} = \dfrac{\frac{9}{2}}{\sqrt{3}} = \dfrac{9}{2\sqrt{3}} \cdot \dfrac{\sqrt{3}}{\sqrt{3}} = \dfrac{9\sqrt{3}}{6} = \dfrac{3\sqrt{3}}{2}$,

and $w = 2z = 2\left(\dfrac{3\sqrt{3}}{2}\right) = 3\sqrt{3}$.

2.2 Exercises (page 59)

21. To find the reference angle for $-300°$, sketch this angle in standard position.

The reference angle for $-300°$ is $-300° + 360° = 60°$. Because $-300°$ is in quadrant I, the values of all its trigonometric functions will be positive, so these values will be identical to the trigonometric function values for $60°$. (See the Function Values of Special Angles table that follows Example 5 in Section 2.1.)

$\sin(-300°) = \dfrac{\sqrt{3}}{2}$ $\csc(-300°) = \dfrac{2\sqrt{3}}{3}$

$\cos(-300°) = \dfrac{1}{2}$ $\sec(-300°) = 2$

$\tan(-300°) = \sqrt{3}$ $\cot(-300°) = \dfrac{\sqrt{3}}{3}$

57. The reference angle for $115°$ is $65°$. Since $115°$ is in quadrant II, the sine is positive. $\sin \theta$ decreases on the interval $(90°, 180°)$ from 1 to 0. Therefore $\sin 115°$ is closest to .9.

2.3 Exercises (page 64)

41. $\sin 10° + \sin 10° \stackrel{?}{=} \sin 20°$

Using a calculator, we get

$\sin 10° + \sin 10° \approx .34729636$

and $\sin 20° \approx .34202014$.

Thus, the statement is false.

55. For parts (a) and (b), $\theta = 3°$, $g = 32.2$, and $f = .14$.

(a) Since 45 mph = 66 ft per sec,

$$R = \frac{V^2}{g(f + \tan \theta)}$$

$$\approx \frac{66^2}{32.2(.14 + \tan 3°)}$$

$$\approx 703 \text{ ft.}$$

(b) Since 70 mph = $\frac{70(5280)}{3600}$ ft per sec = 102.67 ft per sec,

$$R = \frac{V^2}{g(f + \tan \theta)}$$

$$\approx \frac{102.67^2}{32.2(.14 + \tan 3°)}$$

$$\approx 1701 \text{ ft.}$$

(c) Intuitively, increasing θ would make it easier to negotiate the curve at a higher speed much like is done at a race track. Mathematically, a larger value of θ (acute) will lead to a larger value for $\tan \theta$. If $\tan \theta$ increases, then the ratio determining R will *decrease*. Thus, the radius can be smaller and the curve sharper if θ is increased.

$$R = \frac{V^2}{g(f + \tan \theta)}$$

$$\approx \frac{66^2}{32.2(.14 + \tan 4°)}$$

$$\approx 644 \text{ ft}$$

$$R = \frac{V^2}{g(f + \tan \theta)}$$

$$\approx \frac{102.67^2}{32.2(.14 + \tan 4°)}$$

$$\approx 1559 \text{ ft}$$

As predicted, both values are smaller.

2.4 Exercises *(page 72)*

21. Solve the right triangle with $B = 73.00°$, $b = 128$ in., and $C = 90°$.

$$A = 90° - 73.00° = 17.00°$$

$$\tan 73.00° = \frac{128}{a}$$

$$a = \frac{128}{\tan 73.00°} \approx 39.1 \text{ in.} \quad \text{Three significant digits}$$

$$\sin 73.00° = \frac{128}{c}$$

$$c = \frac{128}{\sin 73.00°} \approx 134 \text{ in.} \quad \text{Three significant digits}$$

37. Let x represent the horizontal distance between the two buildings and y represent the height of the portion of the building across the street that is higher than the window.

$$\tan 20.0° = \frac{30.0}{x}$$

$$x = \frac{30.0}{\tan 20.0°} \approx 82.4$$

$$\tan 50.0° = \frac{y}{x}$$

$$y = x \tan 50.0° = \left(\frac{30.0}{\tan 20.0°}\right) \tan 50.0° \approx 98.2$$

$$\text{height} = y + 30.0 = \left(\frac{30.0}{\tan 20.0°}\right) \tan 50.0° + 30.0 \approx 128$$

Three significant digits

The height of the building across the street is about 128 ft.

43. Let h represent the height of the tower.

$$\tan 34.6° = \frac{h}{40.6}$$

$$h = 40.6 \tan 34.6° \approx 28.0$$

Three significant digits

The height of the tower is about 28.0 m.

2.5 Exercises *(page 81)*

17. Let x = the distance between the two ships. The angle between the bearings of the ships is

$$180° - (28° \, 10' + 61° \, 50') = 90°.$$

The triangle formed is a right triangle.

(continued)

Distance traveled at 24.0 mph:

$$(4 \text{ hr})(24.0 \text{ mph}) = 96 \text{ mi}$$

Distance traveled at 28.0 mph:

$$(4 \text{ hr})(28.0 \text{ mph}) = 112 \text{ mi}$$

Applying the Pythagorean theorem gives

$$x^2 = 96^2 + 112^2$$
$$x^2 = 21{,}760$$
$$x \approx 148.$$

The ships are 148 mi apart.

25. Let x = the distance from the closer point on the ground to the base of height h of the pyramid.

In the larger right triangle:

$$\tan 21° \, 10' = \frac{h}{135 + x}$$
$$h = (135 + x) \tan 21° \, 10'$$

In the smaller right triangle:

$$\tan 35° \, 30' = \frac{h}{x}$$
$$h = x \tan 35° \, 30'$$

Substitute for h in this equation, and solve for x.

$$(135 + x) \tan 21° \, 10' = x \tan 35° \, 30'$$
$$135 \tan 21° \, 10' + x \tan 21° \, 10' = x \tan 35° \, 30'$$
$$135 \tan 21° \, 10' = x \tan 35° \, 30' - x \tan 21° \, 10'$$
$$135 \tan 21° \, 10' = x(\tan 35° \, 30' - \tan 21° \, 10')$$
$$\frac{135 \tan 21° \, 10'}{\tan 35° \, 30' - \tan 21° \, 10'} = x$$

Then substitute for x in the equation for the smaller triangle.

$$h = \frac{135 \tan 21° \, 10'(\tan 35° \, 30')}{\tan 35° \, 30' - \tan 21° \, 10'}$$
$$h \approx 114$$

The height of the pyramid is 114 ft.

CHAPTER 3   RADIAN MEASURE AND CIRCULAR FUNCTIONS

3.1 Exercises *(page 97)*

77. (a) In 24 hr, the hour hand will rotate twice around the clock. Since one complete rotation measures 2π radians, the two rotations will measure $2(2\pi) = 4\pi$ radians.

(b) In 4 hr, the hour hand will rotate $\frac{4}{12} = \frac{1}{3}$ of the way around the clock, which will measure $\frac{1}{3}(2\pi) = \frac{2\pi}{3}$ radians.

3.2 Exercises *(page 103)*

23. For the large gear and pedal,

$$s = r\theta = 4.72\pi. \quad 180° = \pi \text{ radians}$$

Thus, the chain moves 4.72π in. Find the angle through which the small gear rotates.

$$\theta = \frac{s}{r} = \frac{4.72\pi}{1.38} \approx 3.42\pi$$

The angle θ for the wheel and for the small gear are the same, so for the wheel,

$$s = r\theta = 13.6(3.42\pi) \approx 146 \text{ in.}$$

The bicycle will move about 146 in.

43. (a)

The triangle formed by the sides of the central angle and the chord is isosceles. Therefore, the bisector of the central angle is also the perpendicular bisector of the chord and divides the larger triangle into two congruent right triangles.

$$\sin 21° = \frac{50}{r}$$
$$r = \frac{50}{\sin 21°} \approx 140 \text{ ft}$$

The radius of the curve is about 140 ft.

(b) $r = \dfrac{50}{\sin 21°}; \theta = 42°$

$$42° = 42\left(\frac{\pi}{180} \text{ radian}\right) = \frac{7\pi}{30} \text{ radian}$$
$$s = r\theta = \frac{50}{\sin 21°} \cdot \frac{7\pi}{30} = \frac{35\pi}{3 \sin 21°} \approx 102 \text{ ft}$$

The length of the arc determined by the 100-ft chord is about 102 ft.

(c) The portion of the circle bounded by the arc and the 100-ft chord is the shaded region in the figure below.

The area of the portion of the circle can be found by subtracting the area of the triangle from the area of the sector. From the figure in part (a),

$$\tan 21° = \frac{50}{h}, \quad \text{so} \quad h = \frac{50}{\tan 21°}.$$
$$A_{\text{sector}} = \frac{1}{2} r^2 \theta$$
$$= \frac{1}{2}\left(\frac{50}{\sin 21°}\right)^2\left(\frac{7\pi}{30}\right) \quad \begin{array}{l}\text{From part (b),}\\ 42° = \frac{7\pi}{30}.\end{array}$$
$$\approx 7135 \text{ ft}^2$$

$$A_{\text{triangle}} = \frac{1}{2} bh = \frac{1}{2}(100)\left(\frac{50}{\tan 21°}\right)$$
$$\approx 6513 \text{ ft}^2$$

$$A_{\text{portion}} = A_{\text{sector}} - A_{\text{triangle}} \approx 7135 \text{ ft}^2 - 6513 \text{ ft}^2$$
$$= 622 \text{ ft}^2$$

The area of the portion is about 622 ft².

45. Use the Pythagorean theorem to find the hypotenuse of the right triangle, which is also the radius of the sector of the circle.

$$r^2 = 30^2 + 40^2 = 900 + 1600 = 2500$$
$$r = \sqrt{2500} = 50$$

$$A_{\text{triangle}} = \frac{1}{2} bh = \frac{1}{2}(30)(40)$$
$$= 600 \text{ yd}^2$$

$$A_{\text{sector}} = \frac{1}{2} r^2 \theta$$

$$= \frac{1}{2}(50)^2 \cdot \frac{\pi}{3} \qquad 60° = \frac{\pi}{3}$$

$$= \frac{1250\pi}{3} \text{ yd}^2$$

$$\text{Total area} = A_{\text{triangle}} + A_{\text{sector}} = 600 \text{ yd}^2 + \frac{1250\pi}{3} \text{ yd}^2$$

$$\approx 1900 \text{ yd}^2$$

3.3 Exercises *(page 113)*

39. cos 2

$\frac{\pi}{2} \approx 1.57$ and $\pi \approx 3.14$, so $\frac{\pi}{2} < 2 < \pi$. Thus, an angle of 2 radians is in quadrant II. (The figure for Exercises 35–38 also shows that 2 radians is in quadrant II.) Because values of the cosine function are negative in quadrant II, cos 2 is negative.

57. $\left[\pi, \frac{3\pi}{2}\right];\quad \tan s = \sqrt{3}$

Recall that $\tan\frac{\pi}{3} = \sqrt{3}$ and in quadrant III tan s is positive. Therefore,

$$\tan\left(\pi + \frac{\pi}{3}\right) = \tan\frac{4\pi}{3} = \sqrt{3},$$

and thus, $s = \frac{4\pi}{3}$.

3.4 Exercises *(page 119)*

25. The hour hand of a clock moves through an angle of 2π radians (one complete revolution) in 12 hr, so

$$\omega = \frac{\theta}{t} = \frac{2\pi}{12} = \frac{\pi}{6} \text{ radian per hr.}$$

35. At 200 revolutions per min, the bicycle tire is moving $200(2\pi) = 400\pi$ radians per min. This is the angular velocity ω. The linear velocity of the bicycle is

$$v = r\omega = 13(400\pi) = 5200\pi \text{ in. per min.}$$

Convert this velocity to miles per hour.

$$v = \frac{5200\pi \text{ in.}}{\text{min}} \cdot \frac{60 \text{ min}}{\text{hr}} \cdot \frac{1 \text{ ft}}{12 \text{ in.}} \cdot \frac{1 \text{ mi}}{5280 \text{ ft}} \approx 15.5 \text{ mph}$$

CHAPTER 4 GRAPHS OF THE CIRCULAR FUNCTIONS

4.1 Exercises *(page 141)*

37. $E = 5 \cos 120\pi t$

(a) The amplitude is $|5| = 5$, and the period is $\frac{2\pi}{120\pi} = \frac{1}{60}$.

(b) Since the period is $\frac{1}{60}$, one cycle is completed in $\frac{1}{60}$ sec. Therefore, in 1 sec, 60 cycles are completed.

(c) For $t = 0$, $E = 5 \cos 120\pi(0) = 5 \cos 0 = 5$.

For $t = .03$, $E = 5 \cos 120\pi(.03) \approx 1.545$.

For $t = .06$, $E = 5 \cos 120\pi(.06) \approx -4.045$.

For $t = .09$, $E \approx -4.045$.

For $t = .12$, $E \approx 1.545$.

(d)

$E = 5 \cos 120\pi t$

4.2 Exercises *(page 152)*

45. $y = \frac{1}{2} + \sin 2\left(x + \frac{\pi}{4}\right)$

This equation has the form $y = c + a \sin b(x - d)$ with $c = \frac{1}{2}$, $a = 1$, $b = 2$, and $d = -\frac{\pi}{4}$. Start with the graph of $y = \sin x$ and modify it to take into account the amplitude, period, and translations required to obtain the desired graph.

Amplitude: $|a| = 1$; Period: $\frac{2\pi}{b} = \frac{2\pi}{2} = \pi$

Vertical translation: $\frac{1}{2}$ unit up

Phase shift (horizontal translation): $\frac{\pi}{4}$ units to the left

4.3 Exercises *(page 165)*

43. $y = -1 + \dfrac{1}{2}\cot(2x - 3\pi)$

$y = -1 + \dfrac{1}{2}\cot 2\left(x - \dfrac{3\pi}{2}\right)$ Rewrite $2x - 3\pi$ as $2\left(x - \frac{3\pi}{2}\right)$.

Period: $\dfrac{\pi}{b} = \dfrac{\pi}{2}$

Vertical translation: 1 unit down

Phase shift (horizontal translation): $\dfrac{3\pi}{2}$ units to the right

Because the function is to be graphed over a two-period interval, locate three adjacent vertical asymptotes. Because asymptotes of the graph of $y = \cot x$ occur at multiples of π, the following equations can be solved to locate asymptotes:

$$2\left(x - \frac{3\pi}{2}\right) = -2\pi, \quad 2\left(x - \frac{3\pi}{2}\right) = -\pi, \quad \text{and}$$

$$2\left(x - \frac{3\pi}{2}\right) = 0.$$

Solve each of these equations.

$$2\left(x - \frac{3\pi}{2}\right) = -2\pi$$

$$x - \frac{3\pi}{2} = -\pi$$

$$x = -\pi + \frac{3\pi}{2} = \frac{\pi}{2}$$

$$2\left(x - \frac{3\pi}{2}\right) = -\pi$$

$$x - \frac{3\pi}{2} = -\frac{\pi}{2}$$

$$x = -\frac{\pi}{2} + \frac{3\pi}{2} = \frac{2\pi}{2} = \pi$$

$$2\left(x - \frac{3\pi}{2}\right) = 0$$

$$x - \frac{3\pi}{2} = 0$$

$$x = \frac{3\pi}{2}$$

Divide the interval $\left(\frac{\pi}{2}, \pi\right)$ into four equal parts to obtain the following key x-values:

first-quarter value: $\dfrac{5\pi}{8}$;   middle value: $\dfrac{3\pi}{4}$;

third-quarter value: $\dfrac{7\pi}{8}$.

Evaluating the given function at these three key x-values gives the following points:

$$\left(\frac{5\pi}{8}, -\frac{1}{2}\right), \quad \left(\frac{3\pi}{4}, -1\right), \quad \left(\frac{7\pi}{8}, -\frac{3}{2}\right).$$

Connect these points with a smooth curve and continue the graph to approach the asymptotes $x = \frac{\pi}{2}$ and $x = \pi$ to complete one period of the graph. Sketch an identical curve between the asymptotes $x = \pi$ and $x = \frac{3\pi}{2}$ to complete a second period of the graph.

$y = -1 + \frac{1}{2}\cot(2x - 3\pi)$

4.4 Exercises *(page 171)*

19. (a) We will use a model of the form $s(t) = a\cos\omega t$ with $a = -3$. Since

$$s(0) = -3\cos(0\omega) = -3\cos 0 = -3 \cdot 1 = -3,$$

using a cosine function rather than a sine function will avoid the need for a phase shift.

Since the frequency $= \frac{6}{\pi}$ cycles per sec, by definition,

$$\frac{6}{\pi} = \frac{\omega}{2\pi}$$

$$\pi\omega = 12\pi \quad \text{Cross products}$$

$$\omega = 12. \quad \text{Divide by } \pi.$$

Therefore, a model for the position of the weight at time t seconds is

$$s(t) = -3\cos 12t.$$

(b) Period $= \dfrac{1}{\frac{6}{\pi}} = \dfrac{\pi}{6}$ sec

CHAPTER 5 TRIGONOMETRIC IDENTITIES

5.1 Exercises *(page 185)*

25. $\cot\theta = \dfrac{4}{3}$, $\sin\theta > 0$

Since $\cot\theta > 0$ and $\sin\theta > 0$, θ is in quadrant I, so all the function values are positive.

$$\tan\theta = \frac{1}{\cot\theta} = \frac{1}{\frac{4}{3}} = \frac{3}{4}$$

$$\sec^2\theta = \tan^2\theta + 1 \qquad \text{Pythagorean identity}$$

$$= \left(\frac{3}{4}\right)^2 + 1 = \frac{9}{16} + \frac{16}{16} = \frac{25}{16}$$

$\sec \theta = \sqrt{\dfrac{25}{16}} = \dfrac{5}{4}$ $\sec \theta > 0$

$\cos \theta = \dfrac{1}{\sec \theta} = \dfrac{1}{\frac{5}{4}} = \dfrac{4}{5}$

$\sin^2 \theta = 1 - \cos^2 \theta$ Alternative form of Pythagorean identity

$\qquad = 1 - \left(\dfrac{4}{5}\right)^2 = \dfrac{9}{25}$

$\sin \theta = \sqrt{\dfrac{9}{25}} = \dfrac{3}{5}$ $\sin \theta > 0$

$\csc \theta = \dfrac{1}{\sin \theta} = \dfrac{1}{\frac{3}{5}} = \dfrac{5}{3}$

Thus, $\sin \theta = \frac{3}{5}$, $\cos \theta = \frac{4}{5}$, $\tan \theta = \frac{3}{4}$, $\sec \theta = \frac{5}{4}$, and $\csc \theta = \frac{5}{3}$.

47. $\csc x = \dfrac{1}{\sin x}$ Alternative form of reciprocal identity

$\qquad = \dfrac{1}{\pm\sqrt{1 - \cos^2 x}}$ Alternative form of Pythagorean identity

$\qquad = \dfrac{\pm 1}{\sqrt{1 - \cos^2 x}}$

$\qquad = \dfrac{\pm 1}{\sqrt{1 - \cos^2 x}} \cdot \dfrac{\sqrt{1 - \cos^2 x}}{\sqrt{1 - \cos^2 x}}$ Rationalize the denominator.

$\qquad = \dfrac{\pm\sqrt{1 - \cos^2 x}}{1 - \cos^2 x}$

59. $\sec \theta - \cos \theta = \dfrac{1}{\cos \theta} - \cos \theta$

$\qquad = \dfrac{1}{\cos \theta} - \dfrac{\cos^2 \theta}{\cos \theta}$ Get a common denominator.

$\qquad = \dfrac{1 - \cos^2 \theta}{\cos \theta}$ Subtract fractions.

$\qquad = \dfrac{\sin^2 \theta}{\cos \theta}$ $1 - \cos^2 \theta = \sin^2 \theta$

65. Since $\cos x = \frac{1}{5}$, which is positive, x is in quadrant I or quadrant IV.

$\sin x = \pm\sqrt{1 - \cos^2 x} = \pm\sqrt{1 - \left(\dfrac{1}{5}\right)^2}$

$\qquad = \pm\sqrt{\dfrac{24}{25}} = \pm\dfrac{2\sqrt{6}}{5}$

$\tan x = \dfrac{\sin x}{\cos x} = \dfrac{\pm\frac{2\sqrt{6}}{5}}{\frac{1}{5}} = \pm 2\sqrt{6}$

$\sec x = \dfrac{1}{\cos x} = \dfrac{1}{\frac{1}{5}} = 5$

Quadrant I:

$\dfrac{\sec x - \tan x}{\sin x} = \dfrac{5 - 2\sqrt{6}}{\frac{2\sqrt{6}}{5}} = \dfrac{5(5 - 2\sqrt{6})}{2\sqrt{6}}$

$\qquad = \dfrac{25 - 10\sqrt{6}}{2\sqrt{6}} \cdot \dfrac{\sqrt{6}}{\sqrt{6}} = \dfrac{25\sqrt{6} - 60}{12}$

Quadrant IV:

$\dfrac{\sec x - \tan x}{\sin x} = \dfrac{5 - (-2\sqrt{6})}{-\frac{2\sqrt{6}}{5}} = \dfrac{5(5 + 2\sqrt{6})}{-2\sqrt{6}}$

$\qquad = \dfrac{25 + 10\sqrt{6}}{-2\sqrt{6}} \cdot \dfrac{-\sqrt{6}}{-\sqrt{6}} = \dfrac{-25\sqrt{6} - 60}{12}$

5.2 Exercises *(page 194)*

11. $\dfrac{1}{1 + \cos x} - \dfrac{1}{1 - \cos x} = \dfrac{1(1 - \cos x) - 1(1 + \cos x)}{(1 + \cos x)(1 - \cos x)}$

$\qquad = \dfrac{1 - \cos x - 1 - \cos x}{1 - \cos^2 x}$

$\qquad = \dfrac{-2 \cos x}{\sin^2 x} = -\dfrac{2 \cos x}{\sin^2 x}$

or $-\dfrac{2 \cos x}{\sin^2 x} = -2\left(\dfrac{\cos x}{\sin x}\right)\left(\dfrac{1}{\sin x}\right)$

$\qquad = -2 \cot x \csc x$

15. $(\sin x + 1)^2 - (\sin x - 1)^2$

$\qquad = (\sin^2 x + 2 \sin x + 1) - (\sin^2 x - 2 \sin x + 1)$

               Square the binomials.

$\qquad = \sin^2 x + 2 \sin x + 1 - \sin^2 x + 2 \sin x - 1$

$\qquad = 4 \sin x$

63. Verify that $\dfrac{\tan^2 t - 1}{\sec^2 t} = \dfrac{\tan t - \cot t}{\tan t + \cot t}$ is an identity.

Work with the right side.

$\dfrac{\tan t - \cot t}{\tan t + \cot t} = \dfrac{\tan t - \dfrac{1}{\tan t}}{\tan t + \dfrac{1}{\tan t}}$ $\cot t = \frac{1}{\tan t}$

$\qquad = \dfrac{\tan t}{\tan t}\left(\dfrac{\tan t - \dfrac{1}{\tan t}}{\tan t + \dfrac{1}{\tan t}}\right)$

Multiply numerator and denominator of the complex fraction by the LCD, $\tan t$.

$\qquad = \dfrac{\tan^2 t - 1}{\tan^2 t + 1}$ Distributive property

$\qquad = \dfrac{\tan^2 t - 1}{\sec^2 t}$ $\tan^2 t + 1 = \sec^2 t$

79. Show that $\sin(\csc s) = 1$ is not an identity.

We need find only one value for which the statement is false. Let $s = 2$. Use a calculator to find that $\sin(\csc 2) \approx .891094$, which is not equal to 1. Thus, $\sin(\csc s) = 1$ is not true for *all* real numbers s, so it is not an identity.

5.3 Exercises *(page 202)*

35. $\sec \theta = \csc\left(\dfrac{\theta}{2} + 20°\right)$

By a cofunction identity, $\sec \theta = \csc(90° - \theta)$. Thus,

$$\csc\left(\dfrac{\theta}{2} + 20°\right) = \csc(90° - \theta) \qquad \text{Substitute.}$$

$$\dfrac{\theta}{2} + 20° = 90° - \theta$$

$$\dfrac{3\theta}{2} = 70° \qquad \text{Add } \theta; \text{ subtract } 20°.$$

$$\theta = \dfrac{2}{3}(70°) = \dfrac{140°}{3}. \quad \text{Multiply by } \tfrac{2}{3}.$$

57. True or false: $\cos \dfrac{\pi}{3} = \cos \dfrac{\pi}{12} \cos \dfrac{\pi}{4} - \sin \dfrac{\pi}{12} \sin \dfrac{\pi}{4}$.

Since $\dfrac{\pi}{3} = \dfrac{\pi}{12} + \dfrac{3\pi}{12} = \dfrac{\pi}{12} + \dfrac{\pi}{4}$,

$$\cos \dfrac{\pi}{3} = \cos\left(\dfrac{\pi}{12} + \dfrac{\pi}{4}\right) \quad \text{Substitute.}$$

$$= \cos \dfrac{\pi}{12} \cos \dfrac{\pi}{4} - \sin \dfrac{\pi}{12} \sin \dfrac{\pi}{4}.$$
$$\text{Cosine sum identity}$$

The given statement is true.

5.4 Exercises *(page 209)*

45. $\cos s = -\dfrac{8}{17}$ and $\cos t = -\dfrac{3}{5}$, s and t in quadrant III

In order to substitute into sum and difference identities, we need to find the values of $\sin s$ and $\sin t$, and also the values of $\tan s$ and $\tan t$. Because s and t are both in quadrant III, the values of $\sin s$ and $\sin t$ will be negative, while $\tan s$ and $\tan t$ will be positive.

$$\sin s = -\sqrt{1 - \cos^2 s} = -\sqrt{1 - \left(-\dfrac{8}{17}\right)^2}$$

$$= -\sqrt{\dfrac{225}{289}} = -\dfrac{15}{17}$$

$$\sin t = -\sqrt{1 - \cos^2 t} = -\sqrt{1 - \left(-\dfrac{3}{5}\right)^2}$$

$$= -\sqrt{\dfrac{16}{25}} = -\dfrac{4}{5}$$

$$\tan s = \dfrac{\sin s}{\cos s} = \dfrac{-\frac{15}{17}}{-\frac{8}{17}} = \dfrac{15}{8}; \ \tan t = \dfrac{\sin t}{\cos t} = \dfrac{-\frac{4}{5}}{-\frac{3}{5}} = \dfrac{4}{3}$$

(a) $\sin(s + t) = \sin s \cos t + \cos s \sin t$

$$= \left(-\dfrac{15}{17}\right)\left(-\dfrac{3}{5}\right) + \left(-\dfrac{8}{17}\right)\left(-\dfrac{4}{5}\right)$$

$$= \dfrac{45}{85} + \dfrac{32}{85} = \dfrac{77}{85}$$

(b) $\tan(s + t) = \dfrac{\tan s + \tan t}{1 - \tan s \tan t} = \dfrac{\frac{15}{8} + \frac{4}{3}}{1 - \left(\frac{15}{8}\right)\left(\frac{4}{3}\right)}$

$$= \dfrac{\frac{45}{24} + \frac{32}{24}}{1 - \frac{60}{24}} = \dfrac{\frac{77}{24}}{-\frac{36}{24}} = -\dfrac{77}{36}$$

(c) From parts (a) and (b), $\sin(s + t) > 0$ and $\tan(s + t) < 0$. The only quadrant in which values of sine are positive and values of tangent are negative is quadrant II.

51. $\tan \dfrac{11\pi}{12} = \tan\left(\pi - \dfrac{\pi}{12}\right)$

$$= \dfrac{\tan \pi - \tan \frac{\pi}{12}}{1 + \tan \pi \tan \frac{\pi}{12}} \quad \text{Tangent difference identity}$$

$$= \dfrac{0 - \tan \frac{\pi}{12}}{1 + 0 \cdot \tan \frac{\pi}{12}} \quad \tan \pi = 0$$

$$= -\tan \dfrac{\pi}{12}$$

Now use a difference identity to find $\tan \frac{\pi}{12}$.

$$\tan \dfrac{\pi}{12} = \tan\left(\dfrac{\pi}{4} - \dfrac{\pi}{6}\right)$$

$$= \dfrac{\tan \frac{\pi}{4} - \tan \frac{\pi}{6}}{1 + \tan \frac{\pi}{4} \cdot \tan \frac{\pi}{6}}$$

$$= \dfrac{1 - \frac{\sqrt{3}}{3}}{1 + 1 \cdot \frac{\sqrt{3}}{3}} \quad \tan \tfrac{\pi}{4} = 1; \tan \tfrac{\pi}{6} = \tfrac{\sqrt{3}}{3}$$

$$= \dfrac{\frac{3 - \sqrt{3}}{3}}{\frac{3 + \sqrt{3}}{3}} \quad \begin{array}{l}\text{Add and subtract with the}\\ \text{common denominator.}\end{array}$$

$$= \dfrac{3 - \sqrt{3}}{3 + \sqrt{3}} \quad \begin{array}{l}\text{Multiply numerator and}\\ \text{denominator by 3.}\end{array}$$

$$= \dfrac{3 - \sqrt{3}}{3 + \sqrt{3}} \cdot \dfrac{3 - \sqrt{3}}{3 - \sqrt{3}} \quad \text{Rationalize the denominator.}$$

$$= \dfrac{9 - 6\sqrt{3} + 3}{9 - 3}$$

Square of a binomial in numerator; product of sum and difference of two terms in denominator

$$= \dfrac{12 - 6\sqrt{3}}{6}$$

$$= \dfrac{6(2 - \sqrt{3})}{6} \quad \text{Factor the numerator.}$$

$$= 2 - \sqrt{3} \quad \text{Lowest terms}$$

Thus,

$$\tan \frac{11\pi}{12} = -\tan \frac{\pi}{12} = -\left(2 - \sqrt{3}\right) = -2 + \sqrt{3}.$$

63. Verify that $\dfrac{\sin(x - y)}{\sin(x + y)} = \dfrac{\tan x - \tan y}{\tan x + \tan y}$ is an identity.

Work with the left side.

$$\frac{\sin(x - y)}{\sin(x + y)} = \frac{\sin x \cos y - \cos x \sin y}{\sin x \cos y + \cos x \sin y}$$

Sine sum and difference identities

$$= \frac{\dfrac{\sin x \cos y}{\cos x \cos y} - \dfrac{\cos x \sin y}{\cos x \cos y}}{\dfrac{\sin x \cos y}{\cos x \cos y} + \dfrac{\cos x \sin y}{\cos x \cos y}}$$

Divide numerator and denominator by $\cos x \cos y$.

$$= \frac{\dfrac{\sin x}{\cos x} \cdot 1 - 1 \cdot \dfrac{\sin y}{\cos y}}{\dfrac{\sin x}{\cos x} \cdot 1 + 1 \cdot \dfrac{\sin y}{\cos y}}$$

$$= \frac{\tan x - \tan y}{\tan x + \tan y}$$

Tangent quotient identity

5.5 Exercises (page 218)

25. $\dfrac{1}{4} - \dfrac{1}{2} \sin^2 47.1° = \dfrac{1}{4}(1 - 2 \sin^2 47.1°)$ Factor out $\frac{1}{4}$.

$$= \frac{1}{4} \cos 2(47.1°)$$

$$\cos 2A = 1 - 2 \sin^2 A$$

$$= \frac{1}{4} \cos 94.2°$$

31. $\tan 3x = \tan(2x + x)$

$$= \frac{\tan 2x + \tan x}{1 - \tan 2x \tan x}$$

Tangent sum identity

$$= \frac{\dfrac{2 \tan x}{1 - \tan^2 x} + \tan x}{1 - \dfrac{2 \tan x}{1 - \tan^2 x} \cdot \tan x}$$

Tangent double-angle identity

$$= \frac{\dfrac{2 \tan x + (1 - \tan^2 x) \tan x}{1 - \tan^2 x}}{\dfrac{1 - \tan^2 x - 2 \tan^2 x}{1 - \tan^2 x}}$$

Add and subtract using the common denominator.

$$= \frac{2 \tan x + \tan x - \tan^3 x}{1 - \tan^2 x - 2 \tan^2 x}$$

Multiply numerator and denominator by $1 - \tan^2 x$.

$$= \frac{3 \tan x - \tan^3 x}{1 - 3 \tan^2 x}$$

Combine terms.

43. Verify that $\sin 4x = 4 \sin x \cos x \cos 2x$ is an identity.

Work with the left side.

$$\sin 4x = \sin 2(2x)$$

$$= 2 \sin 2x \cos 2x \qquad \text{Sine double-angle identity}$$

$$= 2(2 \sin x \cos x)\cos 2x \quad \text{Sine double-angle identity}$$

$$= 4 \sin x \cos x \cos 2x$$

5.6 Exercises (page 224)

21. Find $\tan \dfrac{\theta}{2}$, given $\sin \theta = \dfrac{3}{5}$, with $90° < \theta < 180°$.

To find $\tan \frac{\theta}{2}$, we need the values of $\sin \theta$ and $\cos \theta$. We know $\sin \theta = \frac{3}{5}$.

$$\cos \theta = \pm\sqrt{1 - \sin^2 \theta} \qquad \text{Fundamental identity}$$

$$= \pm\sqrt{1 - \left(\frac{3}{5}\right)^2} \qquad \text{Substitute.}$$

$$= \pm\sqrt{\frac{16}{25}}$$

$$\cos \theta = -\frac{4}{5} \qquad \theta \text{ is in quadrant II.}$$

Thus,

$$\tan \frac{\theta}{2} = \frac{\sin \theta}{1 + \cos \theta} \qquad \text{Half-angle identity}$$

$$= \frac{\frac{3}{5}}{1 - \frac{4}{5}} = 3. \quad \text{Simplify.}$$

49. Verify that $\sec^2 \dfrac{x}{2} = \dfrac{2}{1 + \cos x}$ is an identity.

Work with the left side.

$$\sec^2 \frac{x}{2} = \frac{1}{\cos^2 \dfrac{x}{2}} \qquad \text{Reciprocal identity}$$

$$= \frac{1}{\left(\pm\sqrt{\dfrac{1 + \cos x}{2}}\right)^2} \qquad \begin{array}{l}\text{Cosine half-angle}\\ \text{identity}\end{array}$$

$$= \frac{1}{\dfrac{1 + \cos x}{2}} = \frac{2}{1 + \cos x}$$

CHAPTER 6   INVERSE CIRCULAR FUNCTIONS AND TRIGONOMETRIC EQUATIONS

6.1 Exercises *(page 246)*

71. $\sin\left(2\cos^{-1}\dfrac{1}{5}\right)$

Let $\theta = \cos^{-1}\frac{1}{5}$, so $\cos\theta = \frac{1}{5}$. The inverse cosine function yields values only in quadrants I and II, and since $\frac{1}{5}$ is positive, θ is in quadrant I. Sketch θ in quadrant I, and label the sides of a right triangle. By the Pythagorean theorem, the length of the side opposite θ will be $\sqrt{5^2 - 1^2} = \sqrt{24} = 2\sqrt{6}$.

From the figure, $\sin\theta = \dfrac{2\sqrt{6}}{5}$. Then,

$$\sin\left(2\cos^{-1}\frac{1}{5}\right) = \sin 2\theta$$

$$= 2\sin\theta\cos\theta$$
Sine double-angle identity

$$= 2\left(\frac{2\sqrt{6}}{5}\right)\left(\frac{1}{5}\right) = \frac{4\sqrt{6}}{25}.$$

77. $\sin\left(\sin^{-1}\dfrac{1}{2} + \tan^{-1}(-3)\right)$

Let $\sin^{-1}\frac{1}{2} = A$ and $\tan^{-1}(-3) = B$; then $\sin A = \frac{1}{2}$ and $\tan B = -3$. Sketch angle A in quadrant I and angle B in quadrant IV.

$$\sin\left(\sin^{-1}\frac{1}{2} + \tan^{-1}(-3)\right) = \sin(A + B)$$

$$= \sin A\cos B + \cos A\sin B$$
Sine sum identity

$$= \frac{1}{2}\cdot\frac{1}{\sqrt{10}} + \frac{\sqrt{3}}{2}\cdot\frac{-3}{\sqrt{10}}$$

$$= \frac{1 - 3\sqrt{3}}{2\sqrt{10}} = \frac{\sqrt{10} - 3\sqrt{30}}{20}$$

6.2 Exercises *(page 254)*

15. $\tan^2 x + 3 = 0$,   so $\tan^2 x = -3$.

The square of a real number cannot be negative, so this equation has no solution. Solution set: $\emptyset$

25.
$$2\sin\theta - 1 = \csc\theta$$

$$2\sin\theta - 1 = \frac{1}{\sin\theta} \quad \text{Reciprocal identity}$$

$$2\sin^2\theta - \sin\theta = 1 \quad \text{Multiply by } \sin\theta.$$

$$2\sin^2\theta - \sin\theta - 1 = 0 \quad \text{Subtract 1.}$$

$$(2\sin\theta + 1)(\sin\theta - 1) = 0 \quad \text{Factor.}$$

$$2\sin\theta + 1 = 0 \quad \text{or} \quad \sin\theta - 1 = 0$$
$$\text{Zero-factor property}$$

$$\sin\theta = -\frac{1}{2} \quad \text{or} \quad \sin\theta = 1$$

Over the interval $[0°, 360°)$, the equation $\sin\theta = -\frac{1}{2}$ has two solutions, the angles in quadrants III and IV that have reference angle $30°$. These are $210°$ and $330°$. In the same interval, the only angle θ for which $\sin\theta = 1$ is $90°$.

Solution set: $\{90°, 210°, 330°\}$

49.
$$\frac{2\tan\theta}{3 - \tan^2\theta} = 1$$

$$2\tan\theta = 3 - \tan^2\theta$$

$$\tan^2\theta + 2\tan\theta - 3 = 0$$

$$(\tan\theta - 1)(\tan\theta + 3) = 0$$

$$\tan\theta = 1 \quad \text{or} \quad \tan\theta = -3$$

Over the interval $[0°, 360°)$, the equation $\tan\theta = 1$ has two solutions, $45°$ and $225°$. Over the same interval, the equation $\tan\theta = -3$ has two solutions that are approximately $-71.6° + 180° = 108.4°$ and $-71.6° + 360° = 288.4°$.

The period of the tangent function is $180°$, so all solutions of the given equation are $45° + n\cdot 180°$ and $108.4° + n\cdot 180°$, where n is any integer.

6.3 Exercises *(page 260)*

23. $\cos 2x + \cos x = 0$

We choose the identity for $\cos 2x$ that involves only the cosine function.

$$\cos 2x + \cos x = 0$$

$$2\cos^2 x - 1 + \cos x = 0 \quad \text{Cosine double-angle identity}$$

$$2\cos^2 x + \cos x - 1 = 0 \quad \text{Standard quadratic form}$$

$$(2\cos x - 1)(\cos x + 1) = 0 \quad \text{Factor.}$$

$$2\cos x - 1 = 0 \quad \text{or} \quad \cos x + 1 = 0 \quad \text{Zero-factor property}$$

$$2\cos x = 1 \quad \text{or} \quad \cos x = -1$$

$$\cos x = \frac{1}{2} \quad \text{or} \quad x = \pi$$

$$x = \frac{\pi}{3} \text{ or } \frac{5\pi}{3}$$

Solution set: $\left\{\dfrac{\pi}{3}, \pi, \dfrac{5\pi}{3}\right\}$

31.

$$2 \sin \theta = 2 \cos 2\theta$$

$$\sin \theta = \cos 2\theta \qquad \text{Divide by 2.}$$

$$\sin \theta = 1 - 2 \sin^2 \theta$$

   Cosine double-angle identity

$$2 \sin^2 \theta + \sin \theta - 1 = 0$$

$$(2 \sin \theta - 1)(\sin \theta + 1) = 0$$

$$2 \sin \theta - 1 = 0 \quad \text{or} \quad \sin \theta + 1 = 0$$

$$\sin \theta = \frac{1}{2} \quad \text{or} \quad \sin \theta = -1$$

Over the interval $[0°, 360°)$, the equation $\sin \theta = \frac{1}{2}$ has two solutions, $30°$ and $150°$. Over the same interval, the equation $\sin \theta = -1$ has one solution, $270°$.

Solution set: $\{30°, 150°, 270°\}$

6.4 Exercises *(page 265)*

15. $\quad y = \cos(x + 3)$

$$x + 3 = \arccos y$$

$$x = -3 + \arccos y$$

33. $\arccos x + 2 \arcsin \dfrac{\sqrt{3}}{2} = \pi$

$$\arccos x = \pi - 2 \arcsin \frac{\sqrt{3}}{2}$$

$$\arccos x = \pi - 2\left(\frac{\pi}{3}\right) \quad \arcsin \frac{\sqrt{3}}{2} = \frac{\pi}{3}$$

$$\arccos x = \pi - \frac{2\pi}{3}$$

$$\arccos x = \frac{\pi}{3}$$

$$x = \cos \frac{\pi}{3} = \frac{1}{2}$$

Solution set: $\left\{\dfrac{1}{2}\right\}$

37. $\cos^{-1} x + \tan^{-1} x = \dfrac{\pi}{2}$

$$\cos^{-1} x = \frac{\pi}{2} - \tan^{-1} x$$

$$x = \cos\left(\frac{\pi}{2} - \tan^{-1} x\right)$$

   Definition of $\cos^{-1} x$

$$x = \cos \frac{\pi}{2} \cdot \cos(\tan^{-1} x)$$

$$+ \sin \frac{\pi}{2} \cdot \sin(\tan^{-1} x)$$

   Cosine difference identity

$$x = 0 \cdot \cos(\tan^{-1} x) + 1 \cdot \sin(\tan^{-1} x)$$

   $\cos \frac{\pi}{2} = 0; \sin \frac{\pi}{2} = 1$

$$x = \sin(\tan^{-1} x)$$

Let $u = \tan^{-1} x$, so $\tan u = x$.

From the triangle, we find that $\sin u = \dfrac{x}{\sqrt{1 + x^2}}$, so the equation $x = \sin(\tan^{-1} x)$ becomes $x = \dfrac{x}{\sqrt{1 + x^2}}$. Solve this equation.

$$x = \frac{x}{\sqrt{1 + x^2}}$$

$$x\sqrt{1 + x^2} = x \quad \text{Multiply by } \sqrt{1 + x^2}.$$

$$x\sqrt{1 + x^2} - x = 0$$

$$x\left(\sqrt{1 + x^2} - 1\right) = 0 \quad \text{Factor.}$$

$x = 0 \quad \text{or} \quad \sqrt{1 + x^2} - 1 = 0 \quad$ Zero-factor property

$$\sqrt{1 + x^2} = 1 \quad \text{Isolate the radical.}$$

$$1 + x^2 = 1 \quad \text{Square both sides.}$$

$$x^2 = 0$$

$$x = 0$$

Solution set: $\{0\}$

CHAPTER 7   APPLICATIONS OF TRIGONOMETRY AND VECTORS

7.1 Exercises *(page 282)*

31. We cannot find θ directly because the length of the side opposite angle θ is not given. Redraw the triangle shown in the figure and label the third angle as α.

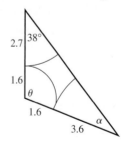

$$\frac{\sin \alpha}{1.6 + 2.7} = \frac{\sin 38°}{1.6 + 3.6} \quad \begin{array}{l}\text{Alternative form of the} \\ \text{law of sines}\end{array}$$

$$\frac{\sin \alpha}{4.3} = \frac{\sin 38°}{5.2}$$

$$\sin \alpha = \frac{4.3 \sin 38°}{5.2} \approx .50910468$$

$$\alpha \approx 31°$$

Then
$$\theta = 180° - 38° - 31°$$

$$\theta \approx 111°.$$

41. To find the area of the triangle, use $\mathscr{A} = \frac{1}{2}bh$, with $b = 1$ and $h = \sqrt{2}$.

$$\mathscr{A} = \frac{1}{2}(1)\left(\sqrt{2}\right) = \frac{\sqrt{2}}{2}$$

Now use $\mathscr{A} = \frac{1}{2}ab \sin C$, with $a = 2$, $b = 1$, and $C = 45°$.

$$\mathscr{A} = \frac{1}{2}(2)(1)\sin 45° = \sin 45° = \frac{\sqrt{2}}{2}$$

Both formulas show that the area is $\frac{\sqrt{2}}{2}$.

7.2 Exercises *(page 291)*

11. $\dfrac{\sin B}{b} = \dfrac{\sin A}{a}$ Alternative form of the law of sines

$\dfrac{\sin B}{2} = \dfrac{\sin 60°}{\sqrt{6}}$ Substitute values from the figure.

$\sin B = \dfrac{2 \sin 60°}{\sqrt{6}}$ Multiply by 2.

$\sin B = \dfrac{2 \cdot \frac{\sqrt{3}}{2}}{\sqrt{6}}$ $\sin 60° = \frac{\sqrt{3}}{2}$

$\sin B = \dfrac{\sqrt{3}}{\sqrt{6}} = \sqrt{\dfrac{1}{2}} = \dfrac{\sqrt{2}}{2}$

$B = 45°$ $\sin 45° = \frac{\sqrt{2}}{2}$

There is another angle between $0°$ and $180°$ whose sine is $\frac{\sqrt{2}}{2}$: $180° - 45° = 135°$. However, this is too large because $A = 60°$ and $60° + 135° = 195° > 180°$, so there is only one solution, $B = 45°$.

19. $A = 142.13°$, $b = 5.432$ ft, $a = 7.297$ ft

$\dfrac{\sin B}{b} = \dfrac{\sin A}{a}$ Alternative form of the law of sines

$\sin B = \dfrac{b \sin A}{a}$

$\sin B = \dfrac{5.432 \sin 142.13°}{7.297}$ Substitute given values.

$\sin B \approx .45697580$

$B \approx 27.19°$ Use the inverse sine function.

Because angle A is obtuse, angle B must be acute, so this is the only possible value for B and there is one triangle with the given measurements.

$C = 180° - A - B$ Sum of the angles of any triangle is $180°$.

$= 180° - 142.13° - 27.19°$

$C = 10.68°$

Thus, $B \approx 27.19°$ and $C \approx 10.68°$.

7.3 Exercises *(page 298)*

21. $C = 45.6°$, $b = 8.94$ m, $a = 7.23$ m

First find c.

$c^2 = a^2 + b^2 - 2ab \cos C$ Law of cosines

$c^2 = 7.23^2 + 8.94^2 - 2(7.23)(8.94)\cos 45.6°$
                    Substitute given values.

$c^2 \approx 41.7493$

$c \approx 6.46$

Find A next since angle A is smaller than angle B (because $a < b$), and thus angle A must be acute.

$\dfrac{\sin A}{a} = \dfrac{\sin C}{c}$ Alternative form of the law of sines

$\sin A = \dfrac{a \sin C}{c}$

$\sin A = \dfrac{7.23 \sin 45.6°}{6.46}$

$\sin A \approx .79963428$

$A \approx 53.1°$

Finally, find B.

$B = 180° - C - A$

$= 180° - 45.6° - 53.1°$

$B = 81.3°$

Thus, $c \approx 6.46$ m, $A \approx 53.1°$, and $B \approx 81.3°$.

41. Find AC, or b, in this figure.

Angle $1 = 180° - 128° \, 40' = 51° \, 20'$

Angles 1 and 2 are alternate interior angles formed when two parallel lines (the north lines) are cut by a transversal, line BC, so angle $2 =$ angle $1 = 51° \, 20'$.

angle $ABC = 90° -$ angle $2 = 90° - 51° \, 20' = 38° \, 40'$
                   Complementary angles

$b^2 = a^2 + c^2 - 2ac \cos B$
                Law of cosines

$b^2 = 359^2 + 450^2 - 2(359)(450)\cos 38° \, 40'$
                Substitute values from the figure.

$b^2 \approx 79{,}106$

$b \approx 281$

C is about 281 km from A.

7.4 Exercises *(page 312)*

19. Use the figure to find the components of **a** and **b**:
a = ⟨−8, 8⟩ and **b** = ⟨4, 8⟩.

   (a) **a** + **b** = ⟨−8, 8⟩ + ⟨4, 8⟩ = ⟨−8 + 4, 8 + 8⟩ = ⟨−4, 16⟩

   (b) **a** − **b** = ⟨−8, 8⟩ − ⟨4, 8⟩ = ⟨−8 − 4, 8 − 8⟩

         = ⟨−12, 0⟩

   (c) −**a** = −⟨−8, 8⟩ = ⟨8, −8⟩

47. **v** = ⟨a, b⟩ = ⟨5 cos(−35°), 5 sin(−35°)⟩ ≈ ⟨4.0958, −2.8679⟩

81. First write the given vectors in component form.

$$3\mathbf{i} + 4\mathbf{j} = \langle 3, 4 \rangle; \quad \mathbf{j} = \langle 0, 1 \rangle$$

$$\cos \theta = \frac{\langle 3, 4 \rangle \cdot \langle 0, 1 \rangle}{|\langle 3, 4 \rangle| \, |\langle 0, 1 \rangle|}$$

$$= \frac{3(0) + 4(1)}{\sqrt{9 + 16} \cdot \sqrt{0 + 1}}$$

$$= \frac{4}{5 \cdot 1} = \frac{4}{5} = .8$$

$$\theta = \cos^{-1} .8 \approx 36.87°$$

7.5 Exercises *(page 318)*

5. Use the parallelogram rule. In the figure, **x** represents the second force and **v** is the resultant.

$$\alpha = 180° − 78° \, 50' = 101° \, 10'$$
$$\beta = 78° \, 50' − 41° \, 10' = 37° \, 40'$$

$$\frac{|\mathbf{x}|}{\sin 41° \, 10'} = \frac{176}{\sin 37° \, 40'} \quad \text{Law of sines}$$

$$|\mathbf{x}| = \frac{176 \sin 41° \, 10'}{\sin 37° \, 40'} \approx 190$$

$$\frac{|\mathbf{v}|}{\sin \alpha} = \frac{176}{\sin 37° \, 40'} \quad \text{Law of sines}$$

$$|\mathbf{v}| = \frac{176 \sin 101° \, 10'}{\sin 37° \, 40'} \approx 283$$

Thus, the magnitude of the second force is about 190 lb and the magnitude of the resultant is about 283 lb.

27. Let **v** represent the airspeed vector.

The groundspeed is $\dfrac{400 \text{ mi}}{2.5 \text{ hr}} = 160$ mph.

angle $BAC = 328° − 180° = 148°$

$$|\mathbf{v}|^2 = 11^2 + 160^2 − 2(11)(160) \cos 148°$$
$$\qquad\qquad\qquad\qquad\qquad \text{Law of cosines}$$
$$|\mathbf{v}|^2 \approx 28{,}706$$
$$|\mathbf{v}| \approx 170$$

The airspeed must be approximately 170 mph.

$$\frac{\sin B}{11} = \frac{\sin 148°}{170} \quad \text{Law of sines}$$

$$\sin B = \frac{11 \sin 148°}{170} \approx .034289$$

$$B \approx 2°$$

The bearing must be approximately 360° − 2° = 358°.

CHAPTER 8   COMPLEX NUMBERS, POLAR EQUATIONS, AND PARAMETRIC EQUATIONS

8.1 Exercises *(page 338)*

55. $−i − 2 − (6 − 4i) − (5 − 2i)$

$$= (−2 − 6 − 5) + [−1 − (−4) − (−2)]i$$
$$= −13 + 5i$$

73. $(2 + i)(2 − i)(4 + 3i)$

$$= [(2 + i)(2 − i)](4 + 3i) \quad \text{Associative property}$$
$$= (2^2 − i^2)(4 + 3i) \qquad \text{Product of the sum and}$$
$$\qquad\qquad\qquad\qquad\qquad\qquad \text{difference of two terms}$$
$$= [4 − (−1)](4 + 3i) \quad i^2 = −1$$
$$= 5(4 + 3i)$$
$$= 20 + 15i \qquad\qquad \text{Distributive property}$$

85. $\dfrac{1}{i^{-11}} = i^{11} = i^8 \cdot i^3 = (i^4)^2 \cdot i^3 = 1(−i) = −i$

101. $\left(\dfrac{\sqrt{2}}{2} + \dfrac{\sqrt{2}}{2}i\right)^2 = \left(\dfrac{\sqrt{2}}{2}\right)^2 + 2 \cdot \dfrac{\sqrt{2}}{2} \cdot \dfrac{\sqrt{2}}{2}i + \left(\dfrac{\sqrt{2}}{2}i\right)^2$

$$\qquad\qquad\qquad\qquad\qquad\qquad \text{Square of a binomial}$$

$$= \frac{2}{4} + 2 \cdot \frac{2}{4}i + \frac{2}{4}i^2$$

$$= \frac{1}{2} + i + \frac{1}{2}i^2$$

$$= \frac{1}{2} + i + \frac{1}{2}(−1) \quad i^2 = −1$$

$$= \frac{1}{2} + i − \frac{1}{2}$$

$$= i$$

Thus, $\dfrac{\sqrt{2}}{2} + \dfrac{\sqrt{2}}{2}i$ is a square root of i.

8.2 Exercises *(page 345)*

29. 3 cis 150° = 3(cos 150° + i sin 150°)

$$= 3\left(−\frac{\sqrt{3}}{2} + i \cdot \frac{1}{2}\right) \quad \begin{array}{l} \cos 150° = −\frac{\sqrt{3}}{2}; \\ \sin 150° = \frac{1}{2} \end{array}$$

$$= −\frac{3\sqrt{3}}{2} + \frac{3}{2}i \qquad \text{Rectangular form}$$

43. $-5 - 5i$

Sketch the graph of $-5 - 5i$ in the complex plane.

Since $x = -5$ and $y = -5$,

$$r = \sqrt{x^2 + y^2} = \sqrt{(-5)^2 + (-5)^2} = \sqrt{50} = 5\sqrt{2}$$

and

$$\tan \theta = \frac{y}{x} = \frac{-5}{-5} = 1.$$

Since $\tan \theta = 1$, the reference angle for θ is $45°$. The graph shows that θ is in quadrant III, so $\theta = 180° + 45° = 225°$. Therefore,

$$-5 - 5i = 5\sqrt{2}(\cos 225° + i \sin 225°).$$

8.3 Exercises *(page 350)*

7. $[4(\cos 60° + i \sin 60°)][6(\cos 330° + i \sin 330°)]$

$= 4 \cdot 6[\cos(60° + 330°) + i \sin(60° + 330°)]$
                           Product theorem

$= 24(\cos 390° + i \sin 390°)$ Multiply and add.

$= 24(\cos 30° + i \sin 30°)$ 390° and 30° are
                               coterminal angles.

$= 24\left(\frac{\sqrt{3}}{2} + i \cdot \frac{1}{2}\right)$ $\cos 30° = \frac{\sqrt{3}}{2}; \sin 30° = \frac{1}{2}$

$= 12\sqrt{3} + 12i$

21. $\dfrac{-i}{1 + i}$

Numerator: $-i = 0 - 1i$

$r = \sqrt{0^2 + (-1)^2} = 1$

$\theta = 270°$ since $\cos 270° = 0$ and $\sin 270° = -1$, so $-i = 1 \operatorname{cis} 270°$.

Denominator: $1 + i = 1 + 1i$

$$r = \sqrt{1^2 + 1^2} = \sqrt{2}$$

$$\tan \theta = \frac{y}{x} = \frac{1}{1} = 1$$

Since x and y are both positive, θ is in quadrant I, so $\theta = \tan^{-1} 1 = 45°$. Thus, $1 + i = \sqrt{2} \operatorname{cis} 45°$.

$$\frac{-i}{1 + i} = \frac{1 \operatorname{cis} 270°}{\sqrt{2} \operatorname{cis} 45°}$$

$= \frac{1}{\sqrt{2}} \operatorname{cis}(270° - 45°)$ Quotient theorem

$= \frac{\sqrt{2}}{2} \operatorname{cis} 225°$

$= \frac{\sqrt{2}}{2}(\cos 225° + i \sin 225°)$

$$= \frac{\sqrt{2}}{2}\left(-\frac{\sqrt{2}}{2} - i \cdot \frac{\sqrt{2}}{2}\right) \quad \begin{array}{l} \cos 225° = -\frac{\sqrt{2}}{2}; \\ \sin 225° = -\frac{\sqrt{2}}{2} \end{array}$$

$= -\frac{1}{2} - \frac{1}{2}i$ Rectangular form

8.4 Exercises *(page 356)*

11. $(-2 - 2i)^5$

First write $-2 - 2i$ in trigonometric form.

$$r = \sqrt{(-2)^2 + (-2)^2} = \sqrt{8} = 2\sqrt{2}$$

$$\tan \theta = \frac{y}{x} = \frac{-2}{-2} = 1$$

Because x and y are both negative, θ is in quadrant III, so $\theta = 225°$.

$$-2 - 2i = 2\sqrt{2}(\cos 225° + i \sin 225°)$$

$$(-2 - 2i)^5 = \left[2\sqrt{2}(\cos 225° + i \sin 225°)\right]^5$$

$= \left(2\sqrt{2}\right)^5[\cos(5 \cdot 225°) + i \sin(5 \cdot 225°)]$
                         De Moivre's theorem

$= 32 \cdot 4\sqrt{2}(\cos 1125° + i \sin 1125°)$

$= 128\sqrt{2}(\cos 1125° + i \sin 1125°)$

$= 128\sqrt{2}(\cos 45° + i \sin 45°)$
                  1125° and 45° are coterminal.

$= 128\sqrt{2}\left(\frac{\sqrt{2}}{2} + i \cdot \frac{\sqrt{2}}{2}\right)$
                      $\cos 45° = \frac{\sqrt{2}}{2}; \sin 45° = \frac{\sqrt{2}}{2}$

$= 128 + 128i$ Rectangular form

41. $x^3 - \left(4 + 4i\sqrt{3}\right) = 0$

$$x^3 = 4 + 4i\sqrt{3}$$

$$r = \sqrt{4^2 + \left(4\sqrt{3}\right)^2} = \sqrt{16 + 48} = \sqrt{64} = 8$$

$$\tan \theta = \frac{4\sqrt{3}}{4} = \sqrt{3}$$

θ is in quadrant I, so $\theta = 60°$.

$$x^3 = 4 + 4i\sqrt{3}$$

$$x^3 = 8\left(\frac{1}{2} + i\frac{\sqrt{3}}{2}\right)$$

$$r^3(\cos 3\alpha + i \sin 3\alpha) = 8(\cos 60° + i \sin 60°)$$

$r^3 = 8$, so $r = 2$.

$\alpha = \dfrac{60°}{3} + \dfrac{360° \cdot k}{3}$, k any integer nth root theorem

$\alpha = 20° + 120° \cdot k$, k any integer

If $k = 0$, then $\alpha = 20° + 0° = 20°$.

If $k = 1$, then $\alpha = 20° + 120° = 140°$.

If $k = 2$, then $\alpha = 20° + 240° = 260°$.

Solution set: $\{2(\cos 20° + i \sin 20°),$
$2(\cos 140° + i \sin 140°), 2(\cos 260° + i \sin 260°)\}$

8.5 Exercises *(page 367)*

53.

$$r = 2 \sin \theta$$

$r^2 = 2r \sin \theta$	Multiply by r.
$x^2 + y^2 = 2y$	$r^2 = x^2 + y^2, r \sin \theta = y$
$x^2 + y^2 - 2y = 0$	Subtract $2y$.
$x^2 + y^2 - 2y + 1 = 1$	Add 1 to complete the square on y.
$x^2 + (y - 1)^2 = 1$	Factor the perfect square trinomial.

The graph is a circle with center $(0, 1)$ and radius 1.

$r = 2 \sin \theta$
$x^2 + (y - 1)^2 = 1$

59.

$$r = 2 \sec \theta$$

$r = \dfrac{2}{\cos \theta}$	Reciprocal identity
$r \cos \theta = 2$	Multiply by $\cos \theta$.
$x = 2$	$r \cos \theta = x$

The graph is the vertical line through $(2, 0)$.

$r = 2 \sec \theta$
$x = 2$

8.6 Exercises *(page 376)*

9. $x = t^3 + 1, y = t^3 - 1$,   for t in $(-\infty, \infty)$

(a)

t	x	y
-2	-7	-9
-1	0	-2
0	1	-1
1	2	0
2	9	7
3	28	26

$x = t^3 + 1$
$y = t^3 - 1$
for t in $(-\infty, \infty)$

(b)

$x = t^3 + 1$	
$y = t^3 - 1$	
$x - y = 2$	Subtract equations to eliminate t.
$y = x - 2$	Solve for y.

The rectangular equation is $y = x - 2$, for x in $(-\infty, \infty)$.
The graph is a line with slope 1 and y-intercept -2.

13. $x = 3 \tan t, y = 2 \sec t$,   for t in $\left(-\dfrac{\pi}{2}, \dfrac{\pi}{2} \right)$

(a)

t	x	y
$-\dfrac{\pi}{3}$	$-3\sqrt{3} \approx -5.2$	4
$-\dfrac{\pi}{6}$	$-\sqrt{3} \approx -1.7$	$\dfrac{4\sqrt{3}}{3} \approx 2.3$
0	0	2
$\dfrac{\pi}{6}$	$\sqrt{3} \approx 1.7$	$\dfrac{4\sqrt{3}}{3} \approx 2.3$
$\dfrac{\pi}{3}$	$3\sqrt{3} \approx 5.2$	4

$x = 3 \tan t$
$y = 2 \sec t$
for t in $\left(-\dfrac{\pi}{2}, \dfrac{\pi}{2} \right)$

(b) $x = 3 \tan t$, so $\dfrac{x}{3} = \tan t$.

$y = 2 \sec t$, so $\dfrac{y}{2} = \sec t$.

$1 + \tan^2 t = \sec^2 t$	Pythagorean identity
$1 + \left(\dfrac{x}{3} \right)^2 = \left(\dfrac{y}{2} \right)^2$	Substitute expressions for $\tan t$ and $\sec t$.
$1 + \dfrac{x^2}{9} = \dfrac{y^2}{4}$	
$y^2 = 4\left(1 + \dfrac{x^2}{9} \right)$	Multiply by 4.
$y = 2\sqrt{1 + \dfrac{x^2}{9}}$	Use the positive square root because $y > 0$ in the given interval for t.

The rectangular equation is $y = 2\sqrt{1 + \dfrac{x^2}{9}}$, for x in $(-\infty, \infty)$. The graph is the upper half of a hyperbola.

CHAPTER 9   EXPONENTIAL AND LOGARITHMIC FUNCTIONS

9.1 Exercises (page 399)

7. $g(x) = \left(\dfrac{1}{4}\right)^x$

$g(-2) = \left(\dfrac{1}{4}\right)^{-2} = 4^2 = 16$

55. $\left(\dfrac{1}{e}\right)^{-x} = \left(\dfrac{1}{e^2}\right)^{x+1}$

$(e^{-1})^{-x} = (e^{-2})^{x+1}$ Definition of negative exponent

$e^x = e^{-2(x+1)}$ $(a^m)^n = a^{mn}$

$e^x = e^{-2x-2}$ Distributive property

$x = -2x - 2$ Property (b)

$3x = -2$ Add $2x$.

$x = -\dfrac{2}{3}$ Divide by 3.

Solution set: $\left\{-\dfrac{2}{3}\right\}$

57. $\left(\sqrt{2}\right)^{x+4} = 4^x$

$(2^{1/2})^{x+4} = (2^2)^x$ Definition of $a^{1/n}$; write both sides as powers of a common base.

$2^{(1/2)(x+4)} = 2^{2x}$ $(a^m)^n = a^{mn}$

$2^{(1/2)x+2} = 2^{2x}$ Distributive property

$\dfrac{1}{2}x + 2 = 2x$ Property (b)

$2 = \dfrac{3}{2}x$ Subtract $\frac{1}{2}x$.

$\dfrac{4}{3} = x$ Multiply by $\frac{2}{3}$.

Solution set: $\left\{\dfrac{4}{3}\right\}$

59. $\dfrac{1}{27} = b^{-3}$

$3^{-3} = b^{-3}$

$b = 3$

Solution set: $\{3\}$

Alternate solution:

$\dfrac{1}{27} = b^{-3}$

$\dfrac{1}{27} = \dfrac{1}{b^3}$

$27 = b^3$

$b = \sqrt[3]{27}$

$b = 3$

Solution set: $\{3\}$

9.2 Exercises (page 410)

23. $\log_x 25 = -2$

$x^{-2} = 25$ Write in exponential form.

$(x^{-2})^{-1/2} = 25^{-1/2}$ Raise both sides to the same power.

$x = \dfrac{1}{25^{1/2}}$

$x = \dfrac{1}{5}$

Solution set: $\left\{\dfrac{1}{5}\right\}$

67. $-\dfrac{2}{3}\log_5(5m^2) + \dfrac{1}{2}\log_5(25m^2)$

$= \log_5(5m^2)^{-2/3} + \log_5(25m^2)^{1/2}$ Power property

$= \log_5[(5m^2)^{-2/3} \cdot (25m^2)^{1/2}]$ Product property

$= \log_5(5^{-2/3}m^{-4/3} \cdot 5m)$ Properties of exponents

$= \log_5(5^{-2/3} \cdot 5^1 \cdot m^{-4/3} \cdot m^1)$

$= \log_5(5^{1/3} \cdot m^{-1/3})$ $a^m \cdot a^n = a^{m+n}$

$= \log_5 \dfrac{5^{1/3}}{m^{1/3}}$ or $\log_5 \sqrt[3]{\dfrac{5}{m}}$

73. $\log_{10}\sqrt{30} = \log_{10} 30^{1/2}$ Definition of $a^{1/n}$

$= \dfrac{1}{2}\log_{10} 30$ Power property

$= \dfrac{1}{2}\log_{10}(10 \cdot 3)$

$= \dfrac{1}{2}(\log_{10} 10 + \log_{10} 3)$ Product property

$= \dfrac{1}{2}(1 + .4771)$ $\log_a a = 1$; Substitute.

$= \dfrac{1}{2}(1.4771) \approx .7386$

9.3 Exercises (page 421)

39. $\log_{\sqrt{13}} 12 = \dfrac{\ln 12}{\ln \sqrt{13}}$ Change-of-base theorem

$= \dfrac{\ln 12}{\ln 13^{1/2}}$ Definition of $a^{1/n}$

$= \dfrac{\ln 12}{.5 \ln 13}$ Power property

≈ 1.9376 Use a calculator; round answer to four decimal places.

The required logarithm can also be found by entering $\ln \sqrt{13}$ into the calculator directly:

$\log_{\sqrt{13}} 12 = \dfrac{\ln 12}{\ln \sqrt{13}}$ Change-of-base theorem

≈ 1.9376 Use a calculator; round answer to four decimal places.

83.
$$\log(x + 25) = 1 + \log(2x - 7)$$
$$\log(x + 25) - \log(2x - 7) = 1$$

$$\log \frac{x + 25}{2x - 7} = 1 \quad \text{Quotient property}$$

$$\frac{x + 25}{2x - 7} = 10^1$$

Write in exponential form; $\log x = \log_{10} x$.

$$x + 25 = 10(2x - 7)$$

Multiply by $2x - 7$.

$$x + 25 = 20x - 70$$

Distributive property

$$95 = 19x$$

Add 70; subtract x.

$$5 = x \quad \text{Divide by 19.}$$

Solution set: $\{5\}$

93. $\log_2(\log_2 x) = 1$

$\log_2 x = 2^1$ Write in exponential form.

$x = 2^2$ Write in exponential form.

$x = 4$ Evaluate.

Solution set: $\{4\}$

95.
$$\log x^2 = (\log x)^2$$

$$2 \log x = (\log x)^2 \quad \text{Power property}$$

$$(\log x)^2 - 2 \log x = 0 \quad \text{Subtract } 2 \log x; \text{ rewrite.}$$

$$\log x(\log x - 2) = 0 \quad \text{Factor.}$$

$$\log x = 0 \quad \text{or} \quad \log x - 2 = 0$$

Zero-factor property

$$\log x = 2$$

$$x = 10^0 \quad \text{or} \qquad\qquad x = 10^2$$

Write in exponential form.

$$x = 1 \quad \text{or} \qquad\qquad x = 100$$

Both solutions check.

Solution set: $\{1, 100\}$

Answers to Selected Exercises

To The Student

In this section we provide the answers that we think most students will obtain when they work the exercises using the methods explained in the text. If your answer does not look exactly like the one given here, it is not necessarily wrong. In many cases there are equivalent forms of the answer. For example, if the answer section shows $\frac{3}{4}$ and your answer is .75, you have obtained the correct answer but written it in a different (yet equivalent) form. Unless the directions specify otherwise, .75 is just as valid an answer as $\frac{3}{4}$. In general, if your answer does not agree with the one given in the text, see whether it can be transformed into the other form. If it can, then it is the correct answer. If you still have doubts, talk with your instructor.

If you need further help with trigonometry, you may want to obtain a copy of the *Student's Solution Manual* that goes with this book. Your college bookstore either has this manual or can order it for you.

CHAPTER 1   TRIGONOMETRIC FUNCTIONS

1.1 Exercises *(page 6)*

3. $45°$ **5. (a)** $60°$ **(b)** $150°$ **7. (a)** $45°$ **(b)** $135°$ **9. (a)** $36°$ **(b)** $126°$ **11.** $150°$ **13.** $70°; 110°$ **15.** $55°; 35°$

17. $80°; 100°$ **19.** $(90 - x)°$ **21.** $(x - 360)°$ **23.** $83° \, 59'$ **25.** $23° \, 49'$ **27.** $38° \, 32'$ **29.** $17° \, 1' \, 49''$

31. $20.900°$ **33.** $91.598°$ **35.** $274.316°$ **37.** $31° \, 25' \, 47''$ **39.** $89° \, 54' \, 1''$ **41.** $178° \, 35' \, 58''$ **45.** $320°$

47. $235°$ **49.** $179°$ **51.** $130°$ **53.** $30° + n \cdot 360°$ **55.** $135° + n \cdot 360°$ **57.** $-90° + n \cdot 360°$

Angles other than those given are possible in Exercises 61–67.

61.   **63.**   **65.**   **67.**

435°; −285°;
quadrant I

534°; −186°;
quadrant II

660°; −60°;
quadrant IV

299°; −421°;
quadrant IV

69. $3\sqrt{2}$ **71.** $\sqrt{34}$ **73.** 4 **75.** $\dfrac{3}{4}$ **77.** $1800°$

79. 12.5 rotations per hr   **81.** 4 sec

1.2 Exercises *(page 14)*

1. vertical angles   **3.** $51°; 51°$ **5.** $50°; 60°; 70°$ **7.** $60°; 60°; 60°$ **9.** $65°; 115°$ **11.** $49°; 49°$ **13.** $48°; 132°$

15. $91°$ **17.** $2° \, 29'$ **19.** $25.4°$ **23.** Answers are given in numerical order: $55°; 65°; 60°; 65°; 60°; 120°; 60°; 60°; 55°;$
$55°$ **25.** right; scalene   **27.** acute; equilateral   **29.** right; scalene   **31.** right; isosceles   **33.** obtuse; scalene

35. acute; isosceles   **41.** A and P; B and Q; C and R; AC and PR; BC and QR; AB and PQ **43.** A and C; E and D; ABE and CBD; EB and DB; AB and CB; AE and CD **45.** $Q = 42°$; $B = R = 48°$ **47.** $B = 106°$; $A = M = 44°$

49. $X = M = 52°$ **51.** $a = 20; b = 15$ **53.** $a = 6; b = 7\dfrac{1}{2}$ **55.** $x = 6$ **57.** 30 m   **59.** 500 m; 700 m

61. 112.5 ft   **63.** $x = 110$ **65.** $c \approx 111.1$ **67.** The unknown side in the first quadrilateral is 40 cm; The unknown sides

in the second quadrilateral are 27 cm and 36 cm.   **69.** $x = 10; y = 2$ **71. (a)** about $\dfrac{1}{4}$ **(b)** about 30 arc degrees

73. (a) about 1,830,000 mi **(b)** yes

1.3 Exercises *(page 24)*

1.   **3.**

In Exercises 5–11 and 25–31, we give, in order, sine, cosine, tangent, cotangent, secant, and cosecant.

5. $\dfrac{4}{5}; -\dfrac{3}{5}; -\dfrac{4}{3}; -\dfrac{3}{4}; -\dfrac{5}{3}; \dfrac{5}{4}$ **7.** $1; 0;$ undefined; $0;$ undefined; 1 **9.** $\dfrac{\sqrt{3}}{2}; \dfrac{1}{2}; \sqrt{3}; \dfrac{\sqrt{3}}{3}; 2; \dfrac{2\sqrt{3}}{3}$

11. $0; -1; 0;$ undefined; $-1;$ undefined   **15.** 0 **17.** negative   **19.** negative   **21.** positive   **23.** positive

25. $-\dfrac{2\sqrt{5}}{5}; \dfrac{\sqrt{5}}{5}; -2; -\dfrac{1}{2}; \sqrt{5}; -\dfrac{\sqrt{5}}{2}$ **27.** $\dfrac{6\sqrt{37}}{37}; -\dfrac{\sqrt{37}}{37}; -6; -\dfrac{1}{6}; -\sqrt{37}; \dfrac{\sqrt{37}}{6}$

29. $-\dfrac{4\sqrt{65}}{65}; \dfrac{7\sqrt{65}}{65}; \dfrac{4}{7}; \dfrac{7}{4}; -\dfrac{\sqrt{65}}{7}; -\dfrac{\sqrt{65}}{4}$ **31.** $1; 0;$ undefined; $0;$ undefined; 1 **33.** -3

35. -3 **37.** 5 **39.** 1 **41.** -1 **43.** 0 **45.** 0 **47.** They are equal.   **49.** They are negatives of each other.

51. $45°; 225°$ **53.** $\theta = 143.13°$ or $\theta = 216.87°$ (Other answers are possible.)   **55.** about $.940;$ about $.342$ **57.** $35°$

59. decrease; increase   **61.** area $= 3x^2 \sin \theta$

1.4 Exercises *(page 33)*

1. $1; \theta = 90°$ **3.** $-.4$ **5.** $\dfrac{1}{3}$ **7.** $\dfrac{\sqrt{15}}{15}$ **9.** $.70069071$ **15.** $-\dfrac{1}{3}$ **17.** $2\sqrt{6}$ **19.** 2.5 **21.** $2°$ **23.** $3°$

27. II   **29.** I or III   **31.** II or III   **33.** $+; +; +$ **35.** $+; -; -$ **37.** $+; +; +$ **39.** $-; +; -$ **41.** $\tan 30°$

43. $\sec 33°$ **45.** $\cos 26°$ **47.** impossible   **49.** possible   **51.** possible   **53.** possible   **55.** $-2\sqrt{2}$

57. $-\dfrac{\sqrt{5}}{2}$ **59.** $-\dfrac{\sqrt{15}}{4}$ **61.** 3.44701905 **63.** yes

In Exercises 65–71, we give, in order, sine, cosine, tangent, cotangent, secant, and cosecant.

65. $\dfrac{15}{17}; -\dfrac{8}{17}; -\dfrac{15}{8}; -\dfrac{8}{15}; -\dfrac{17}{8}; \dfrac{17}{15}$ **67.** $\dfrac{\sqrt{5}}{7}; \dfrac{2\sqrt{11}}{7}; \dfrac{\sqrt{55}}{22}; \dfrac{2\sqrt{55}}{5}; \dfrac{7\sqrt{11}}{22}; \dfrac{7\sqrt{5}}{5}$ **69.** $\dfrac{8\sqrt{67}}{67}; \dfrac{\sqrt{201}}{67}; \dfrac{8\sqrt{3}}{3}; \dfrac{\sqrt{3}}{8}; \dfrac{\sqrt{201}}{3};$

$\dfrac{\sqrt{67}}{8}$ **71.** $.164215; -.986425; -.166475; -6.00691; -1.01376; 6.08958$ **75.** false; $\sin \theta \le 1$ for all θ.

77. (a) 13 prism diopters **(b)** $\tan \theta = \dfrac{5}{12}$ **79.** $+$ **81.** $-$

Chapter 1 Review Exercises *(page 39)*

1. complement: $55°$; supplement: $145°$ **3.** $186°$ **5.** $270° + n \cdot 360°$ **7.** $9360°$ **9.** $119.134°$ **11.** $275° \, 6' \, 2''$

13. $40°; 60°; 80°$ **15.** $.25$ km   **17.** $N = 12°; R = 82°; M = 86°$ **19.** $p = 7; q = 7$ **21.** $k = 14$ **23.** 12 ft

In Exercises 25–31, we give, in order, sine, cosine, tangent, cotangent, secant, and cosecant.

25. $-\dfrac{\sqrt{3}}{2}; \dfrac{1}{2}; -\sqrt{3}; -\dfrac{\sqrt{3}}{3}; 2; -\dfrac{2\sqrt{3}}{3}$ **27.** $-\dfrac{4}{5}; \dfrac{3}{5}; -\dfrac{4}{3}; -\dfrac{3}{4}; \dfrac{5}{3}; -\dfrac{5}{4}$ **29.** $\dfrac{15}{17}; -\dfrac{8}{17}; -\dfrac{15}{8}; -\dfrac{8}{15}; -\dfrac{17}{8}; \dfrac{17}{15}$

31. $-\dfrac{1}{2}; \dfrac{\sqrt{3}}{2}; -\dfrac{\sqrt{3}}{3}; -\sqrt{3}; \dfrac{2\sqrt{3}}{3}; -2$ **33.** tangent and secant

35.

37. -6 **39. (a)** impossible **(b)** possible **(c)** possible

In Exercises 41–45 we give, in order, sine, cosine, tangent, cotangent, secant, and cosecant.

41. $-\dfrac{\sqrt{39}}{8}; -\dfrac{5}{8}; \dfrac{\sqrt{39}}{5}; \dfrac{5\sqrt{39}}{39}; -\dfrac{8}{5}; -\dfrac{8\sqrt{39}}{39}$ **43.** $\dfrac{2\sqrt{5}}{5}; -\dfrac{\sqrt{5}}{5}; -2; -\dfrac{1}{2}; -\sqrt{5}; \dfrac{\sqrt{5}}{2}$

45. $-\dfrac{3}{5}; \dfrac{4}{5}; -\dfrac{3}{4}; -\dfrac{4}{3}; \dfrac{5}{4}; -\dfrac{5}{3}$ **49.** approximately 9500 ft

Chapter 1 Test *(page 42)*

1. $74.2983°$ **2.** $203°$ **3.** $2700°$ **4.** $x = 30°; y = 30°$ **5.** 40 yd   **6.** $\sin\theta = -\dfrac{5\sqrt{29}}{29}; \cos\theta = \dfrac{2\sqrt{29}}{29};$

$\tan\theta = -\dfrac{5}{2}$ **7.** III   **8.** $\sin\theta = -\dfrac{3}{5}; \tan\theta = -\dfrac{3}{4}; \cot\theta = -\dfrac{4}{3}; \sec\theta = \dfrac{5}{4}; \csc\theta = -\dfrac{5}{3}$ **10. (a)** $+$ **(b)** $-$

11. $-\dfrac{1}{10}$ or $-.1$ **12.** $\dfrac{1}{\tan\theta}; \dfrac{\cos\theta}{\sin\theta}; \pm\sqrt{\csc^2\theta - 1}$

CHAPTER 2   ACUTE ANGLES AND RIGHT TRIANGLES

2.1 Exercises *(page 51)*

In Exercises 1 and 3, we give, in order, sine, cosine, and tangent.

1. $\dfrac{21}{29}; \dfrac{20}{29}; \dfrac{21}{20}$ **3.** $\dfrac{n}{p}; \dfrac{m}{p}; \dfrac{n}{m}$ **5.** C   **7.** B   **9.** E

In Exercises 11 and 13, we give, in order, the unknown side, sine, cosine, tangent, cotangent, secant, and cosecant.

11. $c = 13; \dfrac{12}{13}; \dfrac{5}{13}; \dfrac{12}{5}; \dfrac{5}{12}; \dfrac{13}{5}; \dfrac{13}{12}$ **13.** $b = \sqrt{13}; \dfrac{\sqrt{13}}{7}; \dfrac{6}{7}; \dfrac{\sqrt{13}}{6}; \dfrac{6\sqrt{13}}{13}; \dfrac{7}{6}; \dfrac{7\sqrt{13}}{13}$

15. $\sin A = \cos(90° - A); \cos A = \sin(90° - A); \tan A = \cot(90° - A); \cot A = \tan(90° - A); \sec A = \csc(90° - A);$

$\csc A = \sec(90° - A)$ **17.** $\csc 51°$ **19.** $\tan(100° - \theta)$ **21.** $\cos 51.3°$ **23.** $40°$ **25.** $20°$ **27.** $12°$

29. true   **31.** false   **33.** true   **35.** $\dfrac{\sqrt{3}}{3}$ **37.** $\dfrac{1}{2}$ **39.** $\dfrac{2\sqrt{3}}{3}$ **41.** $\sqrt{2}$ **43.** $\dfrac{\sqrt{2}}{2}$

45.   **46.**   **47.** the legs; $(2\sqrt{2}, 2\sqrt{2})$ **48.** $(1, \sqrt{3})$ **49.** $\sin x; \tan x$ **51.** $60°$

53. $\left(\dfrac{\sqrt{2}}{2}, \dfrac{\sqrt{2}}{2}\right); 45°$ **55.** $y = \dfrac{\sqrt{3}}{3}x$ **57.** $60°$ **59. (a)** $60°$ **(b)** k **(c)** $k\sqrt{3}$ **(d)** $2; \sqrt{3}; 30°; 60°$

61. $x = \dfrac{9\sqrt{3}}{2}; y = \dfrac{9}{2}; z = \dfrac{3\sqrt{3}}{2}; w = 3\sqrt{3}$ **63.** $p = 15; r = 15\sqrt{2}; q = 5\sqrt{6}; t = 10\sqrt{6}$ **65.** $A = \dfrac{s^2}{2}$

2.2 Exercises *(page 59)*

1. C   **3.** A   **5.** D   **11.** $\dfrac{\sqrt{2}}{2}; \dfrac{\sqrt{2}}{2}; \sqrt{2}; \sqrt{2}$ **13.** $-\dfrac{1}{2}; -\dfrac{\sqrt{3}}{3}; -2$ **15.** $\dfrac{1}{2}; -\sqrt{3}; -\dfrac{2\sqrt{3}}{3}$

17. $\sqrt{3}; \dfrac{\sqrt{3}}{3}$

In Exercises 19–33, we give, in order, sine, cosine, tangent, cotangent, secant, and cosecant.

19. $-\dfrac{\sqrt{2}}{2}; \dfrac{\sqrt{2}}{2}; -1, -1; \sqrt{2}; -\sqrt{2}$ **21.** $\dfrac{\sqrt{3}}{2}; \dfrac{1}{2}; \sqrt{3}; \dfrac{\sqrt{3}}{3}; 2; \dfrac{2\sqrt{3}}{3}$ **23.** $\dfrac{\sqrt{3}}{2}; -\dfrac{1}{2}; -\sqrt{3}; -\dfrac{\sqrt{3}}{3}; -2; \dfrac{2\sqrt{3}}{3}$

25. $-\dfrac{1}{2}; -\dfrac{\sqrt{3}}{2}; \dfrac{\sqrt{3}}{3}; \sqrt{3}; -\dfrac{2\sqrt{3}}{3}; -2$ **27.** $-\dfrac{\sqrt{2}}{2}; -\dfrac{\sqrt{2}}{2}; 1; 1; -\sqrt{2}; -\sqrt{2}$ **29.** $\dfrac{1}{2}; -\dfrac{\sqrt{3}}{2}; -\dfrac{\sqrt{3}}{3}; -\sqrt{3}; -\dfrac{2\sqrt{3}}{3}; 2$

31. $-\dfrac{1}{2}; -\dfrac{\sqrt{3}}{2}; \dfrac{\sqrt{3}}{3}; \sqrt{3}; -\dfrac{2\sqrt{3}}{3}; -2$ **33.** $\dfrac{1}{2}; -\dfrac{\sqrt{3}}{2}; -\dfrac{\sqrt{3}}{3}; -\sqrt{3}; -\dfrac{2\sqrt{3}}{3}; 2$ **35.** $-\dfrac{\sqrt{3}}{2}$ **37.** $\dfrac{\sqrt{3}}{2}$ **39.** true

41. false; $\dfrac{1}{2} \neq \sqrt{3}$ **43.** true   **45.** $\left(-3\sqrt{3}, 3\right)$ **47.** no   **49.** positive   **51.** positive   **53.** negative   **57.** .9

59. 45°; 225°   **61.** 30°; 150°   **63.** 120°; 300°   **65.** 45°; 315°

2.3 Exercises *(page 64)*

1. sin; 1   **3.** reciprocal; reciprocal   **5.** .6252427   **7.** 1.0273488   **9.** 15.055723   **11.** 1.4887142   **13.** .6743024

15. .9999905   **17.** .4327386   **19.** .2308682   **21.** .0825664   **23.** 55.845496°   **25.** 16.166641°

27. 38.491580°   **29.** 68.673241°   **33.** .3746065934   **35.** −1   **37.** 1   **41.** false   **43.** true   **45.** false

47. false   **49.** 65.96 lb   **51.** −2.87°   **53.** 2.866°   **55.** (a) 703 ft (b) 1701 ft (c) R would decrease. 644 ft, 1559 ft

57. (a) 2×10^8 m per sec (b) 2×10^8 m per sec   **59.** 48.7°   **61.** (a) approximately 155 ft (b) approximately 194 ft

Connections *(page 72)*

Steps 1 and 2 compare to his second step. The first part of Step 3 (solving the equation) compares to his third step, and the last part of Step 3 (checking) compares to his fourth step.

2.4 Exercises *(page 72)*

1. 20,385.5 to 20,386.5   **3.** 8958.5 to 8959.5   **7.** .05   **9.** $B = 53° 40'$; $a = 571$ m; $b = 777$ m   **11.** $M = 38.8°$;

$n = 154$ m; $p = 198$ m   **13.** $A = 47.9108°$; $c = 84.816$ cm; $a = 62.942$ cm   **19.** $B = 62.00°$; $a = 8.17$ ft; $b = 15.4$ ft

21. $A = 17.00°$; $a = 39.1$ in.; $c = 134$ in.   **23.** $c = 85.9$ yd; $A = 62° 50'$; $B = 27° 10'$ **25.** $b = 42.3$ cm; $A = 24° 10'$;

$B = 65° 50'$ **27.** $B = 36° 36'$; $a = 310.8$ ft; $b = 230.8$ ft   **29.** $A = 50° 51'$; $a = .4832$ m; $b = .3934$ m

31. The angle of elevation from X to Y is 90° whenever Y is directly above X.   **35.** 9.35 m   **37.** 128 ft   **39.** 26.92 in.

41. 22°   **43.** 28.0 m   **45.** 13.3 ft   **47.** 146 m   **49.** (a) 29,008 ft (b) shorter

2.5 Exercises *(page 81)*

1. It should be shown as an angle measured clockwise from due north.   **3.** A sketch is important to show the relationships

among the given data and the unknowns.   **5.** 270°; N 90° W or S 90° W   **7.** 315°; N 45° W   **9.** $y = \dfrac{\sqrt{3}}{3}x, x \leq 0$

11. 220 mi   **13.** 47 nautical mi   **15.** 120 mi   **17.** 148 mi   **19.** $x = \dfrac{b}{a - c}$ **21.** $y = (\tan 35°)(x - 25)$

23. 433 ft   **25.** 114 ft   **27.** 5.18 m   **29.** (a) $d = \dfrac{b}{2}\left(\cot\dfrac{\alpha}{2} + \cot\dfrac{\beta}{2}\right)$ (b) 345.3951 cm   **31.** 10.8 ft

33. (a) 323 ft (b) $R\left(1 - \cos\dfrac{\theta}{2}\right)$ **35.** (a) 23.4 ft (b) 48.3 ft (c) The faster the speed, the more land needs to be cleared

inside the curve.

Chapter 2 Review Exercises *(page 88)*

In Exercises 1, 13, and 15, we give, in order, sine, cosine, tangent, cotangent, secant, and cosecant.

1. $\dfrac{60}{61}; \dfrac{11}{61}; \dfrac{60}{11}; \dfrac{11}{60}; \dfrac{61}{11}; \dfrac{61}{60}$ **3.** 10°   **5.** 7°   **7.** true   **9.** true   **13.** $-\dfrac{\sqrt{3}}{2}; \dfrac{1}{2}; -\sqrt{3}; -\dfrac{\sqrt{3}}{3}; 2; -\dfrac{2\sqrt{3}}{3}$

15. $-\dfrac{1}{2}; \dfrac{\sqrt{3}}{2}; -\dfrac{\sqrt{3}}{3}; -\sqrt{3}; \dfrac{2\sqrt{3}}{3}; -2$ **17.** 120°; 240°   **19.** 150°; 210°   **21.** $3 - \dfrac{2\sqrt{3}}{3}$

23. (a) $-\dfrac{\sqrt{2}}{2}; -\dfrac{\sqrt{2}}{2}; 1$ (b) $-\dfrac{\sqrt{3}}{2}; \dfrac{1}{2}; -\sqrt{3}$ **25.** −1.3563417   **27.** 1.0210339   **29.** .20834446   **31.** 55.673870°

33. 12.733938°   **35.** 63.008286°   **37.** 47.1°; 132.9°   **39.** false; 1.4088321 $\neq$ 1   **41.** true   **45.** III

47. $B = 31° 30'$; $a = 638$; $b = 391$ **49.** $B = 50.28°$; $a = 32.38$ m; $c = 50.66$ m   **51.** 73.7 ft   **53.** 18.75 cm

55. 1200 m   **57.** 140 mi   **61. (a)** 716 mi **(b)** 1104 mi

Chapter 2 Test *(page 91)*

1. $\sin A = \dfrac{12}{13}$; $\cos A = \dfrac{5}{13}$; $\tan A = \dfrac{12}{5}$; $\cot A = \dfrac{5}{12}$; $\sec A = \dfrac{13}{5}$; $\csc A = \dfrac{13}{12}$ **2.** $x = 4$; $y = 4\sqrt{3}$; $z = 4\sqrt{2}$; $w = 8$

3. $15°$ **4.** $16.16664145°$ **5.** $135°$; $225°$ **6.** Take the reciprocal of $\tan \theta$ to get $\cot \theta = .5960011896$.   **7. (a)** true

(b) false; For $0° \leq \theta \leq 90°$, cosine is decreasing. **(c)** true   **8.** $-\sqrt{3}$ **9. (a)** .97939940 **(b)** -1.9056082 **(c)** 1.9362132

10. $B = 31° 30'$; $c = 877$; $b = 458$ **11.** $67.1°$ or $67° 10'$ **12.** 15.5 ft   **13.** 92 km   **14.** 448 m

CHAPTER 3   RADIAN MEASURE AND CIRCULAR FUNCTIONS

3.1 Exercises *(page 97)*

1. 1   **3.** 3   **5.** $\dfrac{\pi}{3}$ **7.** $\dfrac{\pi}{2}$ **9.** $\dfrac{5\pi}{6}$ **11.** $\dfrac{5\pi}{3}$ **13.** $\dfrac{5\pi}{2}$ **21.** $60°$ **23.** $315°$ **25.** $330°$ **27.** $-30°$

29. $126°$ **31.** $48°$ **33.** $153°$ **35.** .68   **37.** .742   **39.** 2.43   **41.** 1.122   **43.** .9847   **45.** .832391

47. $114° 35'$ **49.** $99° 42'$ **51.** $19° 35'$ **53.** $287° 6'$ **57.** $\dfrac{\sqrt{3}}{2}$ **59.** 1   **61.** $\dfrac{2\sqrt{3}}{3}$ **63.** 1   **65.** $-\sqrt{3}$

67. $\dfrac{1}{2}$ **69.** -1 **71.** $-\dfrac{\sqrt{3}}{2}$ **73.** $\dfrac{1}{2}$ **75.** We begin the answers with the blank next to $30°$, and then proceed

counterclockwise from there: $\dfrac{\pi}{6}$; 45; $\dfrac{\pi}{3}$; 120; 135; $\dfrac{5\pi}{6}$; π; $\dfrac{7\pi}{6}$; $\dfrac{5\pi}{4}$; 240; 300; $\dfrac{7\pi}{4}$; $\dfrac{11\pi}{6}$.   **77. (a)** 4π **(b)** $\dfrac{2\pi}{3}$

79. (a) 5π **(b)** $\dfrac{8\pi}{3}$

Connections *(page 102)*

Answers will vary. The longitude at Greenwich is $0°$.

3.2 Exercises *(page 103)*

1. 2π **3.** 8   **5.** 1   **7.** 25.8 cm   **9.** 5.05 m   **11.** The length is doubled.   **13.** 3500 km   **15.** 5900 km

17. $44°$ N   **19. (a)** 11.6 in. **(b)** $37° 5'$ **21.** $38.5°$ **23.** 146 in.   **25.** .20 km   **27.** 6π **29.** 1.5

31. 1116.1 m^2 **33.** 706.9 ft^2 **35.** 114.0 cm^2 **37.** 1885.0 mi^2 **39.** 3.6   **41. (a)** $13\dfrac{1}{3}°$; $\dfrac{2\pi}{27}$ **(b)** 480 ft

(c) $\dfrac{160}{9} \approx 17.8$ ft **(d)** approximately 672 ft^2 **43. (a)** 140 ft **(b)** 102 ft **(c)** 622 ft^2 **45.** 1900 yd^2 **47.** radius: 3947 mi;

circumference: 24,800 mi   **49.** $r = \dfrac{L}{\theta}$ **50.** $h = r \cos \dfrac{\theta}{2}$ **51.** $d = r\left(1 - \cos \dfrac{\theta}{2}\right)$ **52.** $d = \dfrac{L}{\theta}\left(1 - \cos \dfrac{\theta}{2}\right)$

53. The area is quadrupled.   **55.** $V = \dfrac{r^2 \theta h}{2}$ (θ in radians)

Connections *(page 113)*

$PQ = y = \dfrac{y}{1} = \sin \theta$; $OQ = x = \dfrac{x}{1} = \cos \theta$; $AB = \dfrac{AB}{1} = \dfrac{AB}{AO} = \dfrac{y}{x} = \tan \theta$ (by similar triangles)

3.3 Exercises *(page 113)*

1. (a) 1 **(b)** 0 **(c)** undefined   **3. (a)** 0 **(b)** 1 **(c)** 0   **5. (a)** 0 **(b)** -1 **(c)** 0   **7.** $-\dfrac{1}{2}$ **9.** -1 **11.** -2 **13.** $-\dfrac{1}{2}$

15. $\dfrac{\sqrt{2}}{2}$ **17.** $\dfrac{\sqrt{3}}{2}$ **19.** $\dfrac{2\sqrt{3}}{3}$ **21.** $-\dfrac{\sqrt{3}}{3}$ **23.** .5736   **25.** .4068   **27.** 1.2065   **29.** 14.3338

31. -1.0460 **33.** -3.8665 **35.** .7   **37.** 4   **39.** negative   **41.** negative   **43.** positive

45. $\sin \theta = \dfrac{\sqrt{2}}{2}$; $\cos \theta = \dfrac{\sqrt{2}}{2}$; $\tan \theta = 1$; $\cot \theta = 1$; $\sec \theta = \sqrt{2}$; $\csc \theta = \sqrt{2}$

47. $\sin \theta = -\dfrac{12}{13}$; $\cos \theta = \dfrac{5}{13}$; $\tan \theta = -\dfrac{12}{5}$; $\cot \theta = -\dfrac{5}{12}$; $\sec \theta = \dfrac{13}{5}$; $\csc \theta = -\dfrac{13}{12}$ **49.** .2095   **51.** 1.4426   **53.** .3887

55. $\dfrac{5\pi}{6}$ **57.** $\dfrac{4\pi}{3}$ **59.** $\dfrac{7\pi}{4}$ **61.** $(-.8011, .5985)$ **63.** $(.4385, -.8987)$ **65.** I   **67.** II   **69.** .9846

71. 8.6 hr; 15.4 hr   **73. (a)** 30° **(b)** 60° **(c)** 75° **(d)** 86° **(e)** 86° **(f)** 60°

3.4 Exercises *(page 119)*

1. 2π sec   **3. (a)** $\dfrac{\pi}{2}$ radians **(b)** 10π cm **(c)** $\dfrac{5\pi}{3}$ cm per sec   **5.** 2π radians   **7.** $\dfrac{3\pi}{32}$ radian per sec   **9.** $\dfrac{6}{5}$ min

11. .180311 radian per sec   **13.** 8π m per sec   **15.** $\dfrac{9}{5}$ radians per sec   **17.** 1.83333 radians per sec   **19.** 18π cm

21. 12 sec   **23.** $\dfrac{3\pi}{32}$ radian per sec   **25.** $\dfrac{\pi}{6}$ radian per hr   **27.** $\dfrac{\pi}{30}$ radian per min   **29.** $\dfrac{7\pi}{30}$ cm per min

31. 168π m per min   **33.** 1500π m per min   **35.** 15.5 mph   **37. (a)** $\dfrac{2\pi}{365}$ radian **(b)** $\dfrac{\pi}{4380}$ radian per hr **(c)** 66,700 mph

39. (a) .24 radian per sec **(b)** 3.11 cm per sec   **41.** 3.73 cm   **43.** 523.6 radians per sec

Chapter 3 Review Exercises *(page 124)*

3. Three of many possible answers are $1 + 2\pi$, $1 + 4\pi$, and $1 + 6\pi$.   **5.** $\dfrac{\pi}{4}$ **7.** $\dfrac{35\pi}{36}$ **9.** $\dfrac{40\pi}{9}$ **11.** 225°

13. 480°   **15.** $-110°$ **17.** π in.   **19.** 12π in.   **21.** 35.8 cm   **23.** 7.683 cm   **25.** 273 m²   **27.** 4500 km

29. $\dfrac{3}{4}$; 1.5 sq units   **31. (a)** $\dfrac{\pi}{3}$ radians **(b)** 2π in.   **33.** $\sqrt{3}$ **35.** $-\dfrac{1}{2}$ **37.** 2   **39.** $\tan 1$ **41.** $\sin 2$

43. .8660   **45.** .9703   **47.** 1.9513   **49.** .3898   **51.** .5148   **53.** 1.1054   **55.** $\dfrac{\pi}{4}$ **57.** $\dfrac{7\pi}{6}$ **59.** $\dfrac{15}{32}$ sec

61. $\dfrac{\pi}{20}$ radian per sec   **63.** 285.3 cm   **65.** $\dfrac{\pi}{36}$ radian per sec   **67. (b)** $\dfrac{\pi}{6}$ **(c)** less ultraviolet light when $\theta = \dfrac{\pi}{3}$

Chapter 3 Test *(page 128)*

1. $\dfrac{2\pi}{3}$ **2.** $-\dfrac{\pi}{4}$ **3.** .09   **4.** 135°   **5.** $-210°$ **6.** 229.18°   **7. (a)** $\dfrac{4}{3}$ **(b)** 15,000 cm²   **8.** 2 radians

9. $\dfrac{\sqrt{2}}{2}$ **10.** $-\dfrac{\sqrt{3}}{2}$ **11.** undefined   **12.** -2 **13.** 0   **14.** 0   **15.** $\sin \dfrac{7\pi}{6} = -\dfrac{1}{2}$; $\cos \dfrac{7\pi}{6} = -\dfrac{\sqrt{3}}{2}$;

$\tan \dfrac{7\pi}{6} = \dfrac{\sqrt{3}}{3}$ **16.** $\left\{ s \,\middle|\, s \neq (2n + 1)\dfrac{\pi}{2}, \text{ where } n \text{ is any integer} \right\}$ **17. (a)** .9716 **(b)** $\dfrac{\pi}{3}$ **18. (a)** $\dfrac{2\pi}{3}$ radians

(b) 40π cm **(c)** 5π cm per sec   **19.** approximately 8.278 mi per sec   **20. (a)** 75 ft **(b)** $\dfrac{2\pi}{90}$ radian per sec

CHAPTER 4   GRAPHS OF THE CIRCULAR FUNCTIONS

Connections *(page 141)*

1. $X = -.4161468$, $Y = .90929743$; X is $\cos 2$ and Y is $\sin 2$.   **2.** $X = 1.9$, $Y = .94630009$; $\sin 1.9 = .94630009$

3. $X = 1.9$, $Y = -.3232896$; $\cos 1.9 = -.3232896$

4.1 Exercises *(page 141)*

1. G   **3.** E   **5.** B   **7.** F   **9.** 2   **11.** $\dfrac{2}{3}$ **13.** 1   **15.** 2

17. 4π; 1   **19.** $\dfrac{8\pi}{3}$; 1   **21.** $\dfrac{2\pi}{3}$; 1   **23.** 8π; 2   **25.** $\dfrac{2\pi}{3}$; 2

27. 2; 1   **29.** 1; 2   **31.** 4; $\dfrac{1}{2}$ **33.** **(a)** 80°; 50° **(b)** 15° **(c)** about 35,000 yr **(d)** downward

35. **(a)** about 2 hr **(b)** 1 yr   **37.** **(a)** 5; $\dfrac{1}{60}$ **(b)** 60 **(c)** 5; 1.545; -4.045; -4.045; 1.545

(d)

39. **(a)** $L(x) = .022x^2 + .55x + 316$ $+ 3.5 \sin 2\pi x$ **(b)** maximums: $x = \dfrac{1}{4}, \dfrac{5}{4}, \dfrac{9}{4}, \dots$; minimums: $x = \dfrac{3}{4}, \dfrac{7}{4}, \dfrac{11}{4}, \dots$

41. **(a)** 1° **(b)** 19° **(c)** 53° **(d)** 58° **(e)** 49° **(f)** 13°   **43.** 24 hr   **45.** approximately 6:00 P.M.; approximately .2 ft

47. approximately 2:00 A.M.; approximately 2.6 ft   **49.** 1; 240° or $\dfrac{4\pi}{3}$

4.2 Exercises *(page 152)*

1. D   **3.** H   **5.** B   **7.** F   **9.** B   **11.** C   **13.** right   **15.** 2; 2π; none; π to the right   **17.** 4; 4π; none; π to

the left   **19.** 3; 4; none; $\dfrac{1}{2}$ to the right   **21.** 1; $\dfrac{2\pi}{3}$; up 2; $\dfrac{\pi}{15}$ to the right   **23.**   **25.**

27.   **29.**   **31.**   **33.**   **35.**

37.

39.

41.

43.

45.

47. **(a)** yes 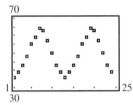   **(b)** It represents the average yearly temperature.

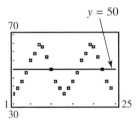

(c) 14; 12; 4.2   **(d)** $f(x) = 14 \sin\left[\dfrac{\pi}{6}(x - 4.2)\right] + 50$

(e) The function gives an excellent model for the given data.

$f(x) = 14 \sin\left[\dfrac{\pi}{6}(x - 4.2)\right] + 50$

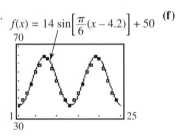

(f)
```
SinReg
y=a*sin(bx+c)+d
a=13.21
b=.52
c=-2.18
d=49.68
```
TI-83 Plus fixed to the
nearest hundredth

4.3 Exercises *(page 165)*

1. B   **3.** E   **5.** D   **7.**   **9.**   **11.**   **13.**

15.   **17.**   **19.**   **21.**   **23.**

25.   **27.**   **29.**   **31.**   **33.**

35.

$y = \cot\left(3x + \frac{\pi}{4}\right)$

37.

$y = 1 + \tan x$

39.

$y = 1 - \cot x$

41.

$y = -1 + 2 \tan x$

43.

$y = -1 + \frac{1}{2} \cot (2x - 3\pi)$

45.

$y = 1 - 2 \cot 2\left(x + \frac{\pi}{2}\right)$

47. true   **49.** true   **51.** false; $\tan(-x) = -\tan x$ for all x in the domain.   **53.** four

55. domain: $\left\{ x \mid x \neq (2n + 1)\dfrac{\pi}{4}, \text{ where } n \text{ is an integer} \right\}$; range: $(-\infty, \infty)$

57. (a) 0 m (b) -2.9 m (c) -12.3 m (d) 12.3 m (e) It leads to $\tan\dfrac{\pi}{2}$, which is undefined.

61. The display is for $Y_1 + Y_2$ at $X = \dfrac{\pi}{6}$.

63. π **64.** $\dfrac{5\pi}{4}$ **65.** $y = \dfrac{5\pi}{4} + n\pi$

66. approximately .3217505544   **67.** approximately 3.463343208   **68.** $\{x \mid x = .3217505544 + n\pi\}$

Summary Exercises on Graphing Circular Functions *(page 168)*

1.

$y = 2 \sin \pi x$

2.

$y = 4 \cos 1.5x$

3.

$y = -2 + .5 \cos \dfrac{\pi}{4}x$

4.

$y = 3 \sec \dfrac{\pi x}{2}$

5.

$y = -4 \csc .5x$

6.

$y = 3 \tan\left(\dfrac{\pi x}{2} + \pi\right)$

7.

$y = -5 \sin \dfrac{x}{3}$

8.

$y = 10 \cos\left(\dfrac{x}{4} + \dfrac{\pi}{2}\right)$

9.

$y = 3 - 4 \sin (2.5x + \pi)$

10.

$y = 2 - \sec[\pi(x - 3)]$

4.4 Exercises *(page 171)*

1. (a) $s(t) = 2 \cos 4\pi t$ (b) $s(1) = 2$; The weight is neither moving upward nor downward. At $t = 1$, the motion of the weight is changing from up to down.   **3.** (a) $s(t) = -3 \cos 2.5\pi t$ (b) $s(1) = 0$; upward

5. $s(t) = .21 \cos 55\pi t$

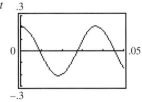

7. $s(t) = .14 \cos 110\pi t$

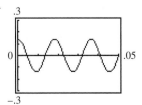

9. (a) $s(t) = -4 \cos \dfrac{2\pi}{3}t$ (b) 3.46 units (c) $\dfrac{1}{3}$ **11.** (a) $s(t) = 2 \sin 2t$; amplitude: 2; period: π; frequency: $\dfrac{1}{\pi}$

11. (b) $s(t) = 2 \sin 4t$; amplitude: 2; period: $\dfrac{\pi}{2}$; frequency: $\dfrac{2}{\pi}$ **13.** $\dfrac{8}{\pi^2}$ ft   **15. (a)** 4 in.   **(b)** after $\dfrac{1}{8}$ sec

(c) 4 cycles per sec; $\dfrac{1}{4}$ sec   **17. (a)** 5 in.   **(b)** 2 cycles per sec; $\dfrac{1}{2}$ sec   **(c)** after $\dfrac{1}{4}$ sec   **(d)** approximately 4; After 1.3 sec, the

weight is about 4 in. above the equilibrium position.   **19. (a)** $s(t) = -3 \cos 12t$ **(b)** $\dfrac{\pi}{6}$ sec   **21.** 0; π; They are the same.

Chapter 4 Review Exercises *(page 175)*

1. B   **3.** sine, cosine, tangent, cotangent   **5.** 2; 2π; none; none   **7.** $\dfrac{1}{2}$; $\dfrac{2\pi}{3}$; none; none   **9.** 2; 8π; 1 up; none

11. 3; 2π; none; $\dfrac{\pi}{2}$ to the left   **13.** not applicable; π; none; $\dfrac{\pi}{8}$ to the right   **15.** not applicable; $\dfrac{\pi}{3}$; none; $\dfrac{\pi}{9}$ to the

right   **17.** tangent   **19.** cosine   **21.** cotangent   **25.**   **27.**

29.   **31.**   **33.**   **35.**   **37.**

39.   **41.**   **45. (b)**   **47. (a)** 30°   **(b)** 60°   **(c)** 75°   **(d)** 86°   **(e)** 86°

(f) 60°   **49. (a)** 100   **(b)** 258   **(c)** 122   **(d)** 296   **51.** amplitude: 4; period: 2; frequency: $\dfrac{1}{2}$ **53.** The frequency is the

number of cycles in one unit of time; -4; 0; $-2\sqrt{2}$

Chapter 4 Test *(page 178)*

1. (a) π **(b)** 6   **(c)** $[-3, 9]$ **(d)** -3 **(e)** $\dfrac{\pi}{4}$ to the left $\left(\text{that is, } -\dfrac{\pi}{4} \right)$ **2.**   **3.**

8. (a)

$f(x) = 17.5 \sin\left[\frac{\pi}{6}(x - 4)\right] + 67.5$

(b) 17.5; 12; 4 to the right; 67.5 up **(c)** approximately 52°F

(d) 50°F in January; 85°F in July

(e) approximately 67.5°; This is the vertical translation.

9. (a) 4 in. **(b)** after $\frac{1}{8}$ sec **(c)** 4 cycles per sec; $\frac{1}{4}$ sec

CHAPTER 5   TRIGONOMETRIC IDENTITIES

5.1 Exercises *(page 185)*

1. -2.6 **3.** .625   **5.** $\frac{\sqrt{7}}{4}$ **7.** $-\frac{2\sqrt{5}}{5}$ **9.** $-\frac{\sqrt{105}}{11}$ **12.** $-\sin x$ **13.** odd   **14.** $\cos x$ **15.** even

16. $-\tan x$ **17.** odd   **19.** $f(-x) = -f(x)$

21. $\cos\theta = -\frac{\sqrt{5}}{3}$; $\tan\theta = -\frac{2\sqrt{5}}{5}$; $\cot\theta = -\frac{\sqrt{5}}{2}$; $\sec\theta = -\frac{3\sqrt{5}}{5}$; $\csc\theta = \frac{3}{2}$

23. $\sin\theta = -\frac{\sqrt{17}}{17}$; $\cos\theta = \frac{4\sqrt{17}}{17}$; $\cot\theta = -4$; $\sec\theta = \frac{\sqrt{17}}{4}$; $\csc\theta = -\sqrt{17}$

25. $\sin\theta = \frac{3}{5}$; $\cos\theta = \frac{4}{5}$; $\tan\theta = \frac{3}{4}$; $\sec\theta = \frac{5}{4}$; $\csc\theta = \frac{5}{3}$

27. $\sin\theta = -\frac{\sqrt{7}}{4}$; $\cos\theta = \frac{3}{4}$; $\tan\theta = -\frac{\sqrt{7}}{3}$; $\cot\theta = -\frac{3\sqrt{7}}{7}$; $\csc\theta = -\frac{4\sqrt{7}}{7}$ **29.** B   **31.** E   **33.** A   **35.** A

37. D   **41.** $\sin\theta = \frac{\pm\sqrt{2x + 1}}{x + 1}$ **43.** $\sin x = \pm\sqrt{1 - \cos^2 x}$ **45.** $\tan x = \pm\sqrt{\sec^2 x - 1}$

47. $\csc x = \frac{\pm\sqrt{1 - \cos^2 x}}{1 - \cos^2 x}$ **49.** $\cos\theta$ **51.** $\cot\theta$ **53.** $\cos^2\theta$ **55.** $\sec\theta - \cos\theta$ **57.** $\cot\theta - \tan\theta$

59. $\tan\theta \sin\theta$ **61.** $\cos^2\theta$ **63.** $\sec^2\theta$ **65.** $\frac{25\sqrt{6} - 60}{12}$; $\frac{-25\sqrt{6} - 60}{12}$ **67.** $-\sin(2x)$ **68.** It is the negative of

$\sin(2x)$.   **69.** $\cos(4x)$ **70.** It is the same function.   **71. (a)** $-\sin(4x)$ **(b)** $\cos(2x)$ **(c)** $5\sin(3x)$ **73.** not an identity
75. identity

5.2 Exercises *(page 194)*

1. $\csc\theta \sec\theta$ or $\frac{1}{\sin\theta\cos\theta}$ **3.** $1 + \sec s$ **5.** 1   **7.** 1   **9.** $2 + 2\sin t$ **11.** $-\frac{2\cos x}{\sin^2 x}$ or $-2\cot x \csc x$

13. $(\sin\theta + 1)(\sin\theta - 1)$ **15.** $4\sin x$ **17.** $(2\sin x + 1)(\sin x + 1)$ **19.** $(\cos^2 x + 1)^2$
21. $(\sin x - \cos x)(1 + \sin x \cos x)$ **23.** $\sin\theta$ **25.** 1   **27.** $\tan^2\beta$ **29.** $\tan^2 x$ **31.** $\sec^2 x$

71. $(\sec\theta + \tan\theta)(1 - \sin\theta) = \cos\theta$ **73.** $\frac{\cos\theta + 1}{\sin\theta + \tan\theta} = \cot\theta$ **75.** identity   **77.** not an identity

83. It is true when $\sin x \geq 0$.   **85. (a)** $P = 16k\cos^2(2\pi t)$ **(b)** $P = 16k[1 - \sin^2(2\pi t)]$

5.3 Exercises *(page 202)*

1. F   **3.** E   **5.** $\frac{\sqrt{6} - \sqrt{2}}{4}$ **7.** $\frac{\sqrt{2} - \sqrt{6}}{4}$ **9.** $\frac{\sqrt{2} - \sqrt{6}}{4}$ **11.** 0   **13.** The calculator gives a value of 0 for the

expression.   **15.** $\cot 3°$ **17.** $\sin\frac{5\pi}{12}$ **19.** $\sec 104° \, 24'$ **21.** $\cos\left(-\frac{\pi}{8}\right)$ **23.** $\csc(-56° \, 42')$

25. $\tan(-86.9814°)$ **27.** $\tan$ **29.** $\cos$ **31.** $\csc$

For Exercises 33–37, other answers are possible.

33. $15°$ **35.** $\dfrac{140°}{3}$ **37.** $20°$ **39.** $\cos\theta$ **41.** $-\cos\theta$ **43.** $\cos\theta$ **45.** $-\cos\theta$ **47.** $\dfrac{4-6\sqrt{6}}{25}$; $\dfrac{4+6\sqrt{6}}{25}$

49. $\dfrac{16}{65}$; $-\dfrac{56}{65}$ **51.** $\dfrac{2\sqrt{638}-\sqrt{30}}{56}$; $\dfrac{2\sqrt{638}+\sqrt{30}}{56}$ **53.** true   **55.** false   **57.** true   **59.** true   **61.** false

68. $\dfrac{-\sqrt{6}-\sqrt{2}}{4}$ **69.** $\dfrac{-\sqrt{6}-\sqrt{2}}{4}$ **70. (a)** $\dfrac{\sqrt{2}-\sqrt{6}}{4}$ **(b)** $\dfrac{-\sqrt{6}-\sqrt{2}}{4}$ **71. (a)** 3 **(b)** 163 and -163; no

5.4 Exercises *(page 209)*

3. C   **5.** E   **7.** B   **9.** $\dfrac{\sqrt{6}+\sqrt{2}}{4}$ **11.** $2-\sqrt{3}$ **13.** $\dfrac{-\sqrt{6}-\sqrt{2}}{4}$ **15.** $\dfrac{\sqrt{2}}{2}$ **17.** -1 **19.** 0   **21.** 1

23. $\dfrac{\sqrt{3}\cos\theta-\sin\theta}{2}$ **25.** $\dfrac{\cos\theta-\sqrt{3}\sin\theta}{2}$ **27.** $\dfrac{\sqrt{2}(\sin x-\cos x)}{2}$ **29.** $\dfrac{\sqrt{3}\tan\theta+1}{\sqrt{3}-\tan\theta}$ **31.** $\dfrac{\sqrt{2}(\cos x+\sin x)}{2}$

33. $-\cos\theta$ **35.** $-\tan\theta$ **37.** $-\tan\theta$ **41. (a)** $\dfrac{63}{65}$ **(b)** $\dfrac{63}{16}$ **(c)** I   **43. (a)** $\dfrac{4\sqrt{2}+\sqrt{5}}{9}$ **(b)** $\dfrac{-8\sqrt{5}-5\sqrt{2}}{20-2\sqrt{10}}$ (Other

forms are possible.) **(c)** II   **45. (a)** $\dfrac{77}{85}$ **(b)** $-\dfrac{77}{36}$ **(c)** II   **47.** $\dfrac{\sqrt{6}-\sqrt{2}}{4}$ **49.** $\dfrac{-\sqrt{6}-\sqrt{2}}{4}$ **51.** $-2+\sqrt{3}$

53. $\sin\left(\dfrac{\pi}{2}+x\right)=\cos x$ **55.** $\tan\left(\dfrac{\pi}{2}+x\right)=-\cot x$ **67.** $180°-\beta$ **68.** $\theta=\beta-\alpha$

69. $\tan\theta=\dfrac{\tan\beta-\tan\alpha}{1+\tan\beta\tan\alpha}$ **71.** $18.4°$ **72.** $80.8°$ **73. (a)** 425 lb **(c)** $0°$ **75.** $-20\cos\dfrac{\pi t}{4}$

5.5 Exercises *(page 218)*

1. C   **3.** B   **5.** C   **7.** $\cos\theta=\dfrac{2\sqrt{5}}{5}$; $\sin\theta=\dfrac{\sqrt{5}}{5}$ **9.** $\cos\theta=-\dfrac{\sqrt{42}}{12}$; $\sin\theta=\dfrac{\sqrt{102}}{12}$ **11.** $\cos2\theta=\dfrac{17}{25}$;

$\sin2\theta=-\dfrac{4\sqrt{21}}{25}$ **13.** $\cos2x=-\dfrac{3}{5}$; $\sin2x=\dfrac{4}{5}$ **15.** $\cos2\theta=\dfrac{39}{49}$; $\sin2\theta=-\dfrac{4\sqrt{55}}{49}$ **17.** $\dfrac{\sqrt{3}}{2}$ **19.** $\dfrac{\sqrt{3}}{2}$

21. $-\dfrac{\sqrt{2}}{2}$ **23.** $\dfrac{1}{2}\tan102°$ **25.** $\dfrac{1}{4}\cos94.2°$ **27.** $-\cos\dfrac{4\pi}{5}$ **29.** $\cos3x=4\cos^3x-3\cos x$

31. $\tan3x=\dfrac{3\tan x-\tan^3x}{1-3\tan^2x}$ **33.** $\cos^4x-\sin^4x=\cos2x$ **35.** $\dfrac{2\tan x}{2-\sec^2x}=\tan2x$ **55.** $\sin160°-\sin44°$

57. $\dfrac{5}{2}\cos5x+\dfrac{5}{2}\cos x$ **59.** $-2\sin3x\sin x$ **61.** $-2\sin11.5°\cos36.5°$ **63.** $2\cos6x\cos2x$ **65.** $a=-885.6$;

$c=885.6$; $\omega=240\pi$

5.6 Exercises *(page 224)*

1. $-$ **3.** $+$ **5.** C   **7.** D   **9.** F   **11.** $\dfrac{\sqrt{2+\sqrt{2}}}{2}$ **13.** $-\dfrac{\sqrt{2+\sqrt{3}}}{2}$ **15.** $-\dfrac{\sqrt{2+\sqrt{3}}}{2}$ **19.** $\dfrac{\sqrt{10}}{4}$

21. 3   **23.** $\dfrac{\sqrt{50-10\sqrt{5}}}{10}$ **25.** $-\sqrt{7}$ **27.** $\dfrac{\sqrt{5}}{5}$ **29.** $-\dfrac{\sqrt{42}}{12}$ **31.** .127   **33.** $\sin20°$ **35.** $\tan73.5°$

37. $\tan29.87°$ **39.** $\cos9x$ **41.** $\tan4\theta$ **43.** $\cos\dfrac{x}{8}$ **45.** $\dfrac{\sin x}{1+\cos x}=\tan\dfrac{x}{2}$ **47.** $\dfrac{\tan\frac{x}{2}+\cot\frac{x}{2}}{\cot\frac{x}{2}-\tan\frac{x}{2}}=\sec x$

59. $106°$ **61.** 2   **63.** They are both radii of the circle.   **64.** It is the supplement of a $30°$ angle.   **65.** Their sum is

$180°-150°=30°$, and they are equal.   **66.** $2+\sqrt{3}$ **68.** $\dfrac{\sqrt{6}+\sqrt{2}}{4}$ **69.** $\dfrac{\sqrt{6}-\sqrt{2}}{4}$ **70.** $2-\sqrt{3}$

Chapter 5 Review Exercises *(page 231)*

1. B **3.** C **5.** D **7.** 1 **9.** $\dfrac{1}{\cos^2 \theta}$ **11.** $-\dfrac{\cos \theta}{\sin \theta}$ **13.** $\sin x = -\dfrac{4}{5}$; $\tan x = -\dfrac{4}{3}$; $\cot(-x) = \dfrac{3}{4}$

15. $\sin 165° = \dfrac{\sqrt{6} - \sqrt{2}}{4}$; $\cos 165° = \dfrac{-\sqrt{6} - \sqrt{2}}{4}$; $\tan 165° = -2 + \sqrt{3}$; $\csc 165° = \sqrt{6} + \sqrt{2}$; $\sec 165° = -\sqrt{6} + \sqrt{2}$;

$\cot 165° = -2 - \sqrt{3}$ **17.** E **19.** J **21.** I **23.** H **25.** G

In Exercises 27–31, other forms are possible for tan($x + y$).

27. $\dfrac{4 + 3\sqrt{15}}{20}$; $\dfrac{4\sqrt{15} + 3}{20}$; $\dfrac{4 + 3\sqrt{15}}{4\sqrt{15} - 3}$; I **29.** $\dfrac{2 + 3\sqrt{7}}{10}$; $\dfrac{2\sqrt{3} + \sqrt{21}}{10}$; $\dfrac{2 + 3\sqrt{7}}{2\sqrt{3} - \sqrt{21}}$; II

31. $\dfrac{4 - 9\sqrt{11}}{50}$; $\dfrac{12\sqrt{11} - 3}{50}$; $\dfrac{4 - 9\sqrt{11}}{12\sqrt{11} + 3}$; IV **33.** $\sin \theta = \dfrac{\sqrt{14}}{4}$; $\cos \theta = \dfrac{\sqrt{2}}{4}$ **35.** $\sin 2x = \dfrac{3}{5}$; $\cos 2x = -\dfrac{4}{5}$

37. $\dfrac{1}{2}$ **39.** $\dfrac{\sqrt{5} - 1}{2}$ **41.** .5 **43.** $-\dfrac{\sin 2x + \sin x}{\cos 2x - \cos x} = \cot \dfrac{x}{2}$ **45.** $\dfrac{\sin x}{1 - \cos x} = \cot \dfrac{x}{2}$

47. $\dfrac{2(\sin x - \sin^3 x)}{\cos x} = \sin 2x$ **71. (a)** $D = \dfrac{v^2 \sin 2\theta}{32}$ **(b)** approximately 35 ft

Chapter 5 Test *(page 233)*

1. $\sin x = -\dfrac{5\sqrt{61}}{61}$; $\cos x = \dfrac{6\sqrt{61}}{61}$ **2.** -1 **3.** $\sin(x + y) = \dfrac{2 - 2\sqrt{42}}{15}$; $\cos(x - y) = \dfrac{4\sqrt{2} - \sqrt{21}}{15}$;

$\tan(x + y) = \dfrac{2\sqrt{2} - 4\sqrt{21}}{8 + \sqrt{42}}$ **4.** $\dfrac{-\sqrt{2 - \sqrt{2}}}{2}$ **5.** $\sec x - \sin x \tan x = \cos x$ **6.** $\cot \dfrac{x}{2} - \cot x = \csc x$

11. (a) $-\sin \theta$ **(b)** $-\sin \theta$ **12. (a)** $V = 163 \cos\left(\dfrac{\pi}{2} - \omega t\right)$ **(b)** 163 volts; $\dfrac{1}{240}$ sec

CHAPTER 6 INVERSE CIRCULAR FUNCTIONS AND TRIGONOMETRIC EQUATIONS

6.1 Exercises *(page 246)*

1. one-to-one **3.** domain **5.** π **7. (a)** $[-1, 1]$ **(b)** $\left[-\dfrac{\pi}{2}, \dfrac{\pi}{2}\right]$ **(c)** increasing **(d)** -2 is not in the domain.

9. (a) $(-\infty, \infty)$ **(b)** $\left(-\dfrac{\pi}{2}, \dfrac{\pi}{2}\right)$ **(c)** increasing **(d)** no **11.** $\cos^{-1} \dfrac{1}{a}$ **13.** 0 **15.** π **17.** $-\dfrac{\pi}{2}$ **19.** 0 **21.** $\dfrac{\pi}{2}$

23. $\dfrac{\pi}{4}$ **25.** $\dfrac{5\pi}{6}$ **27.** $\dfrac{3\pi}{4}$ **29.** $-\dfrac{\pi}{6}$ **31.** $\dfrac{\pi}{6}$ **33.** $-45°$ **35.** $-60°$ **37.** $120°$ **39.** $120°$

41. $-7.6713835°$ **43.** $113.500970°$ **45.** $30.987961°$ **47.** .83798122 **49.** 2.3154725 **51.** 1.1900238

53. **55.** **57.** **59.** The domain of $y = \tan^{-1} x$ is $(-\infty, \infty)$.

60. In both cases, the result is x. In each case, the graph is a straight line bisecting quadrants I and III (i.e., the line $y = x$).

61. It is the graph of $y = x$.

62. It does not agree because the range of the inverse tangent function is $\left(-\dfrac{\pi}{2}, \dfrac{\pi}{2}\right)$, not $(-\infty, \infty)$, as was the case in

Exercise 61.

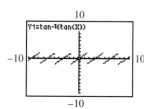

63. $\dfrac{\sqrt{7}}{3}$ **65.** $\dfrac{\sqrt{5}}{5}$ **67.** $\dfrac{120}{169}$ **69.** $-\dfrac{7}{25}$ **71.** $\dfrac{4\sqrt{6}}{25}$ **73.** 2

75. $\dfrac{63}{65}$ **77.** $\dfrac{\sqrt{10} - 3\sqrt{30}}{20}$ **79.** .894427191   **81.** .1234399811   **83.** $\sqrt{1 - u^2}$ **85.** $\sqrt{1 - u^2}$ **87.** $\dfrac{\sqrt{u^2 - 4}}{u}$

89. $\dfrac{u\sqrt{2}}{2}$ **91.** $\dfrac{2\sqrt{4 - u^2}}{4 - u^2}$ **93. (a)** $45°$ **(b)** $\theta = 45°$ **95. (a)** $18°$ **(b)** $18°$ **(c)** $15°$ **(e)** 1.4142151 m (Note: Due to the

computational routine, there may be a discrepancy in the last few decimal places.)   **(f)** $\sqrt{2}$

$y = \tan^{-1}\left(\dfrac{x}{x^2 + 2}\right)$

Radian mode

6.2 Exercises *(page 254)*

1. Solve the linear equation for $\cot x$.   **3.** Solve the quadratic equation for $\sec x$ by factoring.   **5.** Solve the quadratic

equation for $\sin x$ using the quadratic formula.   **7.** Use an identity to rewrite as an equation with one trigonometric function.

11. $\left\{\dfrac{3\pi}{4}, \dfrac{7\pi}{4}\right\}$ **13.** $\left\{\dfrac{\pi}{6}, \dfrac{5\pi}{6}\right\}$ **15.** $\emptyset$ **17.** $\left\{\dfrac{\pi}{4}, \dfrac{2\pi}{3}, \dfrac{5\pi}{4}, \dfrac{5\pi}{3}\right\}$ **19.** $\{\pi\}$ **21.** $\left\{\dfrac{7\pi}{6}, \dfrac{3\pi}{2}, \dfrac{11\pi}{6}\right\}$

23. $\{30°, 210°, 240°, 300°\}$ **25.** $\{90°, 210°, 330°\}$ **27.** $\{45°, 135°, 225°, 315°\}$ **29.** $\{45°, 225°\}$

31. $\{0°, 30°, 150°, 180°\}$ **33.** $\{0°, 45°, 135°, 180°, 225°, 315°\}$ **35.** $\{53.6°, 126.4°, 187.9°, 352.1°\}$

37. $\{149.6°, 329.6°, 106.3°, 286.3°\}$ **39.** $\emptyset$ **41.** $\{57.7°, 159.2°\}$ **43.** $.9 + 2n\pi, 2.3 + 2n\pi, 3.6 + 2n\pi, 5.8 + 2n\pi,$

where n is any integer   **45.** $\dfrac{\pi}{3} + 2n\pi, \dfrac{2\pi}{3} + 2n\pi, \dfrac{4\pi}{3} + 2n\pi, \dfrac{5\pi}{3} + 2n\pi,$ where n is any integer   **47.** $33.6° + 360° n,$

$326.4° + 360° n,$ where n is any integer   **49.** $45° + 180° n, 108.4° + 180° n,$ where n is any integer

51. $\{.6806, 1.4159\}$ **53. (a)** .00164 and .00355 **(b)** $[.00164, .00355]$ **(c)** outward   **55. (a)** $\dfrac{1}{4}$ sec **(b)** $\dfrac{1}{6}$ sec **(c)** .21 sec

57. (a) One such value is $\dfrac{\pi}{3}$. **(b)** One such value is $\dfrac{\pi}{4}$.

6.3 Exercises *(page 260)*

1. $\left\{\dfrac{\pi}{3}, \pi, \dfrac{4\pi}{3}\right\}$ **3.** $\{60°, 210°, 240°, 310°\}$ **7.** $\left\{\dfrac{\pi}{12}, \dfrac{11\pi}{12}, \dfrac{13\pi}{12}, \dfrac{23\pi}{12}\right\}$ **9.** $\left\{\dfrac{\pi}{2}, \dfrac{7\pi}{6}, \dfrac{11\pi}{6}\right\}$

11. $\left\{\dfrac{\pi}{18}, \dfrac{7\pi}{18}, \dfrac{13\pi}{18}, \dfrac{19\pi}{18}, \dfrac{25\pi}{18}, \dfrac{31\pi}{18}\right\}$ **13.** $\left\{\dfrac{3\pi}{8}, \dfrac{5\pi}{8}, \dfrac{11\pi}{8}, \dfrac{13\pi}{8}\right\}$ **15.** $\left\{\dfrac{\pi}{2}, \dfrac{3\pi}{2}\right\}$ **17.** $\left\{0, \dfrac{\pi}{3}, \pi, \dfrac{5\pi}{3}\right\}$ **19.** $\emptyset$

21. $\left\{\dfrac{\pi}{2}\right\}$ **23.** $\left\{\dfrac{\pi}{3}, \pi, \dfrac{5\pi}{3}\right\}$ **25.** $\{15°, 45°, 135°, 165°, 255°, 285°\}$ **27.** $\{0°\}$ **29.** $\{120°, 240°\}$

31. $\{30°, 150°, 270°\}$ **33.** $0° + 360°\,n, 30° + 360°\,n, 150° + 360°\,n, 180° + 360°\,n,$ where n is any integer

35. $60° + 360°\,n, 300° + 360°\,n,$ where n is any integer   **37.** $11.8° + 360°\,n, 78.2° + 360°\,n, 191.8° + 360°\,n,$

$258.2° + 360°\,n,$ where n is any integer   **39.** $30° + 360°\,n, 90° + 360°\,n, 150° + 360°\,n, 210° + 360°\,n, 270° + 360°\,n,$

$330° + 360°\,n,$ where n is any integer   **41.** $\{1.2802\}$

43. (a) For $x = t,$

$P(t) = .003 \sin 220\pi t +$

$\dfrac{.003}{3} \sin 660\pi t +$

$\dfrac{.003}{5} \sin 1100\pi t +$

$\dfrac{.003}{7} \sin 1540\pi t$

(b) The graph is periodic, and the wave has "jagged square" tops and bottoms.

(c) This will occur when t is in one of these intervals: $(.0045, .0091), (.0136, .0182),$ $(.0227, .0273).$

45. (a) For $x = t,$

$P(t) = \dfrac{1}{2}\sin[2\pi(220)t] +$

$\dfrac{1}{3}\sin[2\pi(330)t] +$

$\dfrac{1}{4}\sin[2\pi(440)t]$

(b) $.0007576, .009847, .01894, .02803$ **(c)** 110 Hz **(d)** For $x = t,$

$P(t) = \sin[2\pi(110)t] +$

$\dfrac{1}{2}\sin[2\pi(220)t] +$

$\dfrac{1}{3}\sin[2\pi(330)t] +$

$\dfrac{1}{4}\sin[2\pi(440)t]$

47. $.001$ sec   **49.** $.004$ sec

6.4 Exercises *(page 265)*

1. C   **3.** C   **5.** $x = \arccos \dfrac{y}{5}$ **7.** $x = \dfrac{1}{3}\,\text{arccot}\,2y$ **9.** $x = \dfrac{1}{2}\arctan\dfrac{y}{3}$ **11.** $x = 4\arccos\dfrac{y}{6}$

13. $x = \dfrac{1}{5}\arccos\left(-\dfrac{y}{2}\right)$ **15.** $x = -3 + \arccos y$ **17.** $x = \arcsin(y + 2)$ **19.** $x = \arcsin\left(\dfrac{y + 4}{2}\right)$

23. $\left\{-2\sqrt{2}\right\}$ **25.** $\{\pi - 3\}$ **27.** $\left\{\dfrac{3}{5}\right\}$ **29.** $\left\{\dfrac{4}{5}\right\}$ **31.** $\{0\}$ **33.** $\left\{\dfrac{1}{2}\right\}$ **35.** $\left\{-\dfrac{1}{2}\right\}$ **37.** $\{0\}$

39. $Y = \arcsin X - \arccos X - \dfrac{\pi}{6}$

41. $\{4.4622\}$

43. (a) $A \approx .00506, \phi \approx .484; P = .00506 \sin(440\pi t + .484)$

(b) The two graphs are the same. For $x = t,$

$P(t) = .00506 \sin(440\pi t + .484)$

$P_1(t) + P_2(t) = .0012 \sin(440\pi t + .052) +$

$.004 \sin(440\pi t + .61)$

45. (a) $\tan \alpha = \dfrac{x}{z}$; $\tan \beta = \dfrac{x+y}{z}$ **(b)** $\dfrac{x}{\tan \alpha} = \dfrac{x+y}{\tan \beta}$ **(c)** $\alpha = \arctan\left(\dfrac{x \tan \beta}{x+y}\right)$ **(d)** $\beta = \arctan\left(\dfrac{(x+y) \tan \alpha}{x}\right)$

47. (a) $t = \dfrac{1}{2\pi f} \arcsin \dfrac{e}{E_{max}}$ **(b)** .00068 sec   **49. (a)** $t = \dfrac{3}{4\pi} \arcsin 3y$ **(b)** .27 sec

Chapter 6 Review Exercises *(page 271)*

1. false; The range of the inverse sine function is $\left[-\dfrac{\pi}{2}, \dfrac{\pi}{2}\right]$, while that of the inverse cosine is $[0, \pi]$.   **3.** false;

$\arcsin\left(-\dfrac{1}{2}\right) = -\dfrac{\pi}{6}$, not $\dfrac{11\pi}{6}$.   **5.** $\dfrac{\pi}{4}$ **7.** $-\dfrac{\pi}{3}$ **9.** $\dfrac{3\pi}{4}$ **11.** $\dfrac{2\pi}{3}$ **13.** $\dfrac{3\pi}{4}$ **15.** $-60°$ **17.** $60.67924514°$

19. $36.4895081°$ **21.** $73.26220613°$ **23.** -1 **25.** $\dfrac{3\pi}{4}$ **27.** $\dfrac{\pi}{4}$ **29.** $\dfrac{\sqrt{7}}{4}$ **31.** $\dfrac{\sqrt{3}}{2}$ **33.** $\dfrac{294 + 125\sqrt{6}}{92}$

35. $\dfrac{1}{u}$ **37.** $\{.463647609, 3.605240263\}$ **39.** $\left\{\dfrac{\pi}{4}, \dfrac{3\pi}{4}, \dfrac{5\pi}{4}, \dfrac{7\pi}{4}\right\}$ **41.** $\left\{\dfrac{\pi}{8}, \dfrac{3\pi}{8}, \dfrac{5\pi}{8}, \dfrac{7\pi}{8}, \dfrac{9\pi}{8}, \dfrac{11\pi}{8}, \dfrac{13\pi}{8}, \dfrac{15\pi}{8}\right\}$

43. $\dfrac{\pi}{3} + 2n\pi$, $\pi + 2n\pi$, $\dfrac{5\pi}{3} + 2n\pi$, where n is any integer   **45.** $\{270°\}$ **47.** $\{45°, 90°, 225°, 270°\}$

49. $\{70.5°, 180°, 289.5°\}$ **51.** $x = \arcsin 2y$ **53.** $x = \left(\dfrac{1}{3} \arctan 2y\right) - \dfrac{2}{3}$ **55.** $\emptyset$ **57.** $\left\{-\dfrac{1}{2}\right\}$

59. (b) 8.6602567 ft; There may be a discrepancy in the final digits.

$f(x) = \arctan\left(\dfrac{15}{x}\right) - \arctan\left(\dfrac{5}{x}\right)$

61. The light beam is completely underwater.   **63.**

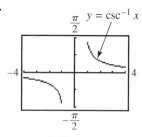

Radian mode

Chapter 6 Test *(page 273)*

1. $[-1, 1]$; $\left[-\dfrac{\pi}{2}, \dfrac{\pi}{2}\right]$ **2. (a)** $\dfrac{2\pi}{3}$ **(b)** $-\dfrac{\pi}{3}$ **(c)** 0 **(d)** $\dfrac{2\pi}{3}$ **3. (a)** $\dfrac{\sqrt{5}}{3}$ **(b)** $\dfrac{4\sqrt{2}}{9}$

6. $\dfrac{u\sqrt{1-u^2}}{1-u^2}$ **7.** $\{90°, 270°\}$ **8.** $\{18.4°, 135°, 198.4°, 315°\}$ **9.** $\left\{0, \dfrac{2\pi}{3}, \dfrac{4\pi}{3}\right\}$

10. $\dfrac{2\pi}{3} + 4n\pi, \dfrac{4\pi}{3} + 4n\pi$, where n is any integer   **11. (a)** $x = \dfrac{1}{3}\arccos y$ **(b)** $\left\{\dfrac{4}{5}\right\}$ **12.** $\dfrac{5}{6}$ sec, $\dfrac{11}{6}$ sec, $\dfrac{17}{6}$ sec

CHAPTER 7 APPLICATIONS OF TRIGONOMETRY AND VECTORS

Connections *(page 281)*

1. house: $X_H = 1131.8$ ft, $Y_H = 4390.2$ ft; fire: $X_F = 2277.5$ ft, $Y_F = -2596.2$ ft   **2.** 7079.7 ft

7.1 Exercises *(page 282)*

1. C   **3.** $\sqrt{3}$ **5.** $C = 95°$, $b = 13$ m, $a = 11$ m   **7.** $B = 37.3°$, $a = 38.5$ ft, $b = 51.0$ ft   **9.** $C = 57.36°$, $b = 11.13$ ft, $c = 11.55$ ft   **11.** $B = 18.5°$, $a = 239$ yd, $c = 230$ yd   **13.** $A = 56°\ 00'$, $AB = 361$ ft, $BC = 308$ ft   **15.** $B = 110.0°$, $a = 27.01$ m, $c = 21.37$ m   **17.** $A = 34.72°$, $a = 3326$ ft, $c = 5704$ ft   **19.** $C = 97°\ 34'$, $b = 283.2$ m, $c = 415.2$ m   **25.** 118 m   **27.** 17.8 km   **29.** 10.4 in.   **31.** 111°   **33.** first location: 5.1 mi; second location: 7.2 mi   **35.** about 419,000 km, which compares favorably to the actual value   **37.** approximately 6600 ft   **39.** $\dfrac{\sqrt{3}}{2}$ **41.** $\dfrac{\sqrt{2}}{2}$ **43.** 46.4 m²   **45.** 356 cm²   **47.** 722.9 in.²   **49.** 65.94 cm²   **51.** 100 m²   **53.** $a = \sin A$, $b = \sin B$, $c = \sin C$ **55.** $x = \dfrac{d \sin \alpha \sin \beta}{\sin(\beta - \alpha)}$

7.2 Exercises *(page 291)*

1. A   **3. (a)** $4 < h < 5$ **(b)** $h = 4$ or $h > 5$ **(c)** $h < 4$ **5.** 1   **7.** 2   **9.** 0   **11.** 45°   **13.** $B_1 \approx 49.1°$, $C_1 \approx 101.2°$, $B_2 \approx 130.9°$, $C_2 \approx 19.4°$ **15.** $B \approx 26°\ 30'$, $A \approx 112°\ 10'$ **17.** no such triangle   **19.** $B \approx 27.19°$, $C \approx 10.68°$ **21.** $B \approx 20.6°$, $C \approx 116.9°$, $c \approx 20.6$ ft   **23.** no such triangle   **25.** $B_1 \approx 49°\ 20'$, $C_1 \approx 92°\ 00'$, $c_1 \approx 15.5$ km, $B_2 \approx 130°\ 40'$, $C_2 \approx 10°\ 40'$, $c_2 \approx 2.88$ km   **27.** $B \approx 37.77°$, $C \approx 45.43°$, $c \approx 4.174$ ft   **29.** $A_1 \approx 53.23°$, $C_1 \approx 87.09°$, $c_1 \approx 37.16$ m, $A_2 \approx 126.77°$, $C_2 \approx 13.55°$, $c_2 \approx 8.719$ m   **31.** 1; 90°; a right triangle   **35.** It cannot exist.

Connections *(page 298)*

1.–3. All three methods give the area as 9.5 sq units.

7.3 Exercises *(page 298)*

1. (a) SAS **(b)** law of cosines   **3. (a)** SSA **(b)** law of sines   **5. (a)** ASA **(b)** law of sines   **7. (a)** ASA **(b)** law of sines   **9.** 7   **11.** 30°   **13.** $a \approx 5.4$, $B \approx 40.7°$, $C \approx 78.3°$ **15.** $A \approx 22.3°$, $B \approx 108.2°$, $C \approx 49.5°$ **17.** $A \approx 33.6°$, $B \approx 50.7°$, $C \approx 95.7°$ **19.** $a \approx 2.60$ yd, $B \approx 45.1°$, $C \approx 93.5°$ **21.** $c \approx 6.46$ m, $A \approx 53.1°$, $B \approx 81.3°$ **23.** $A \approx 82°$, $B \approx 37°$, $C \approx 61°$ **25.** $C \approx 102°\ 10'$, $B \approx 35°\ 50'$, $A \approx 42°\ 00'$ **27.** $C \approx 84°\ 30'$, $B \approx 44°\ 40'$, $A \approx 50°\ 50'$ **29.** $a \approx 156$ cm, $B \approx 64°\ 50'$, $C \approx 34°\ 30'$ **31.** $b \approx 9.529$ in., $A \approx 64.59°$, $C \approx 40.61°$ **33.** $a \approx 15.7$ m, $B \approx 21.6°$, $C \approx 45.6°$ **35.** $A \approx 30°$, $B \approx 56°$, $C \approx 94°$ **37.** The value of $\cos \theta$ will be greater than 1; your calculator will give you an error message (or a complex number) when using the inverse cosine function.   **39.** 257 m   **41.** 281 km   **43.** 10.8 mi   **45.** 40.5°   **47.** 26.4° and 36.3°   **49.** second base: 66.8 ft; first and third bases: 63.7 ft   **51.** 39.2 km   **53.** 350°   **55.** approximately 47.5 ft   **57.** 163.5°   **59.** 22 ft   **61.** 16.26°   **63.** $24\sqrt{3}$ **65.** 78 m²   **67.** 12,600 cm²   **69.** 3650 ft²   **71.** 25.24983 mi   **73.** 392,000 mi²   **75.** Area and perimeter are both 36.   **77. (a)** 87.8° and 92.2° both appear possible. **(b)** 92.2° **(c)** With the law of cosines we are required to find the inverse cosine of a negative number. Therefore, we know that angle C is greater than 90°.

7.4 Exercises *(page 312)*

1. **m** and **p**; **n** and **r**   **3.** **m** and **p** equal 2**t**, or **t** is one-half **m** or **p**; also **m** = 1**p** and **n** = 1**r**

5. −**b**   **7.**   **9.**   **11.**   **13.**

3**a**

15.

17. Yes, it appears that vector addition is associative (and this is true, in general).

19. (a) ⟨−4, 16⟩ (b) ⟨−12, 0⟩ (c) ⟨8, −8⟩   **21.** (a) ⟨8, 0⟩ (b) ⟨0, 16⟩ (c) ⟨−4, −8⟩

23. (a) ⟨0, 12⟩ (b) ⟨−16, −4⟩ (c) ⟨8, −4⟩   **25.** (a) 4**i** (b) 7**i** + 3**j** (c) −5**i** + **j**

27. (a) ⟨−2, 4⟩ (b) ⟨7, 4⟩ (c) ⟨6, −6⟩

29.   **31.**   **33.** 17; 331.9°   **35.** 8; 120°   **37.** 47, 17

39. 38.8, 28.0   **41.** 123, 155   **43.** $\left\langle \dfrac{5\sqrt{3}}{2}, \dfrac{5}{2} \right\rangle$ **45.** ⟨3.0642, 2.5712⟩   **47.** ⟨4.0958, −2.8679⟩   **49.** 530 newtons

51. 88.2 lb   **53.** 94.2 lb   **55.** 24.4 lb   **57.** ⟨a + c, b + d⟩   **59.** ⟨2, 8⟩   **61.** ⟨8, −20⟩   **63.** ⟨−30, −3⟩

65. ⟨8, −7⟩   **67.** −5**i** + 8**j**   **69.** 2**i** or 2**i** + 0**j**   **71.** 7   **73.** −3   **75.** 20   **77.** 135°   **79.** 90°   **81.** 36.87°

83. −6   **85.** −24   **87.** orthogonal   **89.** not orthogonal   **91.** not orthogonal   **93.** magnitude: 9.5208; direction angle: 119.0647°   **94.** ⟨−4.1042, 11.2763⟩   **95.** ⟨−.5209, −2.9544⟩   **96.** ⟨−4.6252, 8.3219⟩

97. magnitude: 9.5208; direction angle: 119.0647°   **98.** They are the same. Preference of method is an individual choice.

7.5 Exercises *(page 318)*

1. 2640 lb at an angle of 167.2° with the 1480-lb force   **3.** 93.9°   **5.** 190 lb and 283 lb, respectively   **7.** 18°

9. 2.4 tons   **11.** 17.5°   **13.** weight: 64.8 lb; tension: 61.9 lb   **15.** 13.5 mi; 50.4°   **17.** 39.2 km   **19.** current: 3.5 mph; motorboat: 19.7 mph   **21.** bearing: 237°; groundspeed: 470 mph   **23.** groundspeed: 161 mph; airspeed: 156 mph

25. bearing: 74°; groundspeed: 202 mph   **27.** bearing: 358°; airspeed: 170 mph   **29.** groundspeed: 230 km per hr; bearing: 167°   **31.** (a) $|\mathbf{R}| = \sqrt{5} \approx 2.2$, $|\mathbf{A}| = \sqrt{1.25} \approx 1.1$; About 2.2 in. of rain fell. The area of the opening of the rain gauge is about 1.1 in.². (b) $V = 1.5$; the volume of rain was 1.5 in.³. (c) **R** and **A** should be parallel and point in opposite directions.

Chapter 7 Review Exercises *(page 325)*

1. 63.7 m   **3.** 41.7°   **5.** 54° 20′ or 125° 40′   **9.** (a) $b = 5, b \geq 10$ (b) $5 < b < 10$ (c) $b < 5$

11. 19.87° or 19° 52′   **13.** 55.5 m   **15.** 19 cm   **17.** $B \approx 17.3°$, $C \approx 137.5°$, $c \approx 11.0$ yd   **19.** $c \approx 18.7$ cm, $A \approx 91° 40′, B \approx 45° 50′$ **21.** 153,600 m²   **23.** .234 km²   **25.** 58.6 ft   **27.** 13 m   **29.** 53.2 ft   **31.** 115 km

33. 25   **35.**   **37.** (a) true (b) false   **39.** 207 lb   **41.** 869; 418   **43.** 15; 126.9°

45. $-9; 142.1°$ **47.** $\left\langle \dfrac{5}{13}, \dfrac{12}{13} \right\rangle$ **49.** 29 lb   **51.** bearing: 306°; speed: 524 mph   **53.** $AB = 1978.28$ ft; $BC = 975.05$ ft

55. Both expressions equal $\dfrac{1 - \sqrt{3}}{2}$.

Chapter 7 Test *(page 329)*

1. $137.5°$ **2.** 180 km   **3.** $49.0°$ **4.** 4300 km² **5. (a)** $b > 10$ **(b)** none **(c)** $b \le 10$ **6.** 264 sq units

7. $|\mathbf{v}| = 10; \theta \approx 126.9°$ **8.**   **9. (a)** $\langle 1, -3 \rangle$ **(b)** $\langle -6, 18 \rangle$ **(c)** -20 **(d)** $\sqrt{10}$ **10.** 2.7 mi

11. $\langle -346, 451 \rangle$ **12.** 1.91 mi   **13.** 28.59 ft, 37.85 ft

14. 30.4 lb

CHAPTER 8 COMPLEX NUMBERS, POLAR EQUATIONS, AND PARAMETRIC EQUATIONS

8.1 Exercises *(page 338)*

1. true   **3.** true   **5.** false; *Every* real number is a complex number.   **7.** real, complex   **9.** complex, pure imaginary

11. complex   **13.** real, complex   **15.** complex, pure imaginary   **17.** $5i$ **19.** $i\sqrt{10}$ **21.** $12i\sqrt{2}$ **23.** $-3i\sqrt{2}$

25. $\{\pm 4i\}$ **27.** $\{\pm 2i\sqrt{3}\}$ **29.** $\left\{ -\dfrac{2}{3} \pm \dfrac{\sqrt{2}}{3}i \right\}$ **31.** $\{3 \pm i\sqrt{5}\}$ **33.** $\left\{ \dfrac{1}{2} \pm \dfrac{\sqrt{6}}{2}i \right\}$ **35.** $\left\{ -\dfrac{1}{2} \pm \dfrac{\sqrt{3}}{2}i \right\}$

37. -13 **39.** $-2\sqrt{6}$ **41.** $\sqrt{3}$ **43.** $i\sqrt{3}$ **45.** $\dfrac{1}{2}$ **47.** -2 **49.** $7 - i$ **51.** 2 **53.** $1 - 10i$

55. $-13 + 5i$ **57.** $8 - i$ **59.** $-14 + 2i$ **61.** $5 - 12i$ **63.** $-8 - 6i$ **65.** 13 **67.** 7 **69.** $25i$

71. $12 + 9i$ **73.** $20 + 15i$ **75.** i **77.** -1 **79.** $-i$ **81.** 1 **83.** $-i$ **85.** $-i$ **89.** $2 - 2i$

91. $\dfrac{3}{5} - \dfrac{4}{5}i$ **93.** $-1 - 2i$ **95.** $-5i$ **97.** $-8i$ **99.** $-\dfrac{2}{3}i$ **103.** $E = 30 + 60i$ **105.** $Z = \dfrac{233}{37} + \dfrac{119}{37}i$

107. $110 + 32i$

8.2 Exercises *(page 345)*

1. length (magnitude)   **3.**   **5.**   **7.**   **9.**

11. $1 - 4i$ **13.** $3 - i$ **15.** $3 - 3i$ **17.** $-3 + 3i$ **19.** $2 + 4i$ **21.** $7 + 9i$ **23.** $\sqrt{2} + i\sqrt{2}$ **25.** $10i$

27. $-2 - 2i\sqrt{3}$ **29.** $-\dfrac{3\sqrt{3}}{2} + \dfrac{3}{2}i$ **31.** $\dfrac{5}{2} - \dfrac{5\sqrt{3}}{2}i$ **33.** $-\sqrt{2}$ **35.** $2\sqrt{3} - 2i$

37. $3\sqrt{2}(\cos 315° + i \sin 315°)$ **39.** $6(\cos 240° + i \sin 240°)$ **41.** $2(\cos 330° + i \sin 330°)$

43. $5\sqrt{2}(\cos 225° + i \sin 225°)$ **45.** $2\sqrt{2}(\cos 45° + i \sin 45°)$ **47.** $5(\cos 90° + i \sin 90°)$

49. $4(\cos 180° + i \sin 180°)$ **51.** $\sqrt{13}(\cos 56.31° + i \sin 56.31°)$ **53.** $-1.0261 - 2.8191i$

55. $12(\cos 90° + i \sin 90°)$ **57.** $\sqrt{34}(\cos 59.04° + i \sin 59.04°)$ **59.** the circle of radius 1 centered at the origin

61. the vertical line $x = 1$ **63.** yes   **67.** B   **69.** A

8.3 Exercises *(page 350)*

1. multiply; add   **3.** $-3\sqrt{3} + 3i$ **5.** $-4i$ **7.** $12\sqrt{3} + 12i$ **9.** $-\dfrac{15\sqrt{2}}{2} + \dfrac{15\sqrt{2}}{2}i$ **11.** $-3i$ **13.** $\sqrt{3} - i$

15. -2 **17.** $-\dfrac{1}{6} - \dfrac{\sqrt{3}}{6}i$ **19.** $2\sqrt{3} - 2i$ **21.** $-\dfrac{1}{2} - \dfrac{1}{2}i$ **23.** $\sqrt{3} + i$ **25.** $.6537 + 7.4715i$

27. $30.8580 + 18.5414i$ **29.** $.2091 + 1.9890i$ **31.** $-3.7588 - 1.3681i$ **33.** 2   **34.** $w = \sqrt{2}$ cis 135°;
$z = \sqrt{2}$ cis 225°   **35.** 2 cis 0°   **36.** 2; It is the same.   **37.** $-i$ **38.** cis$(-90°)$ **39.** $-i$; It is the same.
43. $1.18 - .14i$ **45.** approximately $27.43 + 11.5i$

8.4 Exercises *(page 356)*

1. $27i$ **3.** 1   **5.** $\dfrac{27}{2} - \dfrac{27\sqrt{3}}{2}i$ **7.** $-16\sqrt{3} + 16i$ **9.** $-128 + 128i\sqrt{3}$ **11.** $128 + 128i$

13. (a) $\cos 0° + i \sin 0°$, $\cos 120° + i \sin 120°$, $\cos 240° + i \sin 240°$ **(b)**

15. (a) 2 cis 20°, 2 cis 140°, 2 cis 260° **(b)**

17. (a) $2(\cos 90° + i \sin 90°)$, $2(\cos 210° + i \sin 210°)$, $2(\cos 330° + i \sin 330°)$ **(b)**

19. (a) $4(\cos 60° + i \sin 60°)$, $4(\cos 180° + i \sin 180°)$, $4(\cos 300° + i \sin 300°)$ **(b)**

21. (a) $\sqrt[3]{2}(\cos 20° + i \sin 20°)$, $\sqrt[3]{2}(\cos 140° + i \sin 140°)$, $\sqrt[3]{2}(\cos 260° + i \sin 260°)$ **(b)**

23. (a) $\sqrt[3]{4}(\cos 50° + i \sin 50°)$, $\sqrt[3]{4}(\cos 170° + i \sin 170°)$, $\sqrt[3]{4}(\cos 290° + i \sin 290°)$ **(b)**

25.

$\cos 0° + i \sin 0°$,
$\cos 180° + i \sin 180°$

27.

$\cos 0° + i \sin 0°$, $\cos 60° + i \sin 60°$,
$\cos 120° + i \sin 120°$, $\cos 180° + i \sin 180°$,
$\cos 240° + i \sin 240°$, $\cos 300° + i \sin 300°$

29. $\cos 45° + i \sin 45°, \cos 225° + i \sin 225°$

31. $\{\cos 0° + i \sin 0°, \cos 120° + i \sin 120°, \cos 240° + i \sin 240°\}$ **33.** $\{\cos 90° + i \sin 90°, \cos 210° + i \sin 210°,$
$\cos 330° + i \sin 330°\}$ **35.** $\{2(\cos 0° + i \sin 0°), 2(\cos 120° + i \sin 120°), 2(\cos 240° + i \sin 240°)\}$
37. $\{\cos 45° + i \sin 45°, \cos 135° + i \sin 135°, \cos 225° + i \sin 225°, \cos 315° + i \sin 315°\}$ **39.** $\{\cos 22.5° + i \sin 22.5°,$
$\cos 112.5° + i \sin 112.5°, \cos 202.5° + i \sin 202.5°, \cos 292.5° + i \sin 292.5°\}$ **41.** $\{2(\cos 20° + i \sin 20°),$
$2(\cos 140° + i \sin 140°), 2(\cos 260° + i \sin 260°)\}$ **43.** $1, -\dfrac{1}{2} + \dfrac{\sqrt{3}}{2}i, -\dfrac{1}{2} - \dfrac{\sqrt{3}}{2}i$ **45.** $\cos 2\theta + i \sin 2\theta$

46. $(\cos^2 \theta - \sin^2 \theta) + i(2 \cos \theta \sin \theta) = \cos 2\theta + i \sin 2\theta$ **47.** $\cos 2\theta = \cos^2 \theta - \sin^2 \theta$ **48.** $\sin 2\theta = 2 \cos \theta \sin \theta$
49. (a) yes **(b)** no **(c)** yes   **51.** $1, .30901699 + .95105652i, -.809017 + .58778525i, -.809017 - .5877853i,$
$.30901699 - .9510565i$ **53.** $-4, 2 - 2i\sqrt{3}$ **55.** $\{.87708 + .94922i, -.63173 + 1.1275i, -1.2675 - .25240i,$
$-.15164 - 1.28347i, 1.1738 - .54083i\}$ **57.** false

8.5 Exercises *(page 367)*

1. (a) II   **(b)** I   **(c)** IV   **(d)** III

Graphs for Exercises 3(a), 5(a), 7(a), 9(a), 11(a)

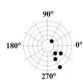

Answers may vary in Exercises 3(b)–11(b).

3. (b) $(1, 405°), (-1, 225°)$ **(c)** $\left(\dfrac{\sqrt{2}}{2}, \dfrac{\sqrt{2}}{2}\right)$ **5. (b)** $(-2, 495°), (2, 315°)$ **(c)** $\left(\sqrt{2}, -\sqrt{2}\right)$ **7. (b)** $(5, 300°), (-5, 120°)$

(c) $\left(\dfrac{5}{2}, -\dfrac{5\sqrt{3}}{2}\right)$ **9. (b)** $(-3, 150°), (3, -30°)$ **(c)** $\left(\dfrac{3\sqrt{3}}{2}, -\dfrac{3}{2}\right)$ **11. (b)** $\left(3, \dfrac{11\pi}{3}\right), \left(-3, \dfrac{2\pi}{3}\right)$ **(c)** $\left(\dfrac{3}{2}, -\dfrac{3\sqrt{3}}{2}\right)$

Graphs for Exercises 13(a), 15(a), 17(a), 19(a), 21(a)

Answers may vary in Exercises 13(b)–21(b).

13. (b) $\left(\sqrt{2}, 315°\right), \left(-\sqrt{2}, 135°\right)$ **15. (b)** $(3, 90°), (-3, 270°)$ **17. (b)** $(2, 45°), (-2, 225°)$

19. (b) $\left(\sqrt{3}, 60°\right), \left(-\sqrt{3}, 240°\right)$ **21. (b)** $(3, 0°), (-3, 180°)$ **23.** $r = \dfrac{4}{\cos \theta - \sin \theta}$

$x - y = 4$
$r = \dfrac{4}{\cos \theta - \sin \theta}$

25. $r = 4$ or $r = -4$ **27.** $r = \dfrac{5}{2 \cos \theta + \sin \theta}$ **29.** $r \sin \theta = k$ **30.** $r = \dfrac{k}{\sin \theta}$

$x^2 + y^2 = 16$
$r = 4$ or
$r = -4$

$2x + y = 5$
$r = \dfrac{5}{2 \cos \theta + \sin \theta}$

31. $r = k \csc \theta$ **32.**

$r = 3 \csc \theta$ $y = 3$ **33.** $r \cos \theta = k$ **34.** $r = \dfrac{k}{\cos \theta}$ **35.** $r = k \sec \theta$

36.

$r = 3 \sec \theta$ $x = 3$ **37.** C   **39.** A   **41.** cardioid

$r = 2 + 2 \cos \theta$

43. limaçon

$r = 3 + \cos \theta$

45. four-leaved rose

$r = 4 \cos 2\theta$

47. lemniscate

$r^2 = 4 \cos 2\theta$

49. cardioid

$r = 4 - 4 \cos \theta$

51.

$r = 2 \sin \theta \tan \theta$

53. $x^2 + (y - 1)^2 = 1$

$r = 2 \sin \theta$
$x^2 + (y - 1)^2 = 1$

55. $y^2 = 4(x + 1)$

$r = \dfrac{2}{1 - \cos \theta}$
$y^2 = 4(x + 1)$

57. $(x + 1)^2 + (y + 1)^2 = 2$

$r = -2 \cos \theta - 2 \sin \theta$
$(x + 1)^2 + (y + 1)^2 = 2$

59. $x = 2$

$r = 2 \sec \theta$
$x = 2$

61. $x + y = 2$

$r = \dfrac{2}{\cos \theta + \sin \theta}$
$x + y = 2$

63.

$r = \theta$

65. $r = \dfrac{2}{2 \cos \theta + \sin \theta}$

67. **(a)** $(r, -\theta)$ **(b)** $(r, \pi - \theta)$ or $(-r, -\theta)$ **(c)** $(r, \pi + \theta)$ or $(-r, \theta)$

69.

$r = \theta, 0 \leq \theta \leq 4\pi$

71.

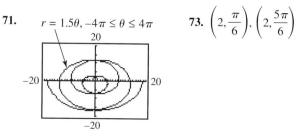

$r = 1.5\theta, -4\pi \leq \theta \leq 4\pi$

73. $\left(2, \dfrac{\pi}{6}\right), \left(2, \dfrac{5\pi}{6}\right)$

75. $\left(\dfrac{4 + \sqrt{2}}{2}, \dfrac{\pi}{4}\right), \left(\dfrac{4 - \sqrt{2}}{2}, \dfrac{5\pi}{4}\right)$ **77. (a)**

(b)

(c) no

Earth is closest to the sun.

8.6 Exercises *(page 376)*

1. C   **3.** A   **5. (a)**

(b) $y = x^2 - 4x + 4$, for x in $[1, 3]$

7. (a)

(b) $y = 3x^2 - 4$, for x in $[0, 2]$ **9. (a)**

(b) $y = x - 2$, for x in $(-\infty, \infty)$

11. (a)

(b) $x^2 + y^2 = 4$, for x in $[-2, 2]$ **13. (a)**

(b) $y = 2\sqrt{1 + \dfrac{x^2}{9}}$, for x in $(-\infty, \infty)$

15. (a)

(b) $y = \dfrac{1}{x}$, for x in $(0, 1]$ **17. (a)**

(b) $y = \sqrt{x^2 + 2}$, for x in $(-\infty, \infty)$

19. (a)

(b) $(x - 2)^2 + (y - 1)^2 = 1$, for x in $[1, 3]$ **21. (a)**

(b) $y = \dfrac{1}{x}$, for x in $(-\infty, 0) \cup (0, \infty)$

23. (a)

(b) $y = x - 6$, for x in $(-\infty, \infty)$ **25.**

$x^2 + y^2 = 9$

27. $\dfrac{x^2}{9} + \dfrac{y^2}{4} = 1$

Answers may vary for Exercises 29 and 31.

29. $x = t$, $y = (t + 3)^2 - 1$, for t in $(-\infty, \infty)$; $x = t - 3$, $y = t^2 - 1$, for t in $(-\infty, \infty)$

31. $x = t$, $y = t^2 - 2t + 3$, for t in $(-\infty, \infty)$; $x = t + 1$, $y = t^2 + 2$, for t in $(-\infty, \infty)$

33.   **35.**   **37.**

$x = 2 \cos t$, $y = 3 \sin 2t$, for t in $[0, 6.5]$ $x = 3 \sin 4t$, $y = 3 \cos 3t$, for t in $[0, 6.5]$

39. **(a)** $x = 24t$, $y = -16t^2 + 24\sqrt{3}t$ **(b)** $y = -\dfrac{1}{36}x^2 + \sqrt{3}x$ **(c)** 2.6 sec; 62 ft

41. **(a)** $x = (88 \cos 20°)t$, $y = 2 - 16t^2 + (88 \sin 20°)t$ **(b)** $y = 2 - \dfrac{x^2}{484 \cos^2 20°} + (\tan 20°)x$ **(c)** 1.9 sec; 161 ft

43. **(a)** $y = -\dfrac{1}{256}x^2 + \sqrt{3}x + 8$; parabolic path **(b)** approximately 7 sec; approximately 448 ft

45. **(a)** $x = 32t$, $y = 32\sqrt{3}t - 16t^2 + 3$ **(b)** about 112.6 ft **(c)** 51 ft maximum height; The ball had traveled horizontally about 55.4 ft. **(d)** yes   **47.** $x = 56.56530965t$ **(b)** 50.0° **(c)** $x = (88 \cos 50.0°)t$, $y = -16t^2 + (88 \sin 50.0°)t$

$y = -16t^2 + 67.41191099t$

49. Many answers are possible; for example, $y = a(t - h)^2 + k$, $x = t$ and $y = at^2 + k$, $x = t + h$.

51. Many answers are possible; for example, $x = a \sin t$, $y = b \cos t$ and $x = t$, $y^2 = b^2\left(1 - \dfrac{t^2}{a^2}\right)$.

55. The graph is translated horizontally by c units.

Chapter 8 Review Exercises *(page 382)*

1. $3i$ **3.** $\{\pm 9i\}$ **5.** $-2 - 3i$ **7.** $5 + 4i$ **9.** $29 + 37i$ **11.** $-32 + 24i$ **13.** $-2 - 2i$

15. $\dfrac{8}{5} + \dfrac{6}{5}i$ **17.** $-\dfrac{3}{26} + \dfrac{11}{26}i$ **19.** i **21.** $-30i$ **23.** $-\dfrac{1}{8} + \dfrac{\sqrt{3}}{8}i$ **25.** $8i$ **27.** $-\dfrac{1}{2} - \dfrac{\sqrt{3}}{2}i$

29.   **31.**   **33.** $2\sqrt{2}(\cos 135° + i \sin 135°)$ **35.** $-\sqrt{2} - i\sqrt{2}$

37. $\sqrt{2}(\cos 315° + i \sin 315°)$ **39.** $4(\cos 270° + i \sin 270°)$

41. the line $y = -x$

43. $\sqrt[3]{2}(\cos 105° + i \sin 105°)$, $\sqrt[3]{2}(\cos 225° + i \sin 225°)$, $\sqrt[3]{2}(\cos 345° + i \sin 345°)$ **45.** none

47. $\{2(\cos 45° + i \sin 45°), 2(\cos 135° + i \sin 135°), 2(\cos 225° + i \sin 225°), 2(\cos 315° + i \sin 315°)\}$ **49.** $(2, 120°)$

51. quadrantal   **53.** circle

$r = 4 \cos \theta$

55. eight-leaved rose

$r = 2 \sin 4\theta$

57. $y^2 = -6\left(x - \dfrac{3}{2}\right)$ or $y^2 + 6x - 9 = 0$ **59.** $x^2 + y^2 = 4$ **61.** $r = \tan \theta \sec \theta$ or $r = \dfrac{\tan \theta}{\cos \theta}$ **63.** A   **65.** B

67. $r = 2 \sec \theta$ or $r = \dfrac{2}{\cos \theta}$ **69.** $r = \dfrac{4}{\cos \theta + 2 \sin \theta}$ **71.**

$x = t + \cos t$
$y = \sin t$
for t in $[0, 2\pi]$

73. $y = \sqrt{x^2 + 1}$, for x in $[0, \infty)$

75. $y = 3\sqrt{1 + \dfrac{x^2}{25}}$, for x in $(-\infty, \infty)$ **77.** $x = 3 + 5 \cos t$, $y = 4 + 5 \sin t$, for t in $[0, 2\pi]$

79. (a) $x = (118 \cos 27°)t$, $y = 3.2 - 16t^2 + (118 \sin 27°)t$ **(b)** $y = 3.2 - \dfrac{4x^2}{3481 \cos^2 27°} + (\tan 27°)x$ **(c)** 3.4 sec; 358 ft

Chapter 8 Test *(page 385)*

1. (a) $7 - 3i$

(b) $-3 - 5i$ **(c)** $14 - 18i$ **(d)** $\dfrac{3}{13} - \dfrac{11}{13}i$ **2. (a)** $-i$ **(b)** $2i$ **3.** $\left\{\dfrac{1}{4} \pm \dfrac{\sqrt{31}}{4}i\right\}$

4. (a) $3(\cos 90° + i \sin 90°)$ **(b)** $\sqrt{5}$ cis $63.43°$ **(c)** $2(\cos 240° + i \sin 240°)$ **5. (a)** $\dfrac{3\sqrt{3}}{2} + \dfrac{3}{2}i$ **(b)** $3.06 + 2.57i$ **(c)** $3i$

6. (a) $16(\cos 50° + i \sin 50°)$ **(b)** $2\sqrt{3} + 2i$ **(c)** $4\sqrt{3} + 4i$ **7.** 2 cis $67.5°$, 2 cis $157.5°$, 2 cis $247.5°$, 2 cis $337.5°$

8. Answers may vary. **(a)** $(5, 90°)$, $(5, -270°)$ **(b)** $\left(2\sqrt{2}, 225°\right)$, $\left(2\sqrt{2}, -135°\right)$ **9. (a)** $\left(\dfrac{3\sqrt{2}}{2}, -\dfrac{3\sqrt{2}}{2}\right)$ **(b)** $(0, -4)$

10. cardioid

$r = 1 - \cos \theta$

11. three-leaved rose

$r = 3 \cos 3\theta$

12. (a) $x - 2y = -4$

$x - 2y = -4$

(b) $x^2 + y^2 = 36$

$x^2 + y^2 = 36$

13.

$(13, 16)$
$(-15, 9)$
$(-3, 0)$ $x = 4t - 3$
$y = t^2$
for t in $[-3, 4]$

14.

$x = 2 \cos 2t$
$y = 2 \sin 2t$
for t in $[0, 2\pi]$

15. $z^2 - 1 = -1 - 2i$; $r = \sqrt{5}$ and $\sqrt{5} > 2$

CHAPTER 9 EXPONENTIAL AND LOGARITHMIC FUNCTIONS

9.1 Exercises *(page 399)*

1. 9 **3.** $\frac{1}{9}$ **5.** $\frac{1}{16}$ **7.** 16 **9.** 5.196 **11.** .039 **13.** **15.**

17. **19.** **21.** **23.** **25.**

27. **29.** **31.** **33.** 2.3 **35.** .75 **37.** .31

39. $f(x) = \dfrac{e^x - e^{-x}}{2}$ **41.** $f(x) = x \cdot 2^x$ **43.** $\left\{ \dfrac{1}{2} \right\}$ **45.** $\{-2\}$ **47.** $\{0\}$ **49.** $\left\{ \dfrac{1}{2} \right\}$

51. $\left\{ \dfrac{1}{5} \right\}$ **53.** $\{-7\}$ **55.** $\left\{ -\dfrac{2}{3} \right\}$ **57.** $\left\{ \dfrac{4}{3} \right\}$ **59.** $\{3\}$ **61.** $\{-8, 8\}$ **63. (a)** $13,891.16; $4984.62

(b) $13,968.24; $5061.70 **65.** $21,223.33 **67.** $3528.81 **69.** 4.5% **71.** Bank A (even though it has the highest stated rate) **73. (a)** **(b)** exponential **(c)**

(d) $P(1500) \approx 828$ mb; $P(11,000) \approx 232$ mb **75. (a)** about 63,000 **(b)** about 42,000 **(c)** about 21,000

9.2 Exercises *(page 410)*

1. (a) C **(b)** A **(c)** E **(d)** B **(e)** F **(f)** D **3.** $\log_3 81 = 4$ **5.** $\log_{2/3} \dfrac{27}{8} = -3$ **7.** $6^2 = 36$ **9.** $\left(\sqrt{3} \right)^8 = 81$

13. $\{-4\}$ **15.** $\{-3\}$ **17.** $\left\{ \dfrac{1}{4} \right\}$ **19.** $\{8\}$ **21.** $\{9\}$ **23.** $\left\{ \dfrac{1}{5} \right\}$ **25.** $\{64\}$ **27.** $\left\{ \dfrac{2}{3} \right\}$ **29.** $\left\{ \dfrac{1}{3} \right\}$

33.

35.

37.

39. E   **41.** B   **43.** F

45.

47.

49.

51.

53. $\log_2 6 + \log_2 x - \log_2 y$ **55.** $1 + \dfrac{1}{2}\log_5 7 - \log_5 3$ **57.** cannot be simplified

59. $\dfrac{1}{2}(\log_m 5 + 3\log_m r - 5\log_m z)$ **61.** $\log_a \dfrac{xy}{m}$ **63.** $\log_m \dfrac{a^2}{b^6}$ **65.** $\log_a [(z-1)^2(3z+2)]$ **67.** $\log_5 \dfrac{5^{1/3}}{m^{1/3}}$ or

$\log_5 \sqrt[3]{\dfrac{5}{m}}$ **69.** .7781   **71.** .3522   **73.** .7386   **75. (a)**   **77. (a)** -4 **(b)** 6

9.3 Exercises *(page 421)*

1. increasing   **3.** $f^{-1}(x) = \log_5 x$ **5.** natural; common   **7.** There is no power of 2 that yields a result of 0.

9. $\log 8 = .90308999$ **11.** 1.5563   **13.** -1.3768 **15.** 4.3010   **17.** 3.5835   **19.** -3.1701 **21.** 4.6931

23. 3.2   **25.** 1.8   **27.** 2.0×10^{-3} **29.** 1.6×10^{-5} **31.** poor fen   **33.** rich fen   **35.** 2.3219   **37.** $-.2537$

39. 1.9376   **41.** -1.4125 **43. (a)** 3 **(b)** 5^2 or 25 **(c)** $\dfrac{1}{e}$ **45. (a)** 5 **(b)** $\ln 3$ **(c)** $2\ln 3$ or $\ln 9$ **47. (a)** 20 **(b)** 30

(c) 50 **(d)** 60 **(e)** about 3 decibels   **49. (a)** 3 **(b)** 6 **(c)** 8   **51.** about $126{,}000{,}000I_0$ **53.** about 70 million visitors;

We must assume that the rate of increase continues to be logarithmic.   **55. (a)** 2 **(b)** 2 **(c)** 2 **(d)** 1   **57.** 1

59. $\log_7 19;\ \dfrac{\log 19}{\log 7};\ \dfrac{\ln 19}{\ln 7}$ **61.** $\log_{1/2} 12;\ \dfrac{\log 12}{\log\left(\dfrac{1}{2}\right)};\ \dfrac{\ln 12}{\ln\left(\dfrac{1}{2}\right)}$ **63.** $\{1.6309\}$ **65.** $\{-.0803\}$ **67.** $\{2.2694\}$

69. $\{2.3863\}$ **71.** $\{-.1227\}$ **73.** $\emptyset$ **75.** $\{2\}$ **77.** $\left\{\dfrac{1}{3}\right\}$ **79.** $\emptyset$ **81.** $\{1\}$ **83.** $\{5\}$ **85.** $\{5\}$

87. $\left\{\dfrac{9}{2}\right\}$ **89.** $\{-17.5314\}$ **91.** $\{8\}$ **93.** $\{4\}$ **95.** $\{1, 100\}$ **97.** during 2011   **99. (a)** about 24% **(b)** 1963

Chapter 9 Review Exercises *(page 428)*

1. B   **3.** C   **5.** $\log_2 32 = 5$ **7.** $\log_{3/4}\dfrac{4}{3} = -1$ **9.** $\log_3 4$ **10.** 2   **11.** 3   **12.** It lies between 2 and 3.

13. By the change-of-base theorem, $\log_3 16 = \dfrac{\log 16}{\log 3} = \dfrac{\ln 16}{\ln 3} \approx 2.523719014.$ **14.** $-1; 0$ **15.** It lies between -1 and 0;

$\log_5 .68 = \dfrac{\log .68}{\log 5} = \dfrac{\ln .68}{\ln 5} \approx -.2396255723$ **17.** $10^{.5378} \approx 3.45$ **19.** 3   **21.** $\log_3 m + \log_3 n - \log_3 5 - \log_3 r$

23. cannot be simplified   **25.** -1.3862 **27.** 11.8776 **29.** 1.1592 **31.** $\left\{\dfrac{3}{2}\right\}$ **33.** $\{-.3138\}$ **35.** $\{4\}$

37. $\{3\}$ **39.** $\{5\}$ **41. (a)** about $200{,}000{,}000 I_0$ **(b)** about $13{,}000{,}000 I_0$ **(c)** The 1906 earthquake had a magnitude almost 16 times greater than the 1989 earthquake.   **43.** 5.1% **45.** $\$60{,}602.77$ **47.** 17.3 yr   **49.** 2007

Chapter 9 Test *(page 431)*

1. (a) B **(b)** A **(c)** C **(d)** D   **2.** $\left\{\dfrac{1}{2}\right\}$ **3.** $\log_4 8 = \dfrac{3}{2}$ **4.** $8^{2/3} = 4$ **5.** They are inverses.

6. $2\log_7 x + \dfrac{1}{4}\log_7 y - 3\log_7 z$ **7.** 2.3755 **8.** -3.0640 **9.** 1.1674 **10.** -2.6038 **11.** $\{5\}$ **12.** $\left\{\dfrac{5}{2}\right\}$

13. $\{2\}$ **14.** $\{4.7833\}$ **15.** $\{20.1246\}$ **17.** 10 sec   **18. (a)** about 18.9 yr **(b)** about 18.8 yr

Appendix A Exercises *(page 440)*

1. true   **3.** false   **5.** B   **7.** $\{-4\}$ **9.** $\{1\}$ **11.** $\left\{-\dfrac{2}{7}\right\}$ **13.** $\left\{-\dfrac{7}{8}\right\}$ **15.** $\{-1\}$ **17.** $\{10\}$ **19.** $\{75\}$

21. $\{0\}$ **23.** identity; {all real numbers}   **25.** conditional equation; $\{0\}$ **27.** contradiction; $\emptyset$ **29. (a)** E **(b)** C **(c)** A

(d) B **(e)** D   **31.** B; $\left\{\dfrac{-5 \pm \sqrt{7}}{2}\right\}$ **33.** $\{2, 3\}$ **35.** $\left\{-\dfrac{2}{5}, 1\right\}$ **37.** $\left\{-\dfrac{3}{4}, 1\right\}$ **39.** $\{\pm 4\}$ **41.** $\{\pm 3\sqrt{3}\}$

43. $\{-5 \pm 2\sqrt{10}\}$ **45.** $\left\{\dfrac{1 \pm 2\sqrt{3}}{3}\right\}$ **47.** $\left\{\dfrac{1 \pm \sqrt{5}}{2}\right\}$ **49.** $\{3 \pm \sqrt{2}\}$ **51.** $\left\{\dfrac{2 \pm \sqrt{10}}{2}\right\}$ **53.** $\left\{\dfrac{-1 \pm \sqrt{97}}{4}\right\}$

55. $\left\{\dfrac{1 \pm \sqrt{13}}{2}\right\}$ **57. (a)** F **(b)** J **(c)** A **(d)** H **(e)** I **(f)** D **(g)** B **(h)** G **(i)** E **(j)** C

59. $(-\infty, 4]$;   **61.** $[-1, \infty)$;   **63.** $(-\infty, 6]$;

65. $(-\infty, 4)$;   **67.** $\left[-\dfrac{11}{5}, \infty\right)$;   **69.** $(-5, 3)$;

71. $[3, 6]$;   **73.** $(4, 6)$;   **75.** $[-9, 9]$;

Appendix B Exercises *(page 447)*

1.–7.   **9.** II   **11.** III   **15. (a)** I **(b)** IV **(c)** III **(d)** IV   **17.** yes   **19.** no   **21.** 58.02 ft

23. 31.6 ft   **25.** 6 **27. (a)** $8\sqrt{2}$ **(b)** $(-9, -3)$ **29. (a)** $\sqrt{34}$ **(b)** $\left(\dfrac{11}{2}, \dfrac{7}{2}\right)$ **31. (a)** $\sqrt{202}$ **(b)** $\left(-\dfrac{5}{2}, -\dfrac{1}{2}\right)$

33. (a) $\sqrt{133}$ **(b)** $\left(2\sqrt{2}, \dfrac{3\sqrt{5}}{2}\right)$ **35.** $5, -1$ **37.** $9 + \sqrt{119}, 9 - \sqrt{119}$ **39.** $x^2 + y^2 = 25$

41. 74.2%; This is 1.1% less than the actual percent of 75.3.

Other ordered pairs are possible in Exercises 43–47.

43. (a)

x	y
0	−2
4	0
2	−1

(b)

45. (a)

x	y
0	0
1	1
−2	4

(b)

47. (a)

x	y
0	0
−1	−1
2	8

(b)

49. (a) $x^2 + y^2 = 36$ **(b)**

51. (a) $(x - 2)^2 + y^2 = 36$ **(b)**

53. (a) $(x + 2)^2 + (y - 5)^2 = 16$ **(b)**

55. $(x - 3)^2 + (y - 2)^2 = 4$

Appendix C Exercises *(page 457)*

3. The graphs in (a), (c), and (d) represent functions. The graph in (b) fails the vertical line test; it is not the graph of a function.

5. function   **7.** not a function   **9.** function   **11.** not a function; domain: $\{0, 1, 2\}$; range: $\{-4, -1, 0, 1, 4\}$

13. function; domain: $\{2, 5, 11, 17\}$; range: $\{1, 7, 20\}$ **15.** function; domain: $(-\infty, \infty)$; range: $(-\infty, \infty)$ **17.** not a function; domain: $[3, \infty)$; range: $(-\infty, \infty)$ **19.** not a function; domain: $[-4, 4]$; range: $[-3, 3]$ **21.** function; domain: $(-\infty, \infty)$; range: $[0, \infty)$ **23.** not a function; domain: $[0, \infty)$; range: $(-\infty, \infty)$ **25.** function; domain: $(-\infty, \infty)$; range: $(-\infty, \infty)$

27. function; domain: $[0, \infty)$; range: $[0, \infty)$ **29.** function; domain: $\left[-\dfrac{1}{2}, \infty \right)$; range: $[0, \infty)$ **31.** function; domain: $(-\infty, 9) \cup (9, \infty)$; range: $(-\infty, 0) \cup (0, \infty)$ **33.** B   **35.** 4   **37.** −11   **39.** $-3p + 4$ **41.** $3x + 4$ **43.** $-3a - 8$

45. (a) 2 **(b)** 3   **47. (a)** 3 **(b)** −3   **49. (a)** 0 **(b)** 4 **(c)** 2 **(d)** 4   **51. (a)** 11 **(b)** 9 **(c)** −2 **(d)** 2, 7, 8 **(e)** $\sqrt{104}$

53. (a) $[4, \infty)$ **(b)** $(-\infty, -1]$ **(c)** $[-1, 4]$ **55. (a)** $(-\infty, 4]$ **(b)** $[4, \infty)$ **(c)** none   **57. (a)** none **(b)** $(-\infty, -2]$; $[3, \infty)$

(c) $(-2, 3)$ **59. (a)** yes **(b)** $[0, 24]$ **(c)** 1200 megawatts **(d)** at 17 hr or 5 P.M.; at 4 A.M. **(e)** $f(12) = 2000$; At 12 noon, electricity use is 2000 megawatts. **(f)** increasing from 4 A.M. to 5 P.M.; decreasing from midnight to 4 A.M. and from 5 P.M. to midnight

Appendix D Exercises *(page 465)*

1. (a) B **(b)** E **(c)** F **(d)** A **(e)** D **(f)** C   **3. (a)** B **(b)** A **(c)** G **(d)** C **(e)** F **(f)** D **(g)** H **(h)** E

5.   **7.**   **9.**   **11.**   **13. (a)** $(8, 3)$ **(b)** $(8, 48)$

15.   **17.**   **19.** y-axis   **21.** x-axis, y-axis, origin   **23.** origin   **25.** none of these

27.

$y = x^2 - 1$

29.

$y = x^2 + 2$

31.

$y = (x - 1)^2$

33.

$y = (x + 2)^2$

35.

$(-3, -4)$
$y = (x + 3)^2 - 4$

37.

$y = 2x^2 - 1$

39.

$f(x) = 2(x - 2)^2 - 4$

41. It is the graph of $f(x) = |x|$ translated 1 unit to the left, reflected across the x-axis, and translated 3 units up. The equation is $y = -|x + 1| + 3$.

43. It is the graph of $g(x) = \sqrt{x}$ translated 4 units to the left, stretched vertically, and translated 4 units down. The equation is $y = 2\sqrt{x + 4} - 4$.

45. $g(x) = 2x + 13$

Index of Applications

Astronomy
Angle of elevation of the sun, 72, 87, 112–113, 116
Angle the celestial North Pole moves, 42
Angular and linear speeds of Earth, 121
Circumference of Earth, 107
Diameter of the moon, 107
Diameter of the sun, 75
Distance between the sun and a star, 27
Distance to the moon, 285
Eclipse on a planet, 20
Fluctuation in the solar constant, 144–145
Height of a satellite, 90–91
Hours in a Martian day, 121
Orbital speed of Jupiter, 129
Orbits of a space vehicle, 99
Phase angle of the moon, 126
Planetary orbits, 370
Rotation and roll of a spacecraft, 234
Satellites, 119
Simulating gravity on the moon, 377
Sizes and distances in the sky, 20
Solar eclipse, 13–14
Velocity of a star, 321
Viewing field of a telescope, 8

Automotive
Accident reconstruction, 255
Area cleaned by a windshield wiper, 106
Braking distance, 68
Car's speed at collision, 68
Force needed to keep a car parked, 318
Grade resistance, 63, 66
Measuring speed by radar, 65
Pickup truck speedometer, 104–105
Rotating tire, 8, 42
Rotation of a gas gauge arrow, 128
Speed limit on a curve, 67
Stopping distance on a curve, 85

Biology and Life Sciences
Activity of a nocturnal animal, 143
Deer population, 402
Diversity of species, 417, 424
Fish's view of the world, 67, 68
Lynx and hare populations, 177
Movement of an arm, 268
Position of a moving arm, 143
World population growth, 401

Business
Cost from inflation, 430
Employee training, 402

Chemistry
Distance between atoms, 284
pH, 414

Construction
Carpentry technique, 16–17
Distance between a beam and a cable, 301
Distance between points on a crane, 300
Height of a building, 19
Layout for a playhouse, 303
Length of a brace, 326
Required amount of paint, 304
Roof trusses, 296
Square corners, 448
Surveying, 8, 79, 91, 292, 294
Truss construction, 300

Education
Aging of college freshmen, 449
Average annual public university costs, 425
Two-year college enrollment, 449
Volunteerism among college freshmen, 423–424

Energy
Alternating electric current, 201–202, 262, 267, 340
Amperage, wattage, and voltage, 220, 233
Area of a solar cell, 26
Electricity usage, 457
Land required for a solar-power plant, 106
Wattage consumption, 216, 220

Engineering
Angle formed by radii of gears, 284
Angular speed of a pulley, 118, 121
Antenna mast guy wire, 91
Communications satellite coverage, 249
Distance across a canyon, 283, 325
Distance across a lake or river, 74, 278–279, 283, 284, 300
Distance of a rotating beacon, 166, 167
Distance through a tunnel, 84
Electrical current, 204, 352
Electromotive force, 255
Engine specifications, 127
Error in measurement, 77
Force needed to hold a monolith, 319
Gears, 101

Ground distance measured by aerial photography, 285
Height of a mountain, 72, 76–77, 83
Height of a tree, 18, 79–80, 124, 326
Height of an object or a structure, 13, 18, 19, 36, 44, 74, 76, 83, 90, 92, 284, 425
Highway curves, 66–67, 85
Highway grades, 66
Impedance, 340, 352
Latitudes, 104
Length of a sag curve, 61
Length of a structure, 301
Linear speed of a belt, 118, 121, 123
Linear speed of a flywheel, 126
Longest vehicular tunnel, 73
Longitude, 102
Measurement using triangulation, 302
Pipeline position, 326
Programming language, 267
Pulley raising a weight, 104
Radio direction finders, 329
Radio towers and broadcasting patterns, 370
Radio tuners, 193, 196–197
Railroad curves, 106, 226
Railroad engineering, 98
Revolutions, 8
Rope wound around a drum, 100
Rotating pulley, 8, 39, 99
Rotating wheels, 104
Transistors on computer chips, 430–431
Voltage, 143, 212, 234, 255
Wires supporting a flagpole, 330

Environment
Atmospheric carbon dioxide, 144, 397–398
Atmospheric effect on sunlight, 127
Atmospheric pressure, 401
Classifying wetlands, 414
Forest fire location, 279
Pollution trends, 177
Ruggedness of a coastline, 386

Finance
Company loans, 401
Compound interest, 394–395, 397
Continuous compounding, 396
Future value, 400, 401
Growth of an account, 429, 430, 432
Interest rate, 400, 401, 412, 429
Present value, 395, 400

Index